T0213208

Texts in Computer Science

Series Editors

David Gries, Department of Computer Science, Cornell University, Ithaca, NY, USA

Orit Hazzan ⓘ, Faculty of Education in Technology and Science, Technion—Israel Institute of Technology, Haifa, Israel

More information about this series at https://link.springer.com/bookseries/3191

Rudolf Kruse • Sanaz Mostaghim •
Christian Borgelt • Christian Braune •
Matthias Steinbrecher

Computational Intelligence

A Methodological Introduction

Third Edition

With contributions from Frank Klawonn
and Christian Moewes

 Springer

Rudolf Kruse
Faculty of Computer Science
Otto-von-Guericke-University
Magdeburg, Sachsen-Anhalt, Germany

Sanaz Mostaghim 🆔
Faculty of Computer Science
Otto-von-Guericke-University
Magdeburg, Sachsen-Anhalt, Germany

Christian Borgelt
Department of Artificial Intelligence
and Human Interfaces
Paris-Lodron-University Salzburg
Salzburg, Austria

Christian Braune
Faculty of Computer Science
Otto-von-Guericke-University
Magdeburg, Sachsen-Anhalt, Germany

Matthias Steinbrecher
D4L data4life gGmbH
Potsdam, Germany

ISSN 1868-0941   ISSN 1868-095X   (electronic)
Texts in Computer Science
ISBN 978-3-030-42229-5   ISBN 978-3-030-42227-1   (eBook)
https://doi.org/10.1007/978-3-030-42227-1

Preface

Computational Intelligence comprises concepts, paradigms, algorithms, and implementations of systems that are supposed to exhibit intelligent behavior in complex environments. It relies heavily on nature-analog or at least nature-inspired methods.

The choice of topics in this book reflects the most important fields in the area of Computational Intelligence. Classical fields such as Artificial Neural Networks, Evolutionary Algorithms, and Fuzzy Systems are described in considerable detail. However, related methods such as Multi-objective Optimization and Probabilistic Graphical Models are discussed as well, although a complete coverage of all approaches and developments is clearly impossible to achieve in a single volume. Rather than to strive for completeness, our goal is to give a methodical introduction to the area of Computational Intelligence. Hence, we try not only to present fundamental concepts and their implementations, but also explain the theoretical background of proposed problem solutions. In addition, we hope to convey to a reader what is necessary in order to apply these methods successfully.

In the third edition, we paid tribute to recent developments in computational intelligence. We also considered the helpful remarks of our readers while revising the book. You can find several new chapters, e.g., Deep Learning, Computational Swarm Intelligence, Approximate Reasoning, and Learning Bayesian Networks.

This textbook is primarily meant as a companion book for lectures on the covered topics in the area of Computational Intelligence. However, it may also be used for self-study by students and practitioners from industry and commerce. This book is based on notes of lectures, exercise lessons, and seminars that have been given by the authors for many years. On the book's website

http://www.computational-intelligence.eu/

a lot of additional material for lectures on Neural Networks, Evolutionary Algorithms, Fuzzy Systems and Bayesian Networks can be found, including lecture slides and exercises.

Magdeburg, Germany
January 2022

Rudolf Kruse
Sanaz Mostaghim
Christian Borgelt
Christian Braune
Matthias Steinbrecher

Contents

Introduction

1.1 Intelligent Systems

Complex problem settings in widely differing technical, commercial, and financial fields evoke an increasing need for computer applications that must show "intelligent behavior." These applications are desired to support decision- making, to control processes, to recognize and interpret patterns or to maneuver vehicles or robots autonomously in unknown environments. Novel approaches, methods, tools, and programming environments have been developed to accomplish such tasks.

Seen from a higher level of abstraction, the general requirements for developing such an "intelligent system" are ultimately always the same, namely simulating intelligent thinking and actions in a certain field of application. For this purpose, the knowledge about this field must be represented and processed. The quality of the resulting system mainly depends on how well the model is chosen in the development process. Sometimes a data-driven method for finding a good model is appropriate, sometimes a logic-based knowledge representation is more suitable, in real applications, in most cases, hybrid models are used. So one has to sift through many available approaches to find those that fit the application area of the intelligent system best.

The mechanisms and processes that underlie intelligent behavior are examined in the research area of intelligent systems. Similar to most other areas of computer science (or science in general), theoretical aspects (how and why do these systems work?) and application-oriented aspects (where and when can these systems be used?) are studied. At the beginning of the development of intelligent systems, researchers are often focused on the idea of seeing "a human being as a machine"—an idea that stems from the age of enlightenment. They ventured to create an (artificial) intelligence that can both think creatively and solve problems in the way a human can. This intelligence was also meant to exhibit forms of both consciousness and emotions. In the infancy of the field, the typical way to design an "artificial intelligence" was to describe a symbolic basis of the relevant mechanisms. This includes the top-down perspective of problem-solving, which mainly addresses the question of why these systems work. The answer to this question is usually given with the help

© Springer Nature Switzerland AG 2022
R. Kruse et al., *Computational Intelligence*, Texts in Computer Science,
https://doi.org/10.1007/978-3-030-42227-1_1

of a symbolic representation and a logic-based inference mechanism. Techniques that are examples of these approaches include rule-based expert systems, automatic theorem provers, and many operations research techniques which underlie modern planning and scheduling software. Although these traditional approaches have been very successful in some cases, they do have limitations, especially when it comes to scalability. A tiny complication of the problem to solve often causes an increase of complexity that cannot be handled feasibly.

For certain types of problems, techniques that are biologically or linguistically motivated proved successful in theory, design, application, and development of intelligent systems. The corresponding research area is called *Computational Intelligence* (CI). CI comprises concepts, paradigms, algorithms, and implementations to develop systems that exhibit intelligent behavior in complex environments. These novel methods have demonstrated their usefulness in many application areas, in most cases in combination with traditional methods.

1.2 Computational Intelligence

Computational Intelligence is based traditionally on the three main pillars Neural Networks, Evolutionary Computation, and Fuzzy Systems. However, with time, many nature computing paradigms have evolved. Over the last few years, there has been an explosion of research on Deep Learning, in particular, deep convolutional neural networks.

Many ideas and principles in the area of artificial neural networks are inspired by neuroscience. Artificial neural networks are information-processing systems whose structure and functionality simulates the nervous systems and particularly the brain of animals and human beings. They consist of a large number of fairly simple processing units, the so-called neurons, which work in parallel. These neurons send information in the form of action potentials via directed links to other neurons. Based on knowledge about the functionality of biological neural networks, one tries to model and mimic them, especially to achieve learning capability.

Evolutionary algorithms draw on ideas from biological evolution, in which organisms, over many generations, get adapted to environmental conditions. They address certain classes of optimization problems and belong to the family of meta-heuristics, which offers algorithms to approximately solve may types of optimization problems. Meta-heuristics are defined by an abstract sequence of steps that are applicable to more or less arbitrary problem descriptions. However, every single step must be implemented in a problem-specific fashion. Meta-heuristics are often applied to problems for which no efficient solution algorithm is known. Although finding an optimal solution is usually not guaranteed and thus a found solution can, in principle, be arbitrarily bad compared to the optimal solution, meta-heuristics offer the possibility to obtain (sufficiently) good solutions in a reasonable time frame. Here biologically inspired methods such as computational swarm intelligence, multi-objective optimization, and so on, have proven themselves to be noteworthy candidates to employ in theory and practice.

Using the human language as a source of inspiration, fuzzy systems model linguistic imprecision and solve problems based on the generalization of traditional logic, which enables us to perform approximate reasoning. Fuzzy approaches can also be used to derive inference mechanisms, thus giving rise to approximate reasoning methods. Fuzzy systems are routinely applied in control engineering because, in many application scenarios, a precise and complete modeling of the system is impractical or even impossible.

Uncertainty is a common feature of many everyday decisions. Uncertainty typically arises in a situation that has limited or incalculable information about the predicted outcomes of behavior. Bayesian networks are means to efficiently store and reason with uncertain knowledge in complex application areas. Formally, a Bayesian network is a probabilistic graphical model that represents a set of random variables and their conditional dependencies by a directed acyclic graph. Due to the probabilistic representation, one can easily draw inferences based on new information. In addition, Bayesian networks are well-suited for learning from data.

1.3 About the Third Edition of this Book

Our main objective with this textbook is to give a methodical introduction to the field of computational intelligence. Therefore, we focus on fundamental concepts and their implementation and strive to explain the theoretical background of proposed solutions to certain problems. We hope to convey to a reader all that is necessary for a profound application of the discussed methods. This book requires only fairly basic knowledge of mathematics, as we tried to introduce all necessary concepts and tools in this book in order to make it as self-contained as possible. Furthermore, the four parts about artificial neural networks, evolutionary algorithms, fuzzy systems, and Bayesian networks can be studied independently of each other as we tried to avoid dependencies between the parts or requiring prerequisites from earlier parts.

This book would not be the same without the many contributions by Frank Klawonn and Christian Moewes for the parts on fuzzy systems and evolutionary algorithms from the first edition. We would like to thank Jonas Schulze for his careful proofreading.

This book is intended as an accompanying book for lectures about the field of computational intelligence and is fundamentally based on written notes about lectures that the authors have given periodically for students of different faculties. On the website

http://www.computational-intelligence.eu

lecture slides matching the book, exercise sheets, and additional material can be found for all four parts of the book.

Part I
Neural Networks

Introduction to Artificial Neural Networks

2

(Artificial) neural networks are information processing systems, whose structure and operation principles are inspired by the nervous system and the brain of animals and humans. They consist of a large number of fairly simple units, the so-called *neurons*, which are working in parallel. These neurons communicate by sending information in the form of activation signals, along directed connections, to each other.

A commonly used synonym for "neural network" is the term "connectionist model." The research area that is devoted to the study of connectionist models is called "connectionism." Furthermore, the expression "parallel distributed processing" can often be found in relation to (artificial) neural networks.

2.1 Motivation

(Artificial) neural networks are studied for various reasons: in (neuro-)biology and (neuro-)physiology, but also in psychology, one is mainly interested in their similarity to biological nervous systems. In these areas, (artificial) neural networks are used as computational models with which one tries to simulate and thus to understand the mechanisms of nerve and brain functions. Especially in computer science, but also in other engineering sciences, one tries to mimic certain cognitive powers of humans (especially learning ability) by using functional elements of the nervous system and the brain. In physics, certain mathematical models that are analogous to (artificial) neural networks are employed to describe specific physical phenomena. An example are models of magnetism, for instance, the Ising model [7].

As can already be seen from this brief list, the study of (artificial) neural networks is a highly interdisciplinary research area. However, in this book, we widely neglect the use of (artificial) neural networks in physics (even though we draw on examples from physics to explain certain network models) and consider their biological basis only very briefly (see the next section). Rather we focus on the mathematical and

© Springer Nature Switzerland AG 2022
R. Kruse et al., *Computational Intelligence*, Texts in Computer Science,
https://doi.org/10.1007/978-3-030-42227-1_2

engineering aspects, particularly the use of (artificial) neural networks in the area of computer science that is commonly called "artificial intelligence."

While the reasons why biologists study (artificial) neural networks are fairly obvious, we may have to justify why neural networks are (or should be) studied in artificial intelligence. The reason is that the paradigm of classical artificial intelligence (sometimes called, in a somewhat pejorative manner, GOFAI—"good old-fashioned artificial intelligence") is based on a very strong hypothesis about how machines can be made to behave "intelligently." This hypothesis says that the essential requirement for intelligent behavior is the ability to manipulate symbols and symbol structures that are represented by physical structures. Here, *symbol* means a token that refers to an object or a situation. This relation is interpreted in an operational manner: the system can perceive and/or manipulate the object referred to. This hypothesis was first formulated explicitly by [10]:

Physical Symbol System Hypothesis: A physical-symbol system has the necessary and sufficient means for general intelligent action.

As a matter of fact, classical artificial intelligence concentrated, based on this hypothesis, on symbolic forms of representing knowledge and in particular on propositional and predicate logic. (Artificial) neural networks, on the other hand, are no physical symbol systems, since they do not process *symbols*, but rather much more elementary *signals*, which, taken individually, rarely have a (clear) meaning. As a consequence, (artificial) neural networks are often called "sub-symbolic." However, if the ability to process symbols is necessary to produce intelligent behavior, then it is unnecessary to study (artificial) neural networks in artificial intelligence.

There is no doubt that classical artificial intelligence has achieved remarkable successes: with its methods, computers can automatically solve many types of puzzles and brain-twisters and can play games like chess and Reversi (also known as Othello) on an extremely high level. However, when it comes to perception (seeing, hearing etc.), computers usually perform fairly poorly compared to humans—at least if symbolic representations are relied upon: here computers are often too slow, too inflexible, and too little tolerant to noise and faults. We may conjecture that the problem is that in order to recognize patterns—a core task of perception—symbolic representations are not very well suited, because there are no adequate symbols on this level of processing. Rather "raw" (measurement) data needs to be structured and summarized before symbolic methods can effectively be applied. Hence, it appears to be reasonable to examine the mechanisms of sub-symbolic information processing in natural intelligent systems—that is, animals and humans—in more detail and possibly to exploit these mechanisms to mimic intelligent behavior.

Additional arguments why studying (artificial) neural networks may be beneficial arise from the following observations:

- Expert systems that use symbolic representations usually become slower with a larger knowledge base, because larger sets of rules need to be traversed. Human experts, however, usually become faster. Maybe a non-symbolic representation (as it is used in natural neural networks) is more efficient.

• Despite the fairly long switching time of natural neurons (in the order of several milliseconds) essential cognitive tasks (like recognizing an object) are solved in a fraction of a second. If neural processing were sequential, only about 100 switching operations could be performed ("100 step rule"). Hence, high parallelization must be present, which is easy to achieve with neural networks, but much more difficult to implement with other approaches.

• There is a large number of successful applications of (artificial) neural networks in industry, commerce, and finance.

Currently, we even experience—with approaches that have become known under the label **deep learning**—a new "golden age" of (artificial) neural networks, with intensified research and extremely promising advances. One may even go so far as to claim that if one wants to be at the current frontier of research in artificial intelligence, one has to work (at least: also) on (artificial) neural networks.

Historically, this can be seen as the third such "golden age". After the first trainable neuron-inspired computational models were developed at the end of the 1950s [11], it was originally assumed that the core principles of automated learning had been discovered. This led to huge excitement and expectations of great progress: the first "golden age" of (artificial) neural networks started. However, the frenzy was curtailed fairly quickly. After a thorough mathematical analysis of these models revealed that their computational capabilities are actually fairly limited [9], while the used learning rule could not be extended to allow for learning what was actually needed, (artificial) neural networks were even regarded as a "dead end" of artificial intelligence research. Thus, the so-called "dark age" of (artificial) neural networks was ushered in.

Even though a new learning paradigm able to overcome the identified obstacles was already discovered soon after [19], it took until 1986, when the approach was independently rediscovered [14, 15], before it gained widespread attention. However, once that happened, the interest in (artificial) neural networks was keenly revived. This period, which is sometimes referred to as the second "golden age" of (artificial) neural networks, saw many theoretical advances as well as a plethora of applications of (artificial) neural networks. It lasted well into the 1990s and was not ended by discoveries of further limitations or any other fundamental technical problems like the first "golden age", because it could be shown that (artificial) neural networks were universal function approximators [6]. Rather, it was the development of strong competitors—especially for classification tasks—in the form of **support vector machines** [4, 16–18] and **random forests** [3, 5] that shifted some attention away from (artificial) neural networks. However, the somewhat reduced attention during this period can hardly be called a second "dark age".

(Artificial) neural networks jumped back to the center of attention when the ImageNet Large Scale Visual Recognition Challenge (ILSVRC), which was established in 2010, was won in 2012 by AlexNet, a deep neural network [8]. This network improved over the winner of the previous year by more than 10 percentage points, thus cutting the top5 error rate almost in half. The result was renewed intense research into using (artificial) neural networks for recognition tasks, especially for image classification and object recognition in images. This third "golden age" lasts to this day and has not reached its culmination yet.

2.2 Biological Background

As already mentioned, (artificial) neural networks are inspired by the structure and the operation principles of the nervous system and particularly the brain of animals and humans. In fact, the neural network models that we study in this book are not very close to their biological original, since they are too simplified to model the characteristics of natural neural networks correctly. Nevertheless, we briefly consider natural neural networks here, because they formed the starting point for investigating artificial neural networks. The description follows [1].

The nervous system of animals consists of the brain (in so-called "lower" life forms often only referred to as the "central nervous system"), the different sensory systems, which collect information from the different body parts (visual, auditory, olfactory, gustatory, thermal, tactile, etc), and the motor system, which controls movements. The greater part of information processing happens in the brain/central nervous system, although the amount of preprocessing outside the brain can be considerable, for example, in the retina of the eye.

With respect to processing information, the neurons are the most important components of the nervous system.[1] According to common estimates, there are about 100 billion (10^{11}) neurons in the human brain, of which a fairly large part is active in parallel. Neurons process information mainly by interacting with each other.

A **neuron** is a cell that collects and transmits electrical activity. Neurons exist in many different shapes and sizes. Nevertheless, one can derive a "prototypical" neuron that resembles all kinds of neurons to some degree (although this is a fairly severe simplification). This prototype is shown schematically in Fig. 2.1. The **cell body** of the neuron, which contains the **nucleus**, is also called **soma**. It has a diameter of about 5 to 100 μm (micrometer, 1 μm = 10^{-6} m). From the cell, body extend several short, heavily ramified branches that are called **dendrites**. In addition, it has a long extension called **axon**. The axon can be between a few millimeters and one meter long. Axon and dendrites differ in the structure and the properties of the **cell membrane** . In particular, the axon is often covered by a **myelin sheath**.

The axons are the fixed paths along which neurons communicate with each other. The axon of a neuron leads to the dendrites of other neurons. At its end, the axon is heavily ramified and possesses at the ends of these branches **terminal buttons**. Each terminal button almost touches a dendrite or the cell body of another neuron. The gap between a terminal button and a dendrite is usually between 10 and 50 nm

---

[1] The nervous system consists not only of neurons, not even for the largest part. Besides neurons, there are various other cells, for instance, the so-called glia cells , which have a supporting function.

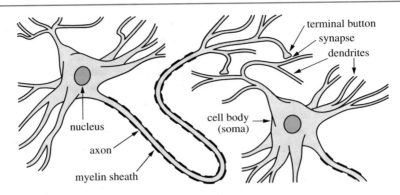

Fig. 2.1 Prototypical structure of biological neurons

(nanometer; 1 nm = 10^{-9} m) wide. Such a place, at which an axon and a dendrite almost touch each other, is called **synapse**.

The most common form of communication between neurons is that a terminal button of the axon releases certain chemicals, the so-called **neurotransmitters**, which act on the membrane of the receiving dendrite and change its polarization (its electrical potential). Usually, the inside of the cell membrane, which encloses the whole neuron, is about 70 mV (millivolts; 1 mV = 10^{-3} V) more negative than its outside, because the concentration of negative ions is greater on the inside, while the concentration of positive ions is greater on the outside. Depending on the type of the released neurotransmitter, the potential difference may be reduced or increased on the side of the dendrite. Synapses that reduce the potential difference are called **excitatory**, while those that increase it are called **inhibitory**.

In an adult human, all connections between neurons are completely established and no new connections are created (again this is a severe simplification). An average neuron possesses between 1000 and 10,000 connections to other neurons. The change of the electrical potential that is caused by a single synapse is fairly small, but the individual excitatory and inhibitory effects can accumulate (counting the excitatory influences as positive and the inhibitory ones as negative). If the excitatory net input is large enough, the potential difference in the cell body can be significantly reduced. If the reduction is large enough, the axon's base is depolarized. This depolarization is caused by positive sodium ions entering the cell. As a consequence, the inside of the cell becomes temporarily (for about one millisecond) more positive than its outside. Afterward, the potential difference is rebuilt by positive potassium ions leaving the cell. Finally, the original distribution of sodium and potassium ions is reestablished by special ion pumps in the cell membrane.

The sudden, temporary change of the electrical potential, which is called **action potential**, propagates along the axon. The propagation speed lies between 0.5 and 130 m/s, depending on the properties of the axon. In particular, it depends on how heavily the axon is covered with a myelin sheath (the more myelin, the faster the action potential is propagated). When this nerve impulse reaches the end of the axon, it causes neurotransmitters to be released at the terminal buttons, thus passing the signal on to the next cell, where the process is repeated.

In summary: changes of the electrical potential are accumulated at the cell body of a neuron and, if they reach a certain threshold, are propagated along the axon. This nerve impulse causes that neurotransmitters are released by the terminal buttons at the end of the axon, thus inducing a change of the electrical potential in the receiving neuron. Even though this description is heavily simplified, it captures the essentials of neural information processing on the level of individual neurons.

In the human nervous system, information is encoded by continuously changing quantities, primarily two: the electrical potential of the neuron's membrane and the number of nerve impulses that a neuron transmits per second. The latter is also called the **firing rate** of the neuron. It is commonly assumed that the number of impulses is more important than their shape (in the sense of a change of the electrical potential), although competing theories of **neural coding** exist. A neuron can emit 100 or even more impulses per second. The higher the firing rate, the higher the influence a neuron has on connected neurons. However, in artificial neural networks, this frequency coding of information is usually not emulated.

References

1. Anderson JR (1995) Cognitive psychology and its implications, 4th edn. Freeman, New York, NY, USA
2. Boden MA (ed) (1990) The philosophy of artificial intelligence. Oxford University Press, Oxford, United Kingdom
3. Breiman L (2001) Random forests. Mach Learn 45(1):5-32. Kluwer, Dordrecht, Netherlands
4. Cristianini N, Shawe-Taylor J (2004) Kernel methods for pattern analysis. Cambridge University Press, Cambridge, Great Britain
5. Ho TK (1995) Random decision forests. In: Proceedings of 3rd international conference on document analysis and recognition (ICDAR 1995, Montreal, Canada). IEEE Press, Piscataway, NJ, USA, pp 278–282
6. Hornik K (1991) Approximation capabilities of multilayer feedforward networks. Neural Netw 4(2):251–257. Elsevier, Amsterdam, Netherlands
7. Ising E (1925) Beitrag zur Theorie des Ferromagnetismus. Zeitschrift für Physik 31(253)
8. Krishevsky A, Sutskever I, Hinton GE (2012) ImageNet classification with deep convolutional neural networks. In: Proceedings of 25th international conference on neural information processing systems (NIPS 2012). Curran Associates Inc., Red Hook, NY, USA, pp 1097–1105
9. Minsky LM, Papert S (1969) Perceptrons. MIT Press, Cambridge, MA, USA
10. Newell A, Simon HA (1976) Computer science as empirical enquiry: symbols and search. In: Communications of the association for computing machinery 19. Association for Computing Machinery, New York, NY, USA. Reprinted in [Boden 1990], pp 105–132
11. Rosenblatt F (1958) The perceptron: a probabilistic modell for information storage and organization in the brain. Psychol Rev 65:386-408. USA
12. Rosenblatt F (1962) Principles of neurodynamics. Spartan Books, New York, NY, USA
13. Rumelhart DE, McClelland JL (eds) (1986) Parallel distributed processing: explorations in the microstructures of cognition, vol 1: foundations. MIT Press, Cambridge, MA, USA

14. Rumelhart DE, Hinton GE, Williams RJ (1986) learning internal representations by error propagation. In: [Rumelhart und McClelland 1986], pp 318–362
15. Rumelhart DE, Hinton GE, Williams RJ (1986) Learning representations by back-propagating errors. Nature 323:533–536. Nature Publishing Group, New York, NY, USA
16. Schölkopf B, Smola A (2002) Learning with kernels: support vector machines, regularization, optimization, and beyond. MIT Press, Cambridge, MA, USA
17. Vapnik V (1995) The nature of statistical learning theory. Springer, New York, NY, USA
18. Vapnik V (1998) Statistical learning theory. Wiley, Chichester, Great Britain
19. Werbos PJ (1974) Beyond Regression: New Tools for Prediction and Analysis in the Behavioral Sciences. PhD Thesis, Harvard University, Cambridge, MA, USA

Threshold Logic Units

The description of biological neural networks in the preceding chapter makes it natural to model neurons as **threshold logic units**: if a neuron receives enough excitatory input that is not compensated by equally strong inhibitory input, it becomes active and sends a signal to other neurons. Such a model was already examined very early in much detail by [4]. As a consequence, threshold logic units are also known as **McCulloch–Pitts neurons**. Another name that is commonly used for a threshold logic unit is **perceptron**, even though the processing units that [10,11] called "perceptrons" are actually somewhat more complex than simple threshold logic units.[1]

3.1 Definition and Examples

Definition 3.1 A **threshold logic unit** is a simple processing unit for real-valued numbers with n inputs $x_1, \ldots, x_n$ and one output y. The unit as a whole possesses a **threshold** θ. To each input x_i a **weight** w_i is assigned. A threshold logic unit computes the function

$$y = \begin{cases} 1 \text{ if } \sum_{i=1}^{n} w_i x_i \geq \theta, \\ 0 \text{ otherwise.} \end{cases}$$

The inputs are often combined into an input vector $\mathbf{x} = (x_1, \ldots, x_n)^\top$ and the weights into a weight vector $\mathbf{w} = (w_1, \ldots, w_n)^\top$. With the help of the scalar product the condition tested by a threshold logic unit may then also be written as $\mathbf{w}^\top \mathbf{x} \geq \theta$.

---

[1] In a perceptron there is, besides the actual threshold logic unit, an input layer that executes additional operations on the input signals. However, this input layer consists of immutable functional elements, and therefore, is often neglected.

15

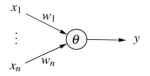

Fig. 3.1 Representation of a threshold logic unit

We depict a threshold logic unit as shown in Fig. 3.1. That is, we draw it as a circle, in which the threshold θ is recorded. Each input is drawn as an arrow that points to the circle and that is labeled with the weight of the input. The output of the threshold logic unit is shown as an arrow that points away from the circle.

To illustrate how threshold logic units work and to demonstrate their capabilities, we consider a couple of simple examples. Figure 3.2 shows on the left a threshold logic unit with two inputs x_1 and x_2, which carry the weights $w_1 = 3$ and $w_2 = 2$, respectively. The threshold is $\theta = 4$. If we assume that the input variables can only have values 0 and 1, we obtain the table shown in Fig. 3.2 on the right. Clearly, this threshold logic unit computes the conjunction of its inputs: only if both inputs are active (that is, equal to 1), it becomes active itself and outputs a 1.

Figure 3.3 shows another threshold logic unit with two inputs, which differs from the one shown in Fig. 3.2 by a negative threshold $\theta = -1$ and one negative weight $w_2 = -2$. Due to the negative threshold, it is active (that is, outputs a 1) even if both inputs are inactive (that is, are equal to 0). Intuitively, the negative weight corresponds to an inhibitory synapse: if the corresponding input becomes active (that is, equal to 1), the threshold logic unit is deactivated and its output becomes 0. We can also observe here that positive weights correspond to excitatory synapses: even if the input x_2 inhibits the threshold logic unit, (that is, if $x_2 = 1$), it can become active, namely if it is "excited" by an active input x_1 (that is, by $x_1 = 1$). In summary,

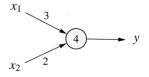

x_1	x_2	$3x_1 + 2x_2$	y
0	0	0	0
1	0	3	0
0	1	2	0
1	1	5	1

Fig. 3.2 A threshold logic unit for the conjunction $x_1 \wedge x_2$

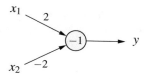

x_1	x_2	$2x_1 - 2x_2$	y
0	0	0	1
1	0	2	1
0	1	-2	0
1	1	0	1

Fig. 3.3 A threshold logic unit for the implication $x_2 \rightarrow x_1$

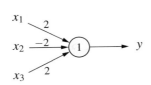

x_1	x_2	x_3	$\sum_i w_i x_i$	y
0	0	0	0	0
1	0	0	2	1
0	1	0	-2	0
1	1	0	0	0
0	0	1	2	1
1	0	1	4	1
0	1	1	0	0
1	1	1	2	1

Fig. 3.4 A threshold logic unit for $(x_1 \wedge \overline{x_2}) \vee (x_1 \wedge x_3) \vee (\overline{x_2} \wedge x_3)$

this threshold logic unit computes the function shown in the table in Fig. 3.3 on the right, that is, the implication $y = x_2 \rightarrow x_1$.

An example of a threshold logic unit with three inputs is shown in Fig. 3.4 on the left. This threshold logic unit already computes a fairly complex function, namely the logical expression $y = (x_1 \wedge \overline{x_2}) \vee (x_1 \wedge x_3) \vee (\overline{x_2} \wedge x_3)$. The truth table of this function and the computations that are carried out by the threshold logic unit for the different input vectors are shown in Fig. 3.4 on the right. This and the preceding threshold logic unit may lead us to presume that negations (in logical expressions) are (often) represented by negative weights.

3.2 Geometric Interpretation

The condition that is tested by a threshold logic unit in order to decide whether it should output a 0 or a 1 is very similar to the equation of a straight line (cf. Sect. 10.1). Indeed, the computation of a threshold logic unit can easily be interpreted geometrically if we turn this condition into a line, plane or hyperplane equation, that is, if we consider the equation

$$\sum_{i=1}^{n} w_i x_i = \theta \quad \text{or} \quad \sum_{i=1}^{n} w_i x_i - \theta = 0.$$

(Note that the line equation differs from the actual condition, in that, it uses "=" instead of "≥." This is taken care of below, where the inequality is reinstated). The resulting geometric interpretation is illustrated in Figs. 3.5, 3.6 and 3.8.

Figure 3.5 repeats on the left the threshold logic unit for the conjunction considered above. In the diagram on the right, the input space of this threshold logic unit is shown. The input vectors, which are listed in the table in Fig. 3.2 on the right, are marked according to the output of the threshold logic unit: a filled circle indicates that the threshold logic unit yields output 1 for this point, while an empty circle indicates that it yields output 0. In addition, the diagram shows the straight line described by the equation $3x_1 + 2x_2 = 4$, which corresponds to the **decision border** of the threshold logic unit. It is easy to verify that the threshold logic unit yields output 1 for all points

Fig. 3.5 Geometry of the threshold logic unit for $x_1 \wedge x_2$. The straight line shown in the right diagram has the equation $3x_1 + 2x_2 = 4$; for the gray half-plane it is $3x_1 + 2x_2 \geq 4$

Fig. 3.6 Geometry of the threshold logic unit for $x_2 \rightarrow x_1$: The straight line shown in the right diagram has the equation $2x_1 - 2x_2 = -1$; for the gray half-plane it is $2x_1 - 2x_2 \geq -1$

to the right of this line and output 0 for all points to the left of it, even if we allow for other input values than 0 and 1.

On which side the output of the threshold logic unit is 1 and on which it is 0 can also be read from the line equation: it is well known that the coefficients of x_1 and x_2 are the elements of a normal vector of the line (cf. also Sect. 10.1, which collects some important facts about straight lines and their equations.) The side of the line to which this normal vector points if it is attached to a point on the line is the side on which the output is 1. Indeed, the normal vector $\mathbf{n} = (3, 2)$ that can be read from the equation $3x_1 + 2x_2 = 4$ points to the right top and thus to the side on which the point $(1, 1)$ is located. Hence, the threshold logic unit yields output 1 in the **half-plane** $3x_1 + 2x_2 \geq 4$, which is shown in gray in Fig. 3.5.

Analogously, Fig. 3.6 shows the threshold logic unit computing the implication $x_2 \rightarrow x_1$ and its input space. The straight line drawn into this input space corresponds to its decision border: it separates the points of the input space for which the output is 0 from those for which the output is 1. Since the normal vector $\mathbf{n} = (2, -2)$ that can be read from the equation $2x_1 - 2x_2 = -1$ points to the bottom right, the output is 1 for all points below the line and 0 for all points above it. This coincides with the computations listed in the table in Fig. 3.3, which are represented by filled and empty circles in the diagram in Fig. 3.6. Generally, the threshold logic unit yields output 1 for points in the gray half-plane shown in Fig. 3.6.

Naturally, the computations of threshold logic units with more than two inputs can be interpreted geometrically as well. However, due to the limited spatial imagination of humans, we have to confine ourselves to threshold logic units with no more than three inputs. With three inputs the **separating line** turns into a **separating plane**. We illustrate this by depicting the input space of a threshold logic unit with three

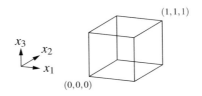

Fig. 3.7 Graphical representation of ternary Boolean functions

Fig. 3.8 Geometry of the threshold logic unit for the function $(x_1 \wedge \overline{x_2}) \vee (x_1 \wedge x_3) \vee (\overline{x_2} \wedge x_3)$: The plane shown in the left diagram is described by the equation $2x_1 - 2x_2 + 2x_3 = 1$

inputs as a unit cube as it is shown in Fig. 3.7. With this diagram, let us reconsider the example of a threshold logic unit with three inputs studied in the preceding section, which is repeated in Fig. 3.8. Into the unit cube on the right of this figure, the plane with the equation $2x_1 - 2x_2 + 2x_3 = 1$ is drawn in gray, which corresponds to the decision rule of this threshold logic unit. In addition, all input vectors, for which the table in Fig. 3.4 lists an output of 1, are marked with a filled circle. For all other corners of the unit cube, the output is 0. Like in the two-dimensional case, we can again read the side of the plane on which the output is 1 from the normal vector of the plane: from the plane equation we derive the normal vector $\mathbf{n} = (2, -2, 2)$, which points out of the drawing plane to the top right.

3.3 Limitations

The examples studied in the preceding section—especially the threshold logic unit with three inputs—may lead us to presume that threshold logic units are fairly powerful processing units. Unfortunately, though, *single* threshold logic units are severely limited in their expressive and computational power. We know from the geometric interpretation of their computations that threshold logic units can represent only functions that are, as one says, **linearly separable**, that is, functions for which the points with output 1 can be separated from the points with output 0 by a linear function—that is, by a line, plane or hyperplane.

Definition 3.2 Two sets of points in a Euclidean space are called **linearly separable** iff there exists at least one point, line, plane or hyperplane (depending on the dimension of the space), such that all points of the one set lie on one side and all

x_1	x_2	y
0	0	1
1	0	0
0	1	0
1	1	1

Fig. 3.9 The biimplication problem: there is no separating straight line

points of the other set lie on the other side of this point, line, plane or hyperplane (or on it). That is, the point sets can be separated by a **linear decision function**.

More formally, two sets of points $X, Y \subset \mathbb{R}^m$ are linearly separable if and only if there exist a vector $\mathbf{w} \in \mathbb{R}^m$ and a value $\theta \in \mathbb{R}$ such that

$$\forall \mathbf{x} \in X : \quad \mathbf{w}^\top \mathbf{x} \geq \theta \quad \text{and} \quad \forall \mathbf{y} \in Y : \quad \mathbf{w}^\top \mathbf{y} < \theta.$$

Especially Boolean functions (as studied above) define two points sets, namely the set of points that are mapped to the function value 0 and the set of points that are mapped to 1. Hence, the term "linearly separable" can be transferred to Boolean functions, or actually to any function that yields only two different values.

Unfortunately, though, not all such functions are linearly separable. A very simple example of a Boolean function that is not linearly separable is the biimplication (that is, $x_1 \leftrightarrow x_2$), the truth table of which is shown in Fig. 3.9 on the left. From the graphical representation of this function, which is shown in the same figure on the right, we can already see that there is no separating line. As a consequence, there cannot be any threshold logic unit computing this function.

The formal proof is executed by the common method of *reductio ad absurdum* (proof by contradiction). We assume that there exists a threshold logic unit with weights w_1 and w_2 and threshold θ that computes the biimplication. Then it is

$$
\begin{aligned}
\text{due to } (0,0) \mapsto 1: && 0 && \geq \theta, && (1)\\
\text{due to } (1,0) \mapsto 0: && w_1 && < \theta, && (2)\\
\text{due to } (0,1) \mapsto 0: && w_2 && < \theta, && (3)\\
\text{due to } (1,1) \mapsto 1: && w_1 + w_2 && \geq \theta. && (4)
\end{aligned}
$$

From (2) and (3), it follows $w_1 + w_2 < 2\theta$, which together with (4) yields $2\theta > \theta$, or $\theta > 0$. However, this contradicts (1). We conclude that our assumption is impossible. Rather, there is no threshold logic unit that computes the biimplication.

The fact that only linearly separable functions can be represented may appear to be, at first sight, a small and bearable restriction, since only two of the 16 possible Boolean functions of two variables are *not* linearly separable (namely the biimplication and the exclusive or). However, if the number of inputs is increased, the fraction of all Boolean functions that are linearly separable drops rapidly (see Table 3.1). For a larger number of inputs (single), threshold logic units can, therefore, compute "almost no" Boolean functions (relative to all possible ones).

For checking whether two sets of points (or a function yielding only two values) are linearly separable, the following theorem is useful:

Table 3.1 The number of all Boolean functions of n inputs and the number of them that are linearly separable (On-Line Encyclopedia of Integer Sequences oeis.org, A001146 and A00609)

Inputs	Boolean functions	Linearly separable functions
1	4	4
2	16	14
3	256	104
4	65,536	1,774
5	4,294,967,296	94,572
6	18,446,744,073,709,551,616	15,028,134
n	$2^{(2^n)}$	No general formula known

Definition 3.3 A set of points in a Euclidean space is called **convex** if it is non-empty and connected (that is, if it is a *region*) and for every pair of points in it, every point on the straight line segment connecting the points of the pair is also in the set.

Definition 3.4 The **convex hull** of a set of points X in a Euclidean space is the smallest convex set of points that contains X. Alternatively, the **convex hull** of a set of points X is the intersection of all convex sets that contain X.

Theorem 3.1 *Two sets of points in a Euclidean space are* **linearly separable** *if and only if their convex hulls are disjoint (that is, have no point in common).*

To illustrate this theorem (which we do not prove here), we apply it to the biimplication problem. Here, the convex hulls of the two point sets are the diagonal line segments, connecting $(0, 0)$ and $(1, 1)$ for output 1 and connecting $(1, 0)$ and $(0, 1)$ for output 0. These two line segments share their intersection point and are thus not disjoint. Therefore, the biimplication is not linearly separable.

3.4 Networks of Threshold Logic Units

As demonstrated in the preceding section, threshold logic units are severely limited. However, up to now, we only considered *single* threshold logic units. The powers of threshold logic units can be increased considerably if we combine several threshold logic units, that is, if we consider networks of threshold logic units.

As an example, we consider a possible solution for the biimplication problem with the help of three threshold logic units that are organized into two layers. This solution exploits the logical equivalence

$$x_1 \leftrightarrow x_2 \equiv (x_1 \to x_2) \wedge (x_2 \to x_1),$$

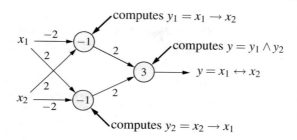

Fig. 3.10 Combining several threshold logic units

which splits the biimplication into three functions. From Figs. 3.3 and 3.6, we already know that the implication $x_2 \to x_1$ is linearly separable. In the implication $x_1 \to x_2$, the variables are merely exchanged, so it is linearly separable as well. Finally, we know from Figs. 3.2 and 3.5 that the conjunction of two Boolean variables is linearly separable. As a consequence, we only have to connect the corresponding threshold logic units, see Fig. 3.10. Thus, we obtain a network with two layers, corresponding to the nested structure of the logical expression.

Intuitively, the two threshold logic units on the left (first layer) compute new Boolean coordinates y_1 and y_2 for the input vectors, so that the transformed input vectors in the input space of the threshold logic unit on the right (second layer) become linearly separable. This is illustrated geometrically with the two diagrams shown in Fig. 3.11. The separating line g_1 corresponds to the upper threshold logic unit and describes the implication $y_1 = x_1 \to x_2$: for all points above this line the output is 1, for all points below it, the output is 0. The separating line g_2 belongs to the lower threshold logic unit and describes the implication $y_2 = x_2 \to x_1$: for all points above this line the output is 0, for all points below it the output is 1.

The threshold logic units on the left assign the new coordinates $(y_1, y_2) = (1, 0)$ to the input vector $b \mathrel{\hat{=}} (x_1, x_2) = (1, 0)$ and the new coordinates $(y_1, y_2) = (0, 1)$ to the input vector $c \mathrel{\hat{=}} (x_1, x_2) = (0, 1)$, while they assign the coordinates $(y_1, y_2) = (1, 1)$ to both the input vector $a \mathrel{\hat{=}} (x_1, x_2) = (0, 0)$ and the input vector $d \mathrel{\hat{=}} (x_1, x_2) =$

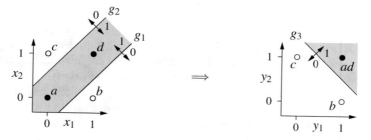

Fig. 3.11 Geometric interpretation of combining multiple threshold logic units into a network to compute the biimplication

(1, 1) (see Fig. 3.11 on the right). After this transformation, the input vectors for which the output is 1 can easily be separated from those for which the output is 0, for instance, by the line g_3 that is shown in the diagram in Fig. 3.11 on the right.

It can be shown that all Boolean functions with an arbitrary number of inputs can be computed by networks of threshold logic units, simply by exploiting logical equivalences to divide these functions in such a way that all occurring sub-functions are linearly separable. With the help of the disjunctive normal form (or, analogously, the conjunctive normal form) one can even show that the networks only need to have two layers, regardless of the Boolean function to represent:

Algorithm 3.1 (Representation of Boolean Functions)
Let $y = f(x_1, \ldots, x_n)$ be a Boolean function of n variables.

1. Represent the Boolean function $f(x_1, \ldots, x_n)$ in disjunctive normal form. That is, determine $D_f = K_1 \vee \cdots \vee K_m$, where all K_j are conjunctions of n literals, that is, $K_j = l_{j1} \wedge \cdots \wedge l_{jn}$ with $l_{ji} = x_i$ (positive literal) or $l_{ji} = \neg x_i$ (negative literal).

2. For each conjunction K_j of the disjunctive normal form create one neuron (with n inputs—one input for each of the variables) where

$$w_{ji} = \begin{cases} 2 \text{ if } l_{ji} = x_i, \\ -2 \text{ if } l_{ji} = \neg x_i, \end{cases} \quad \text{and} \quad \theta_j = n - 1 + \frac{1}{2} \sum_{i=1}^{n} w_{ji}.$$

3. Create one output neuron (with m inputs—one input for each of the neurons created in step 2) where

$$w_{(n+1)k} = 2, \quad k = 1, \ldots, m, \quad \text{and} \quad \theta_{n+1} = 1.$$

In the network constructed in this way, every neuron created in step 2 computes a conjunction and the output neuron computes their disjunction.

Intuitively, each neuron in the first layer describes a hyperplane that separates the corner of the hypercube, for which the conjunction is 1, from the rest of the unit hypercube. The equation of this hyperplane is easy to determine: its normal vector points from the center of the unit hypercube to the corner that is cut off and thus is 1 in all components in which the position vector of the corner has value 1, and it is -1 in all components in which the position vector of the corner has the value 0. We multiply this normal vector with 2 in order to obtain an integer threshold (only for aesthetics, no significance attached). The threshold has to be determined in such a way that it is exceeded only if all inputs that carry a weight of 2 are 1 and all other inputs are 0. The formula stated in step 2 yields such a value.

To compute the disjunction of the outputs of the neurons created in step 2, we have to separate, in an m-dimensional unit hypercube of the conjunctions, the corner $(0, \ldots, 0)$, for which the output is 0, from all other corners, for which the output is 1. This can be achieved, for instance, with the hyperplane that possesses the normal vector $(1, \ldots, 1)$ and the support vector $(\frac{1}{2}, 0, \ldots, 0)$. (As an illustration, consider the three-dimensional case again.) The parameters of the output neuron stated in step 3 are then simply read from the corresponding equation.

x_1	x_2	x_3	y	C_j
0	0	0	0	
1	0	0	1	$x_1 \wedge \overline{x_2} \wedge \overline{x_3}$
0	1	0	0	
1	1	0	0	
0	0	1	0	
1	0	1	0	
0	1	1	1	$\overline{x_1} \wedge x_2 \wedge x_3$
1	1	1	1	$x_1 \wedge x_2 \wedge x_3$

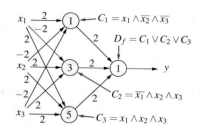

Fig. 3.12 A ternary Boolean function and a network computing it as it results from Algorithm 3.1

First layer (conjunctions):

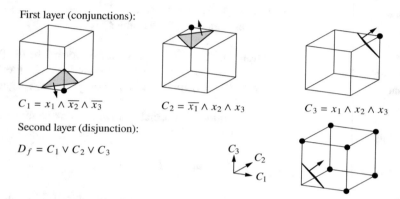

$C_1 = x_1 \wedge \overline{x_2} \wedge \overline{x_3}$ $C_2 = \overline{x_1} \wedge x_2 \wedge x_3$ $C_3 = x_1 \wedge x_2 \wedge x_3$

Second layer (disjunction):

$D_f = C_1 \vee C_2 \vee C_3$

Fig. 3.13 Geometric interpretation of the network shown in Fig. 3.12 (cf. also Fig. 3.7)

To illustrate this algorithm, we apply it to ternary Boolean function, the truth table of which is shown on the left in Fig. 3.12. In order to construct the disjunctive normal form of this function, we create one conjunction for each row of the truth table where the output y is 1 with literals according to input values (positive literal if input is 1, negative literal if it is 0). The needed conjunctions are shown in the rightmost column of the table. They give rise to the three threshold logic units in the first layer (left) of the network shown on the right in Fig. 3.12. The outputs of these three units have to be disjunctively combined, which is achieved with the threshold logic unit in the second layer (right). It is easy to check that this network actually computes the desired function. Its geometric interpretation is shown in Fig. 3.13.

3.5 Training the Parameters

By interpreting the computations of a threshold logic unit geometrically, as shown in Sect. 3.2, we have (at least for functions with 2 and 3 variables) a simple method to find, for a given linearly separable function, a threshold logic unit that computes it: we determine a line, plane or hyperplane that separates the points for which the output is 1 from those for which the output is 0. In doing so, we take care that the normal vector of that line, plane or hyperplane points toward the points for which the output is one. From the equation describing this line, plane or hyperplane, we can then easily read the weights and the threshold.

However, this method becomes difficult and eventually infeasible if the function to compute has more than three arguments, because we cannot imagine the input space of such a function. Furthermore, it is impossible to automate this method, because we find a suitable separating line or plane by "visual inspection" of the point sets to separate. This "visual inspection" cannot be mimicked directly by a computer. In order to be able to determine the parameters of a threshold logic unit with a computer, so that it computes a given function, we need a different approach. The principle consists in starting with randomly chosen values for the weights and the threshold and then changing these values step by step until the desired function is computed. The slow, stepwise adaptation of the weights and the threshold is also called **learning** or—in order to avoid confusions with the much more complex human learning process—the **training** of the threshold logic unit.

In order to find a method to adapt the weights and the threshold, we start from the following consideration: depending on the values of the weights and the threshold, the computation of the threshold logic unit will be more or less correct. Therefore, we can define an error function $e(w_1, \ldots, w_n, \theta)$, which states how well, for given weights and threshold, the computed function coincides with the desired one. Our objective is, of course, to determine the weights and the threshold in such a way that the error vanishes, that is, that the error function becomes 0. To achieve this, we try to reduce the value of the error function in every step.

We illustrate this procedure with the help of a very simple example, namely a threshold logic unit with only one input. The parameters of this unit are to be determined in such a way that it computes the negation. Such a threshold logic unit is shown in Fig. 3.14 together with the two training examples for the negation: if the input is 0, the output should be 1, if the input is 1, the output should be 0.

The error function we define first, as it appears to be natural, as the absolute value of the difference between the desired and the actual output. This function is shown in Fig. 3.15. The left diagram shows the error for the input $x = 0$, for which an output of 1 is desired. Since the threshold logic unit computes a 1 if $xw \geq \theta$, the error is 0 for a negative threshold and 1 for a positive threshold. (Obviously, the weight does not have any influence, because it is multiplied with the input, which is 0.) The middle diagram shows the error for the input $x = 1$, for which an output of 0 is desired. Here, both the weight and the threshold have an influence. If the weight is less than the threshold, we have $xw < \theta$ and thus the output and consequently the

Fig. 3.14 A threshold logic unit with a single input and training examples for the negation

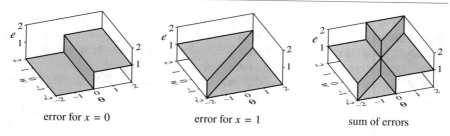

error for $x = 0$ error for $x = 1$ sum of errors

Fig. 3.15 Error of computing the negation w.r.t. the threshold

error is 0. The diagram on the right shows the sum of these individual errors. Clearly, our objective is to make this error sum vanish.

From the right diagram, a human can now easily read how the weight and the threshold have to be chosen so that the threshold logic unit computes the negation: the values of these parameters must lie in the triangle in the lower left of the w-θ plane, in which the error is 0. However, it is not yet possible to automatically adapt the parameters with this function, because the global "visual inspection" of the error function, which a human relies on to find the solution, cannot be mimicked directly in a computer. Rather we would have to be able to read from the shape of the function at the point, which is given by the current weight and the current threshold, in which directions we have to change the weight and the threshold so that the error is reduced. With this error function, however, this is impossible, because it consists of "plateaus" or "terraces." In "almost all" points (the "edges" of the error function are the only exceptions) the error stays the same in all directions.[2]

In order to circumvent this problem, we modify the error function. Where the threshold logic unit produces the wrong output, we consider how far the threshold is exceeded (for a desired output of 0) or how far it is underrun (for a desired output of 1). Intuitively, we may say that the computation is "the more wrong" the farther the threshold is exceeded for a desired output of 0, or the farther the threshold is underrun for a desired output of 1. The modified error function is shown in Fig. 3.16. Again the left diagram shows the error for the input $x = 0$, the middle diagram the error for the input $x = 1$, and the right diagram the sum of these individual errors.

If a threshold logic unit now produces a wrong output, we adapt the weight and the threshold in such a way that the error is reduced. That is, we try to "descend in the error landscape". With the modified error function, this is possible, because we can read from it "locally" (that is, without a visual inspection of the whole error function, but merely by looking at the shape of the error function at the point that is given by the current values of the weight and the threshold) in which directions we have to change the weight and the threshold: we simply move in the direction in which the error function has the strongest downward slope. Intuitively, we follow the

---

[2] The somewhat imprecise notion "almost all points" can be made mathematically precise by drawing on measure theory: the set of points at which the error function changes has measure 0.

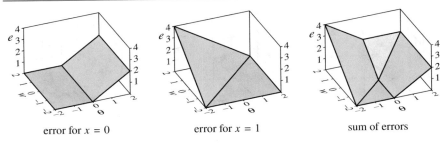

error for $x = 0$ error for $x = 1$ sum of errors

Fig. 3.16 Error of computing the negation w.r.t. how far the threshold is exceeded or underrun

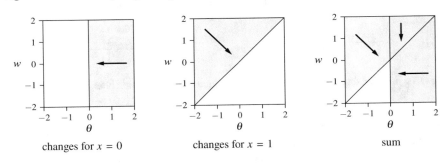

changes for $x = 0$ changes for $x = 1$ sum

Fig. 3.17 Directions of the weight and threshold changes

common scout advice on how to find water: always go downhill. The directions that result from this rule are shown schematically in Fig. 3.17. The arrows indicate how the weight and the threshold should be adapted in different regions of the parameter space. In those regions, in which no arrows are drawn, weight and threshold are left unchanged, because there is no error.

The adaptation rules that are shown in Fig. 3.17 can be applied in two different ways. In the first place, we may consider the inputs $x = 0$ and $x = 1$ alternatingly and may adapt the weight and the threshold according to the corresponding rules. That is, first we adapt the weight and the threshold according to the left diagram, then we adapt them according to the middle diagram, then we adapt them again according to the left diagram and so forth until the error vanishes. This way of training a neural network is called **online learning** or **online training**, since with every training example that becomes available, a training step can be carried out.

The second option consists of aggregating the changes over all training examples without applying them immediately. Only at the end of a **(learning/training) epoch**, that is, after all training examples have been traversed, the aggregated changes are applied. Then the training examples are traversed again and, at the end, the weight and the threshold are adapted and so forth until the error vanishes. This way of training is called **(full) batch learning** or **(full) batch training**, since all training examples

Fig. 3.18 Training processes with initial values $\theta = \frac{3}{2}$, $w = 2$ and learning rate 1

Fig. 3.19 Learned threshold logic unit for the negation and its geometric interpretation

have to be available together (in a *batch*), and corresponds to adapting weight and threshold according to the diagram on the right.[3]

Figure 3.18 shows the training processes for the initial values $\theta = \frac{3}{2}$ and $w = 2$. Both online training (left) and batch training (middle) use a **learning rate** of 1. The learning rate states by how much the weight and the threshold are changed, and thus how "fast" the training is. (However, the learning rate should also not be chosen arbitrarily large, see Chap. 5.) If the learning rate is 1, the weight and the threshold are increased or reduced by 1. In order to illustrate the "descent in the error landscape", the batch training is repeated in a three-dimensional diagram in Fig. 3.18 on the right. The final, fully trained threshold logic unit (with $\theta = -\frac{1}{2}$ and $w = -1$) is shown, together with its geometric interpretation, in Fig. 3.19.

In this simple example, we derived the adaptation rules directly from a visual inspection of the error function. An alternative way of obtaining these adaptation rules are the following considerations: if a threshold logic unit produces an output of 1 instead of a desired 0, then the threshold is too small and/or the weights are too large. Hence, we should increase the threshold and reduce the weights. Of course, the latter is reasonable only if the corresponding input is 1, as otherwise the weight has no influence on the output.) Contrariwise, if a threshold logic unit produces an output of 0 instead of a desired 1, then the threshold is too large and/or the weights are too small. Hence, the threshold should be reduced and the weights should be increased (provided, of course, that the corresponding input is 1).

For our simple threshold logic unit, the changes shown in Fig. 3.17 have exactly these effects. However, the considerations above have the advantage that they can be

---

[3] The reason for the qualifier "(full)" instead of merely calling this approach "batch training" or "batch learning" will be discussed later in the context of stochastic gradient descent.

Fig. 3.20 Turning the threshold into a weight

applied to threshold logic units with more than one input. Therefore, we can define the following general training method for threshold logic units:

Definition 3.5 Let $\mathbf{x} = (x_1, \ldots, x_n)^\top$ be an input vector of a threshold logic unit, o the desired output for this input vector and y the actual output of the threshold logic unit. If $y \neq o$, then, in order to reduce the error, the threshold θ and the weight vector $\mathbf{w} = (w_1, \ldots, w_n)^\top$ are adapted as follows:

$$\theta^{(\text{new})} = \theta^{(\text{old})} + \Delta\theta \quad \text{with } \Delta\theta = -\eta(o - y),$$
$$\forall i \in \{1, \ldots, n\} : w_i^{(\text{new})} = w_i^{(\text{old})} + \Delta w_i \text{ with } \Delta w_i = \eta(o - y)x_i,$$

where η is a parameter that is called **learning rate**. It determines the severity of the weight and threshold changes. This method is called the **delta rule** or **Widrow–Hoff procedure** [17].

In this definition, we have to distinguish between an adaptation of the threshold and adaptations of the weights, because the directions of these changes are opposite to each other (opposite signs for $\eta(t - y)$ and $\eta(t - y)x_i$, respectively). However, we can unify the adaptation rules by turning the threshold into a weight. The rationale of this transformation is illustrated in Fig. 3.20: the threshold is fixed to 0 and to compensate this reduction in the number of parameters, an additional (imaginary) input x_0 is introduced, which has a fixed value of 1. This input is weighted with the negated threshold. The two threshold logic units are clearly equivalent, since the left one tests the condition $\sum_{i=1}^{n} w_i x_i \geq \theta$, the right one tests $\sum_{i=1}^{n} w_i x_i - \theta \geq 0$, in order to determine the output. Note that the same effect is produced if the additional input has the fixed value -1 and is weighted with the threshold directly (shown in blue). This alternative way of turning the threshold into a weight is also very common.

Regardless of how the threshold is turned into a weight, we obtain the same adaptation directions for all parameters. With a negated threshold: if the output is 1 instead of 0, then both the w_i as well as $-\theta$ should be reduced. If the output is 0 instead of 1, both the w_i as well as $-\theta$ should be increased. On the other hand, with an unnegated θ and a fixed input of -1, we obtain a uniform rule, because the needed negative sign is produced by the input. Therefore, we can determine the adaptation direction of all parameters by simply subtracting the actual from the desired output.

Thus, we can formulate the delta rule as follows: Let $\mathbf{x} = (x_0 = 1, x_1, \ldots, x_n)^\top$ be an extended input vector of a threshold logic unit (note the additional input $x_0 = 1$), o the desired output for this input vector and y the actual output of the threshold logic unit. If $y \neq o$, then, in order to reduce the error, the extended weight vector $\mathbf{w} = (w_0 = -\theta, w_1, \ldots, w_n)^\top$ (note the added weight $w_0 = -\theta$) is adapted as follows:

$$\forall i \in \{0, 1, \ldots, n\} : \quad w_i^{(\text{new})} = w_i^{(\text{old})} + \Delta w_i \quad \text{with} \quad \Delta w_i = \eta(o - y)x_i.$$

We point out this possibility here because it can often be used, for example, to simplify derivations (see, for instance, Sect. 5.5). However, for the sake of clarity, we maintain the distinction of weight and threshold for the remainder of this chapter.

With the help of the delta rule, we can now state two algorithms for training a threshold logic unit: an online version and a batch version. In order to formulate these algorithms, we assume that we are given a set $L = \{(\mathbf{x}_1, o_1), \ldots, (\mathbf{x}_m, o_m)\}$ of training examples, each of which consist of an input vector $\mathbf{x}_i \in \mathbb{R}^n$ and the desired output $o_i \in \{0, 1\}$ for this input vector, $i = 1, \ldots, m$. Furthermore, let arbitrary initial weights $\mathbf{w}$ and an arbitrary initial threshold θ be given (e.g., chosen randomly). We consider online training first:

Algorithm 3.2 (Online Training of a Threshold Logic Unit)

```
procedure online_training (varw, var θ, L, η);
var y, e;                              (* output, sum of errors *)
begin
  repeat
    e := 0;                            (* initialize the sum of errors *)
    for all (x, o) ∈ L do begin        (* traverse the examples *)
      if (wᵀx ≥ θ) then   y := 1;      (* compute the output of *)
                  else   y := 0;       (* the threshold logic unit *)
      if (y ≠ o) then begin            (* if the output is wrong *)
        θ := θ − η(o − y);             (* adapt the threshold *)
        w := w + η(o − y)x;            (* and the weights *)
        e := e + |o − y|;              (* sum the errors *)
      end;
    end;
  until (e ≤ 0);                       (* repeat the computations *)
end;                                   (* until the error vanishes *)
```

Obviously, this algorithm repeatedly applies the delta rule until the sum of errors over all training examples vanishes. Note that, in this algorithm, the weight adaptation is written in vector form, which, however, is clearly equivalent to an adaptation of the individual weights. Let us now turn to the batch version

Algorithm 3.3 (Batch Training of a Threshold Logic Unit)

procedure batch_training (**var w**, **var** θ, L, η);
var y, e,   (* output, sum of errors *)
    θ_c, $\mathbf{w}_c$;   (* aggregated changes *)
begin
  repeat
    $e := 0$; $\theta_c := 0$; $\mathbf{w}_c := \mathbf{0}$;   (* initializations *)
    for all $(\mathbf{x}, o) \in L$ **do begin**   (* traverse the examples *)
      if $(\mathbf{w}^\top \mathbf{x} \geq \theta)$ **then** $y := 1$;   (* compute the output of *)
              else $y := 0$;   (* the threshold logic unit *)
      if $(y \neq o)$ **then begin**   (* if the output is wrong *)
        $\theta_c := \theta_c - \eta(o - y)$;   (* sum the threshold and *)
        $\mathbf{w}_c := \mathbf{w}_c + \eta(o - y)\mathbf{x}$; (* the weight changes *)
        $e := e + |o - y|$;   (* sum the errors *)
      end;
    end;
    $\theta := \theta + \theta_c$;   (* adapt the threshold *)
    $\mathbf{w} := \mathbf{w} + \mathbf{w}_c$;   (* and the weights *)
  until $(e \leq 0)$;   (* repeat the computations *)
end;   (* until the error vanishes *)

In this algorithm, the delta rule is applied in a modified form, since for each traversal of the training examples, the same threshold and the same weights are used. If the output is wrong, the computed changes are not applied directly, but summed in the variables θ_c and $\mathbf{w}_c$. Only after all training examples have been visited, the threshold and the weights are adapted with the help of these variables.

To illustrate the operation of these two algorithms, Table 3.2 shows the online training of the simple threshold logic unit considered above, which is to be trained in such a way that it computes the negation. Like in Fig. 3.18 on p. 28 the initial values are $\theta = \frac{3}{2}$ and $w = 3$. It is easy to check that the online training shown here in tabular form corresponds exactly to the one shown graphically in Fig. 3.18 on the left. Analogously, Table 3.3 shows batch training. It corresponds exactly to the procedure depicted in Fig. 3.18 in the middle or on the right.

As another example, we consider a threshold logic unit with two inputs that are to be trained in such a way that it computes the conjunction of its inputs. Such a threshold logic unit is shown, together with the corresponding training examples, in Fig. 3.21. For this example, we only consider online training. The corresponding training procedure for the initial values $\theta = w_1 = w_2 = 0$ with learning rate 1 is shown in Table 3.4. As for the negation, the training is successful and finally yields the threshold $\theta = 3$ and the weights $w_1 = 2$ and $w_2 = 1$. The resulting threshold logic unit is shown, together with its geometric interpretation, in Fig. 3.22. Note that this threshold logic unit indeed computes the conjunction, even though the point $(1, 1)$ lies on the separating line, because it yields output 1 not only for points to the right of the line, but also for all points *on* the line. This shows that training

Table 3.2 Online training for the negation with initial values $\theta = \frac{3}{2}$, $w = 2$ and learning rate 1

Epoch	x	o	wx	y	e	$\Delta\theta$	Δw	θ	w
								1.5	2
1	0	1	−1.5	0	1	−1	0	0.5	2
	1	0	1.5	1	−1	1	−1	1.5	1
2	0	1	−1.5	0	1	−1	0	0.5	1
	1	0	0.5	1	−1	1	−1	1.5	0
3	0	1	−1.5	0	1	−1	0	0.5	0
	1	0	−0.5	0	0	0	0	−0.5	0
4	0	1	−0.5	0	1	−1	0	−0.5	0
	1	0	0.5	1	−1	1	−1	0.5	−1
5	0	1	−0.5	0	1	−1	0	−0.5	−1
	1	0	−0.5	0	0	0	0	−0.5	−1
6	0	1	0.5	1	0	0	0	−0.5	−1
	1	0	−0.5	0	0	0	0	−0.5	−1

Table 3.3 Batch training for the negation with initial values $\theta = \frac{3}{2}$, $w = 2$ and learning rate 1

Epoch	x	o	wx	y	e	$\Delta\theta$	Δw	θ	w
								1.5	2
1	0	1	−1.5	0	1	−1	0		
	1	0	0.5	1	−1	1	−1	1.5	1
2	0	1	−1.5	0	1	−1	0		
	1	0	−0.5	0	0	0	0	0.5	1
3	0	1	−0.5	0	1	−1	0		
	1	0	0.5	1	−1	1	−1	0.5	0
4	0	1	−0.5	0	1	−1	0		
	1	0	−0.5	0	0	0	0	−0.5	0
5	0	1	0.5	1	0	0	0		
	1	0	0.5	1	−1	1	−1	0.5	−1
6	0	1	−0.5	0	1	−1	0		
	1	0	−1.5	0	0	0	0	−0.5	−1
7	0	1	0.5	1	0	0	0		
	1	0	−0.5	0	0	0	0	−0.5	−1

Fig. 3.21 A threshold logic unit with two inputs and training examples for the conjunction $y = x_1 \wedge x_2$

Table 3.4 Training of a threshold logic unit for the conjunction

Epoch	x_1	x_2	o	$\mathbf{w}^{\top}\mathbf{x}$	y	e	$\Delta\theta$	Δw_1	Δw_2	θ	w_1	w_2
										0	0	0
1	0	0	0	0	1	−1	1	0	0	1	0	0
	0	1	0	−1	0	0	0	0	0	1	0	0
	1	0	0	−1	0	0	0	0	0	1	0	0
	1	1	1	−1	0	1	−1	1	1	0	1	1
2	0	0	0	0	1	−1	1	0	0	1	1	1
	0	1	0	0	1	−1	1	0	−1	2	1	0
	1	0	0	−1	0	0	0	0	0	2	1	0
	1	1	1	−1	0	1	−1	1	1	1	2	1
3	0	0	0	−1	0	0	0	0	0	1	2	1
	0	1	0	0	1	−1	1	0	−1	2	2	0
	1	0	0	0	1	−1	1	−1	0	3	1	0
	1	1	1	−2	0	1	−1	1	1	2	2	1
4	0	0	0	−2	0	0	0	0	0	2	2	1
	0	1	0	−1	0	0	0	0	0	2	2	1
	1	0	0	0	1	−1	1	−1	0	3	1	1
	1	1	1	−1	0	1	−1	1	1	2	2	2
5	0	0	0	−2	0	0	0	0	0	2	2	2
	0	1	0	0	1	−1	1	0	−1	3	2	1
	1	0	0	−1	0	0	0	0	0	3	2	1
	1	1	1	0	1	0	0	0	0	3	2	1
6	0	0	0	−3	0	0	0	0	0	3	2	1
	0	1	0	−2	0	0	0	0	0	3	2	1
	1	0	0	−1	0	0	0	0	0	3	2	1
	1	1	1	0	1	0	0	0	0	3	2	1

does not necessarily yield an "optimal" solution, but merely one that "works", even though we may prefer a separating line between the filled and empty points.

After we have seen two examples of successful training, we naturally face the question whether Algorithms 3.2 and 3.3 always achieve their objective. As a first

Fig. 3.22 Geometry of the learned threshold logic unit for $x_1 \wedge x_2$. The straight line shown on the right is described by the equation $2x_1 + x_2 = 3$

Table 3.5 Training of a threshold logic unit for the biimplication

Epoch	x_1	x_2	o	$\mathbf{w}^\mathsf{T}\mathbf{x}$	y	e	$\Delta\theta$	Δw_1	Δw_2	θ	w_1	w_2
										0	0	0
1	0	0	1	0	1	0	0	0	0	0	0	0
	0	1	0	0	1	-1	1	0	-1	1	0	-1
	1	0	0	-1	0	0	0	0	0	1	0	-1
	1	1	1	-2	0	1	-1	1	1	0	1	0
2	0	0	1	0	1	0	0	0	0	0	1	0
	0	1	0	0	1	-1	1	0	-1	1	1	-1
	1	0	0	0	1	-1	1	-1	0	2	0	-1
	1	1	1	-3	0	1	-1	1	1	1	1	0
3	0	0	1	0	1	0	0	0	0	0	1	0
	0	1	0	0	1	-1	1	0	-1	1	1	-1
	1	0	0	0	1	-1	1	-1	0	2	0	-1
	1	1	1	-3	0	1	-1	1	1	1	1	0

step, we can assert that these algorithms do not terminate if the function to be learned is *not* linearly separable. This is illustrated in Fig. 3.5 with the help of the online training procedure for the biimplication. Epochs 2 and 3 are clearly identical and will thus be repeated indefinitely without a solution ever being found. However, this is not surprising, since the training procedure terminates only if the sum of the errors over all training examples vanishes. Since we know from Sect. 3.3 that there is no threshold logic unit that computes the biimplication, the error can never vanish and thus the algorithm cannot terminate (Table 3.5).

For linearly separable functions, however, that is, for functions that can actually be computed by a threshold logic unit, it is guaranteed that the algorithms find a solution. That is, the following theorem holds:

Theorem 3.2 (Convergence Theorem for the Delta Rule) *Let* $L = \{(\mathbf{x}_1, o_1), \ldots (\mathbf{x}_m, o_m)\}$ *be a set of training examples, each consisting of an input vector* $\mathbf{x}_i \in \mathbb{R}^n$ *and the desired output* $o_i \in \{0, 1\}$ *for this input vector. Furthermore, let* $L_0 =$

$\{(\mathbf{x}, o) \in L \mid o = 0\}$ and $L_1 = \{(\mathbf{x}, o) \in L \mid o = 1\}$. *If L_0 and L_1 are linearly separable, that is, if there exist $\mathbf{w} \in \mathbb{R}^n$ and $\theta \in \mathbb{R}$ such that*

$$\forall(\mathbf{x}, 0) \in L_0 : \quad \mathbf{w}^\top \mathbf{x} < \theta \quad and \quad \forall(\mathbf{x}, 1) \in L_1 : \quad \mathbf{w}^\top \mathbf{x} \geq \theta,$$

then the Algorithms 3.2 and 3.3 terminate.

Proof The proof, which we do not want to spell out here, can be found, for example, in [9] or in [6].

Since both algorithms terminate only when the error vanishes, the computed values for the threshold and the weights must be a solution of the learning problem.

3.6 Variants

All examples that we considered up to now referred to logical functions and we encoded *false* as 0 and *true* as 1. However, this encoding has the disadvantage that with an input of *false* the corresponding weight cannot be changed, because the formula for the weight change contains the input as a factor (see Definition 3.5 on p. 29). This disadvantage can, in certain situations, slow down training unnecessarily, since a weight can only be adapted if the corresponding input is *true*.

To avoid this problem, the ADALINE model (ADAptive LINear Element) relies on the encoding *false* $\,\hat{=}\, - 1$ and *true* $\,\hat{=}\, + 1$. Thus, an input of *false* also leads, provided the output is wrong, to a weight adaptation. Indeed, the delta rule was originally developed for the ADALINE model [17], so that strictly we may only speak of the **delta rule** or the **Widrow–Hoff procedure** if the ADALINE model is employed. Although the procedure is equally applicable (and has the same convergence properties) for the encoding *false* $\,\hat{=}\, 0$ and *true* $\,\hat{=}\, 1$ (see the preceding section), it is sometimes called **error correction procedure** to avoid confusion [7,8]. We ignore this distinction here, because it is due to historical rather than conceptual reasons.

3.7 Training Networks

After simple neuro-computers had been used successfully for pattern recognition tasks at the end of the 1950s (for example, [10]), the simple and fast delta rule training method had been developed by [17] and the perceptron convergence theorem (corresponds to the convergence theorem for the delta rule) had been proven by [11], great expectations were placed in the development of (artificial) neural networks. As already mentioned in the introduction, this started the "first bloom" of neural network research, in which it was believed that one had already discovered the core principles underlying systems that are able to learn automatically.

Only after [5] carried out a careful mathematical analysis of the perceptron and pointed out in all clarity that threshold logic units can compute only linearly separable functions, the limitations of the models and procedures used at the time were properly recognized. Although it was already known from the early works of [4], that the limitations of the expressive and computational power can be lifted by using *networks* of threshold logic units (since, for example, such networks can compute arbitrary Boolean functions), training methods were confined to *single* threshold logic units.

Unfortunately, transferring the training procedures to networks of threshold logic units turned out to be a surprisingly difficult problem. For example, the delta rule derives the weight adaptation from the difference between the actual and the desired output (see Definition 3.5 on p. 29). However, the desired output is available only for the neuron that yields the output of the network as a whole. For all other threshold logic units, which carry out some kind of preprocessing and transmit their outputs only to other threshold logic units, no such desired output can be given. As an example, consider the biimplication problem and the structure of the network that we proposed as a solution for this problem (Fig. 3.10 on p. 22): the training examples do not state desired outputs for the two threshold logic units on the left. One of the main reasons for this is that the necessary coordinate transformation is not uniquely determined: separating lines in the input space may just as well be placed in a completely different way (for example, perpendicular to the bisectrix) or one may direct the normal vectors somewhat differently.

As a consequence, (artificial) neural networks were seen as a "research dead end" and the so-called "dark age" of neural network research began. Only when the training procedure of **error backpropagation** was developed, the area was revived. This procedure was described first in [16], but did not receive any attention. Only when [13,14] independently developed the method again and advertised it in the research community, the modern age ("second bloom") of (artificial) neural network began, which lasts to the present day, with only a temporary diversion of attention to other methods in the 2000s (as already mentioned in the introduction).

We consider error backpropagation only in Chap. 5, since it cannot be applied directly to threshold logic units. It requires that the activation of a neuron does not jump at a crisply defined threshold from 0 to 1, but that the activation rises slowly, according to a differentiable function. For networks consisting of pure threshold logic units, no training method is known still.

References

1. Anderson JA, Rosenfeld E (1988) Neurocomputing: foundations of research. MIT Press, Cambridge, MA, USA
2. Boden MA (ed) (1990) The philosophy of artificial intelligence. Oxford University Press, Oxford, United Kingdom
3. McCulloch WS (1965) Embodiments of mind. MIT Press, Cambridge, MA, USA

4. McCulloch WS, Pitts WH (1943) A logical calculus of the ideas immanent in nervous activity. Bull Math Biophys 5:115–133. USA. Reprinted in [McCulloch 1965], 19–39, in [Anderson und Rosenfeld 1988], 18–28, and in [Boden 1990], 22–39

5. Minsky LM, Papert S (1969) Perceptrons. MIT Press, Cambridge, MA, USA

6. Nauck D, Klawonn F, Kruse R (1997) Foundations of neuro-fuzzy systems. Wiley, Chichester, United Kingdom

7. Nilsson NJ (1965) Learning machines: the foundations of trainable pattern-classifying systems. McGraw-Hill, New York, NY

8. Nilsson NJ (1998) Artificial intelligence: a new synthesis. Morgan Kaufmann, San Francisco, CA, USA

9. Rojas R (1996) Theorie der neuronalen Netze – Eine systematische Einführung. Springer, Berlin, Germany

10. Rosenblatt F (1958) The perceptron: a probabilistic model for information storage and organization in the brain. Psychol Rev 65:386–408. USA

11. Rosenblatt F (1962) Principles of neurodynamics. Spartan Books, New York, NY, USA

12. Rumelhart DE, McClelland JL (eds) (1986) Parallel distributed processing: explorations in the microstructures of cognition, vol 1: foundations

13. Rumelhart DE, Hinton GE, Williams RJ (1986a) Learning internal representations by error propagation. In [Rumelhart und McClelland 1986], pp 318–362

14. Rumelhart DE, Hinton GE, Williams RJ (1986b) Learning representations by back-propagating errors. Nature 323:533–536

15. Wasserman PD (1989) Neural computing: theory and practice. Van Nostrand Reinhold, New York, NY, USA

16. Werbos PJ (1974) Beyond Regression: New Tools for Prediction and Analysis in the Behavioral Sciences. PhD Thesis, Harvard University, Cambridge, MA, USA

17. Widrow B, Hoff ME (1960) Adaptive switching circuits. In: IRE WESCON convention record. Institute of Radio Engineers, New York, NY, USA, pp 96–104

General Neural Networks

4

In this chapter, we introduce a general model of (artificial) neural networks that capture (more or less) all special forms, which we consider in the following chapters. We start by defining the structure of an (artificial) neural network and then describe generally the operation and finally the training of an (artificial) neural network.

4.1 Structure of Neural Networks

In the preceding chapter, we already considered, briefly, networks of threshold logic units. The way in which we represented these networks suggests to describe neural networks with the help of a graph (in the sense of graph theory). Therefore, we first define the notion of a graph and a few useful notions, which we draw on in the following definitions and the subsequent chapters.

Definition 4.1 A (directed) **graph** is a pair $G = (V, E)$ consisting of a (finite) set V of **vertices** or **nodes** and a (finite) set $E \subseteq V \times V$ of **edges**. We say that an edge $e = (u, v) \in E$ is **directed** from the vertex u to the vertex v.

It is, of course, also possible to define undirected graphs (for example, by using unordered pairs $\{u, v\}$ for the edges). However, to describe neural networks, we only need directed graphs, since the connections between neurons are always directed.

Definition 4.2 Let $G = (V, E)$ be a (directed) graph and $u \in V$ a vertex. The vertices of the set

$$\mathrm{pred}(u) = \{v \in V \mid (v, u) \in E\}$$

© Springer Nature Switzerland AG 2022
R. Kruse et al., *Computational Intelligence*, Texts in Computer Science,
https://doi.org/10.1007/978-3-030-42227-1_4

are called the **predecessors** of the vertex u and the vertices of the set

$$\text{succ}(u) = \{v \in V \mid (u, v) \in E\}$$

are called the **successors** of the vertex u.

Definition 4.3 An (artificial) **neural network** is a (directed) graph $G = (U, C)$ whose vertices $u \in U$ are called **neurons** or **units** and whose edges $c \in C$ are called **connections**. The set U of vertices is divided into the set U_{in} of **input neurons**, the set U_{out} of **output neurons** and the set U_{hidden} of **hidden neurons**. It is

$$U = U_{\text{in}} \cup U_{\text{out}} \cup U_{\text{hidden}},$$
$$U_{\text{in}} \neq \emptyset, \qquad U_{\text{out}} \neq \emptyset, \qquad U_{\text{hidden}} \cap (U_{\text{in}} \cup U_{\text{out}}) = \emptyset.$$

Each connection $(v, u) \in C$ carries a **weight** w_{uv} and to each neuron $u \in U$ three (real-valued) quantities are assigned: the **network input** net_u, the **activation** act_u and the **output** out_u. In addition, each input neuron $u \in U_{\text{in}}$ has a fourth (real-valued) quantity, the **external input** ext_u. Each neuron $u \in U$ also possesses three functions:

the **network input function**	$f_{\text{net}}^{(u)} : \mathbb{R}^{2\|\text{pred}(u)\|+\kappa_1(u)} \to \mathbb{R},$	
the **activation function**	$f_{\text{act}}^{(u)} : \mathbb{R}^{\kappa_2(u)} \to \mathbb{R},$	and
the **output function**	$f_{\text{out}}^{(u)} : \mathbb{R} \to \mathbb{R},$	

with which the network input net_u, the activation act_u and the output out_u of the neuron u are computed. $\kappa_1(u)$ and $\kappa_2(u)$ depend on the type and the number of arguments of the functions (see below).

The neurons are divided into input, output, and hidden neurons in order to specify which neurons receive input from the environment (input neurons) and which emit output to the environment (output neurons). The remaining neurons have no contact with the environment (but only with other neurons) and thus are "hidden."

Note that the set U_{in} of input neurons and the set U_{out} of output neurons need not be disjoint: A neuron can be both an input and output neuron. In Chap. 8, we even consider networks (so-called Hopfield networks) in which all neurons are both input and output neurons and there are no hidden neurons.

Note also that, in the index of a weight w_{uv}, the neuron to which the corresponding connection leads is named first. The reason for this order is that the graph of a neural network is often described by an adjacency matrix which instead of the values 1 (connection) and 0 (no connection) contains the weights of the connections (if a weight is zero, the corresponding connection does not exist). Due to reasons that will be studied in more detail in Chap. 5, it is advantageous to collect the weights of the connections that lead to a specific neuron in the same matrix *row* (and not in the same matrix *column*). Since the elements of a matrix are indexed according to the rule "row first, then column", the neuron is named first to which the connections lead. Thus, we obtain the following scheme (with $r = |U|$):

Fig. 4.1   A simple (artificial)
neural network

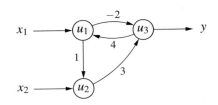

$$\begin{array}{cccc} u_1 & u_2 & \cdots & u_r \end{array}$$

$$\begin{pmatrix} w_{u_1 u_1} & w_{u_1 u_2} & \cdots & w_{u_1 u_r} \\ w_{u_2 u_1} & w_{u_2 u_2} & & w_{u_2 u_r} \\ \vdots & & & \vdots \\ w_{u_r u_1} & w_{u_r u_2} & \cdots & w_{u_r u_r} \end{pmatrix} \begin{array}{c} u_1 \\ u_2 \\ \vdots \\ u_r \end{array}$$

This matrix is to be read from the top to the right: the columns correspond to the neurons, from which the connections emanate, the rows to the neurons to which the connections lead. (Note that neurons may also be connected to themselves—diagonal elements of the above matrix.) This matrix and the corresponding weighted graph (i.e., graph with weighted edges) are called the **network structure**.

According to the network structure, we distinguish two fundamental types of neural networks: if the graph that describes the network structure of a neural network is acyclic—that is, if it does not contain loops[1] and no directed cycles—the network is called a **feed forward network**. However, if the graph contains loops or directed cycles, it is called a **recurrent network**. The reasons for these names are, of course, that in a neural network, information can be transmitted only along the (directed) connections. If the graph is acyclic, there is only one direction, namely forward, from the input neurons to the output neurons. However, if there are loops or directed cycles, outputs can be coupled back to inputs. In the subsequent chapters, we first consider different types of feed forward networks, since they are easier to analyze. In Chaps. 8 and 9, we then turn to recurrent networks.

To illustrate the definition of the structure of a neural network, we consider as an example the network with three neurons (that is, $U = \{u_1, u_2, u_3\}$), that is shown in Fig. 4.1. The neurons u_1 and u_2 are input neurons (that is, $U_{\text{in}} = \{u_1, u_2\}$). They receive the external inputs x_1 and x_2, respectively. The neuron u_3 is the only output neuron (that is, $U_{\text{out}} = \{u_3\}$). It produces the output y of the neural network. This network does not contain any hidden neurons (that is, $U_{\text{hidden}} = \emptyset$).

There is a total of four connections between the neurons (that is, the graph has the edges $C = \{(u_1, u_2), (u_1, u_3), (u_2, u_3), (u_3, u_1)\}$), the weights of which are indicated by the numbers with which the arrows are labeled that represent these connections (for example, $w_{u_3 u_2} = 3$). This network is recurrent, as there are two

---

[1] A loop is an edge/connection from a vertex to this vertex itself, that is, an edge $e = (v, v)$ with a vertex $v \in V$. Neural networks with loops are studied in Chap. 9.

directed cycles (for instance, the cycle $(u_1, u_3), (u_3, u_1)$). If we describe the network structure by a matrix (as explained above), we obtain the 3×3 matrix

$$
\begin{matrix}
u_1 \ u_2 \ u_3 \\
\begin{pmatrix} 0 & 0 & 4 \\ 1 & 0 & 0 \\ -2 & 3 & 0 \end{pmatrix}
\begin{matrix} u_1 \\ u_2 \\ u_3 \end{matrix}
\end{matrix}
$$

Note that the neuron, from which a connection emanates, selects the column, while the neuron, to which a connection leads, selects the row, in which the corresponding connection weight is entered.

4.2 Operation of Neural Networks

To describe the operation of neural networks, we have to specify (1) how a single neuron computes its output from its inputs (that is, the outputs of its predecessors) and (2) how the computations of the different neurons are organized, in particular, how the external input is processed and in what order the neurons are updated.

Let us first consider a single neuron. Each neuron can be seen as a simple processor, the structure of which is shown in Fig. 4.2. If the neuron u is an input neuron, its (initial) activation is set directly from the external input ext_u. Otherwise, the network input function $f_{\text{net}}^{(u)}$ computes the network input net_u from the inputs $\text{in}_{uv_1}, \ldots, \text{in}_{uv_n}$ (which correspond to the outputs $\text{out}_{v_1}, \ldots, \text{out}_{v_n}$ of the predecessors of the neuron u) and the connection weights $w_{uv_1}, \ldots, w_{uv_n}$. This computation may be influenced by additional parameters $\sigma_1, \ldots, \sigma_l$ (see, for instance, Sect. 6.6). From the network input, additional parameters $\theta_1, \ldots, \theta_k$, and possibly a feedback of the current activation of the neuron u (see, for instance, Chap. 9) the activation function $f_{\text{act}}^{(u)}$ computes the new activation act_u of the neuron u. Finally, the output of the neuron u is computed by the output function $f_{\text{out}}^{(u)}$ from its activation.

The number $\kappa_1(u)$ of the additional arguments of the network input function and the number $\kappa_2(u)$ of the arguments of the activation function depend on the type

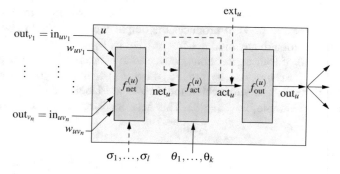

Fig. 4.2 Structure of a generalized neuron

of these functions and the structure of the neuron (for example, whether there is a feedback of the current activation or not). They may differ for each of the neurons of a neural network. Usually, the network input function has only $2|\text{pred}(u)|$ arguments (namely the outputs of the predecessor neurons and the corresponding connection weights), since no other parameters enter it. The activation function usually has two arguments: the network input and a parameter, which may be, for instance (as in the preceding chapter), a threshold. The output function, on the other hand, has only the activation as its argument and usually serves the purpose to scale the output to a desired output range, most commonly by a linear mapping.

Note that the network input function is often written with vector arguments:

$$f_{\text{net}}^{(u)}(\text{in}_u, \mathbf{w}_u) = f_{\text{net}}^{(u)}(\text{in}_{uv_1}, \dots, \text{in}_{uv_n}, w_{uv_1}, \dots, w_{uv_n})$$
$$= f_{\text{net}}^{(u)}(\text{out}_{v_1}, \dots, \text{out}_{v_n}, w_{uv_1}, \dots, w_{uv_n}).$$

This is analogous to the way in which we have worked with an input vector $\mathbf{x}$ and a weight vector $\mathbf{w}$ in the preceding chapter.

Having clarified the operation of a single neuron, we turn to the neural network as a whole. We divide the computations of a neural network into two phases: the **input phase**, in which the external inputs are fed into the network, and the **work phase**, in which the output of the neural network is computed.

The input phase serves the purpose to initialize the network. In this phase, the activations of the input neurons are set to the values of the corresponding external inputs. The activations of the remaining neurons are initialized arbitrarily, usually by simply setting them to 0. In addition, the output function is applied to the initialized activations, so that all neurons produce initial outputs.

In the work phase, the external inputs are switched off and the activations and outputs of the neurons are recomputed (possibly multiple times). To achieve this, the network input function, the activation function, and the output function are applied as described above. If a neuron does not receive any network input, because it does not have any predecessors, we define that it simply maintains its activation (and thus also its output). Essentially this is only important for the input neurons in a feed forward network. For these input neurons, which do not have predecessors, this definition is meant to guarantee that they always possess a well-defined activation (and output), since the external inputs are switched off in the work phase.

The recomputations are terminated either if the network reaches a stable state, that is, if further recomputations do not change the outputs of the neurons anymore, or if a predetermined number of recomputations have been carried out.

The temporal order of the recomputations is not generally fixed (although there are, depending on the network type, certain orders that suggest themselves). For example, all neurons of a network may recompute their outputs at the same time (synchronous update), drawing on the old outputs of their predecessors. Or we may define an order of the neurons in which they compute their new output one after the other (asynchronous update). In this case, the new outputs of other neurons may already be used as inputs for subsequent computations.

Fig. 4.3 A simple (artificial) neural network. The numbers inside the neurons state the threshold of the activation function, the labels at the arrows the connection weights

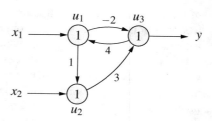

For a feed forward network, the computations usually follow a **topological ordering**[2] of the neurons, as no redundant computations are carried out in this way. Note that for recurrent networks, the final output may depend on the order in which the neurons recompute their outputs (if they are updated asynchronously) as well as on how many recomputations are carried out.

As an example, we reconsider the (artificial) neural network consisting of three neurons that is shown in Fig. 4.1. We assume that all have the weighted sum of their inputs (outputs of their predecessors) as their network input functions. That is,

$$f_{\text{net}}^{(u)}(\text{in}_u, \mathbf{w}_u) = \sum_{v \in \text{pred}(u)} w_{uv} \text{in}_{uv} = \sum_{v \in \text{pred}(u)} w_{uv} \text{out}_v.$$

We also assume that the activation function of all neurons is the threshold function

$$f_{\text{act}}^{(u)}(\text{net}_u, \theta) = \begin{cases} 1, & \text{if net}_u \geq \theta, \\ 0, & \text{otherwise.} \end{cases}$$

If we write the threshold into the neurons, as we did in the preceding chapter, we can represent the neural network as shown in Fig. 4.3. Finally, we assume that the output function of all neurons is the identity, that is,

$$f_{\text{out}}^{(u)}(\text{act}_u) = \text{act}_u.$$

Therefore, we need not distinguish between activation and output.

We consider first how this network operates if it receives the inputs $x_1 = 1$ and $x_2 = 0$ and updates the outputs of the neurons in the order $u_3, u_1, u_2, u_3, u_1, u_2, u_3, \ldots$. The corresponding computations are shown in Table 4.1.

In the input phase, the activations of the input neurons u_1 and u_2 are initialized with the values of the external inputs $\text{ext}_{u_1} = x_1 = 1$ and $\text{ext}_{u_2} = x_2 = 0$, respectively. The activation of the output neuron u_3 is initialized to the (arbitrarily chosen) value 0. Since we assumed that the output function is the identity, we need not carry out any calculations in the input phase, but simply copy the external inputs. The neurons now have the outputs $\text{out}_{u_1} = 1$ and $\text{out}_{u_2} = \text{out}_{u_3} = 0$ (see Table 4.1).

---

[2] A topological ordering is a numbering of the vertices of a directed graph, such that all edges are directed from a vertex with a lower number to a vertex with a higher number. A topological ordering exists only for acyclic graphs, that is, for feed forward networks. For feed forward networks, an update following a topological ordering ensures that all inputs of a neuron are already available (have already been computed), before it (re)computes its own activation and output.

Table 4.1 Computations carried out by the simple neural network shown in Fig. 4.1 for the input $(x_1 = 1, x_2 = 0)$ if the activations are updated in the order $u_3, u_1, u_2, u_3, u_1, u_2, u_3, \ldots$

	u_1	u_2	u_3	
Input phase	1	0	0	
Work phase	1	0	0	$net_{u_3} = -2 < 1$
	0	0	0	$net_{u_1} = 0 \geq 1$
	0	0	0	$net_{u_2} = 0 < 1$
	0	0	0	$net_{u_3} = 0 < 1$
	0	0	0	$net_{u_1} = 0 < 1$

Table 4.2 Computations carried out by the neural network shown in Fig. 4.1 for the input $(x_1 = 1, x_2 = 0)$ if the activations are updated in the order $u_3, u_2, u_1, u_3, u_2, u_1, u_3, \ldots$

	u_1	u_2	u_3	
Input phase	1	0	0	
Work phase	1	0	0	$net_{u_3} = -2 < 1$
	1	1	0	$net_{u_2} = 1 \geq 1$
	0	1	0	$net_{u_1} = 0 < 1$
	0	1	1	$net_{u_3} = 3 \geq 1$
	0	0	1	$net_{u_2} = 0 < 1$
	1	0	1	$net_{u_1} = 4 \geq 1$
	1	0	0	$net_{u_3} = -2 < 1$

The work phase starts with an update of the neuron u_3. Its network input is the weighted sum of the outputs of neurons u_1 and u_2, which are weighted with -2 and 3, respectively, that is, $net_{u_3} = -2 \cdot 1 + 3 \cdot 0 = -2$. Since -2 is less than 1, the activation (and thus the output) of the neuron u_3 is set to 0. In the next step of the work phase, the output of the neuron u_1 is updated. (Note that its external input is no longer available, but has been switched off.) Since it receives the network input 0, its activation (and thus its output) is set to 0. Likewise, the network input of the neuron u_2 is 0 and thus its activation (and its output as well) is also set to 0 in the third step. After two additional steps, it becomes clear that we have reached a stable state, since after the fifth step of the work phase, we have exactly the same situation as after the second step. Therefore, the work phase is terminated and the activation 0 of the output neuron u_3 yields the output $y = 0$ of the neural network.

A stable state is reached due to the fact that the neurons were updated in the order $u_3, u_1, u_2, u_3, u_1, u_2, u_3, \ldots$. If we choose, as an alternative, the order $u_3, u_2, u_1, u_3, u_2, u_1, u_3, \ldots$, we observe a completely different behavior that is shown in Table 4.2. In the seventh step of the work phase, it becomes clear that the outputs of all three neurons oscillate and thus that no stable state can be reached: the situation after the seventh step is identical to the one after the first step and thus the computations will repeat indefinitely. Hence, we cannot terminate the work phase, because a stable state has been reached, but have to choose a different criterion, for example, that a certain number of update steps have been computed. However, in this case, the output of the neural network depends on the step, after which the work phase is terminated. If it is terminated after step k with $(k - 1)$ mod $6 < 3$, the activation of the output neuron u_3 and thus the output of the network is $y = 0$. However, if the work phase is terminated after step k with $(k - 1)$ mod $6 \geq 3$, the activation of the output neuron u_3 and thus the output of the network is $y = 1$.

4.3 Scale Types and Encoding

Up to now, we assumed (sometimes implicitly) that the inputs and outputs of a neural network are (real) numbers. However, the data that we want to process with an (artificial) neural network do not always come in this form. In practice, we meet attributes having different so-called **scale types**, which may require preprocessing. An overview of some important scale types is shown in Table 4.3. Scales types are distinguished by the operations that are meaningful for attribute values.

The most basic operation is a test for the equality of two attribute values. If only such a test is meaningful, an attribute is called **nominal** (or **qualitative** or **categorical**, sometimes also **symbolic**). If it is also meaningful to speak of one of two values being greater than the other (and hence the second value being less than the first), but no other operation can be performed meaningfully, then the attribute is called **ordinal** (or **comparative** or **rank scaled**). Finally, if it is also meaningful to form differences of values or even their ratios, the attribute is called **metric** (or **quantitative** or simply **numerical**). If differences are meaningful, one speaks of an *interval scale*, if also ratios are meaningful, one speaks of a *ratio scale*.

Whether an attribute has an interval or a ratio scale may depend on the units used. For example, the temperature measured in °Celsius only has an interval scale, but not a ratio scale, because it is not meaningful to say that 20° Celsius is twice as warm

Table 4.3 The most important scale types with a few examples

Scale type	Possible operations	Examples
Nominal (qualitative, categorical)	Test for equality	Sex
		Blood type
Ordinal (comparative, rank scaled)	Test for equality	School grade
	Greater/less than	Wind strength
Metric (quantitative, numerical) (interval scale) (ratio scale)	Test for equality	Length
	Greater/less than	Weight
	Difference	Time
	Ratio (optional)	Temperature

as 10° Celsius (since the zero point is arbitrary). Temperature measured in °Kelvin, however, has a ratio scale, because there is a natural origin (absolute zero) and due to this natural origin ratios are meaningful (they relate energy quantities).

Clearly, metric attributes can be processed directly with (artificial) neural networks, since they come directly in the form of real numbers. Other scale types, however, need some preprocessing. For example, if we want to process nominal attributes, we may have to transform their values into numbers. (Note that they may already be encoded as numbers in the data we want to process, but that does not mean that they are metric or can be used directly like that.) Although it may seem natural to simply number the different values of such attributes, this can lead to undesired effects if the numbers do not reflect a natural order of the values. Only ordinal attributes possess such an order. Furthermore, even if there is such a natural order, it may still not be appropriate to choose equal steps between neighboring values, as this requires that differences can be computed meaningfully.

The typical approach to handle nominal attributes is so-called **1-in-n encoding** (also called **1-hot encoding**): each nominal attribute is represented by as many binary variables as it has values, each variable corresponds to one attribute value. The variable that corresponds to the value of the nominal attribute obtaining in a sample case is set to 1, while all other variables that belong to the same attribute are set to 0. That is, only 1 in n variables (where n is the number of attributes values) is set to 1, the others to 0, which explains the name(s) of this encoding. A common alternative is **1-in-(n−1) encoding**, where one attribute value is treated specially and is encoded as all zeros, while only the remaining $n - 1$ values are encoded by setting one of the variables to 1. Since this encoding does not treat all attribute values in the same way (one is treated differently from all others), there should be a good reason to single out the value that is encoded by all zeros.

Ordinal attributes usually pose an unpleasant problem. Of course, one may use 1-in-n encoding for them, treating them like nominal attributes, and this is actually what is most commonly done. However, this discards the ordinal structure of the domain of such an attribute. On the other hand, if we represent the values of an ordinal attribute by numbers that reflect the ordering, they will be treated as if they were metric attributes, implicitly assuming that at least differences can be computed meaningfully. Unfortunately, there seems to be no perfect solution for this problem. One has to choose for the application at hand which drawback is easier to tolerate.

Apart from encoding the data, it is, in practice, advisable to normalize the inputs of a neural network in order to avoid certain numerical problems, which can result from an unequal scaling of the different input variables (e.g., due to the measurement units used). Most commonly, each input variable is scaled in such a way that it has arithmetic mean 0 and variance 1 (so-called z-**score normalization**). To achieve this, one computes from the given data set for each input x_k the arithmetic mean μ_k and the standard deviation σ_k, that is,

$$\mu_k = \frac{1}{m} \sum_{i=1}^{m} x_k^{(i)} \quad \text{and} \quad \sigma_k = \sqrt{\frac{1}{m} \sum_{i=1}^{m} \left(x_k^{(i)} - \mu_k\right)^2},$$

where m is the number of sample cases in the data set and the upper index indicates the sample case referred to.[3] Then the values are transformed according to

$$z_k^{(i)} = \frac{x_k^{(i)} - \mu_k}{\sigma_k}.$$

This normalization can be carried out as a preprocessing step or (in a feed forward network) by the output function of the input neurons.

4.4 Training Neural Networks

One of the most enticing properties of (artificial) neural networks is the possibility to train them for certain tasks with the help of an example data. To some degree, we already considered this possibility in the preceding chapter with the help of the delta rule. Although the delta rule is only applicable for single threshold logic units and cannot be transferred to networks directly, it already illustrates the basic principle: training a neural network consists in adapting the connection weights and possibly some other parameters (like thresholds) such that a certain criterion is optimized.

Depending on the type of the training data and the criterion to optimize, we can distinguish two fundamental learning tasks: fixed and free.

Definition 4.4 A **fixed learning task** L_{fixed} for a neural network with n input neurons, that is, $U_{\text{in}} = \{u_1, \ldots, u_n\}$, and m output neurons, that is, $U_{\text{out}} = \{v_1, \ldots, v_m\}$, is a set of **training patterns** $l = \left(\mathbf{i}^{(l)}, \mathbf{o}^{(l)}\right)$, each consisting of an **input vector** $\mathbf{i}^{(l)} = \left(\text{ext}_{u_1}^{(l)}, \ldots, \text{ext}_{u_n}^{(l)}\right)$ and an **output vector** $\mathbf{o}^{(l)} = \left(o_{v_1}^{(l)}, \ldots, o_{v_m}^{(l)}\right)$.

If we are given a fixed learning task, we desire to train a neural network in such a way that it produces for all training patterns $l \in L_{\text{fixed}}$ the outputs contained in the output vector $\mathbf{o}^{(l)}$ if the external inputs of the corresponding input vector $\mathbf{i}^{(l)}$ are fed into the network. This is also called **supervised learning**.

In practice, this optimum can rarely be achieved and thus one may have to accept a partial or approximate solution. In order to determine how well a neural network solves a fixed learning task, an error function is employed, which measures how well the actual outputs coincide with the desired outputs in the training patterns. Commonly, this error function is defined as the sum of squared deviations of desired

---

[3] The second formula is based on the maximum likelihood estimator for the variance of a normal distribution. In statistics, often the unbiased estimator is preferred, which differs from the one shown above by using $m - 1$ instead of m. For a normalization, this difference is usually negligible.

and actual output over all training patterns and all output neurons. That is, the error of a neural network w.r.t. a fixed learning task L_{fixed} is defined as

$$e = \sum_{l \in L_{\text{fixed}}} e^{(l)} = \sum_{v \in U_{\text{out}}} e_v = \sum_{l \in L_{\text{fixed}}} \sum_{v \in U_{\text{out}}} e_v^{(l)},$$

where

$$e_v^{(l)} = \left(o_v^{(l)} - \text{out}_v^{(l)} \right)^2$$

is the individual error for a training pattern l and an output neuron v.

The square of the deviations of the desired and the actual output is chosen for various reasons. In the first place, it is clear that we cannot simply sum the deviations directly, since then positive and negative deviations could cancel, thus producing a misleading impression of the actual quality of the network. Therefore, we have to sum (at least) the absolute values of the deviations.

However, the square of the deviation of the actual and the desired output has at least two advantages over the absolute value: in the first place it is continuously differentiable everywhere, while the derivative of the absolute value does not exist/is discontinuous at 0. It is desirable that the error function is continuously differentiable, because this simplifies the derivation of the update rules for the weights (see Sect. 5.5). Secondly, large deviations from the desired output are weighted more severely, so that there is a tendency that during training, individual strong deviations (i.e., for individual training patterns) from the desired value are avoided.

The sum of squared deviations is certainly a reasonable error measure if the task is **regression**, that is, if the output (or target) attributes are metric. However, in practice, we often face nominal output (or target) attributes; in this case, the task is **classification**. As discussed above, if we want to use nominal attributes in a neural network, we have to encode their values, for example, with 1-in-n encoding. Once this is done, the sum of squared deviations can be used as an error measure for classification, summing over all variables that were introduced for the encoding.

However, there are some problems with this approach. First, we need to be able to "decode" the output of the neural network. It is not likely that, even after proper training, the output of a neural network is a clean 1-in-n encoding. Rather, we should expect that all values lie somewhere between 0 and 1 (provided we choose appropriate activation and output functions). This is no problem for the error computation, but we have to define which attribute value (or class) is predicted. The standard (and very natural) solution is to interpret the computation of such a neural network as predicting the value for which the greatest output is produced.

However, the fact that all outputs are between 0 and 1 (or at least can be made to lie between 0 and 1) leads to the desire to be able to interpret the outputs as probabilities, with which the corresponding attribute value/class is the obtaining value/class. For the decoding, we could simply normalize the values to sum 1 (provided they are actually all non-negative and there is at least one positive value). Although this is certainly feasible, it is more common to transform the values with the so-called

softmax function[4] : This function takes n values $y_1, \ldots, y_n$, which could be the outputs produced by a neural network, and transforms them according to

$$y_i' = \frac{e^{y_i}}{\sum_{k=1}^{n} e^{y_k}},$$

for all $i \in \{1, \ldots, n\}$. The result is a list of numbers $y_1', \ldots, y_n'$ that are formally a probability distribution over n values. The advantage of this function is that we need not make sure that the input values $y_1, \ldots, y_n$ are non-negative. The softmax function yields a probability distribution for any list $y_1, \ldots, y_n$ of n real numbers.

Having thus ensured that the output of a neural network with a nominal target can be interpreted as a probability distribution over classes, other error measures than the sum of squared deviations suggest themselves. We could draw on measures for the divergence of two probability distributions, where we interpret a clean 1-in-n encoding as an extreme probability distribution that assigns a probability of 1 to exactly one value and a probability of 0 to all other values.
One of the best known measures is [1]:

Definition 4.5 (*Kullback–Leibler Information Divergence*) Let p_1 and p_2 be two strictly positive probability distributions on the same space Ω of events. Then

$$I_{\text{KLdiv}}(p_1, p_2) = \sum_{\omega \in \Omega} p_1(\omega) \log_2 \frac{p_1(\omega)}{p_2(\omega)} = \mathbb{E}_{\omega \sim p_1} \left(\log_2 \frac{p_1(\omega)}{p_2(\omega)} \right)$$

is called the **Kullback–Leibler information divergence** of p_1 and p_2. Here, $\mathbb{E}_{\omega \sim p_1}$ denotes the expected value taken over events ω that are distributed according to p_1.

An error measure can be derived from the Kullback–Leibler information divergence by combining it with Shannon entropy [2] to obtain a relative entropy of one probability distribution w.r.t. another, which is also known as cross entropy:

Definition 4.6 (*Shannon Entropy*) The **Shannon entropy** of a strictly positive probability distribution p on a space Ω of events is

$$H_{\text{Shannon}}(p) = - \sum_{\omega \in \Omega} p(\omega) \log_2 p(\omega) = \mathbb{E}_{\omega \sim p}(- \log_2 p(\omega)).$$

Definition 4.7 (*Cross Entropy*) Let p_1 and p_2 be two strictly positive probability distributions on the same space Ω of events. Then

---

[4] It is quite unfortunate that this name has taken hold in the neural network community, since this function is actually *not* a soft version of the maximum operator. It does *not* yield, in a soft fashion, the value of the greatest element (of some set of elements). Rather, it is a soft version of the argmax operator (which in its crisp/hard form would assign a 1 to the element with the greatest value and a 0 to all other elements). Hence, it would be better to call it softargmax.

$$H_{\text{cross}}(p_1, p_2) = H(p_1) + I_{\text{KLdiv}}(p_1, p_2)$$

$$= -\sum_{\omega \in \Omega} p_1(\omega) \log_2 p_1(\omega) + \sum_{\omega \in \Omega} p_1(\omega) \log_2 \frac{p_1(\omega)}{p_2(\omega)}$$

$$= -\sum_{\omega \in \Omega} p_1(\omega) \log_2 p_2(\omega) = \mathbb{E}_{\omega \sim p_1}(-\log_2 p_2(\omega))$$

is called the **cross entropy** of p_1 and p_2 or the **relative entropy** of p_2 w.r.t. p_1.

Cross entropy is used as an error function for classification tasks by applying it to the computed output distribution as it is derived via the softmax function and the desired output distribution (which is a clean 1-in-n encoding of the obtaining class)[5] :

$$H_{\text{cross}}(L) = \sum_{l \in L_{\text{fixed}}} H_{\text{cross}}(\mathbf{o}^{(l)}, \text{out}^{(l)}) = -\sum_{l \in L_{\text{fixed}}} \sum_{v \in U_{\text{out}}} o_v^{(l)} \log_2 (\text{out}_v^{(l)}).$$

This is actually a fairly intuitive measure. An entropy measure can generally be seen as a quantification of the uncertainty about the obtaining value that is encoded in a probability distribution. The desired output distribution exhibits no uncertainty because the desired output value has probability 1, all other values have probability zero. The computed output distribution, however, is strictly positive and hence exhibits uncertainty about the obtaining value. The higher the amount of uncertainty about the obtaining output value, especially w.r.t. the desired output value (which is the ground truth), the worse the performance of the neural network.

Note that in the above expression for the cross entropy the inner sum reduces to a single term for each training pattern l, because $\mathbf{o}^{(l)}$ is a clean 1-in-n-encoding and hence all values $o_v^{(l)}$ except one are zero. That is,

$$H_{\text{cross}}(L) = -\sum_{l \in L_{\text{fixed}}} \log_2 (\text{out}_{v(l)}^{(l)}),$$

where $v(l)$ is the output neuron representing the desired output value (here: a class) for the training pattern l. At first sight, this makes it look as if the error only depends on a single output value per training pattern, while all the other output values are ignored. However, this is not the case, as can easily be seen if we plug in the softmax function: Let $y_v^{(l)}$ for $v \in U_{\text{out}}$ be the values that enter the softmax function, which are transformed by it into the output values $\text{out}_v^{(l)}$. Then we can write

$$H_{\text{cross}}(L) = -\sum_{l \in L_{\text{fixed}}} \log_2 \left(\frac{e^{y_{v(l)}^{(l)}}}{\sum_{v \in U_{\text{out}}} e^{y_v^{(l)}}} \right) = \sum_{l \in L_{\text{fixed}}} \left(\log_2 \left(\sum_{v \in U_{\text{out}}} e^{y_v^{(l)}} \right) - \frac{y_{v(l)}^{(l)}}{\ln(2)} \right).$$

---

[5] Note that this is actually, formally, impossible because the definition requires the two probability distributions to be strictly positive. This is clearly not the case for the desired output distribution (which consists of a single one and $n - 1$ zeros). This problem is circumvented by simply canceling all terms from the sum for which this (desired) distribution is zero. This is harmless, because in the final form $p_1(e)$ only occurs as a factor; and neither as the argument of a logarithm, nor is there any division by it. Hence, even evaluating all terms works. The computed output distribution, on the other hand, is not a problem, because the softmax function always yields positive outputs for finite inputs. This ensures that in the computation no division by zero is possible.

Finally, note that it is also common to refer to an error measure as a **loss function**, because whenever an error is committed, there is a loss of performance.

Let us now turn to free learning tasks, also called **unsupervised learning**.

Definition 4.8   A **free learning task** L_{free} for a neural network with n input neurons, that is, $U_{\text{in}} = \{u_1, \ldots, u_n\}$, is a set of **training patterns** $l = \left(\mathbf{i}^{(l)}\right)$, each of which consists of an **input vector** $\mathbf{i}^{(l)} = \left(\text{ext}_{u_1}^{(l)}, \ldots, \text{ext}_{u_n}^{(l)}\right)$.

While the training patterns of a fixed learning task contain a desired output, which allows us to compute an error or loss, free learning tasks need a different criterion in order to assess how well a neural network solves the task. In principle, with a free learning task for a neural network, we ask for a training result that "produces similar outputs for similar inputs", where the outputs can be chosen by the training method. The objective of the training can be, for example, to group the input vectors into clusters of similar vectors (*clustering* or *cluster analysis*), so that for all vectors in a cluster the same output is produced (see, for instance, Sect. 7.2).

If we are given a free learning task, the most important aspect for training a neural network is how the similarity between the training patterns is measured. This may be defined, for example, with the help of a distance function (details about distance functions can be found in Sect. 6.1). The outputs that are produced for a group of similar input vectors are then often assigned by choosing representatives or by forming prototypes (see Chap. 7 for details).

References

1. Kullback S, Leibler RA (1951) On information and sufficiency. Ann Math Stat 22:79–86. Institute of Mathematical Statistics, Hayward, CA, USA
2. Shannon CE (1948) The mathematical theory of communication. Bell Syst Tech J 27:379–423. Bell Laboratories, USA

Multi-layer Perceptrons

5

Having described the structure, the operation and the training of (artificial) neural networks in a general fashion in the preceding chapter, we turn in this and the subsequent chapters to specific forms of (artificial) neural networks. We start with the best known and most widely used form, the so-called **multi-layer perceptron** (MLP), which is closely related to the networks of threshold logic units we studied in Chap. 3. They exhibit a strictly layered structure (see the definition below) and may employ other activation functions than a step at a crisp threshold.

5.1 Definition and Examples

Definition 5.1 An **r-layer perceptron** is a neural network with a graph $G = (U, C)$ that satisfies the following restrictions:

1. $U_{\text{in}} \cap U_{\text{out}} = \emptyset$,
2. $U_{\text{hidden}} = U_{\text{hidden}}^{(1)} \cup \cdots \cup U_{\text{hidden}}^{(r-2)}, \quad \forall 1 \leq i < j \leq r-2 : \quad U_{\text{hidden}}^{(i)} \cap U_{\text{hidden}}^{(j)} = \emptyset$,
3. $C \subseteq \left(U_{\text{in}} \times U_{\text{hidden}}^{(1)}\right) \cup \left(\bigcup_{i=1}^{r-3} U_{\text{hidden}}^{(i)} \times U_{\text{hidden}}^{(i+1)}\right) \cup \left(U_{\text{hidden}}^{(r-2)} \times U_{\text{out}}\right)$
   or, if there are no hidden neurons ($r = 2, U_{\text{hidden}} = \emptyset$), $\quad C \subseteq U_{\text{in}} \times U_{\text{out}}$.

The network input function of each hidden and each output neuron is the weighted sum (weighted with the connection weights) of the inputs, that is,

$$\forall u \in U_{\text{hidden}} \cup U_{\text{out}} : \qquad f_{\text{net}}^{(u)}(\text{in}_u, \boldsymbol{w}_u) = \boldsymbol{w}_u^\top \text{in}_u = \sum_{v \in \text{pred}(u)} w_{uv} \text{out}_v.$$

© Springer Nature Switzerland AG 2022
R. Kruse et al., *Computational Intelligence*, Texts in Computer Science,
https://doi.org/10.1007/978-3-030-42227-1_5

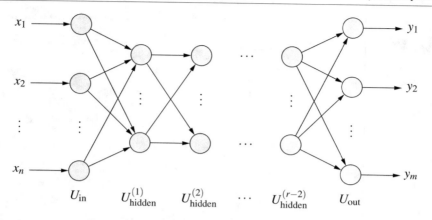

Fig. 5.1 General structure of an r-layered perceptron

The activation function of each neuron is a so-called **sigmoid function**, that is, a monotonically non-decreasing function with

$$f : \mathbb{R} \to [0, 1] \quad \text{with} \quad \lim_{x \to -\infty} f(x) = 0 \quad \text{and} \quad \lim_{x \to \infty} f(x) = 1.$$

The activation function of each output neuron is either also a sigmoid function or a linear function $f_{\text{act}}(\text{net}, \theta) = \alpha \text{net} - \theta$.

Intuitively, the restrictions of the graph in this definition mean that a multi-layer perceptron consists of an input and an output layer (the neurons of the sets U_{in} and U_{out}) and none, one, or several hidden layers (the neurons in the sets $U_{\text{hidden}}^{(i)}$) between them. Connections exist only between the neurons of consecutive layers, that is, between the input layer and the first hidden layer, between consecutive hidden layers and between the last hidden layer and the output layer (see Fig. 5.1). Note that according to this definition a multi-layer perceptron has always at least two layers, namely the input and the output layer.

Examples of sigmoid activation functions, all of which have a parameter, namely a **bias value** θ, are shown in Fig. 5.2. The threshold logic units we studied in Chap. 3 employ exclusively the (Heaviside or unit) step function as their activation function. The advantages of other activation functions are discussed in Sect. 5.3. Here, we remark only that instead of the listed **unipolar** sigmoid functions ($\lim_{x \to -\infty} f(x) = 0$) sometimes **bipolar** sigmoid functions ($\lim_{x \to -\infty} f(x) = -1$) are employed. An example of such a function is the hyperbolic tangent (see Fig. 5.3), which is closely related to the logistic function. In addition, it should be clear that any unipolar sigmoid function can be turned into a bipolar one by simply multiplying it by 2 and subtracting 1. Using bipolar sigmoid activation functions does not cause any fundamental differences. In this book, we therefore confine ourselves to unipolar sigmoid activation functions. All considerations and derivations of the subsequent sections can easily be transferred to bipolar functions.

While sigmoid functions are the classical activation functions for multi-layer perceptrons, more recent developments, which led to what is commonly referred to

(Heaviside or unit) step function:

$$f_{act}(net, \theta) = \begin{cases} 1 & \text{if } net \geq \theta, \\ 0 & \text{otherwise.} \end{cases}$$

semi-linear function:

$$f_{act}(net, \theta) = \begin{cases} 1 & \text{if } net > \theta + \frac{1}{2}, \\ 0 & \text{if } net < \theta - \frac{1}{2}, \\ (net - \theta) + \frac{1}{2} & \text{otherwise.} \end{cases}$$

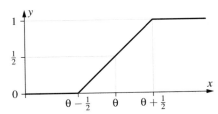

sine up to saturation:

$$f_{act}(net, \theta) = \begin{cases} 1 & \text{if } net > \theta + \frac{\pi}{2}, \\ 0 & \text{if } net < \theta - \frac{\pi}{2}, \\ \frac{\sin(net - \theta) + 1}{2} & \text{otherwise.} \end{cases}$$

logistic function:

$$f_{act}(net, \theta) = \frac{1}{1 + e^{-(net - \theta)}}$$

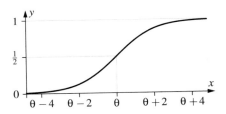

Fig. 5.2 Different unipolar sigmoid activation functions

hyperbolic tangent:

$$f_{act}(net, \theta) = \tanh(net - \theta)$$
$$= \frac{2}{1 + e^{-2(net - \theta)}} - 1$$

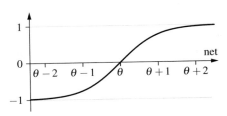

Fig. 5.3 The hyperbolic tangent, a bipolar sigmoid function

as **deep learning**, made other activation functions popular. In particular, the so-called **rectified linear unit (ReLU)** has been suggested in this area, which employs as its activation function either a **rectified maximum** or **ramp function** (see Fig. 5.4 on the left) or its smooth approximation by the so-called **softplus function** (see Fig. 5.4 on the right). The reasons for introducing these alternative activation functions are examined in Sect. 5.12, where deep learning is discussed.

rectified maximum/ramp function:

$$f_{act}(net, \theta) = \max\{0, net-\theta\}$$

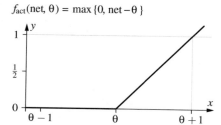

softplus function:

$$f_{act}(net, \theta) = \ln(1 + e^{net-\theta})$$

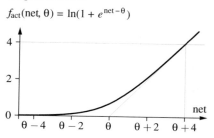

Fig. 5.4 Activation functions of rectified linear units. (Note the different scales of the diagrams!)

Although we initially confine our discussion to sigmoid activation functions, we mention these alternatives here already, because knowing them is needed, for example, for the discussion of weight initialization schemes in Sect. 5.10. Note that common variants of ReLUs are the **leaky ReLU**, which has the activation function

$$f_{act}(net, \theta) = \begin{cases} net - \theta & \text{if } net \geq \theta, \\ v(net - \theta) & \text{otherwise,} \end{cases}$$

with $v = 0.01$. With other values for v, one also speaks of a **parametric ReLU**. Note that for values $v < 1$ the activation function is equivalent to $f_{act}(net, \theta) = \max(net - \theta, v(net - \theta))$. Leaky/parametric ReLUs offer advantages for training over standard ReLUs, since their activation derivative does not vanish anywhere.

Modern frameworks for neural networks offer even more activation functions, which, however, we cannot list exhaustively. An interested reader is referred, for instance, to the documentation of the PyTorch and TensorFlow frameworks.[1]

The strictly layered structure of a multi-layer perceptron and the special network input function of the hidden as well as the output neurons suggest to describe the network structure with the help of a weight matrix, as already discussed in Chap. 4. In this way, the computations carried out by a multi-layer perceptron can be written in a simpler way, using vector and matrix notation. However, for this purpose we do not use a weight matrix for the network as a whole (although this would be possible as well), but one matrix for the connections between one layer and the next: let $U_1 = \{v_1, \ldots, v_m\}$ and $U_2 = \{u_1, \ldots, u_n\}$ be the neurons of two layers of a multi-layer perceptron, where U_2 follows U_1. We construct an $n \times m$ matrix

$$\mathbf{W} = \begin{pmatrix} w_{u_1 v_1} & w_{u_1 v_2} & \cdots & w_{u_1 v_m} \\ w_{u_2 v_1} & w_{u_2 v_2} & \cdots & w_{u_2 v_m} \\ \vdots & \vdots & & \vdots \\ w_{u_n v_1} & w_{u_n v_2} & \cdots & w_{u_n v_m} \end{pmatrix}$$

[1] https://www.pytorch.org and https://www.tensorflow.org, respectively.

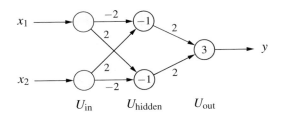

Fig. 5.5 A three-layer perceptron for the biimplication

of the weights of the connections between these two layers, setting $w_{u_i v_j} = 0$ if there is no connection between neuron v_j and neuron u_i. The advantage of such a matrix is that it allows us to write the network input of the neurons of the layer U_2 as

$$\text{net}_{U_2} = \mathbf{W} \cdot \text{in}_{U_2} = \mathbf{W} \cdot \text{out}_{U_1}$$

where $\text{net}_{U_2} = (\text{net}_{u_1}, \dots, \text{net}_{u_n})^\top$ and $\text{in}_{U_2} = \text{out}_{U_1} = (\text{out}_{v_1}, \dots, \text{out}_{v_m})^\top$ (the superscript $\top$ means that the vector is transposed, that is, that it is turned from a row vector into a column vector—as if it were a $n \times 1$ or $m \times 1$ matrix).

The placement of the weights in the matrix is determined by the convention that matrix-vector equations are written with column vectors as well as the common rules of matrix-vector multiplication. This explains why we fixed the order of the indices of the weights in Definition 4.3 on page 40 so that the neuron to which the connection leads is written first, the neuron from which it comes second.

As a first example of a multi-layer perceptron, we reconsider the network of threshold logic units studied in Sect. 3.4 that computes the biimplication. This network is shown in Fig. 5.5 as a three-layer perceptron. Note that compared to Fig. 3.10 on page 22 there are two additional neurons, namely the two input neurons. Formally these two neurons are needed, because our definition of a neural network only allows us to assign weights to the edges of the graph, but not directly to the inputs. Hence, we need the input neuron so that we have edges to the neurons of the hidden layer to which we can assign the input weights. (Note, however, that the input neurons may also transform the input quantities if they possess a suitable output function. For example, if the logarithm of an input is to be used for the computations of a neural network, we simply choose $f_{\text{out}}(\text{act}) \equiv \log(\text{act})$ for the corresponding input neuron.)

To illustrate the matrix notation of the weights, we describe the connection weights of this network by two matrices. We obtain

$$\mathbf{W}_1 = \begin{pmatrix} -2 & 2 \\ 2 & -2 \end{pmatrix} \quad \text{and} \quad \mathbf{W}_2 = \begin{pmatrix} 2 & 2 \end{pmatrix},$$

where the matrix $\mathbf{W}_1$ contains the weights of the connections from the input layer to the hidden layer and the matrix $\mathbf{W}_2$ contains the weights of the connections from the hidden layer to the output layer. If we denote the vector of output values of a layer by out_U, we can write the vector of network inputs for the hidden layer as

$$\text{net}_{U_{\text{hidden}}} = \mathbf{W}_1 \cdot \text{out}_{U_{\text{in}}}$$

s	0	0	0	0	1	1	1	1
x_1	0	0	1	1	0	0	1	1
x_2	0	1	0	1	0	1	0	1
y_1	0	0	1	1	0	1	0	1
y_2	0	1	0	1	0	0	1	1

Fig. 5.6 The Fredkin gate or controlled switch gate [6]

Fig. 5.7 Geometric interpretation of the function that is computed by a Fredkin gate (ignoring the input s which is simply passed through). Marked corner: output is 1, unmarked corner: output is 0

and analogously for the vector of network inputs for the output layer

$$\text{net}_{U_{\text{out}}} = \mathbf{W}_2 \cdot \text{out}_{U_{\text{hidden}}}.$$

As another example we consider the **Fredkin gate**, which plays an important role in so-called **conservative logic**[2] [6]. This gate has three inputs: s, x_1 and x_2, and three outputs: s, y_1 and y_2 (see Fig. 5.6). The "switch variable" s is always passed through without change. The inputs x_1 and x_2 are connected either parallel or crossed to the two outputs y_1 and y_2, depending on whether the switch variable s has value 0 or value 1. The function that is computed by a Fredkin gate is shown in Fig. 5.6 as a table and geometrically in Fig. 5.7.

Figure 5.8 shows a three-layer perceptron that computes the function of the Fredkin gate (ignoring the switch variable s, which is merely passed through without change). Actually, this network consists of two separate three-layer perceptrons, since there are no connections from any of the neurons in the hidden layer to *both* output neurons. This is, of course, not always the case for multi-layer perceptron with more than one output, but a result of the special function of the Fredkin gate.

To illustrate the matrix notation of the weights, we write the weights of this network in two matrices. We obtain

$$\mathbf{W}_1 = \begin{pmatrix} 2 & -2 & 0 \\ 2 & 2 & 0 \\ 0 & 2 & 2 \\ 0 & -2 & 2 \end{pmatrix} \quad \text{and} \quad \mathbf{W}_2 = \begin{pmatrix} 2 & 0 & 2 & 0 \\ 0 & 2 & 0 & 2 \end{pmatrix},$$

---

[2] Conservative logic is a mathematical model for computations and computational powers of computers, in which the fundamental physical principles that govern computing machines are explicitly taken into account. Among these principles are, for instance, that the speed with which information can travel as well as the amount of information that can be stored in the state of a finite system are both finite [6].

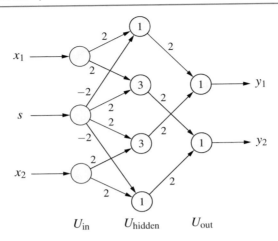

Fig. 5.8 A three-layer perceptron that computes the function of the Fredkin gate (see Fig. 5.6)

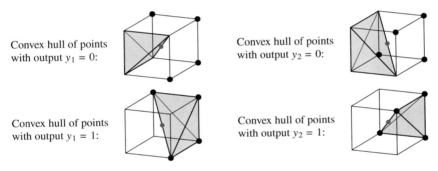

Fig. 5.9 Neither function of the Fredkin gate is linearly separable: the convex hull of the points with output 0 and the convex hull of the points with output 1 intersect (red points).

where the matrix $\mathbf{W}_1$ represents the connections from the input layer to the hidden layer and the matrix $\mathbf{W}_2$ the connections from the hidden layer to the output layer. Note that in these matrices zero elements correspond to missing/absent connections.

Note also that one actually needs a network to compute the two outputs of the Fredkin gate. A single neuron per output is not enough, because neither function is linearly separable. This is demonstrated in Fig. 5.9, which shows, for both outputs (y_1 on the left, y_2 on the right) the convex hulls of the set of input points for which an output of 0 or an output of 1, respectively, are desired. For both outputs these convex hulls share the point in the middle of the unit cube. With the Convex Hull Theorem 3.1 on page 21 it follows that both functions are not linearly separable.

5.2 Why Non-linear Activation Functions?

With the help of the matrix notation of the weights it is easy to show why sigmoid or generally non-linear activation functions are decisive for the computational capabilities of a multi-layer perceptron. Suppose all activation and output functions were linear, that is, functions $f_{\text{act}}(\text{net}, \theta) = \alpha \text{net} - \theta$. Then such a multi-layer perceptron can always be reduced to a two-layer perceptron (only input and output layer):

As mentioned above, we have for two consecutive layers U_1 and U_2

$$\text{net}_{U_2} = \mathbf{W} \cdot \text{in}_{U_2} = \mathbf{W} \cdot \text{out}_{U_1}.$$

If all activation functions are linear, then the activations of the neurons of the layer U_2 can also be determined by a matrix-vector calculation, namely by

$$\text{act}_{U_2} = \mathbf{D}_{\text{act}} \cdot \text{net}_{U_2} - \boldsymbol{\theta},$$

where $\text{act}_{U_2} = (\text{act}_{u_1}, \ldots, \text{act}_{u_n})^\top$ is the vector of activations of the neurons of layer U_2, $\mathbf{D}_{\text{act}}$ is an $n \times n$ diagonal matrix of the factors α_{u_i}, $i = 1, \ldots, n$, and $\boldsymbol{\theta} = (\theta_{u_1}, \ldots, \theta_{u_n})^\top$ is a bias vector. If the output function is a linear function as well, we have analogously

$$\text{out}_{U_2} = \mathbf{D}_{\text{out}} \cdot \text{act}_{U_2} - \boldsymbol{\xi},$$

where $\text{out}_{U_2} = (\text{out}_{u_1}, \ldots, \text{out}_{u_n})^\top$ is the vector of outputs of the neurons of layer U_2, $\mathbf{D}_{\text{out}}$ is again an $n \times n$ diagonal matrix of factors and finally $\boldsymbol{\xi} = (\xi_{u_1}, \ldots, \xi_{u_n})^\top$ is again a bias vector. Therefore, we can write the computation of the outputs of the neurons of layer U_2 from the outputs of the neurons of the preceding layer U_1 as

$$\text{out}_{U_2} = \mathbf{D}_{\text{out}} \cdot \left(\mathbf{D}_{\text{act}} \cdot \left(\mathbf{W} \cdot \text{out}_{U_1} \right) - \boldsymbol{\theta} \right) - \boldsymbol{\xi},$$

which can be simplified to

$$\text{out}_{U_2} = \mathbf{A}_{12} \cdot \text{out}_{U_1} + \boldsymbol{b}_{12},$$

with an $n \times m$ matrix $\mathbf{A}_{12}$ and an n-dimensional vector $\boldsymbol{b}_{12}$. Analogously, we obtain for the computations of outputs of the neurons of a layer U_3, which follows layer U_2, from the outputs of the neurons of layer U_2

$$\text{out}_{U_3} = \mathbf{A}_{23} \cdot \text{out}_{U_2} + \boldsymbol{b}_{23},$$

and therefore for computing the outputs of the neurons of layer U_3 from the outputs of the neurons of layer U_1

$$\text{out}_{U_3} = \mathbf{A}_{13} \cdot \text{out}_{U_1} + \boldsymbol{b}_{13},$$

where $\mathbf{A}_{13} = \mathbf{A}_{23} \cdot \mathbf{A}_{12}$ and $\boldsymbol{b}_{13} = \mathbf{A}_{23} \cdot \boldsymbol{b}_{12} + \boldsymbol{b}_{23}$. As a consequence, the computations of two consecutive layers can be reduced to a single layer. It should be clear that by iterating this result we can incorporate the computations of arbitrarily many layers. Therefore, multi-layer perceptrons can compute only affine transformations if the activation and output functions of all neurons are linear. For more complex tasks non-linear activation functions are needed. However, polynomials as activation functions would also be problematic, because then it can be shown, with an argument completely analogous to the one presented above, that a neural network using such activation functions can only compute polynomial functions.

5.3 Function Approximation

In this section, we study in more detail what we gain compared to threshold logic units (that is, neurons with the (Heaviside or unit) step function as their activation function) if we allow for other activation function.[3] In a first step, we demonstrate that all Riemann-integrable functions can be approximated by four-layer perceptrons with arbitrary accuracy, provided that the output neuron has the identity, instead of a step function, as its activation function.

The principle is illustrated in Figs. 5.10 and 5.11 for a unary function: the function to compute is approximated by a step function (see Fig. 5.10). For each step border x_i, we create a neuron in the first hidden layer of a multi-layer perceptron with a total of four layers (see Fig. 5.11). This neuron serves the purpose to determine on which side of the step border an input values lies.

In the second hidden layer we create one neuron for each step, which receives input from the two neurons in the first hidden layer that refer to the values x_i and x_{i+1} marking the borders of this step (see Fig. 5.11). The weights and the threshold are chosen in such a way that the neuron is activated if the input value is no less than x_i, but less than x_{i+1}, that is, if the input value lies in the range of the step. Note that in this way only exactly one neuron in the second hidden layer can be active, namely the one representing the step in which the input value lies.

The connections from the neurons of the second hidden layer to the output neuron are weighted with the function values of the steps that are represented by the neurons. Since only one neuron can be active on the second hidden layer, the output neuron receives as input the height of the step in which the input value lies. Since the activation function of the output neuron is the identity (as we generally assume here), this value is emitted unchanged. As a consequence, the four-layer perceptron shown in Fig. 5.11 computes exactly the step function sketched in Fig. 5.10.[4]

It should be clear that the approximation accuracy can be increased arbitrarily by making the steps sufficiently small. It may help to recall the introduction of the notion of an integral in calculus by upper and lower Riemann sums: for any given error limit $\varepsilon > 0$ there exists a step width $\delta(\varepsilon) > 0$, such that the upper and lower Riemann sum differ by less than ε. Therefore, we can state the following theorem.

---

[3] In the following, we assume implicitly that the output function of all neurons is the identity. Only the activation functions are changed to obtain certain beneficial properties.

[4] Note that there are not even any anomalies at the step borders, because the step border is included in the step to the right of it, but excluded from the step to the left of it.

Fig. 5.10 Approximating a continuous function with step functions

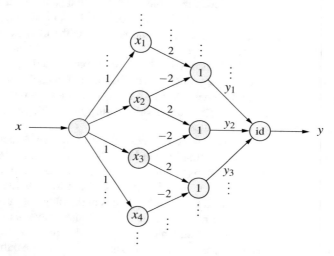

Fig. 5.11 A neural network that computes the step function shown in Fig. 5.10 ("id" instead of a threshold value means that this neuron uses the identity instead of a threshold function.)

Theorem 5.1 *Any Riemann-integrable function can be approximated with arbitrary accuracy by a multi-layer perceptron.*

Note that this theorem only requires that the function to represent is Riemann-integrable. It needs *not* be continuous. That is, the function to represent may have "jumps." However, it may have only finitely many "jumps" of finite height in the region in which it is to be approximated by a multi-layer perceptron. In other words, the function must be continuous "almost everywhere."

Note also that in this theorem the approximation error is measured by the *area* between the function to approximate and the output of the multi-layer perceptron. This area can be made arbitrarily small by increasing the number of neurons (i.e., by increasing the number of steps). However, this does *not* guarantee that for a given multi-layer perceptron, which achieves a certain approximation accuracy in this sense, the difference between its output and the function to approximate is less

Fig. 5.12 Limits of the theorem about approximating a function by a multi-layer perceptron

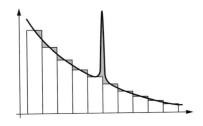

than a certain error bound everywhere. The function could, for instance, possess a very thin spike, which is not captured by any step (see Fig. 5.12). In such a case the area between the function to represent and the output of the multi-layer perceptron is small (because the spike is thin and thus encloses only a small area), but at the location of the spike the deviation of the output from the true function value can nevertheless be considerable.

Naturally, the idea to approximate a given function by a step function can directly be transferred to functions with multiple arguments: the input space is divided—depending on the arity of the function—into rectangles, boxes, or generally hyperboxes, to each of which a function value is assigned. It should be clear that a four-layer perceptron can be constructed again that computes the higher-dimensional "step function." Since we can also increase the approximation accuracy arbitrarily by making the rectangles, boxes or hyperboxes sufficiently small, the above theorem is not limited to unary functions, but holds for functions of arbitrary arity.

Even though the above theorem attests multi-layer perceptrons a high expressive power, one has to concede that it is of little use in practice. The reason is clearly that in order to achieve a sufficiently accurate approximation we have to chose step functions with a very small step width (and thus very many steps). This then forces us to construct multi-layer perceptrons with a possibly huge number of neurons (one neuron for each step and for each step border).

In order to understand how multi-layer perceptrons can approximate functions much better, we consider the case of a unary function in a little more detail. It is easy to see that we can save one layer if we do not use the absolute, but the relative height of a step (that is, the change w.r.t. the preceding step) as the weight of the connection to the output neuron. This idea is illustrated in Figs. 5.13 and 5.14. Every neuron of the hidden layer represents a step border and determines whether an input values lies to the left or to the right of this border. If it lies to the right, the neuron becomes active (outputs a 1). If it lies to the left, the neuron remains inactive (outputs a 0). The output neuron then receives as an additional network input the relative height of the step (that is, the change compared to the preceding step). Since always all those neurons of the hidden layer are active for which the step border lies to the left of the current input value, the weights add up to the height of the step.[5] Note

---

[5] Note that this approach is not easily transferred to functions with multiple arguments. For this to be possible, the influences of the two or more inputs have to be independent in a certain sense.

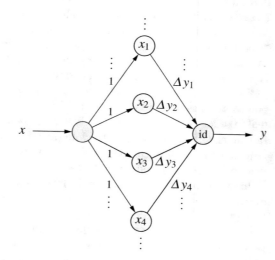

Fig. 5.13 Representing the step function shown in Fig. 5.10 by a weighted sum of (Heaviside) step functions. It is $\Delta y_i = y_i - y_{i-1}$

Fig. 5.14 A neural network that computes the step function shown in Fig. 5.10 as a weighted sum of (Heaviside) step functions, cf. Fig. 5.13. ("id" instead of a threshold means that this neuron has the identity instead of a threshold function as its activation function.)

that the (relative) step heights may as well be negative and therefore the function to approximate need not be monotonically non-decreasing.

Although we saved a layer of neurons in this way, we still need a fairly large number of neurons to achieve a good approximation, since we need sufficiently narrow steps. However, we can improve the accuracy of the approximation not only by reducing the width of the steps, but also by changing the activation functions of the neurons in the hidden layer. For example, if we replace the step functions by semi-linear functions, we can approximate the function by a piecewise linear function.

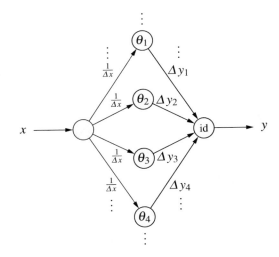

Fig. 5.15 Approximation of a continuous function by a weighted sum of semi-linear functions. It is $\Delta y_i = y_i - y_{i-1}$

Fig. 5.16 A neural network that computes the piecewise linear function shown in Fig. 5.15 by a weighted sum of semi-linear functions. It is $\Delta x = x_{i+1} - x_i$ and $\theta_i = \frac{x_i}{\Delta x}$ ("id" means again that the activation function of the output neuron is the identity.)

This is illustrated in Fig. 5.15. Needless to say that the step heights Δy_i may also be negative. The corresponding three-layer perceptron is shown in Fig. 5.16.

It is clear that with this approach and the same "step width" Δx, the error is much smaller than with step functions. Or the other way round: in order to stay below a given error limit we need considerably fewer neurons in the hidden layer. The number of neurons can be reduced even further if the steps are not made equally wide, but if one employs narrower ones where the function is heavily curved (and thus a linear approximation is poor) and fewer where it is almost linear. By using curved activation functions—like the logistic function—the approximation may be improved further or the same accuracy may be achieved with even fewer neurons.

The principle that we exploited above to eliminate one hidden layer of the multi-layer perceptron may not be directly transferable to functions with multiple arguments, because in two or more dimensions we certainly have to engird, in two steps, the regions for which the weights of the connections to the output layer state the function values. However, with stronger mathematical tools and a few and fairly weak additional assumptions it can be shown that for functions with multiple arguments a single hidden layer suffices as well. To be more specific, the following theorem can be shown [15].

Theorem 5.2 (Universal Approximation Theorem) *Let $\varphi(\cdot)$ be a continuous, bounded and non-constant function, let X denote an arbitrary compact subset of $\mathbb{R}^m$, and let $C(X)$ denote the space of continuous functions on X.*

Given any function $f \in C(X)$ and $\varepsilon > 0$, there exists an integer N, real constants $v_i, \theta_i \in \mathbb{R}$ and real vectors $w_i \in \mathbb{R}^m$, $i = 1, \ldots, N$, such that we may define

$$F(x) = \sum_{i=1}^{N} v_i \, \varphi\!\left(w_i^\top x - \theta_i\right)$$

as an approximate realization of the function f where f is independent of φ. That is,

$$|F(x) - f(x)| < \varepsilon$$

for all $x \in X$. In other words, functions of the form $F(x)$ are dense in $C(X)$.

In this theorem, the function φ corresponds to the activation function of the hidden neurons, the w_i are the weights of the connections from the input neurons to the hidden neurons, the θ_i are the bias values of the hidden neurons, and the v_i are the weights of the connections from the hidden neurons to the output neuron.

Note that with this theorem the approximation computed by a neural network satisfies even the stronger condition that the difference between the output of the multi-layer perceptron and the function to approximate is everywhere smaller than a given error bound ϵ (while Theorem 5.1 only states that the *area* between the output and the actual function can be made arbitrarily small). Note also that it is *not* the shape of the activation function φ, but the layered structure of the network that renders multi-layer perceptrons universal approximators.

An overview of results about the approximation powers of multi-layer perceptrons and a proof of the above theorem can be found, for example, in [23] (and, of course, in the paper originally stating it [15]). Note, however, that such results are relevant only in as far as they ensure that it is not the choice of a structure with only one hidden layer that already rules out the possibility to approximate certain (continuous) functions sufficiently well. That is, they ensure that there are no fundamental obstacles to achieving a good approximation, provided one uses at least one hidden layer and activation functions satisfying the stated conditions. These results do not say anything, though, about how, for a given network structure and particularly a given number of hidden neurons, one can find the parameter values with which the best possible approximation accuracy is achieved. They do not even tell us how many neurons may be needed in the hidden layer.

One should also be careful *not* to read from the above theorem that multi-layer perceptrons with more than one hidden layer are useless, because they do not increase the expressive power of multi-layer perceptrons (even though it is often cited as an argument in this direction). With a second hidden layer it may sometimes be possible to compute the function to represent in a much simpler fashion. Multi-layer perceptrons with two hidden layers may also have advantages when we have to train them. However, since multi-layer perceptrons with more that one hidden layer are much more difficult to analyze, fairly little is known about such possibilities.

5.4 Logistic Regression

Having convinced ourselves that multi-layer perceptrons with general activation functions have considerable expressive and computational powers, we now turn to the task of determining their parameters with the help of training examples. In Chap. 4, we already mentioned that we need an error function for this and that the most common error function is the sum of the squared errors over the output neurons and the training patterns. This sum of squared errors is to be minimized by suitable adaptations of the weights and the parameters of the activation functions. This approach leads to the **method of least squares**, also known as **regression**, which is well known in calculus and statistics, where it is used to determine best fit lines (regression lines) and generally best fit polynomials for a given set of data points (x_i, y_i). The fundamentals of this methods are recalled in Sect. 10.2.

Although we are not interested in best fit lines or polynomials here, the method is worth studying. The reason is that computing a best fit polynomial can also be used to determine other best fit functions, namely if we succeed in finding an appropriate transformation that reduces the problem to the task of finding a regression polynomial. For instance, best fit functions of the form

$$y = ax^b$$

(where a and b are parameters to be identified) can be found by determining a regression line: if we take the logarithm of the equation, we obtain

$$\ln y = \ln a + b \cdot \ln x.$$

This equation can be handled by computing a regression line. We merely have to take the logarithms of the data points (x_i, y_i) and work with the transformed values.[6]

---

[6] Note, however, that with this approach the sum of squared errors is minimized in the transformed space (coordinates $x' = \ln x$ and $y' = \ln y$), but this does not imply that it is also minimized in the original space (coordinates x and y). Nevertheless, this approach usually yields very good results or at least an initial solution that may then be improved by other means.

For (artificial) neural networks it is important that for the **logistic function**

$$y = \frac{Y}{1 + e^{a+bx}},$$

where Y, a, and b are parameters, there also exists a transformation with which the problem can be reduced to the task of computing a regression line (so-called **logistic regression**). The logistic function is very frequently employed as an activation function (see also Sect. 5.5). If we can find a method to determine a logistic regression function, we immediately possess a method to determine the parameters of a two-layered perceptron with a single input, since the value of a is the bias value of the output neuron and the value of b is the weight of the input.

However, how can we "linearize" the logistic function, that is, how can we transform it in such a way that the problem is reduced to the task of finding a regression line? We start by forming the reciprocal value of the logistic equation:

$$\frac{1}{y} = \frac{1 + e^{a+bx}}{Y}.$$

Therefore, it is

$$\frac{Y - y}{y} = e^{a+bx}.$$

Taking the logarithm of this equation yields

$$\ln \left(\frac{Y - y}{y} \right) = a + bx.$$

This equation can easily be handled by finding a regression line if we transform the y-values according to the left-hand side of this equation. (Note that we need to know the value of Y, which effectively describes a scaling, to compute this transformation.) This transformation is commonly known as **logit transformation**. It corresponds to a kind of inverse of the logistic function. By finding a regression line for the data points that are transformed accordingly, we (indirectly) obtain a regression curve for the original data points.[7]

To illustrate the procedure, we consider a simple example. The table below shows a data set consisting of five points $(x_1, y_1), \ldots, (x_5, y_5)$:

x	1	2	3	4	5
y	0.4	1.0	3.0	5.0	5.6

We transform these data points with

$$z = \ln \left(\frac{Y - y}{y} \right), \qquad Y = 6.$$

The transformed data points are

---

[7] Note again that with this procedure the sum of squared errors is minimized in the transformed space (coordinates x and $z = \ln \left(\frac{Y-y}{y} \right)$), but this does not imply that it is also minimized in the original space (coordinates x and y), cf. the preceding footnote.

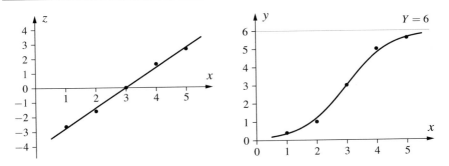

Fig. 5.17 Transformed data (left) and original data (right) as well as the computed regression line (transformed data) and the corresponding regression curve (original data)

x	1	2	3	4	5
z	2.64	1.61	0.00	-1.61	-2.64

To set up the system of normal equations, we compute

$$\sum_{i=1}^{5} x_i = 15, \qquad \sum_{i=1}^{5} x_i^2 = 55, \qquad \sum_{i=1}^{5} z_i = 0, \qquad \sum_{i=1}^{5} x_i z_i \approx -13.775.$$

Thus, we obtain the (linear) equation system (normal equations)

$$
\begin{aligned}
5a + 15b &= 0, \\
15a + 55b &= -13.775,
\end{aligned}
$$

which possesses the solution $a \approx 4.133$ and $b \approx -1.3775$. Hence, the regression line for the transformed data is

$$z \approx 4.133 - 1.3775x$$

and the regression curve for the original data consequently

$$y \approx \frac{6}{1 + e^{4.133 - 1.3775x}}.$$

These two regression functions are shown, together with the (transformed and original, respectively) data points, in Fig. 5.17.

The resulting regression curve for the original data can be computed by a neuron with one input x that has the network input function $f_{\text{net}}(x) \equiv wx$ with $w = b \approx -1.3775$, the logistic activation function $f_{\text{act}}(\text{net}, \theta) \equiv (1 + e^{-(\text{net} - \theta)})^{-1}$ with the parameter $\theta = a \approx 4.133$ and the output function $f_{\text{out}}(\text{act}) \equiv 6\text{act}$.

Note that with the help of logistic regression we can compute not only the parameters of a neuron with a single input, but—in analogy to **multi-linear regression**, see Sect. 10.2—also the parameters of a neuron with multiple inputs.

As a simple illustration, we consider the functions

$$y_1 = f_2(\boldsymbol{x}) = \frac{1}{1 + \exp(-(\,x_1 + x_2 - 4))} = \frac{1}{1 + \exp\big(-((1,\,1)^\top (x_1, x_2) - 4)\big)},$$

$$y_2 = f_2(\boldsymbol{x}) = \frac{1}{1 + \exp(-(2x_1 + x_2 - 6))} = \frac{1}{1 + \exp\left(-((2, 1)^\top(x_1, x_2) - 6)\right)},$$

which are shown in Fig. 5.18 at the top and at the bottom, respectively. Note that the contour lines (i.e., lines of equal function value) of both functions are straight lines and that the normal vector of these lines (as shown by black arrows in the diagrams on the right) are the coefficients of the two input variables in the formulae of the two functions. We know this already from Sect. 3.2, where we studied the geometric interpretation of threshold logic units. Here we have a completely analogous situation, only that the threshold function, which changes from 0 to 1 at a specific line, is replaced by the logistic function, which rises slowly from 0 to 1. The crisp threshold is thus replaced by a "soft" threshold, the midpoint of which (that is, the 0.5 contour line) is singled out by the bias value.

Figure 5.19 shows, as colored dots, data that was generated from the logistic function shown in Fig. 5.18 at the bottom, with some random noise added. This ground truth is indicated by light gray lines in the contour diagram on the right, while the black and dark gray lines indicate that result of a logistic regression. Clearly, this regression result and the ground truth differ significantly. At first one may suspect that this difference is due to the noise that was added to the sampled data points. However, even though this noise certainly also causes a deviation from the ground truth, the main reason for the deviation is that the regression result was computed with the logit transformation approach explained above. As already emphasized, this

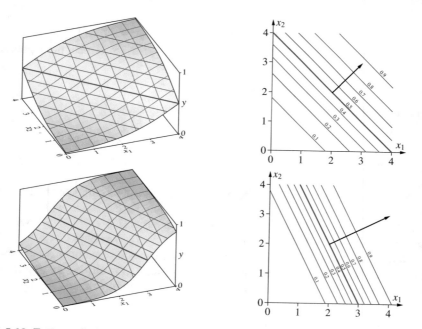

Fig. 5.18 Two two-dimensional logistic functions with different parameters; three-dimensional representation on the left, contour line representation on the right

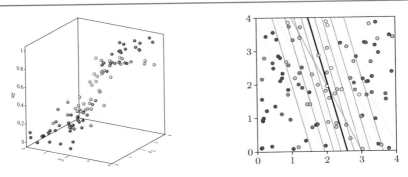

Fig. 5.19 Example of a two-dimensional logistic regression (ground truth: gray, regression: black)

approach only minimizes the error in the transformed space, but thus not necessarily in the original space. Here, due to the fact that the data points were sampled uniformly at random from $[0, 4] \times [0, 4]$, but the ground truth contour lines run diagonally through this square, the minimum error in the transformed space results for a fairly different parameterization than the minimum error in the original space, leading to a significant deviation from the ground truth. In the following section, we study how this problem can be addressed by using gradient descent.

However, the logistic regression approach studied up to now has yet another problem. Since the sum of squared errors can be determined only for output neurons, this method is limited to two-layer perceptrons (that is, with only an input and an output layer, but without any hidden layer). It is not possible to directly transfer it to three- and multi-layer perceptrons. Thus, we face essentially the same problem as in Sect. 3.7, where we could not transfer the delta rule. Therefore we consider in the next section a different method, which can be extended to multi-layer perceptrons.

5.5 Gradient Descent

In the following, we consider the method of gradient descent to determine the parameters of a multi-layer perceptron. In principle, this method relies on the same idea as the procedure studied in Sect. 3.5: depending on the values of the weights and the biases, the output of the multi-layer perceptron will be more or less correct. If we can derive from the error function in which directions we have to change the weights and the bias values in order to reduce the error, we have a way to train the parameters of the network. We simply make a small step into this direction, determine the direction of change again, make another small step and so on—in the same way as we proceeded in Sect. 3.5 (cf. Fig. 3.18 on page 28).

However, in Sect. 3.5 we could not derive the change directions directly from the natural error function (cf. Fig. 3.15 on page 26), but had to invest additional thought in order to modify the error function appropriately. However, this was necessary only,

Fig. 5.20 Intuitive interpretation of the gradient of a real-valued function $z = f(x, y)$ at a point $\boldsymbol{p} = (x_0, y_0)$. It is $\nabla z|_{(x_0, y_0)} = \left(\frac{\partial z}{\partial x}|_{(x_0, y_0)}, \frac{\partial z}{\partial y}|_{(x_0, y_0)} \right)$

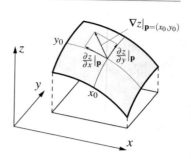

because we used the (Heaviside or unit) step function as the activation function, due to which the error function consisted of "plateaus" or "terraces". In the multi-layer perceptrons we are studying here, however, we have other choices for the activation functions at our disposal (cf. Fig. 5.2 on page 55).

In particular, we should choose a **differentiable activation function**, for example, the logistic function. However, differentiability is not enough. The Heaviside step function is differentiable almost everywhere (except at the threshold value). Nevertheless it is not usable, because its derivative is zero almost everywhere (except at the threshold value, where its derivative does not exist). If the derivative of a function is zero, there is no indication of the direction in which to change the argument of the function in order to increase or decrease the function value. Therefore, we need a differentiable function that has a non-vanishing derivative in at least a sufficiently large region, where it provides direction information. The logistic function in particular has the advantage that it does not vanish for any finite value.

Choosing a differentiable activation function (with a non-zero derivative in a sufficiently large region) has the advantage that the error function is differentiable as well.[8] Hence we can determine the directions, in which the weights and the bias values have to be changed by simply computing the **gradient** of the error function.

Intuitively, the gradient describes the slope of a function (see Fig. 5.20). Formally, computing the gradient yields a **vector field**. That is, the gradient assigns to each point in the domain of the function a vector, the elements of which are the **partial derivatives** of the function w.r.t. its arguments (also known as **direction derivatives**). This vector is often simply called the gradient of the function at the given point (see Fig. 5.20). It points into the direction of the steepest slope of the function at this point. Forming a gradient (at a point or for a whole function) is commonly denoted by the differential operator ∇ (pronounced "nabla" or "del").

As already mentioned above, the principle of gradient methods is this: Starting from a randomly chosen point in the search space, make small steps in the parameter space, always in the direction of the steepest ascent (or descent) of the function to

---

[8] Unless the output function is not differentiable. However, we usually assume (implicitly) that the output function is the identity and thus does not introduce any problems.

optimize, until a (local) maximum (or minimum) is reached (at least approximately). Formulated as a cookbook recipe, gradient descent (or ascent) looks like this:

1. Choose a (random) starting point $\boldsymbol{u}^{(0)} = \left(u_1^{(0)}, \ldots, u_n^{(0)}\right)^{\top}$
2. Compute the gradient of the objective function f at the current point $\boldsymbol{u}^{(i)}$:

$$\nabla_{\boldsymbol{u}} f(\boldsymbol{u})\Big|_{\boldsymbol{u}^{(i)}} = \left(\frac{\partial}{\partial u_1} f(\boldsymbol{u})\Big|_{u_1^{(i)}}, \ldots, \frac{\partial}{\partial u_n} f(\boldsymbol{u})\Big|_{u_n^{(i)}}\right)^{\top}.$$

3. Make a small step in the direction (or against the direction) of the gradient:

$$\boldsymbol{u}^{(i+1)} = \boldsymbol{u}^{(i)} \pm \eta \, \nabla_{\boldsymbol{u}} f(\boldsymbol{u})\Big|_{\boldsymbol{u}^{(i)}}. \qquad \begin{array}{l} + : \text{gradient ascent} \\ - : \text{gradient descent} \end{array}$$

   η is a step width parameter ("learning rate" in artificial neuronal networks).
4. Repeat steps 2 and 3, until some termination criterion is satisfied
   (e.g., a certain number of steps has been executed, current gradient is small).

We illustrate this method with a very simple example, namely how gradient descent can be used to find the minimum of a polynomial function, here specifically

$$f(u) = \frac{5}{6}u^4 - 7u^3 + \frac{115}{6}u^2 - 18u + 6.$$

Although this function has no direct relationship to the error function of a multi-layer perceptron, it is well suited to demonstrate how gradient descent works in principle, as well as certain problems that gradient descent can run into. We start by finding the derivative of the polynomial target function, that is,

$$f'(u) = \frac{10}{3}u^3 - 21u^2 + \frac{115}{3}u - 18,$$

which corresponds to the gradient (the sign indicates the direction of the steepest slope). The computations then proceed according to the scheme

$$u_{i+1} = u_i + \Delta u_i \qquad \text{with} \qquad \Delta u_i = -\eta f'(u_i),$$

where u_0 is a chosen initial value and η is a step width parameter.

We first consider the course of the gradient descent for an initial value $u_0 = 0.2$ and a step width parameter $\eta = 0.01$ as it is shown in Fig. 5.21. Starting at an initial point on the left branch of the function, small steps are carried out towards the minimum. Clearly, the (global) minimum is approached fairly rapidly: after only 10 steps it is almost reached. This is how gradient descent should work ideally.

However, whether it actually does in a certain case depends on the starting point and the learning rate. Especially the value of the learning rate is crucial. Suppose we reduce the step width parameter to $\eta = 0.001$ (i.e., one tenth of the learning rate used in Fig. 5.21). Then, gradient descent proceeds as shown in Fig. 5.22. Although the (global) minimum will finally be reached, this will happen only after a very large number of steps. Obviously, the learning rate was chosen too small in this case, so that the procedure takes very long (or even too long) to complete.

On the other hand, the learning rate should also not be chosen too large, since this can cause oscillations or even chaotic jumps back and forth on the function to

i	u_i	$f(u_i)$	$f'(u_i)$	Δu_i
0	0.200	3.112	-11.147	0.111
1	0.311	2.050	-7.999	0.080
2	0.391	1.491	-6.015	0.060
3	0.451	1.171	-4.667	0.047
4	0.498	0.976	-3.704	0.037
5	0.535	0.852	-2.990	0.030
6	0.565	0.771	-2.444	0.024
7	0.589	0.716	-2.019	0.020
8	0.610	0.679	-1.681	0.017
9	0.626	0.653	-1.409	0.014
10	0.640	0.635		

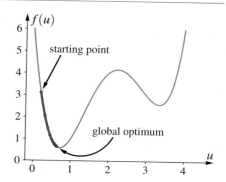

Fig. 5.21 Gradient descent with initial value 0.2 and step width parameter 0.01

i	u_i	$f(u_i)$	$f'(u_i)$	Δu_i
0	0.200	3.112	-11.147	0.011
1	0.211	2.990	-10.811	0.011
2	0.222	2.874	-10.490	0.010
3	0.232	2.766	-10.182	0.010
4	0.243	2.664	-9.888	0.010
5	0.253	2.568	-9.606	0.010
6	0.262	2.477	-9.335	0.009
7	0.271	2.391	-9.075	0.009
8	0.281	2.309	-8.825	0.009
9	0.289	2.233	-8.585	0.009
10	0.298	2.160		

Fig. 5.22 Gradient descent with initial value 0.2 and step width parameter 0.001

minimize. As an illustration consider the gradient descent that is shown in Fig. 5.23 for an initial value $u_0 = 1.5$ and a learning rate $\eta = 0.25$. The process again and again jumps over the minimum and after a few steps values are reached that are even farther away from the minimum than the initial value. If the computations were continued for a few steps more, we could even observe a jump over the local maximum in the middle, thus resulting in values on the right branch of the function. In general, with a large learning rate the behavior is often entirely erratic and thus cannot be expected to ever approach a local, let alone the global minimum.

However, even if a suitable value is chosen for the learning rate, we cannot guarantee that the procedure will be successful. As can be seen in Fig. 5.24, which shows the course of the gradient descent for an initial value $u_0 = 2.6$ and a learning rate $\eta = 0.05$, the closest minimum is approached quickly. However, this minimum is only a local one; the global minimum is missed. This problem is caused mainly by the initial value and thus cannot be solved by changing the learning rate.

Unfortunately, there is no general solution to this problem. This is a fundamental drawback of gradient descent, as of many other heuristic optimization methods: an optimum solution cannot be guaranteed. Often the best one can do is to employ what is referred to as **random restarts**, that is, one repeats the process multiple times starting

i	u_i	$f(u_i)$	$f'(u_i)$	Δu_i
0	1.500	2.719	3.500	−0.875
1	0.625	0.655	−1.431	0.358
2	0.983	0.955	2.554	−0.639
3	0.344	1.801	−7.157	1.789
4	2.134	4.127	0.567	−0.142
5	1.992	3.989	1.380	−0.345
6	1.647	3.203	3.063	−0.766
7	0.881	0.734	1.753	−0.438
8	0.443	1.211	−4.851	1.213
9	1.656	3.231	3.029	−0.757
10	0.898	0.766		

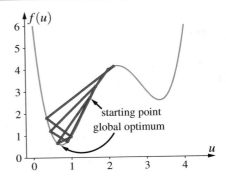

Fig. 5.23 Gradient descent with initial value 1.5 and step width parameter 0.25

i	u_i	$f(u_i)$	$f'(u_i)$	Δu_i
0	2.600	3.816	−1.707	0.085
1	2.685	3.660	−1.947	0.097
2	2.783	3.461	−2.116	0.106
3	2.888	3.233	−2.153	0.108
4	2.996	3.008	−2.009	0.100
5	3.097	2.820	−1.688	0.084
6	3.181	2.695	−1.263	0.063
7	3.244	2.628	−0.845	0.042
8	3.286	2.599	−0.515	0.026
9	3.312	2.589	−0.293	0.015
10	3.327	2.585		

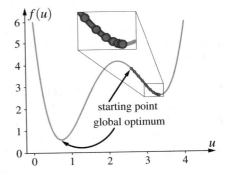

Fig. 5.24 Gradient descent with initial value 2.6 and step width parameter 0.05

from different initializations, and finally selects the best result reached. Even though this does not guarantee that the global optimum is found, it yields better chances of finding it or at least a sufficiently good local optimum.

As a second example, we examine how the solution to the logistic regression problem considered in the preceding section can be improved with gradient descent (cf. Fig. 5.19 on page 71). As we saw, the logit transformation approach followed by a linear regression yields a suboptimal solution. This solution is not useless, though. It can be used as the starting point for a gradient descent scheme.

In order to derive the gradient descent scheme for logistic regression, we start in the same way as for linear or polynomial regression (see Sect. 10.2): We are given a data set $\mathbf{D} = \{(x_1, y_1), \ldots, (x_n, y_n)\}$ with n data points $x_i \in \mathbb{R}^m$, $i = 1, \ldots, n$. To keep things simple, we assume $y_i \in [0, 1]$, $i = 1, \ldots, m$, so that we can work with the standard form of the logistic function, which has $Y = 1$ (cf. Sect. 5.4). We desire to find a parameter vector $a' = (a_1, \ldots, a_m)^\top$ such that

$$f(x, a') = \frac{1}{1 + e^{-a'^\top x_i - a_0}}$$

best fits the given data points. However, to simplify the notation, we use extended vectors $x_i^* = (1, x_{i1}, \ldots, x_{im})^\top$ and $a = (a_0, a_1, \ldots, a_m)^\top$. This is analogous to how the threshold of a threshold logic unit was turned into a weight in Fig. 3.20 on page 29: the leading 1 in x_i^* captures the constant a_0. Next we set up the functional to minimize, namely the sum of squared errors (or deviations):

$$F(a) = \sum_{i=1}^{n} \left(y_i - \frac{1}{1 + e^{-a^\top x_i^*}} \right)^2 \overset{!}{=} \min.$$

A necessary condition for a minimum of this functional is that the gradient of the objective function $F(a)$ w.r.t. the parameter vector a vanishes:

$$\nabla_a F(a) \overset{!}{=} 0.$$

With linear or polynomial regression, this leads to a linear equation system that can be solved with standard methods from linear algebra. Unfortunately, this is not the case for logistic regression. We still obtain an equation system, but due to the logistic function, it is not linear and hence difficult to solve analytically.

There are at least two ways to get around this difficulty: (1) we may use gradient descent on the objective function $F(a)$ (this is the approach we consider in the following) or (2) we may carry out a root search on the gradient $\nabla_a F(a)$. The latter can be done, for example, by applying the Newton–Raphson algorithm (see, e.g., [25]). However, this requires (as the method is applied to a gradient) to compute the second derivative of the objective function (its Hessian matrix in the multi-dimensional case). Due to this computation of the second derivative, such an approach is referred to as a *second order optimization* method. We do not consider this method here, because it is not feasible for multi-layer perceptrons with at least one hidden layer, unless the network is very small.

To obtain the update formula for gradient descent, we start again by deriving the gradient. With the abbreviation $f(z) = \frac{1}{1+e^{-z}}$ for the logistic function it is

$$\nabla_a F(a) = \nabla_a \sum_{i=1}^{n} (y_i - f(a^\top x_i^*))^2 = -2 \sum_{i=1}^{n} (y_i - f(a^\top x_i^*)) \cdot f'(a^\top x_i^*) \cdot x_i^*,$$

The derivative of the logistic functions w.r.t. its argument (indicated by the $'$) is

$$f'(z) = \frac{d}{dz} \left(1 + e^{-z} \right)^{-1} = -\left(1 + e^{-z} \right)^{-2} \left(-e^{-z} \right) \tag{5.1}$$

$$= \frac{1 + e^{-z} - 1}{(1 + e^{-z})^2} = \frac{1}{1 + e^{-z}} \left(1 - \frac{1}{1 + e^{-z}} \right) = f(z) \cdot (1 - f(z)), \tag{5.2}$$

As an illustration, this standardized logistic function (i.e., without a shift by a bias value θ) as well as its derivative are depicted in Fig. 5.25.

Gradient descent now proceeds as follows: Choose as initial values a_0 the result of a logit transform and a linear regression. Update the parameters a according to

$$a_{t+1} = a_t - \frac{\eta}{2} \cdot \nabla_a F(a)|_{a_t}$$

$$= a_t + \eta \cdot \sum_{i=1}^{n} (y_i - f(a_t^\top x_i^*)) \cdot f(a_t^\top x_i^*) \cdot (1 - f(a_t^\top x_i^*)) \cdot x_i^*,$$

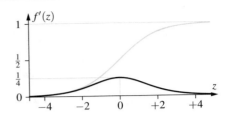

Fig. 5.25 Standardized logistic function (left and in gray on the right) and its derivative (right)

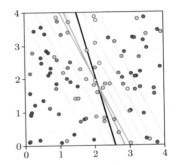

Fig. 5.26 Two-dimensional logistic regression improved with gradient descent (gray: ground truth, black: starting point obtained via logit transform, green: result of gradient descent)

where η is a step width parameter to be chosen by a user (for our example, see Fig. 5.19 on page 71, e.g., $\eta = 0.05$). Repeat the update until convergence, e.g., until $\|a_{t+1} - a_t\| < \tau$ with a chosen threshold τ (for our example: e.g., $\tau = 10^{-6}$).

As a result, we obtain the logistic function that is shown in Fig. 5.26 as a green line (this is the 0.5 contour line). Since this result was obtained by minimizing the sum of squared errors in the original space via gradient descent, we conclude that the result obtained via a logit transform (which was used as a starting point) does not minimize the error in the original space (as was emphasized above). Of course, we cannot expect the result of the gradient descent (green) to coincide with the ground truth (gray), because noise was added to the data points, which has some influence. However, we obtain a much better result.

As a last example, we consider **logistic classification**.[9] Here we confine ourselves to two classes, although logistic classification can be generalized to more than two classes. The idea is to model the conditional probability of one class as a logistic function. For a formal description, we adopt the same notation as above, using a parameter vector $a = (a_0, a_1, \ldots, a_m)^\top$ as well as an extended argument vector $x^* = (1, x_1, \ldots, x_m)^\top$. Then the conditional probability of the target class, which we call c_1, is assumed to be a logistic function:

$$P(C = c_1 \mid X = x) = p_1(x) = p(x; a) = \frac{1}{1 + e^{-a^\top x^*}}.$$

---

[9] In statistics this is often also referred to as logistic regression, which can cause some confusion.

Logistic function with $Y = 1$:

$$y = f(x) = \frac{1}{1 + e^{-a(x-x_0)}}$$

Interpret the logistic function as the probability of one class.

Fig. 5.27 Modeling two-class problems with a logistic function as the probability of one class

With only two classes the conditional probability of the other class c_0 is

$$P(C = c_0 \mid X = x) \; = \; p_1(x) \; = \; 1 - p(x; a).$$

This is illustrated in Fig. 5.27. The conditional probability of the target class is shown in red, the conditional probability of the complement class in blue. Note that this approach is equivalent to an 1-in-$(n - 1)$ encoding of the two classes.

With this approach, we can apply the following classification rule to new data:

$$C = \begin{cases} c_1 & \text{if } p(x; a) \geq \theta, \\ c_0 & \text{if } p(x; a) < \theta, \end{cases} \qquad \theta = 0.5.$$

Of course, the threshold need not be $\theta = 0.5$. This is merely, because in the absence of other information, it is most natural to predict the class that is more likely. If misclassification costs are available, a different threshold θ may be preferable. However, a discussion of misclassification costs is beyond the scope of this book.

Note that this classification approach leads to a **linear classifier**, just like those that can be obtained with threshold logic units. However, it has the additional advantage that it does not merely separate the classes, but also yields a class probability (which a threshold logic unit does not). This probability gets more extreme, the farther away a data point is from the classification boundary.

We now apply this approach to the data shown in Fig. 5.29. Note that the locations of the data points are the same as in Fig. 5.19, but that the target values are now classes, encoded as zero and 1. These classes, encoded as 0 and 1, have been drawn from the conditional probability distributions at each data point, which are described by the logistic function indicated in gray (cf. also Figs. 5.27 and 5.28). We now want to apply the classification model described above to this data.

This classification model allows us to obtain the parameters via maximizing the likelihood of the data. That is, we choose those parameters that make the given data most likely. We do so by setting up a so-called **likelihood function**, which describes the likelihood of the given data depending on the parameters a of the (conjectured) data generating process. Here, we assume that the data was generated by sampling from some distribution on the data space and then classes were assigned according to a probability as given by a logistic function (which is indeed the case). Class $C = 1$ occurs with probability $p_1(x) = f(a^\top x)$, class $C = 0$ with $p_0(x) = 1 - f(a^\top x)$ with $f(z) = \frac{1}{1+e^{-z}}$ and $x^* = (1, x_1, \ldots, x_m)^\top$ and $a = (a_0, a_1, \ldots, a_m)^\top$. Then the **likelihood function** for the data $\mathbf{D} = \{(x_1, y_1), \ldots, (x_n, y_n)\}$ with $y_i \in \{0, 1\}$ is

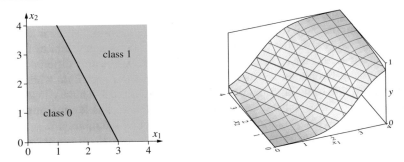

Fig. 5.28 Logistic classification describes a linear classifier (linear classification boundary)

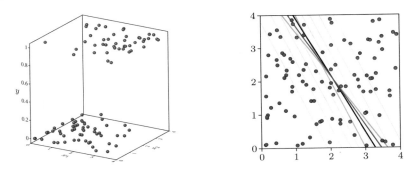

Fig. 5.29 Left: two-dimensional classified data (blue: class $c_0 \stackrel{\wedge}{=} 0$, red: class $c_1 \stackrel{\wedge}{=} 1$). Right: two-dimensional logistic classification improved with gradient descent (gray: ground truth, black: starting point obtained via logit transform, green: result of gradient descent on sum of squared errors, magenta: result of gradient descent on negative log-likelihood function)

$$L(\boldsymbol{a}) = \prod_{i=1}^{n} p_1(\boldsymbol{x}_i)^{y_i} \cdot p_0(\boldsymbol{x}_i)^{1-y_i} = \prod_{i=1}^{n} f(\boldsymbol{a}^\top \boldsymbol{x}_i^*)^{y_i} \cdot (1 - f(\boldsymbol{a}^\top \boldsymbol{x}_i^*))^{1-y_i}.$$

The task to find the parameters that maximize this function can be simplified by considering the **log-likelihood function**. Furthermore, in order to obtain a loss function (which is to be minimized, rather than score function, which is to be maximized), it is common to consider the **negative log-likelihood function**

$$-\ln L(\boldsymbol{a}) = -\sum_{i=1}^{n} \left(y_i \cdot \ln f(\boldsymbol{a}^\top \boldsymbol{x}_i^*) + (1 - y_i) \cdot \ln(1 - f(\boldsymbol{a}^\top \boldsymbol{x}_i^*)) \right)$$

$$= -\sum_{i=1}^{n} \left(y_i \cdot \ln \frac{1}{1 + e^{-(\boldsymbol{a}^\top \boldsymbol{x}_i^*)}} + (1 - y_i) \cdot \ln \frac{e^{-(\boldsymbol{a}^\top \boldsymbol{x}_i^*)}}{1 + e^{-(\boldsymbol{a}^\top \boldsymbol{x}_i^*)}} \right)$$

$$= -\sum_{i=1}^{n} \left((y_i - 1) \cdot \boldsymbol{a}^\top \boldsymbol{x}_i^* - \ln \left(1 + e^{-(\boldsymbol{a}^\top \boldsymbol{x}_i^*)} \right) \right)$$

Note that taking the logarithm of the likelihood function does not change the location of the optimum (because the logarithm is a monotone function), which is all we are interested in. Note also that this negative log-likelihood is equivalent to the cross entropy as it was introduced in Definition 4.7 on page 50. The only difference is that the negative log-likelihood is usually defined via a natural logarithm while for the cross entropy, due to its information-theoretic underpinnings, a binary logarithm is more natural.[10] Clearly, this does not make any difference, as it only introduces a constant factor and hence does not change the location of the minimum.

In principle, we could now proceed as with linear regression: a necessary condition for a minimum of the negative log-likelihood is that its gradient vanishes:

$$\nabla_a - \ln L(a) \overset{!}{=} 0.$$

However, as with the sum of squared errors above, the resulting equation system is not linear and hence difficult to solve analytically. We tackle this problem in the same way as above: we consider a gradient descent.[11]

The gradient of the negative log-likelihood function is (with $f(z) = \frac{1}{1+e^{-z}}$):

$$\nabla_a (-\ln L(a)) = -\nabla_a \sum_{i=1}^{n} \left((y_i - 1) \cdot a^\top x_i^* - \ln\left(1 + e^{-(a^\top x_i^*)}\right) \right)$$

$$= -\sum_{i=1}^{n} \left((y_i - 1) \cdot x_i^* + \frac{e^{-(a^\top x_i^*)}}{1 + e^{-(a^\top x_i^*)}} \cdot x_i^* \right)$$

$$= -\sum_{i=1}^{n} \left((y_i - 1) \cdot x_i^* + \left(1 - f(a^\top x_i^*)\right) \cdot x_i^* \right)$$

$$= -\sum_{i=1}^{n} (y_i - f(a^\top x_i^*)) \cdot x_i^*.$$

It is enlightening to compare this to the gradient of the sum of squared errors:

$$\nabla_a F(a) = -2 \sum_{i=1}^{n} (y_i - f(a^\top x_i^*)) \cdot \underbrace{f(a^\top x_i^*) \cdot (1 - f(a^\top x_i^*))}_{\text{additional factor: derivative of the logistic function}} \cdot x_i^*.$$

Gradient descent now proceeds as follows: Choose as initial values a_0 the result of a logit transform and a linear regression. Note that for this we have to modify the representation of the classes to $y_0' = \epsilon$ and $y_1' = 1 - \epsilon$ in order to be able to compute the logit transform. This is harmless, though, because this step is only supposed to yield a starting point for the gradient descent, not an actual solution. Furthermore,

---

[10] Nevertheless the cross entropy is sometimes also defined via a natural logarithm.
[11] As for the sum of squared errors, applying the Newton–Raphson method to find a root of the gradient is an alternative, which we do not consider here.

the value of ϵ is irrelevant (as long as $\epsilon \in (0, \frac{1}{2})$), since the result is independent of the choice of ϵ. The parameters $\boldsymbol{a}$ are then updated according to

$$\boldsymbol{a}_{t+1} = \boldsymbol{a}_t - \eta \cdot \nabla_{\boldsymbol{a}} (-\ln L(\boldsymbol{a}))|_{\boldsymbol{a}_t} = \boldsymbol{a}_t + \eta \cdot \sum_{i=1}^{n} (y_i - f(\boldsymbol{a}_i^{\top} \boldsymbol{x}_i^*)) \cdot \boldsymbol{x}_i^*,$$

where η is a step width parameter to be chosen by a user (for our example, see Fig. 5.19 on page 71, e.g., $\eta = 0.01$). Repeat the update until convergence, e.g., until $\|\boldsymbol{a}_{t+1} - \boldsymbol{a}_t\| < \tau$ with a chosen threshold τ (for our example: e.g., $\tau = 10^{-6}$). Note that the difference in the step width we use ($\eta = 0.01$ compared to $\eta = 0.05$ on page 74) is due to the missing factor pointed out above, namely the derivative of the logistic function. Since this derivative has a value of at most $\frac{1}{4}$ (see Fig. 5.25), we can use a larger step width if it is present (despite the also added factor 2).

If we apply this gradient descent scheme to the data shown in Fig. 5.29, we obtain the magenta line (this is the 0.5 contour line) in the diagram shown in Fig. 5.29 on the right (while the green line shows, like in Fig. 5.26, the result of a gradient descent on the sum of squared errors, which could be applied here as well). Note that in this case the results of gradient descent, regardless of whether carried out on the sum of squared errors or on the negative log-likelihood, are farther away from the ground truth than the result of a logit transform follows by a linear regression (black line). The reason is that here the influence of the noise (resulting from the random choice of classes) is greater than the influence of whether the objective function is optimized in the transformed or the original space.

With the method of gradient descent training a neural network for a fixed learning task becomes very simple: the weights and the bias values are initialized randomly. Then, the gradient of the error function is computed at the point that is given by these weights and bias values. Since we want to minimize the error, but the gradient points into the direction of the steepest slope, we make a small step in the opposite direction. At the new point (new weights and bias values), we recompute the gradient, etc., until we reach a minimum of the error function.

With these considerations and the above examples, the general procedure should be clear. Therefore, we now turn to a detailed formal derivation of the adaptation rules for the weights and bias values. In order to avoid unnecessary and clumsy distinctions of special cases, we denote in the following the set of neurons of the input layer of an r-layered perceptron with U_0, the sets of neurons of the $r - 2$ hidden layers with U_1 to U_{r-2} and the set of neurons of the output layer (sometimes) with U_{r-1}. We start from the total error of a multi-layer perceptron with output neurons U_{out} w.r.t. a fixed learning task L_{fixed}, which is defined as (cf. Sect. 4.4)

$$e = \sum_{l \in L_{\text{fixed}}} e^{(l)} = \sum_{v \in U_{\text{out}}} e_v = \sum_{l \in L_{\text{fixed}}} \sum_{v \in U_{\text{out}}} e_v^{(l)},$$

that is, as the sum of the individual errors over all output neurons v and all training patterns l. Let u be a neuron of the output layer or a hidden layer, that is, $u \in U_k$, $0 < k < r$. Its predecessors are the neurons $\text{pred}(u) = \{p_1, \ldots, p_n\} \subseteq U_{k-1}$; the corresponding (extended) weight vector is $\boldsymbol{w}_u = (-\theta_u, w_{up_1}, \ldots, w_{up_n})$. Note the additional vector element $-\theta_u$: as shown in Sect. 3.5, a bias value can be turned into a weight, so that all parameters can be treated in a uniform manner (see Fig. 3.20 on page 29)—a simplification that we also drew on in the above examples.

We now compute the gradient of the total error w.r.t. these weights, that is,

$$\nabla_{\boldsymbol{w}_u} e = \frac{\partial e}{\partial \boldsymbol{w}_u} = \left(-\frac{\partial e}{\partial \theta_u}, \frac{\partial e}{\partial w_{up_1}}, \dots, \frac{\partial e}{\partial w_{up_n}} \right).$$

As the total error is the sum of the individual errors over the training patterns, we get

$$\nabla_{\boldsymbol{w}_u} e = \frac{\partial e}{\partial \boldsymbol{w}_u} = \frac{\partial}{\partial \boldsymbol{w}_u} \sum_{l \in L_{\text{fixed}}} e^{(l)} = \sum_{l \in L_{\text{fixed}}} \frac{\partial e^{(l)}}{\partial \boldsymbol{w}_u}.$$

Hence, we can confine ourselves to the error $e^{(l)}$ for a single training pattern l. This error depends on the weights in $\boldsymbol{w}_u$ only via the network input $\text{net}_u^{(l)} = \boldsymbol{w}_u \text{in}_u^{(l)}$ with the (extended) network input vector $\text{in}_u^{(l)} = \left(1, \text{out}_{p_1}^{(l)}, \dots, \text{out}_{p_n}^{(l)} \right)$. We apply the chain rule and obtain

$$\nabla_{\boldsymbol{w}_u} e^{(l)} = \frac{\partial e^{(l)}}{\partial \boldsymbol{w}_u} = \frac{\partial e^{(l)}}{\partial \text{net}_u^{(l)}} \frac{\partial \text{net}_u^{(l)}}{\partial \boldsymbol{w}_u}.$$

Since $\text{net}_u^{(l)} = \boldsymbol{w}_u \text{in}_u^{(l)}$, we get for the second factor immediately

$$\frac{\partial \text{net}_u^{(l)}}{\partial \boldsymbol{w}_u} = \text{in}_u^{(l)}.$$

For the first factor we consider the error $e^{(l)}$ for the training pattern $l = \left(\boldsymbol{i}^{(l)}, \boldsymbol{o}^{(l)} \right)$. If we choose the sum of squared deviations, this error is

$$e^{(l)} = \sum_{v \in U_{\text{out}}} e_u^{(l)} = \sum_{v \in U_{\text{out}}} \left(o_v^{(l)} - \text{out}_v^{(l)} \right)^2,$$

that is, the sum of errors over all output neurons. Therefore, we have

$$\frac{\partial e^{(l)}}{\partial \text{net}_u^{(l)}} = \frac{\partial \sum_{v \in U_{\text{out}}} \left(o_v^{(l)} - \text{out}_v^{(l)} \right)^2}{\partial \text{net}_u^{(l)}} = \sum_{v \in U_{\text{out}}} \frac{\partial \left(o_v^{(l)} - \text{out}_v^{(l)} \right)^2}{\partial \text{net}_u^{(l)}}.$$

Since the actual output $\text{out}_v^{(l)}$ of an output neuron v depends on the network input $\text{net}_u^{(l)}$ of the considered neuron u, it is

$$\frac{\partial e^{(l)}}{\partial \text{net}_u^{(l)}} = \underbrace{-2 \sum_{v \in U_{\text{out}}} \left(o_v^{(l)} - \text{out}_v^{(l)} \right) \frac{\partial \text{out}_v^{(l)}}{\partial \text{net}_u^{(l)}}}_{\delta_u^{(l)}}.$$

This also introduces the abbreviation $\delta_u^{(l)}$ for the sum over the output neurons that occurs here and that plays an important role below.

To determine the sums $\delta_u^{(l)}$ we have to distinguish two cases: if u is an output neuron, we can simplify the expression for $\delta_u^{(l)}$ considerably, because the outputs of all other output neurons are clearly independent of the network input of the neuron u. Therefore, all terms of the sum vanish except the one with $v = u$. We obtain

$$\forall u \in U_{\text{out}} : \quad \delta_u^{(l)} = \left(o_u^{(l)} - \text{out}_u^{(l)} \right) \frac{\partial \text{out}_u^{(l)}}{\partial \text{net}_u^{(l)}}$$

Therefore, the gradient is

$$\forall u \in U_{\text{out}} : \quad \nabla_{\mathbf{w}_u} e_u^{(l)} = \frac{\partial e_u^{(l)}}{\partial \mathbf{w}_u} = -2 \left(o_u^{(l)} - \text{out}_u^{(l)} \right) \frac{\partial \text{out}_u^{(l)}}{\partial \text{net}_u^{(l)}} \, \text{in}_u^{(l)},$$

which implies the general weight change

$$\forall u \in U_{\text{out}} : \quad \Delta \mathbf{w}_u^{(l)} = -\frac{\eta}{2} \nabla_{\mathbf{w}_u} e_u^{(l)} = \eta \left(o_u^{(l)} - \text{out}_u^{(l)} \right) \frac{\partial \text{out}_u^{(l)}}{\partial \text{net}_u^{(l)}} \, \text{in}_u^{(l)}.$$

The negatives sign disappears, because we have to minimize the error and thus have to move in the direction opposite to the gradient, since the gradient points into the direction of the steepest slope of the error function. The constant factor 2 is incorporated into the **learning rate** η.[12] A typical value for the learning rate is $\eta = 0.2$.

Note, however, that this is only the weight change that results from a single training pattern l, since we neglected the sum over the training patterns at the beginning. In other words, this is the adaptation rule for **online training**, in which the weights are adapted after each training pattern (cf. page 27f and Algorithm 3.2 on page 30). For **batch training** we have to sum the changes described by the above formula over all training patterns rather than changing the parameters directly (cf. page 27f and Algorithm 3.3 on page 31f), since the weights are adapted only at the end of a (learning/training) epoch, that is, after all training patterns have been visited.

Of course, a natural generalization of these two schemes is to update the network parameters after a certain number of training patterns that is to be chosen by a user. This scheme is usually referred to as **mini-batch training**, were "mini-batch" refers to the set of training patterns that is evaluated to obtain the parameter changes. Clearly, online training and batch training (which is sometimes also called **full batch training** to distinguish it from mini-batch training) are special cases of mini-batch training where the number k of processed training examples between two updates is $k = 1$ for online training or $k = |L|$ for (full) batch training.

Mini-batch training for neural networks is a special case of the more general method of **stochastic gradient descent**. Generally, stochastic gradient descent means that the function to optimize is composed of several subfunctions, for example, one subfunction per training pattern (as is the case here). In this case a (partial or approximate) gradient can be computed from a (random) subsample of these subfunctions (here: training patterns) and can be used directly to update the parameters. These (random) subsamples are what was called *mini-batches* above. With randomly chosen

---

[12] In order to avoid this factor right from the start, the error of an output neuron is sometimes defined as $e_u^{(l)} = \frac{1}{2} \left(o_u^{(l)} - \text{out}_u^{(l)} \right)^2$. In this way the factor 2 simply cancels in the derivation.

subsamples, the gradient descent is *stochastic*, because it contains the random element of the choice of the subfunctions/training samples. For training neural networks this stochasticity is usually achieved by randomly shuffling the training patterns after each full traversal, so that one can always cut consecutive slices of equal size from the training patterns and use them as mini-batches.

In the above formula for the weight changes we cannot determine the derivative of the output $\text{out}_u^{(l)}$ w.r.t. the network input $\text{net}_u^{(l)}$ generally, because the output is computed from the network input with the help of the output function f_{out} and the activation function f_{act} of the neuron u. That is, we have

$$\text{out}_u^{(l)} = f_{\text{out}}\left(\text{act}_u^{(l)}\right) = f_{\text{out}}\left(f_{\text{act}}\left(\text{net}_u^{(l)}\right)\right),$$

and there are several choices for these two functions.

To simplify matters, we assume here that the activation function does not take any extra parameters.[13] It may be, for instance, the logistic function, which is the most common choice. Furthermore, we assume for the sake of simplicity that the output function f_{out} is the identity and thus that we can neglect it. Then, we obtain

$$\frac{\partial \text{out}_u^{(l)}}{\partial \text{net}_u^{(l)}} = \frac{\partial \text{act}_u^{(l)}}{\partial \text{net}_u^{(l)}} = f'_{\text{act}}\left(\text{net}_u^{(l)}\right),$$

where the prime $'$ means taking the derivative w.r.t. the argument $\text{net}_u^{(l)}$. For the logistic activation function in particular, that is, for

$$f_{\text{act}}(x) = \frac{1}{1 + e^{-x}},$$

we already determined, when we considered logistic regression (see page 76), that

$$f'_{\text{act}}(x) = f_{\text{act}}(x) \cdot (1 - f_{\text{act}}(x))$$

holds and therefore (as we assume that the output function is the identity)

$$f'_{\text{act}}\left(\text{net}_u^{(l)}\right) = f_{\text{act}}\left(\text{net}_u^{(l)}\right) \cdot \left(1 - f_{\text{act}}\left(\text{net}_u^{(l)}\right)\right) = \text{out}_u^{(l)}\left(1 - \text{out}_u^{(l)}\right).$$

We obtain for the weight changes (resulting from one training pattern l)

$$\Delta w_u^{(l)} = \eta \left(o_u^{(l)} - \text{out}_u^{(l)}\right) \text{out}_u^{(l)} \left(1 - \text{out}_u^{(l)}\right) \text{in}_u^{(l)},$$

which makes the computations particularly simple. These weight changes are either applied directly (online training) or aggregated over a certain number of training patterns before they are applied (mini-batch training) or even aggregated over all training pattern with an update only once per epoch (full batch training).

---

[13] Note that the bias value θ_u is already contained in the extended weight vector.

5.6 Error Backpropagation

In the preceding section, we considered the term $\delta_u^{(l)}$ only for output neurons u. That is, the resulting adaptation rule applies only to the weights of the connections from the last hidden layer to the output layer (or, alternatively, only to two-layer perceptrons). In this situation we found ourselves already with the delta rule (see Definition 3.5 on page 28) and faced the problem that the procedure cannot be extended to networks, because we did not have desired outputs for the hidden neurons. The gradient descent approach, however, can be extended to multi-layer perceptrons, since a differentiable activation function (with a sufficiently large region of non-vanishing gradients) allow us to differentiate the output also w.r.t. the weights of the connections from the input layer to the first hidden layer or w.r.t. the weights of the connections between two consecutive hidden layers.

Let u be a neuron of a hidden layer, that is, let $u \in U_k, 0 < k < r - 1$. In this case, the output $\text{out}_v^{(l)}$ of an output neuron v for a training pattern l depends on the network input $\text{net}_u^{(l)}$ of the neuron u only indirectly via the successors of u, that is, $\text{succ}(u) = \{s \in U \mid (u, s) \in C\} = \{s_1, \ldots, s_m\} \subseteq U_{k+1}$, namely via their network input $\text{net}_s^{(l)}$. By applying the chain rule, we obtain

$$\delta_u^{(l)} = \sum_{v \in U_{\text{out}}} \sum_{s \in \text{succ}(u)} (o_v^{(l)} - \text{out}_v^{(l)}) \frac{\partial \text{out}_v^{(l)}}{\partial \text{net}_s^{(l)}} \frac{\partial \text{net}_s^{(l)}}{\partial \text{net}_u^{(l)}}.$$

Since both sums are finite, we can exchange the summations and thus arrive at

$$\delta_u^{(l)} = \sum_{s \in \text{succ}(u)} \left(\sum_{v \in U_{\text{out}}} (o_v^{(l)} - \text{out}_v^{(l)}) \frac{\partial \text{out}_v^{(l)}}{\partial \text{net}_s^{(l)}} \right) \frac{\partial \text{net}_s^{(l)}}{\partial \text{net}_u^{(l)}} = \sum_{s \in \text{succ}(u)} \delta_s^{(l)} \frac{\partial \text{net}_s^{(l)}}{\partial \text{net}_u^{(l)}}.$$

Finally, we have to determine the partial derivative of the network input. It is

$$\text{net}_s^{(l)} = \mathbf{w}_s \text{in}_s^{(l)} = \left(\sum_{p \in \text{pred}(s)} w_{sp} \text{out}_p^{(l)} \right) - \theta_s,$$

where one element of the vector $\text{in}_s^{(l)}$ is the output $\text{out}_u^{(l)}$ of the neuron u. Obviously, $\text{net}_s^{(l)}$ depends on $\text{net}_u^{(l)}$ only via this element $\text{out}_u^{(l)}$. Therefore

$$\frac{\partial \text{net}_s^{(l)}}{\partial \text{net}_u^{(l)}} = \left(\sum_{p \in \text{pred}(s)} w_{sp} \frac{\partial \text{out}_p^{(l)}}{\partial \text{net}_u^{(l)}} \right) - \frac{\partial \theta_s}{\partial \text{net}_u^{(l)}} = w_{su} \frac{\partial \text{out}_u^{(l)}}{\partial \text{net}_u^{(l)}},$$

since all terms vanish except the one with $p = u$. As a consequence we have

$$\delta_u^{(l)} = \left(\sum_{s \in \text{succ}(u)} \delta_s^{(l)} w_{su} \right) \frac{\partial \text{out}_u^{(l)}}{\partial \text{net}_u^{(l)}}.$$

Thus, we arrived at a layer-wise recursion formula for computing the δ-values of the neurons of the hidden layers. If we compare this result to the one obtained in the preceding section for the output neurons, we see that the sum

$$\sum_{s \in \text{succ}(u)} \delta_s^{(l)} w_{su}$$

plays the role of the difference $o_u^{(l)} - \text{out}_u^{(l)}$ of the desired and the actual output of the neuron u for the training pattern l. Therefore we may see it as an error value for a neuron in a hidden layer, like $o_u^{(l)} - \text{out}_u^{(l)}$ is the error value for an output neuron. As a consequence, the error values of any (hidden) layer of a multi-layer perceptron can be computed from the error values of is successor layer. We may also say that an error signal is transmitted from the output layer backwards through the hidden layers. Therefore this method is also called **error backpropagation**. It serves the purpose to compute the gradients for the parameters (weights and bias values) of the hidden layers and thus allow for their adaptation. It can be seen as related to *dynamic programming*, in which results obtained for some subproblems are reused to solve other subproblems. (However, we do not explore this relation here.)

The resulting weight adaptation is

$$\Delta w_u^{(l)} = -\frac{\eta}{2} \nabla_{w_u} e^{(l)} = \eta \, \delta_u^{(l)} \, \text{in}_u^{(l)} = \eta \left(\sum_{s \in \text{succ}(u)} \delta_s^{(l)} w_{su} \right) \frac{\partial \text{out}_u^{(l)}}{\partial \text{net}_u^{(l)}} \, \text{in}_u^{(l)}.$$

Note, however, that this is only the change resulting from a single training pattern l. For mini- or full batch training these changes have to be summed over a (randomly chosen) subset or even all training patterns, respectively.

For the rest of the derivation we again make the simplifying assumption, as in the preceding section, that the output function is the identity. Furthermore, we consider the special case of a logistic activation function. This yields the particularly simple weight adaptation rule (cf. the derivations for output neurons on page 84).

$$\Delta w_u^{(l)} = \eta \left(\sum_{s \in \text{succ}(u)} \delta_s^{(l)} w_{su} \right) \text{out}_u^{(l)} \left(1 - \text{out}_u^{(l)} \right) \text{in}_u^{(l)}.$$

Figure 5.30 combines all formulas that we need to execute and to train a multi-layer perceptron, which employs logistic activation functions in the hidden and the output neurons. The formulas applied are indicated by colors.

5.7 Gradient Descent Examples

To illustrate the gradient descent method, we consider training a two-layer perceptron for the negation, as we used it already in Sect. 3.5. This perceptron and the corresponding training examples are shown in Fig. 5.31. In analogy to Fig. 3.15 on page 26, Fig. 5.32 shows the (sum of) squared errors of computing the negation depending on the values of the weights and the bias value. We assume a logistic activation function, as can clearly be seen from the shape of the error function. Note that the fact that the activation function is now (meaningfully) differentiable, the error function is as well and does no longer consist of "plateaus" or "terraces". Thus, we can now execute a gradient descent on the (unmodified) error function.

The course of this gradient descent, starting with the initial values $\theta = 3$ and $w = \frac{7}{2}$ and proceeding with learning rate 1 is shown in Table 5.1 (online training on

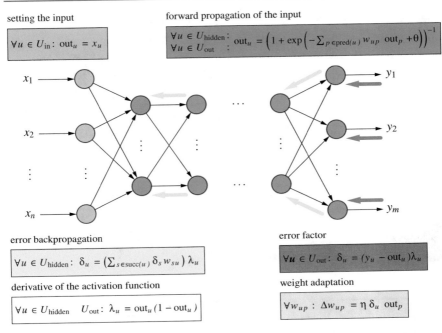

Fig. 5.30 Cookbook recipe for executing (forward propagation) and training (error backpropagation) a multi-layer perceptron with logistic activation functions in the hidden and output neurons

Fig. 5.31 A two-layer perceptron with a single input and training examples for the negation

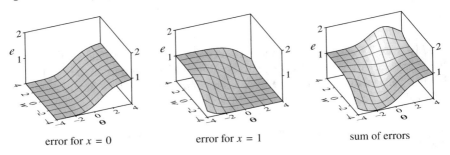

error for $x = 0$ error for $x = 1$ sum of errors

Fig. 5.32 (Sum of) squared errors for computing the negation with a logistic activation function

the left, batch training on the right). The two courses are very similar, which is due to the low number of training examples and the smoothness of the error function. Figure 5.33 shows a graphical representation of the course of the training, with the left and the middle diagram displaying, for the sake of comparison, the regions we

Table 5.1 Training processes with initial values $\theta = 3$, $w = \frac{7}{2}$ and learning rate 1

Epoch	θ	w	error	Epoch	θ	w	Error
0	3.00	3.50	1.307	0	3.00	3.50	1.295
20	3.77	2.19	0.986	20	3.76	2.20	0.985
40	3.71	1.81	0.970	40	3.70	1.82	0.970
60	3.50	1.53	0.958	60	3.48	1.53	0.957
80	3.15	1.24	0.937	80	3.11	1.25	0.934
100	2.57	0.88	0.890	100	2.49	0.88	0.880
120	1.48	0.25	0.725	120	1.27	0.22	0.676
140	−0.06	−0.98	0.331	140	−0.21	−1.04	0.292
160	−0.80	−2.07	0.149	160	−0.86	−2.08	0.140
180	−1.19	−2.74	0.087	180	−1.21	−2.74	0.084
200	−1.44	−3.20	0.059	200	−1.45	−3.19	0.058
220	−1.62	−3.54	0.044	220	−1.63	−3.53	0.044
Online training				Batch training			

online training   batch training   batch training

Fig. 5.33 Training processes with initial values $\theta = 3$, $w = \frac{7}{2}$ and learning rate 1

used in Sect. 3.5 (cf. Figs. 3.17 and 3.18 on page 28). The dots indicate the state of the network every 20 epochs. In the three-dimensional diagram on the right, it can be seen particularly well how the error is slowly reduced and how finally a region is reached in which the error almost vanishes.

For a visual illustration of error backpropagation, we recommend trying the visualization programs wmlp (for Microsoft Windows[tm]) and xmlp (for Unix/Linux), that are available (free of charge, including source code) on the web page

http://www.borgelt.net/mlpd.html

With these programs a three-layer perceptron can be trained in such a way that the biimplication, the exclusive or, or (an approximation of) one of two different real-valued functions are computed. After each training step the computations of the neural network are visualized by drawing the current position of the separating

Fig. 5.34 Screenshot of the demonstration program xmlp (cf. Fig. 3.11 on page 22 and Fig. 5.18 on page 70 for explanations of the diagrams, which show a training result)

lines or the current output of the neurons over their input space. Although this does not allow to follow directly the descent on the error function (which is impossible in principle due to the number of parameters), one gets a good impression of what happens in the course of training. An extensive explanation of these programs can be found on the web page named above, a screenshot is shown in Fig. 5.34.

5.8 Variants of Gradient Descent

One of the problems mentioned above, namely "getting stuck" in a local minimum, cannot be solved in principle. It can only be mitigated by training the network multiple times, each time with different initial values, and choosing the best result. This increases the chances that the global minimum (or at least a good local minimum) is found. However, there is no guarantee that the global minimum will be discovered.

To cope with the other two problems, which concern the learning rate and thus the size of the steps in the parameter space, however, several variants of gradient descent have been developed. Some of these we discuss in the following. We describe them by stating the rules according to which a weight has to be changed based on the gradient of the error function. Since some of these methods draw on gradients or parameter values from earlier training steps, we introduce a parameter t that denotes this training step. For example, $\nabla_w e|_{w_t}$ denotes the gradient of the error function w.r.t. the weight w, evaluated at w_t, the value weight w has in step t. As a comparison, the weight adaptation rule for standard gradient descent is thus written as

$$w_{t+1} = w_t + \Delta w_t \quad \text{with} \quad \Delta w_t = -\frac{\eta}{2} \nabla_w e|_{w_t} = -\frac{\eta}{2} (\nabla_w e)(w_t)$$

(cf. the derivations on page 83 and page 86). We do not distinguish explicitly between full batch, mini-batch and online training, since the difference only consists in how many training patterns are used to compute an (approximate) gradient.

5.8.1 Manhattan Training

In the preceding section, we saw that training can take very long if the learning rate is (too) small. However, even if the learning rate is generally chosen well, training can be slow, namely if it takes place in a region of the parameter space in which the error function is "flat", that is, in which the gradient is fairly small. In order to eliminate this dependence on the magnitude of the gradient, which can slow down training significantly, one may employ so-called Manhattan training, which considers only the sign of the gradient. Thus, the weight adaptation rule becomes

$$\Delta w_t = -\eta \ \mathrm{sgn}(\nabla_w e|_{w_t}).$$

Note that this adaptation rule is also obtained if one uses the sum of the absolute deviations of the actual from the desired output as the error function and completes the derivative at 0 (where it does not exist) in a suitable fashion.

The advantage of this approach is that training proceeds with constant speed (in the sense of a fixed step width), irrespective of the shape of the error function. A disadvantage is, though, that the weights can only assume certain discrete values (taken from a grid with the grid width η), which can make it impossible in principle to get arbitrarily close to the minimum of the error function. In addition, the problem of how to chose the learning rate appropriately still prevails.

5.8.2 Momentum Term

The momentum term procedure [24,31] adds a fraction of the preceding weight change to a normal gradient descent step to speed up training. With this additional term the weight adaptation rule reads

$$\Delta w_t = -\frac{\eta}{2} \nabla_w e|_{w_t} + \alpha \ \Delta w_{t-1},$$

where α is a parameter that has to be less than 1 in order to render the procedure stable. Typically, α is chosen between 0.5 and 0.95. Larger values can lead to increasingly larger weight changes and thus unstable behavior.

The additional term $\beta \ \Delta w_{t-1}$ is called **momentum term**, since its effect is similar to the momentum that is gained by a ball rolling down a slope. The longer the ball rolls into the same direction, the more momentum it gathers and hence the faster it gets. Therefore it tends to continue to roll in the old direction (momentum term), but nevertheless follows (though delayed) the shape of the surface (gradient term).

By introducing a momentum term training can be accelerated in regions of the parameter space in which the error function is fairly flat, but has a uniform slope in one direction. In addition, the problem of how to choose the learning rate is mitigated somewhat, since the momentum term increases or reduces the step width according to the shape of the error function. However, a momentum term cannot fully compensate a learning rate that was chosen much too small, since in case of a constant gradient $\nabla_w e|_{w_t}$ the step width is bounded by $|\Delta w_t| \leq \left| \frac{\eta \nabla_w e|_{w_t}}{2(1-\alpha)} \right|$. Furthermore, a learning rate that is too large can still lead to oscillations and chaotic jumps.

An alternative formulation of the momentum term approach that uses an auxiliary variable to accumulate the gradients (instead of weight changes) is

$$\Delta w_t = -\frac{\eta}{2} m_t \quad \text{with} \quad m_0 = 0;\ m_t = \alpha m_{t-1} + \nabla_w e|_{w_t};\ \alpha \in [0.5, 0.999].$$

This variant is often more convenient for implementations. Note, however, that in this variant the parameter α gets multiplied by $\frac{\eta}{2}$, which is not the case in the original version. This explains the different range of typical values for α.

5.8.3   Nesterov's Accelerated Gradient (NAG)

The idea of Nesterov's accelerated gradient [22] is analogous to a momentum term, but the momentum step is applied to the parameter that is adapted also before computing the gradient, that is,

$$\Delta w_t = -\frac{\eta}{2} \nabla_w e|_{w_t + \alpha \Delta w_{t-1}} + \alpha\ \Delta w_{t-1}.$$

The reason is that the momentum term does not depend on the current gradient, so one can get a better adaptation direction by applying the momentum term also *before* computing the gradient, that is, by computing the gradient at $w_t + \alpha \Delta w_{t-1}$. This point will be reached by the momentum step anyway, so it is advantageous to use the gradient at this point rather than at the current point.

An alternative formulation of Nesterov's accelerated gradient is

$$\Delta w_t = -\frac{\eta}{2} \overline{m}_t \quad \text{with} \quad \begin{aligned} & m_0 = 0;\ m_t = \alpha m_{t-1} + \nabla_w e|_{w_t};\ \alpha \in [0.5, 0.999]; \\ & \overline{m}_t = \alpha m_t \quad\ \ + \nabla_w e|_{w_t}. \end{aligned}$$

Note that this formulation is *not* exactly equivalent to the original formulation above, but leads to analogous training behavior and is easier to implement, because it eliminates the need to first compute the new point at which the gradient is evaluated.

5.8.4   Self-adaptive Error Backpropagation (SuperSAB)

The method of (super) self-adaptive error backpropagation (SuperSAB) [17,34] introduces a separate learning rate $\eta^{(w)}$ for each parameter w of a neural network, that is, for each weight and each bias value. In addition, these learning rates are not constant, but are adapted (prior to their use in a training step) depending on the current and the previous gradient according to

$$\eta_t^{(w)} = \begin{cases} c^- \cdot \eta_{t-1}^{(w)} & \text{if } \nabla_w e|_{w_t} \quad \cdot \nabla_w e|_{w_{t-1}} < 0, \\ c^+ \cdot \eta_{t-1}^{(w)} & \text{if } \nabla_w e|_{w_t} \quad \cdot \nabla_w e|_{w_{t-1}} > 0 \\ & \quad \wedge \nabla_w e|_{w_{t-1}} \cdot \nabla_w e|_{w_{t-2}} \geq 0, \\ \eta_{t-1}^{(w)} & \text{otherwise.} \end{cases}$$

c^- is a shrinkage factor ($0 < c^- < 1$), with which the learning rate is decreased if the current and the previous gradient have opposite signs. The intuitive reason is that in such a case the training step must have leaped over the minimum (as the gradient now indicates the opposite direction to reach it), and thus smaller steps are necessary to actually approach it. Typically c^- is chosen between 0.5 and 0.7.

c^+ is a growth factor ($c^+ > 1$), with which the learning rate is increased if the current and the previous gradient have the same sign. In this case, two training steps were made in the same direction and thus it is plausible to assume that a longer slope of the error function is currently traversed. Hence, the learning rate should be increased in order to traverse it more quickly. (This is similar in spirit to the idea of the momentum term, see above.) Typically c^+ is chosen between 1.05 and 1.2, so that the learning rate grows only slowly (whereas it is shrunk more quickly).

The second condition for applying the growth factor c^+ is meant to prevent that the learning rate is increased again immediately after it has been reduced. This is usually implemented in such a way that the old gradient is set to zero after the learning rate has been reduced, in order to signal the executed reduction. Although this also suppresses a repeated reduction, this procedure eliminates the need to store the gradient $\nabla_w e(t - 2)$ or to create a corresponding marker.

In order to avoid large jumps as well as slow training, it is common to bound the learning rate from above and from below. Furthermore, (super) self-adaptive error backpropagation should only be applied with (full) batch training (or sufficiently large mini-batches), since online training is often unstable with it.

5.8.5 Resilient Backpropagation

Resilient backpropagation (Rprop) can be seen as combining the ideas of Manhattan training and self-adaptive error backpropagation [28,29]. It introduces a *step width* Δw for each parameter of the neural network, that is, for each weight and each bias value, which is adapted according to the following rule, depending on the current and the previous gradient:

$$\Delta w_t = \begin{cases} c^- \cdot \Delta w_{t-1}, & \text{if } \nabla_w e|_{w_t} \cdot \nabla_w e|_{w_{t-1}} < 0, \\ c^+ \cdot \Delta w_{t-1}, & \text{if } \nabla_w e|_{w_t} \cdot \nabla_w e|_{w_{t-1}} > 0 \\ & \quad \wedge \nabla_w e|_{w_{t-1}} \cdot \nabla_w e|_{w_{t-2}} \geq 0, \\ \Delta w_{t-1}, & \text{otherwise.} \end{cases}$$

Like for self-adaptive error backpropagation c^- is a shrinkage factor ($0 < c^- < 1$) and c^+ a growth factor ($c^+ > 1$), with which the step width is reduced or increased, respectively. Applying these factors is justified in essentially the same way as for self-adaptive error backpropagation. The ranges of typical values for these parameters coincide as well, namely $c^- \in [0.5, 0.7]$ and $c^+ \in [1.05, 1.2]$.

Like the value of the learning rate in self-adaptive error backpropagation, the value of the step width is bounded from above and from below in order to avoid large jumps as well as slow training. Furthermore, resilient error backpropagation should only be applied with (full) batch training (or sufficiently large mini-batches), since online training is even less stable as for self-adaptive error backpropagation.

Resilient error backpropagation has proven to be significantly better and faster than many methods (e.g., momentum term and self-adaptive error backpropagation)

Fig. 5.35 Quick propagation relies on a local approximation of the error function by a parabola. m is the true minimum

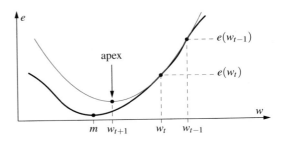

Fig. 5.36 The weight adaptation formula can be computed from the slope triangles of the derivative of the approximation parabola

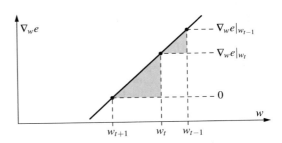

in many applications. For a long time it was one of the most highly recommended training methods, until more recent methods superseded it.

5.8.6   Quick Propagation

The quick propagation method [7] approximates the error function at the location of the current parameter value by a parabola (see Fig. 5.35) and computes from the current and the previous gradient the location of the apex of this parabola. Training then "jumps" directly to this apex, that is, the weight is set to the computed location of the apex. If the error function is "benevolent", training may thus get very close to the actual minimum in a single training step.

The weight adaptation rule can be obtained, for example, from two slope triangles of the derivative of the parabola (see the gray triangles in Fig. 5.36). We have

$$\frac{\nabla_w e|_{w_{t-1}} - \nabla_w e|_{w_t}}{w_{t-1} - w_t} = \frac{\nabla_w e|_{w_t}}{w_t - w_{t-1}}.$$

By solving for $\Delta w_t = w_{t+1} - w_t$ and exploiting $\Delta w_{t-1} = w_t - w_{t-1}$ we arrive at

$$\Delta w_t = \frac{\nabla_w e|_{w_t}}{\nabla_w e|_{w_{t-1}} - \nabla_w e|_{w_t}} \cdot \Delta w_{t-1}.$$

Note, however, that this equation does not distinguish between a parabola that opens upward and one that opens downward, so that it may happen that a (local) maximum of the error function is approached. Although this may be avoided by a test whether

$$\frac{\nabla_w e|_{w_{t-1}} - \nabla_w e|_{t_t}}{\Delta w_{t-1}} < 0$$

(this indicates a parabola that opens upward), implementations often skip this test. Instead a parameter is introduced that limits the increase of the weight change relative to the preceding step. That is, it is made sure that

$$|\Delta w_t| \leq c \cdot |\Delta w_{t-1}|$$

holds, where c is a parameter typically chosen between 1.75 and 2.25. This improves the behavior, but does *not* ensure that the weight is adapted in the correct direction. It merely avoids large jumps and unstable behavior.

Furthermore, implementations often add a standard gradient step to the update rule, provided the gradients $\nabla_w e|_{w_t}$ and $\nabla_w e|_{w_{t-1}}$ have the same sign, that is, provided the minimum does not lie between the current and the preceding weight value.

If the assumptions of the quick-propagation method hold, namely that the error function can be locally approximated by a parabola that opens upward and that the parameters are largely independent, and if batch training is employed, quick propagation can be a very fast training algorithm for multi-layer perceptrons, thus justifying its name. Otherwise it exhibits a tendency toward unstable behavior.

5.8.7 Adaptive Subgradient Descent (AdaGrad)

Up to now we considered only the effect of improvements of gradient descent on the training of a single parameter. However, in a neural network there is a large, sometimes even huge number of parameters that need to be adapted. Naturally, then, the question arises how the updates of different parameters relate to each other. It is to be expected that gradients for different parameters will differ in size, so that training proceeds with different speed in different directions of the parameter space. This can lead to training having advanced very much for some parameters while it is lagging behind for others, leading to undesirable trajectories in the parameter space.

A first and very simple approach to cope with this problem is the method of adaptive subgradient descent (AdaGrad) [5]. It is the first of several methods (more are discussed below) that introduce a kind of normalization of the gradient, which has the additional advantage that it eases the choice of the learning rate (since the influence of the size of the gradient on the training speed is reduced by the normalization). The AdaGrad update rule is defined as

$$\Delta w_t = -\eta \frac{\nabla_w e|_{w_t}}{\sqrt{v_t} + \epsilon} \quad \text{with} \quad v_0 = 0; \ v_t = v_{t-1} + \left(\nabla_w e|_{w_t}\right)^2; \ \epsilon = 10^{-6}; \ \eta = 0.01.$$

Intuitively, the gradient is normalized by the square root of the raw variance[14] of all preceding gradients.[15] This slows down training along dimensions of the parameter space in which parameters already changed substantially (i.e., for which past gradients were large, leading to large steps), and (relatively) speeds up training along

---

[14] In contrast to the variance about the mean, for which the squares of the differences to the mean are summed, the squares of the values themselves are summed for the raw variance.
[15] The added ϵ is only a technical trick for an implementation as it rules out a division by zero.

dimensions in which parameters changed only slightly. In this way, it "stabilizes" the network's representation of common features. However, since the raw variance is a sum of squares, it can only grow. Hence, AdaGrad suffers from the disadvantage that training becomes slower over time and finally stops completely. Due to this it can happen that the initial learning rate (which is commonly chosen as $\eta = 0.01$) is too small to let the adaptations reach a (local) optimum.

5.8.8 Root Mean Squared Gradient Descent (RMSProp)

As stated in the preceding section, the disadvantage of AdaGrad is that the normalization factor can only grow. A simple approach to eliminate this problem is to compute the raw variance not for *all* preceding gradients, but only for those in a history windows of a certain size (i.e., extending a certain number of training steps into the past). However, this is inconvenient to implement, because all gradients would have to pass through a buffer, which is updated in each step by replacing the oldest gradient with the current one. Since such a buffer would have to be maintained for each parameter, the memory consumption would be a (possibly large) multiple of the number of parameters, which is too large to be feasible in practice.

This problem can be circumvented by switching from a history window of a certain size to a scheme of "exponentially forgetting" older gradients. The idea is to form a convex combination of the previous aggregate of the squared gradients and the current one, controlled by a mixing parameter β [33]:

$$\Delta w_t = -\eta \frac{\nabla_w e|_{w_t}}{\sqrt{v_t} + \epsilon} \quad \text{with} \quad \begin{matrix} v_0 = 0; \quad v_t = \beta v_{t-1} + (1 - \beta)\left(\nabla_w e|_{w_t}\right)^2; \\ \beta = 0.999; \quad \epsilon = 10^{-6} \end{matrix}$$

In this way, the weight with which a gradient influences the normalization factor decays geometrically over time (because it is multiplied by a factor $\beta \leq 1$ in each time step). This scheme is called *root mean squared gradient descent (RMSProp)*, because the normalization factor is the root of a sum of squared gradients. No mean is explicitly computed, but with a history window of a fixed size (which is the underlying intuition), the mean computation would only introduce a constant factor (namely the size of the history window), which can be incorporated into the learning rate. This does not affect the switch to a convex combination scheme.

The clear advantage of this scheme is that only a single additional value needs to be stored per parameter (instead of a history window), making it easily feasible for implementations. Furthermore, the normalization factor can shrink again, namely in regions of the parameter space in which the error function is rather flat and hence the gradients are small (while larger gradients from earlier update steps are slowly "forgotten"). This avoids the drawback of AdaGrad that training comes to a halt, because here the normalization factor does not grow uninhibited.

5.8.9 Adaptive Subgradient Descent Over Windows (AdaDelta)

A problem of gradient descent is that it adds a multiple (as controlled by the learning rate η) of the gradient to the current value of a parameter, but a parameter and its change (as computed via a derivative) do not have the same "units". For example, a distance and its derivative, which is a velocity, do not have the same units (m versus m/s). If in physics a numerical update of, say, a location is computed by adding a certain factor times a velocity, then the factor is a time interval and thus have a proper unit that removes the mismatch (e.g., $l' = l + \Delta t \cdot v$, which has units $[m] = [m] + [s] \cdot [m/s]$). However, in gradient descent the learning rate is a unit-less quantity and hence no correction for the "unit mismatch" occurs. Note that this is also the case for AdaGrad or RMSProp: the gradient normalization actually removes all units and makes the parameter change a unit-less quantity.

However, in a Newton–Raphson approach to finding the optimum of a function via a root search on its derivative, the units of a parameter and its change match (due the use of second derivative information/a Hessian matrix). Starting conceptually from such a Newton–Raphson approach is the basic idea of *adaptive subgradient descent over windows (AdaDelta)* [38]. With some simplifying assumptions (see [38] for details), the following rule can be derived:

$$\Delta w_t = -\frac{\sqrt{u_t + \epsilon}}{\sqrt{v_t + \epsilon}} \nabla_w e|_{w_t} \quad \text{with} \quad \begin{aligned} u_0 &= 0; \; u_t = \alpha u_{t-1} + (1-\alpha)(\Delta w_{t-1})^2; \\ v_0 &= 0; \; v_t = \beta v_{t-1} + (1-\beta)\left(\nabla_w e|_{w_t}\right)^2; \\ \alpha &= \beta = 0.95; \epsilon = 10^{-6}. \end{aligned}$$

Note that there is *no learning rate* (no η) in this scheme (as there is no learning rate in the Newton–Raphson method generally).[16] Note also that due to the simplifying assumptions, which remove, for example, interactions between parameters, this is actually *not* a second-order optimization method (as a true Newton–Raphson approach would be), because no second-order derivative is actually computed. Finally, note that the first step of this procedure is effectively identical to Manhattan training (cf. Sect. 5.8.1), while the following steps are somewhat analogous a "normalized" momentum term approach (cf. the alternative formulation of the momentum term approach in Sect. 5.8.2), only that it is applied here to the squared weight changes instead of the weight changes themselves.

5.8.10 Adaptive Moment Estimation (Adam)

One of the most popular gradient descent variants is *adaptive moment estimation (Adam)* [19]. This method combines adding a momentum term (cf. Sect. 5.8.2) and normalizing the gradient as in RMSProp (cf. Sect. 5.8.8). The momentum term is

---

[16] A side issue, which one should be careful about in an implementation, is that the two ϵ (which serve the purpose of preventing a vanishing step size as well as a division by zero) appear *inside* the square root, while in the preceding methods ϵ was added *outside* the square root. This is *not* a typographic error, but can be found like this in the original definitions of these methods.

implemented as in the alternative formulation of this approach, while the normalization follows exactly the RMSProp scheme:

$$\Delta w_t = -\eta \frac{m_t}{\sqrt{v_t} + \epsilon} \quad \text{with} \quad \begin{array}{l} m_0 = 0; \ m_t = \alpha m_{t-1} + (1 - \alpha)\left(\nabla_w e|_{w_t}\right); \quad \alpha = 0.9; \\ v_0 = 0; \ v_t = \beta v_{t-1} + (1 - \beta)\left(\nabla_w e|_{w_t}\right)^2; \quad \beta = 0.999. \end{array}$$

This method is called *adaptive moment estimation*, because m_t is an estimate of the (recent) first moment (mean) of the gradient, while v_t is an estimate of the (recent) raw second moment (raw variance) of the gradient. Again the underlying idea is that of a history window, from which the moments are computed, which for purposes of a memory-friendly implementation is replaced by "exponential forgetting".

A problem of Adam is that in the first training steps there is no or just a very short actual history to draw on. Formally, the values of m_t and v_t thus implicitly rely on a history in which all gradients are zero before the first one that is actually computed (because the formulae start with $m_0 = v_0 = 0$). This introduces a bias toward zero (i.e., the values of m_t and v_t tend to be too small, especially for small t), which can be corrected as follows (Adam with bias correction) [19]:

$$\Delta w_t = -\eta \frac{\widehat{m}_t}{\sqrt{\widehat{v}_t} + \epsilon} \quad \text{with} \quad \begin{array}{ll} \widehat{m}_t = m_t/(1 - \alpha^t) & (m_t \text{ as above}), \\ \widehat{v}_t = v_t/(1 - \beta^t) & (v_t \text{ as above}). \end{array}$$

A natural extension of Adam, which relies on the standard momentum term scheme, is to use Nesterov's accelerated gradient scheme instead. This leads to Adam with Nesterov acceleration (NAdam) [4]:

$$\Delta w_t = -\eta \frac{\alpha m_t + (1 - \alpha)\left(\nabla_w e|_{w_t}\right)}{\sqrt{v_t} + \epsilon} \quad \text{with } m_t \text{ and } v_t \text{ as for Adam.}$$

The idea is, of course, the same as with Nesterov's accelerated gradient in general: since the momentum term modifies the parameter value anyway, independent of the current gradient, it is preferable to apply it before computing the current gradient. In order to ease implementation, the alternative (though not exactly equivalent) formulation of Nesterov's accelerated gradient is used (cf. Sect. 5.8.3).

Of course, one may also add a bias correction to this approach [4]:

$$\Delta w_t = -\eta \frac{\alpha \widehat{m}_t + \frac{1-\alpha}{1-\alpha^t}\left(\nabla_w e|_{w_t}\right)}{\sqrt{\widehat{v}_t} + \epsilon} \quad \begin{array}{l} \text{with } \widehat{m}_t \text{ and } \widehat{v}_t \text{ as for Adam} \\ \text{with bias correction.} \end{array}$$

An alternative to the Adam scheme, which computes the raw variance of previous squared gradients and hence can also be seen as computing an L_2 (or Euclidean) norm of the squared gradients in the history window, one may consider using an L_1 (or maximum) norm instead. This leads to *AdaMax* [19]:

$$\Delta w_t = -\eta \frac{m_t}{\sqrt{v_t} + \epsilon} \quad \text{with} \quad \begin{array}{l} m_0 = 0; \ m_t = \alpha m_{t-1} + (1 - \alpha)\left(\nabla_w e|_{w_t}\right); \quad \alpha = 0.9; \\ v_0 = 0; \ v_t = \max\left(\beta v_{t-1}, \left(\nabla_w e|_{w_t}\right)^2\right); \quad \beta = 0.999. \end{array}$$

Again there is also a version with bias correction [19]:

$$\Delta w_t = -\eta \frac{\widehat{m}_t}{\sqrt{v_t} + \epsilon} \quad \text{with} \quad \begin{array}{ll} \widehat{m}_t = m_t/(1 - \alpha^t) & (m_t \text{ as above}), \\ v_t \text{ does not need a bias correction.} \end{array}$$

Note that in this case only the momentum term needs a bias correction, while the variance term does not. The reason is that the term referring from squared gradients always refers to a (possibly decayed) single squared gradient (which is picked up from some location in the history window, possibly the most recent) and hence the implicit zero terms before the actual start of the history do not matter (much).

5.8.11   Lifting the Derivative of the Activation Function

Often the error function is fairly flat in some regions of the parameter space, because the activation functions are in their saturation region (that is, are evaluated far away from the bias value θ, cf. Fig. 5.2 on page 55), where the gradient is very small or even vanishes completely. In order to speed up training in such a case, the derivative f'_{act} of the activation function may be lifted artificially by a fixed value α, so that sufficiently large training steps are carried out even in these saturation regions [7]. Choosing $\alpha = 0.1$ often leads to good results. This modification of gradient descent is also known as **flat spot elimination**.

Lifting the derivative of the activation function has the additional advantage that it counteracts a weakening of the error signal in the backpropagation procedure. Such a weakening is due to the fact that, for example, the derivative of the logistic function is bounded by $\frac{1}{4}$ (which results for the function value $\frac{1}{2}$, that is, for the location of the bias value, cf. Fig. 5.25 on page 77). As a consequence the error value has a tendency to become smaller and smaller with every layer through which it is propagated, so that training is slower in the front layers of the network.

5.8.12   Weight Decay

It is generally unfavorable if training leads to very large values for the connection weights of a neural network. The reasons are, in the first place, that with large weights one easily reaches the saturation region of a logistic activation function, in which the gradients almost vanish. Secondly, large weights increase the risk of overfitting the network to accidental properties of the training data, so that the performance of the network on new data falls short of what can be achieved otherwise.

The weight decay method [37] serves the purpose to avoid a heavy growth of the weights. To prevent an excessive growth, each weight is reduced in each step by a small factor, for example, with

$$\Delta w(t) = -\frac{\eta}{2}\nabla_w(t) - \xi w(t)$$

for standard gradient descent. Alternatively, each weight may be multiplied before its adaptation by the factor $(1 - \xi)$, which is often easier to implement. The value of ξ should be chosen very small, so that the weights are not kept permanently at low values. Typical values for ξ are in the range between 0.005 to 0.03.

Note that we may obtain the weight decay rule by using an extended error function that penalizes large weights:

$$e^* = e + \frac{\xi}{2} \sum_{u \in U_{\text{out}} \cup U_{\text{hidden}}} \left(\theta_u^2 + \sum_{p \in \text{pred}(u)} w_{up}^2 \right).$$

The derivative of this modified error leads to the weight adaptation stated above.

5.8.13 Batch Normalization

In Sect. 4.3 we recommended to z-score normalize the input data, because depending on the units used for the different inputs they can come from considerably different ranges of values. This is undesirable, because it can cause the training to focus on those inputs that come from a large range of values. (Recall that the input is a factor in the weight update formula, cf. page 84, and thus its magnitude directly influences the update.) This tendency is removed by z-score normalization:

$$z_u^{(l)} = \frac{x_u^{(l)} - \mu_u}{\sqrt{\sigma_u^2}} \quad \text{with} \quad \mu_u = \frac{1}{|L|} \sum_{l \in L} x_u^{(l)}, \quad \sigma_u^2 = \frac{1}{|L| - 1} \sum_{l \in L} \left(x_u^{(l)} - \mu_u \right)^2.$$

Of course, depending on the weights we could encounter a similar situation after the first layer: the outputs of the neurons may cover considerably different ranges of values (even though, with the activation functions we studied up to now, they are all bounded by 0 and 1). Hence, the idea suggests itself to apply such a normalization not only for the input but also after each hidden layer. This is the core idea of the method of *batch normalization* [16].

However, a strict application of such a scheme would require full batch training (to compute the mean and standard deviation over all training patterns). In order to make it feasible for mini-batch training as well, [16] suggest to compute $\mu_u^{(B_i)}$ and $\sigma_u^{2(B_i)}$ from each mini-batch B_i (instead of from all training patterns L) and apply a normalization based on these values to the mini-batch they were computed from. In addition to this normalization, a trainable scale γ_u and shift β_u are introduced, that is, the normalized values are further transformed by $z_u' = \gamma_u \cdot z_u + \beta_u$.

For the final network, the aggregated mean μ_u and and variance σ_u^2 are used:

$$\mu_u = \frac{1}{m} \sum_{i=1}^{m} \mu_u^{(B_i)}, \quad \sigma_u^2 = \frac{1}{m} \sum_{i=1}^{m} \sigma_u^{2(B_i)}. \qquad \begin{array}{l} m\text{: number of} \\ \text{mini-batches} \end{array}$$

The transformation (consisting of normalization, scaling, and shifting) that is computed in the final (trained) network (with trained parameters γ_u and β_u) is:

$$z_u = \frac{x_u - \mu_u}{\sqrt{\sigma_u^2 + \epsilon}}, \quad z_u' = \gamma_u \cdot z_u + \beta_u. \qquad \begin{array}{l} \epsilon\text{: to prevent a} \\ \text{division by zero} \end{array}$$

This form of batch normalization has proven to be very effective in practice.

5.9 Examples for Some Variants

In this section, we illustrate some of the gradient descent variants with some simple examples to emphasize their characteristics. We start with gradient descent with a momentum term, which we apply to the minimization of the polynomial function studied in Sect. 5.7 (see page 73f). With the help of a momentum term the very slow descent of Fig. 5.22 can be accelerated considerably, as shown in Fig. 5.37. However, the very small learning rate cannot be compensated completely. The reason is, as already mentioned in the preceding section, that even for a constant gradient $f'(x)$ the step width is bounded by $\left|\frac{\eta f'(x)}{1-\beta}\right|$, as already mentioned.

By using an adaptive learning rate even the chaotic back and forth jumps can be avoided that can result from a large learning rate, as we observed in Fig. 5.23 on page 75. This is demonstrated in Fig. 5.38, which shows the training procedure for the even larger learning rate $\eta = 0.3$ (compared to $\eta = 0.25$ in Fig. 5.23). Although the initial value is too large, it is quickly corrected by the shrinkage factor, so that the minimum of the function is reached in astonishingly few steps.

i	x_i	$f(x_i)$	$f'(x_i)$	Δx_i
0	0.200	3.112	−11.147	0.011
1	0.211	2.990	−10.811	0.021
2	0.232	2.771	−10.196	0.029
3	0.261	2.488	−9.368	0.035
4	0.296	2.173	−8.397	0.040
5	0.337	1.856	−7.348	0.044
6	0.380	1.559	−6.277	0.046
7	0.426	1.298	−5.228	0.046
8	0.472	1.079	−4.235	0.046
9	0.518	0.907	−3.319	0.045
10	0.562	0.777		

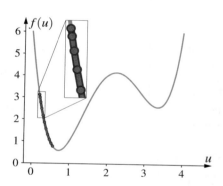

Fig. 5.37 Gradient descent with a momentum term ($\beta = 0.9$) starting from the initial value 0.2 and with a learning rate 0.001

i	x_i	$f(x_i)$	$f'(x_i)$	Δx_i
0	1.500	2.719	3.500	−1.050
1	0.450	1.178	−4.699	0.705
2	1.155	1.476	3.396	−0.509
3	0.645	0.629	−1.110	0.083
4	0.729	0.587	0.072	−0.005
5	0.723	0.587	0.001	0.000
6	0.723	0.587	0.000	0.000
7	0.723	0.587	0.000	0.000
8	0.723	0.587	0.000	0.000
9	0.723	0.587	0.000	0.000
10	0.723	0.587		

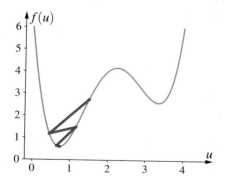

Fig. 5.38 Gradient descent with adaptive learning rate (with $\eta_0 = 0.3$, $c^+ = 1.2$, $c^- = 0.5$) starting from the initial value 1.5

Table 5.2 Training with and without a momentum term ($\beta = 0.9$)

Epoch	θ	w	Error	Epoch	θ	w	Error
0	3.00	3.50	1.295	0	3.00	3.50	1.295
20	3.76	2.20	0.985	10	3.80	2.19	0.984
40	3.70	1.82	0.970	20	3.75	1.84	0.971
60	3.48	1.53	0.957	30	3.56	1.58	0.960
80	3.11	1.25	0.934	40	3.26	1.33	0.943
100	2.49	0.88	0.880	50	2.79	1.04	0.910
120	1.27	0.22	0.676	60	1.99	0.60	0.814
140	−0.21	−1.04	0.292	70	0.54	−0.25	0.497
160	−0.86	−2.08	0.140	80	−0.53	−1.51	0.211
180	−1.21	−2.74	0.084	90	−1.02	−2.36	0.113
200	−1.45	−3.19	0.058	100	−1.31	−2.92	0.073
220	−1.63	−3.53	0.044	110	−1.52	−3.31	0.053
				120	−1.67	−3.61	0.041

Without momentum term   With momentum term

As a second example we consider, in analogy to Sect. 5.7, training a two-layer perceptron for the negation, again starting from the initial values $\theta = 3$ and $w = \frac{7}{2}$. The courses of training without a momentum term and with such a term with the factor $\beta = 0.9$ are shown in Table 5.2 and in Fig. 5.39. Clearly, the training process advances in almost the same way, only that with a momentum term merely about half the number of epochs are needed to obtain the same error value. That is, the momentum term roughly doubled the training speed. If bigger networks are trained with a larger number of training patterns, the difference in training speed can be even much larger, making any method with a momentum term (including methods like Adam or AdaDelta) highly recommendable for training such networks.

To illustrate error backpropagation with a momentum term we also refer again to the programs wmlp and xmlp, which we already mentioned on page 88. They allow to introduce a momentum term, which accelerates the training considerably. It is instructive to observe how much faster a solution for the biimplication or the exclusive or is found and how much more swiftly the network output approximates the given training examples for the two function fitting tasks.

In order to illustrate the behavior of methods with gradient normalization (Ada-Grad, RMSProp, AdaDelta, and especially Adam), we compare standard gradient descent, gradient descent with a momentum term and Adam with bias correction on two simple two-dimensional functions, namely a parabola $y = x_1^2 + 3x_2^3$ and a Rosenbrock function $y = (1 − x_1)^2 + 10(x_2 − x_1^2)^2$ [30], which is a common test case for gradient-based optimization methods. The trajectories in the parameter space that are traversed for these three methods and the two functions are shown in Figs. 5.40 and 5.41, respectively, with a two-dimensional view from the top on without momentum term   with momentum term   with momentum term

Fig. 5.39 Training with and without a momentum term ($\beta = 0.9$); the points show the values of the weight w and the bias value θ every 20 or 10 epochs, respectively

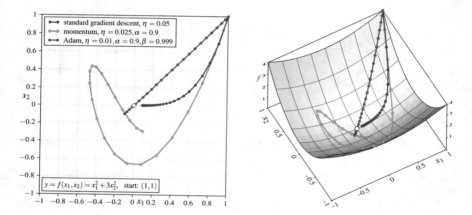

Fig. 5.40 Optimization behavior on a simple parabolic function $y = x_1^2 + 3x_2^2$. Note the balancing of the training speed in the two dimensions by the Adam optimization method

the left and a three-dimensional diagram on the right. Note how Adam (red line) balances the training speed in the two dimensions x_1 and x_2 for the parabola, which is a result of the gradient normalization. In contrast, standard gradient descent (blue line) adapts (at least at the beginning). x_2 much faster than x_1, because the parabola is steeper in x_2- than in x_1-direction. Note also how the momentum term accelerates the training, which can lead (as it does here) to considerable and repeated overshoots before the actual minimum (gray dot) is approached.

The problem of the Rosenbrock function in Fig. 5.41 is that the minimum is located in a narrow curved valley, which makes it very difficult for standard gradient descent (blue line) to approach it, despite a larger learning rate (compared to the other two methods). Gradient descent with a momentum term (green line) and Adam (red line) on the other hand, fare much better, even though the momentum term approach exhibits a stronger tendency to overshoot. (Adam overshoots as well, though, if one computes more update steps, but not as far.)

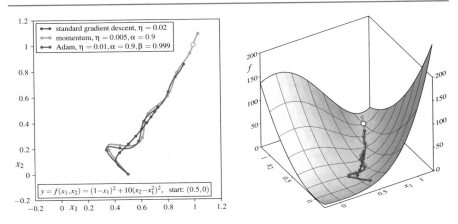

Fig. 5.41 Optimization behavior on a Rosenbrock function [30] $y = (1 - x_1)^2 + 10(x_2 - x_1^2)^2$. Note the very slow training of standard gradient descent despite larger η

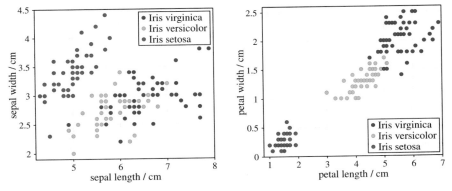

Fig. 5.42 Scatter plots of the iris flower data showing all four measured quantities

As a final example, we consider training a neural network for the iris flower data. This data set, which is arguably the most famous data set in statistics, pattern recognition and data analysis, was orginally collected in part by [1] on the Gaspé Peninsula (Canada). It was completed and first analyzed with statistical methods by [8]. The iris flower data set consist of 150 cases in total, 50 cases for each of three iris flower types: *Iris setosa*, *Iris virginica* and *Iris versicolor*. Each iris flower specimen is described by measurements of the sepal length and width and the petal length and width (in cm, see scatter plots in Fig. 5.42).

In this example, we trained a multi-layer perceptron for the iris data. We used four input neurons (one for each of the four measurements), three output neurons (one for each iris type), and a single hidden layer with three neurons. All hidden and output neurons used the logistic function as their activation functions and the identity as their output functions. The network was trained with sum of squared errors (which

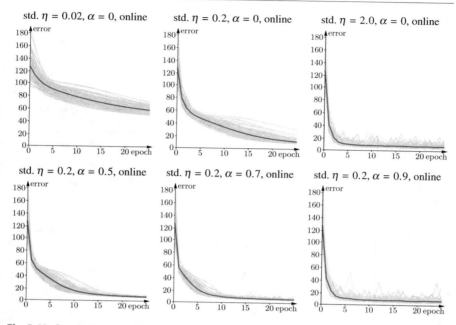

Fig. 5.43 Standard gradient descent without and with a momentum term on the iris data

works well here, even though a cross-entropy loss may be even more appropriate). We tried many different training methods with different learning rates and executed 100 independent training runs, starting from weights that were initialized by sampling uniformly at random from $[-1, 1]$ (bias values were set to 0). The training was conducted with the command line programs that can be found at

http://www.borgelt.net/mlp.html

The results are depicted in Figs. 5.43, 5.44, 5.45 and 5.46. The gray lines indicate to the individual training runs, the red line is an epoch-wise average of these runs.

Figure 5.43 clearly shows the influence of the learning rate and the momentum term parameter α: For both larger values accelerate training, but can easily lead to unstable behavior, as can be seen from the spikes in the error in the diagrams on the right. Figure 5.44 demonstrates how self-adaptive backpropagation and resilient propagation can handle learning rates or initial step width that rather small: they are increased in the course of training and approach, at least after some epochs, the behavior of training runs with much larger learning rates. Even though quickprop mainly jumps to the apex of an approximating parabola, the learning rate still has an influence as it determines from which points the parabola is estimated.

The new variants, which rely on a normalization of the gradient, all outperform the older methods, at least if a proper learning rate is chosen, as can be seen in Figs. 5.45 and 5.46. They reach very low sums of squared errors after very few epochs and are

Fig. 5.44 Super self-adaptive, resilient and quick propagation on the iris data

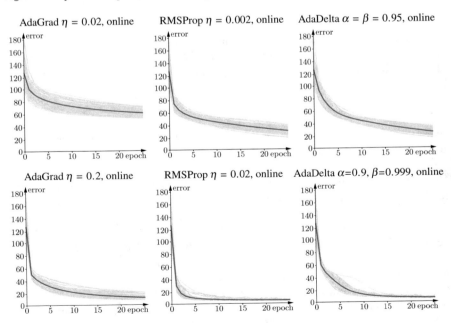

Fig. 5.45 AdaGrad, RMSProp, and AdaDelta on the iris data

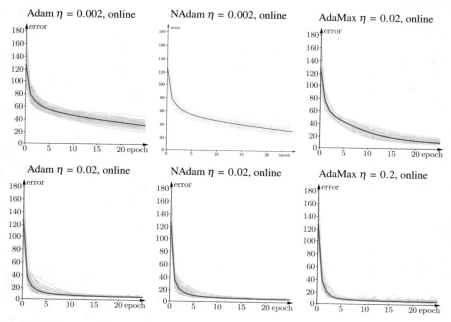

Fig. 5.46 Adam, NAdam, and AdaMax on the iris data

less susceptible to oscillations. Only AdaMax for $\eta = 0.2$ (bottom right diagram in Fig. 5.46) exhibits some instability, which would make a slightly smaller learning rate advisable. At the very least, these example training runs certainly support the popularity of Adam and, in second place, of RMSProp.

5.10 Initializing the Parameters

Up to now we merely said that the parameters, especially the weights, are initialized randomly. However, in practice, bias values are usually set to zero, in order not to bias the computations right from the start. The weights, on the other hand, should not be initialized to zero (or, more generally, should not be initialized to equal values), because this would prevent useful training: all weights of connections leading to the same neuron would receive the same gradient. Although adaptation differences eventually result from different inputs for the connections to the first hidden layer and from different outputs for the connections to the output layer, all hidden sibling neurons (i.e., neurons residing in the same hidden layer) would be trained to compute the same function, which cannot produce a useful training result.

The simplest weight initialization method, which we also employed in the preceding section, is **uniform weight initialization**, where the connection weights are sampled from a uniform distribution on $[-\epsilon, \epsilon]$, with, e.g., $\epsilon \in \left\{1, \frac{1}{p}, \frac{1}{\sqrt{p}}\right\}$. Here, p is the number of predecessors of the neuron to which the connections lead, the weights of which are initialized. Making the weight range dependent on the number of predecessors tries to equalize the (initial) neuron activations across layers. This principle also underlies **Xavier weight initialization** [9],[17] which derives the weight range by assuming normally distributed inputs and that the activation computation occurs in the almost linear region of a logistic function or a hyperbolic tangent. By analyzing how the variance develops under these conditions over consecutive layers, a suitable weight range is determined, so that the variance stays roughly constant, namely $[-\epsilon, \epsilon]$, with $\epsilon = \sqrt{6/(p+s)}$. Here s is the number of neurons in the layer, to which the initialized connections lead. As noted in [11], the conditions derived in [9] can also be met by drawing the weights from a normal distribution with mean 0 and variance $\frac{1}{p}$. Unfortunately, this can cause some ambiguity what is actually meant by "Xavier weight initialization" in an implementation, as both versions can be found.

A similar approach underlies **He et al. / Kaiming weight initialization** [11],[18] which repeats the analysis of [9], but also considers the rectified maximum as activation function. Since this activation function yields an output of zero for half of the network inputs (if these are distributed symmetrically around the initial bias value 0), the weight range should be twice as large. Hence, weights should be drawn from a normal distribution with mean 0 and variance $\frac{2}{p}$.

5.11 Number of Hidden Neurons

When training multi-layer perceptrons in practice, we naturally face the question, how many hidden neurons (often in only a single layer) it should have. A common rule of thumb is that (number of inputs + number of outputs) / 2 should be chosen in case of a single hidden layer, but clearly this can only be a rough guideline.

A better way to determine an appropriate number of hidden neurons is the following procedure, even though it can be computationally fairly costly (depending on the size of the data set): The data set is split into two (roughly) equal size subsets. Multi-layer perceptrons with different numbers of hidden neurons are trained on one part of the data (the training data) and evaluated on the other part (the validation data). That is, the error is computed on the second part of the data, which was *not* used for training the network. This process is repeated with multiple random splits of the data, and the results are averaged per number of hidden neurons. The number of hidden neurons that yields (on average) the best result on the validation data (*not* on the training data!) is chosen. With this number of hidden neurons, a multi-layer perceptron is trained using the full data set (training and validation data together).

---

[17] "Xavier" is the given name of the first author of the paper that proposed this method.
[18] "He" is the family name, "Kaiming" the given name of the first author of the proposing paper.

This procedure exploits the following principle: if the number of neurons in the hidden layer is too small, the multi-layer perceptron may not be able to capture the structure of the relationship between inputs and outputs precisely enough due to a lack of parameters (see the discussion in Sect. 5.3). Therefore, the error will stay larger than necessary. This is usually called **underfitting**.

The more neurons are present in its hidden layer, the better a multi-layer perceptron can capture the dependence between the outputs and the inputs. However, this does not mean that one should use as many neurons as the computing power and memory equipment of the employed computer system permit, in order to obtain the highest expressive power and thus adaptability. In practice, data are always affected by noise, that is, it contains random errors and deviations from the true values, which may pertain to the inputs as well as the outputs. Furthermore, the given data are always merely a finite sample and thus may depict the dependence between outputs and inputs only in a distorted fashion, even if they were collected with a lot of care. Unfortunately, with a larger number of hidden neurons a multi-layer perceptron may adapt not only to the regular dependence between outputs and inputs we try to capture, but also to the accidental specifics (and thus also the errors and deviations) of the training data set. This is usually called **overfitting**.

Overfitting will usually lead to the effect that the error a multi-layer perceptron with many hidden neurons yields on the validation data will be (possibly considerably) greater than the error it yields on the training data. The reason is that the validation data set is likely distorted in a different fashion, since the errors and deviations are random. Therefore, minimizing the error on the validation data by properly choosing the number of hidden neurons prevents both under- and overfitting.

The described method of iteratively splitting the data into training and validation data may be referred to as **cross validation**, although this term is more often used for the following specific procedure: the given data set is split into n parts or subsets (also called *folds*) of about equal size (so-called *n-fold cross validation*). If the output is nominal (categorical/ symbolic), this split is done in such a way that the relative frequencies of the different values of the output attribute in the subsets/folds represent as well as possible the relative frequencies of these values in the data set as a whole. This is also called **stratification** (derived from the Latin *stratum*: layer, level, tier), since it can be achieved by splitting each *stratum* (that is, each set of cases with the same value of the output attribute) separately.

Out of these n data subsets (or folds) n pairs of training and test data sets are formed by using one fold as a validation data set while the remaining $n - 1$ folds are combined into a training data set. The advantage of this method is that one random split of the data yields n different pairs of training and validation data set. Another advantage is that each data point is used exactly once in a validation data set. Thus, it is prevented that some data points have (possibly only due to repeated random splits) a higher influence on the validation result. An obvious disadvantage is that (except for $n = 2$) the size of the training and the test data set are considerably different, which makes the results on the validation data statistically less reliable. It is therefore only recommended for sufficiently large data sets or sufficiently small n, so that the validation data sets are of sufficient size.

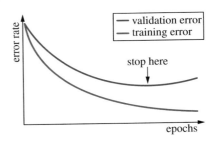

Fig. 5.47 Early stopping the training of a neural network: stop training when the error on the validation data increases again

An alternative way to prevent overfitting is the following approach: during training the performance of the multi-layer perceptron is evaluated after each epoch (or every few epochs) on a validation data set (which is *not* used for training the parameters of the network). While the error on the training data set should always decrease with each epoch, the error on the validation data set should, after decreasing initially as well, increase again as soon as overfitting sets in. This is illustrated in Fig. 5.47, where the training error (red curve) is consistently sloping down, while the validation error (blue curve) starts to rise again at some point. At this moment training is terminated and either the current state or (if available) the state of the multi-layer perceptron, for which the error on the validation data reached a minimum, is reported as the training result. Furthermore there are approaches which derive a stopping criterion for the training from the shape of the error curve on the training data over the training epochs, or which train the network only for a fixed, relatively small number of epochs (also known as **early stopping**).

All of these approaches share the advantage that they need to train only one network (with a specific number of hidden neurons), while the method of cross validation that we described before needs a large number of training runs with different networks and thus is computationally much more costly, especially for larger data sets. Nevertheless, it is preferable to these simpler methods due to essentially one reason: while the simpler methods try to avoid overfitting by stopping the training of a complex network early enough, so that the training could not yet exploit the full expressive power of the network to adapt to the given data, cross validation actually tries to adjust the complexity of the network to the "correct" level. That is, it tried to find the actually most appropriate model for the data.

5.12   Deep Learning

In Sect. 5.3 we saw that, in principle, multi-layer perceptrons with only one hidden layer (and thus, together with the output layer, only two layers that execute computations) can approximate basically any continuous function on a compact part of the $\mathbb{R}^n$ with arbitrary accuracy (see page 66). This theorem is often cited as (allegedly) meaning that we can confine ourselves to multi-layer perceptrons with only one hidden layer, that there is no real need to look at multi-layer perceptrons with more hidden

Fig. 5.48 Multi-layer perceptron for computing the parity function with many hidden layers

layers. However, such a conclusion would be a bit too rash and would overextend considerably on what can actually be inferred. The main reason is that the theorem says nothing about the number of hidden neurons that may be needed to achieve a desired approximation accuracy. Depending on the function to approximate, a very large number of neurons (e.g., exponential in the number of inputs) may be necessary, while allowing for more hidden layers may enable us to achieve the same approximation quality with a significantly lower number of neurons.

A very simple and commonly used example of such a function is the n-bit (even) parity function, which maps n binary inputs to a single binary output. The output is 1 if and only if an even number of inputs are 1 and all other inputs are 0. However, if an odd number of inputs are 1, the output is 0. This function can easily be represented by a multi-layer perceptron with only one hidden layer if we employ Algorithm 3.1 as stated on page 23, adapted to multi-layer perceptrons. However, this algorithm leads to 2^{n-1} hidden neurons, because the disjunctive normal form of the n-bit parity function is a disjunction of 2^{n-1} conjunctions, which represent the 2^{n-1} input combinations with an even number of set bits. That is, the number of hidden neurons grows exponentially with the number of inputs.

However, if we are allowed to use more than one hidden layer, we can construct a multi-layer perceptron that chains together $n - 1$ simple networks for computing the biimplication (like the one shown in Fig. 3.10 on page 22) and the exclusive or (which is analogous to the biimplication network of Fig. 3.10, but with the threshold value and the weights of the output neuron negated). A sketch of the structure of such a network is shown in Fig. 5.48. The first two hidden layers of this network combines two inputs with a biimplication network. Each of the remaining $n - 2$ networks, which compute the exclusive or of their inputs, combines the output of the preceding network with another input. Since each of the networks (regardless of biimplication or exclusive or) requires three neurons (see Fig. 3.10), the final network consists of $n + 3(n - 1) = 4n - 3$ neurons (n input neurons, $3(n - 1) - 1$ hidden neurons, and 1 output neuron). That is, the size of the network grows only linearly with the number of inputs, although at the price of having $2n - 1$ layers (including input and output layer). By using a (binary) tree structure of biimplication and exclusive or networks (using a biimplication network to combine two inputs or the outputs of two networks, and an exclusive or network to combine an input and a network output), the number of layers can be reduced to $O(\log n)$ while the number of needed neurons stays linear in the number of inputs (actually also $4n - 3$).

It is worth noting that the situation we face with the number of hidden neurons needed to represent n-ary Boolean functions is actually fairly similar to the situation w.r.t. linearly separable n-ary Boolean functions as we discussed it in Sect. 3.3 on page 19f. Just like most Boolean functions of n inputs are not linearly separable, only a small fraction of all n-ary Boolean functions can be represented with few hidden neurons by a multi-layer perceptron with only one hidden layer.

To demonstrate this more concretely, suppose that for an arbitrary n-ary Boolean function, there are k_0 input combinations with an output of 0 and k_1 input combinations with an output of 1. Clearly, $k_0 + k_1 = 2^n$. We can guarantee generally that we need at most $2^{\min(k_0, k_1)}$ hidden neurons if we choose between a disjunctive normal form as it is used in Algorithm 3.1 and a conjunctive normal form (for which an analogous algorithm can easily be constructed) based on whether $k_1 \leq k_0$ (prefer disjunctive normal form) or $k_1 > k_0$ (prefer conjunctive normal form). However, since there are $\binom{2^n}{k_0}$ possible functions with k_0 input combinations mapped to 0 and k_1 mapped to 1, we see that a very large part of all possible functions require a substantial number of hidden neurons. Although the number of needed neurons may be reduced with minimization methods like the Quine–McCluskey algorithm [21,26,27], the fundamental problem remains.

It should be noted, though, that in practice the problem is mitigated considerably by the simple fact that training data sets are limited in size. For example, while a complete training data set for an n-ary Boolean function requires 2^n training examples (one for each input configuration), practical problems that require learning functions with n inputs usually contain much fewer sample cases. This leads to many input configurations for which the given data do not prescribe a desired output. As a consequence, the training has the freedom to assign outputs to these input combinations, which allows for a simpler representation (similar to how so-called *don't care symbols*—that is, non-fixed outputs—are exploited in the abovementioned Quine–McCluskey algorithm), that is, a simpler network with (much) fewer neurons.

Nevertheless, using more than one hidden layer promises in many cases, especially for complex functional dependencies and large data sets (like, for example, in handwriting recognition), to reduce the number of needed neurons. This is the focus of the area of **deep learning**, where the "depth" of a neural network means the length of the longest path in the graph underlying the network (and thus the largest number of consecutive computation steps that are executed). Specifically for multi-layer perceptrons, where this longest path is the number of hidden layers plus one (for the output layer), deep learning starts with more than one hidden layer. Once the number of hidden layers reaches 10, one sometimes speaks of *very deep learning*. A general overview of deep learning (which is not restricted to neural networks like this section) can be found, e.g., in [2,3].

The two main problems of deep learning multi-layer perceptrons are (1) overfitting and (2) vanishing gradient. The overfitting problem results mainly from the increased number of adaptable parameters in the additional hidden layers. It can be tackled,

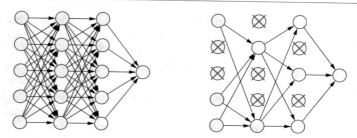

Fig. 5.49 Dropout training randomly deactivates a fraction of the neurons during training

for example, with weight decay (as discussed in Sect. 5.8.12), which prevents overly large weights and thus an overly precise adaptation to (accidental) properties of the data. Furthermore, sparsity constraints can be introduced, either in the form of a restricted number of neurons in the hidden layers or by requiring that only few of the neurons in the hidden layers should be active (on average). The latter is usually achieved by adding a regularization term to the error function, which compares the observed number of activated neurons with the desired low number and pushes the adaptations into a direction that tries to match these numbers. Finally, a training method that has become known as **dropout training** can be applied [32]. With this method some units are randomly omitted from the input layer and the hidden layers during training (for both forward and backward propagation), as indicated in the sketch in Fig. 5.49, which has been shown to be a simple and effective way to prevent overfitting.

The **vanishing gradient** problem may, in principle, affect any gradient descent procedure that is applied to chained functions and thus has to rely on the chain rule. In neural network training particularly it was discovered when trials of training multi-layer perceptrons with several hidden layers led to a performance that fell short of expectations. The core reason was formally exposed in [13] (see also [14]): the farther away a hidden layer is from the output layer, the more slowly it is trained with the standard backpropagation procedure (cf. Sect. 5.6) in the sense that the changes that are applied to the connection weights and the threshold values are usually (much) smaller than those that occur for the output layer or hidden layers that are (very) close to it. This is caused by the fact that the commonly employed logistic activation function

$$f_{\mathrm{act}}(x) = \frac{1}{1 + e^{-x}} \quad \text{has the derivative} \quad f'_{\mathrm{act}}(x) = f_{\mathrm{act}}(x) \cdot (1 - f_{\mathrm{act}}(x)),$$

which is at most $\frac{1}{4}$, namely for $x = 0$ (see Fig. 5.25 on page 77). As a consequence each backpropagation step to a preceding layer (cf. Sect. 5.6) adds a factor, the absolute value of which is usually (considerably) smaller than 1, thus reducing the gradient from layer to layer (hence the term *vanishing gradient*). Although, in principle, the small gradient of the logistic function may be counteracted by a weight (which is also a factor in the backpropagation formula) with a large absolute value, it is usually the case that such a large weight, since it also enters the activation function as an argument, drives the derivative to the saturation regions of the logistic function (that is, far to the left or far to the right). Thus, (the absolute value of) the derivative factor is usually the smaller the larger (the absolute value of) the weights. Furthermore, the connection weights are commonly initialized to a random value in the range from $-\epsilon$ to ϵ with $\epsilon \leq 1$ (see Sect. 5.10), so that the initial training steps are particularly affected, as in them both the gradient of the logistic function as well as the weights provide a factor less than 1. Even though theoretically there can be exceptions to this intuitive and general description, they rarely occur in practice and thus one usually observes a vanishing gradient.

An alternative way to understand the vanishing gradient effect is to observe that the logistic activation function is a *contracting* function. That is, for any two arguments x and y, $x \neq y$, we have $|f_{act}(x) - f_{act}(y)| < |x - y|$, or in words: two arguments are always farther apart than their images under the logistic function—the logistic function contracts its argument range. This is immediately obvious from the fact that the derivative of the logistic function is always less than 1 (actually always $\leq \frac{1}{4}$). If several logistic functions are chained, these contractions (even though, in principle, they may be counteracted by sufficiently large weights) combine and yield an even stronger contraction of the input range. As a consequence, a rather large change of the input values will produce only a rather small change in the output values, and the more so, the more logistic functions are chained together. Therefore, the function that maps the inputs of a multi-layer perceptron to its outputs usually becomes the flatter the more layers the multi-layer perceptron has and consequently the gradient (in the first hidden layer were the inputs are processed) becomes the smaller.

There are various approaches that try to overcome the vanishing gradient problem, which will be discussed below. However, one also has to admit that advances in hardware, especially implementations of multi-layer perceptrons and their training with backpropagation that run on graphics processing units (GPUs, whose specialized processors are well suited for such computations) have alleviated the problem (see, e.g., [18]). By using such special hardware, the computations needed for executing and training a multi-layer perceptron can be accelerated by considerable factors, so that the slow training of the front layers poses much less of a problem, as one can afford a much larger number of training epochs.

Better approaches (which, of course, may exploit the accelarated hardware as well) consist in actually tackling the vanishing gradient problem itself. One of these approaches we already mentioned in Sect. 5.8.11, namely lifting the derivative of the activation function. Alternatively, one may use different activation functions altogether, which do not suffer from the vanishing gradient or contraction problem (so much). In this sense the hyperbolic tangent (see Fig. 5.3 on page 55), which is effectively a scaled form of the logistic function, already fares a little better, because its gradient can be as large as 1 (although only at one point, namely for $x = 0$).

In order to address the problem in a more principled manner, so-called **recti-fied linear units** (**ReLUs**) have been suggested, which were mentioned already in Sect. 5.1 (see page 55). The advantage of the activation function of these units (either the rectified maximum/ramp function or the softplus function, see Fig. 5.4 on page 56) is that their derivative is much larger for a much larger range of values.

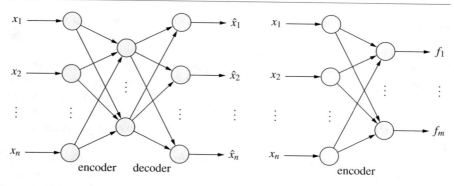

Fig. 5.50 An autoencoder/decoder (left), of which only the encoder part (right) is later used. The x_i are the given inputs, the x_i' the reconstructed inputs, and the y_i the constructed features

The softplus function in particular possesses the logistic function as its derivative, showing that for half of its argument values (i.e., for $x \geq 0$) the derivative is at least $\frac{1}{2}$, with most of these argument values producing a derivative close to 1. This certainly mitigates the vanishing gradient problem, but can sometimes cause the opposite, that is, an *exploding gradient*. This, in turn, may be battled by weight decay, since factors greater than one in the backpropagation procedure (which cause the gradient to explode) stem from weights that have grown too large.

A completely different approach consists in building the multi-layer perceptron layer-by-layer, training only the newly added layer in each step. A very popular specific procedure for this is to build the network as **stacked autoencoders**.

An **autoencoder** is a multi-layer perceptron that maps its inputs to an approximation of these inputs, using a hidden layer (see Fig. 5.50). The hidden layer (and the connections leading to it from the input layer) forms an encoder that transforms the inputs into some form of internal representation, which is then decoded by the output layer (and the connections from the hidden layer to the output layer). Such an autoencoder is trained with standard backpropagation (or one of its variants), which does not suffer much from a vanishing gradient, as there is only one hidden layer.

The rationale of training an autoencoder is that the hidden layer is expected to construct features (and detectors for these features) that capture the information contained in the input data in a compressed form (encoder), so that the input can be well reconstructed from it (decoder). Of course, if the learning task is fixed (see Definition 4.4 on page 48) this implicitly assumes that features that are well suited to represent the inputs in a compressed way are also useful to predict some desired output later. Experience shows that this assumption is often justified, since features constructed by autoencoders often lead to good results on fixed learning tasks.

The main problem of training an autoencoder is how many units should be chosen for the hidden layer and how this layer should be treated. Clearly, if we allow as many (or even more) hidden units as there are inputs, the network may not learn any useful features. Rather it is likely that it will merely pass through its inputs to the output layer (only minimally distorted by the fact that the logistic function is not perfectly linear anywhere; if it were, the input could always be mapped perfectly with appropriate weights and threshold values and an appropriate output scaling).

There are mainly three ways in which this problem is commonly handled: The most straightforward solution is to use (considerably) less neurons in the hidden layer than there are inputs, so that simply passing through the input values is not a feasible option and the autoencoder is forced to learn some actually relevant features. Of course, this leads to the question how many hidden neurons are a good choice. Although in Sect. 5.11 we discussed *cross validation* as a possibility to find a good number of hidden layers, one has to be careful here due to the fact that a number of hidden neurons that equals the number of inputs (or is very close to it) is likely to produce the best results in terms of a minimum error (minimum deviation of the computed outputs from the inputs), but not necessarily the best features.

A second commonly used approach is a sparse activation scheme as we already mentioned it above as a means to avoid overfitting. The number of active neurons in the hidden layer (i.e., neurons with a high activation) is restricted to a small number, which is enforced by either adding a regularization term to the error function that punishes a larger number of active hidden neurons or by explicitly deactivating all but a few neurons with the highest activations (i.e., for all other neurons the output is set to zero). As this also prevents the autoencoder from simply passing through its inputs to its outputs, useful features are likely constructed in the hidden layer.

A third approach is to add noise (that is, random variations) to the input (but not to the copy of the input that is used to evaluate the reconstruction error, that is, the error between the input and the computed output). This leads to so-called **denoising autoencoders** [36], because the autoencoder to be trained is expected to map the input with noise to (a copy of) the input without noise. Sufficient noise also prevents an autoencoder from simply passing through its inputs to its outputs and may also have a beneficial effect in terms of reducing the risk of overfitting, especially if the added noise is changed between training epochs, so that the network cannot adapt itself to specific random features of this noise.

To obtain multi-layer perceptrons with multiple hidden layers, several autoencoders are combined. These autoencoders are trained in a layer-wise fashion as illustrated in Fig. 5.51. In the first step, a simple autoencoder for the input tuples of a given data set is trained. The decoder (output layer) is removed and only the encoder (hidden layer) is kept. With this encoder a new data set is produced by propagating each input tuple through the encoder and recording the activations of the neurons in the hidden layer as a new data tuple. With this new data set (which represents the data in terms of the constructed features) another autoencoder is trained, which is hoped to produce higher-level features from the low-level features found in the first step. For example, in handwriting recognition, the low-level features may be (short) edges, both straight and curved, in specific parts of an input image. The second autoencoder may then combine these features into loops and stems of digits or letters. The procedure may then be repeated to obtain a third or even more autoencoders. Finally, the encoder parts of all trained autoencoders are *stacked*—that is, chained together in the order in which they were trained—to obtain the final network.

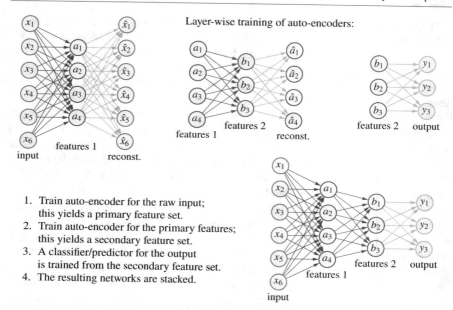

Fig. 5.51 A stacked autoencoder is created by level-wise training of standard autoencoders and then stacking the encoder parts. The final network is fine-tuned with standard gradient descent

If a fixed learning task (see Definition 4.4 on page 48) is to be solved, a final output layer is added, which refers to the actual outputs of the fixed learning task (while for training the autoencoders only the inputs are used). The whole network is then fine-tuned with a final training run that involves all hidden layers (while constructing each autoencoder involved only one hidden layer). Note that in this final training run the vanishing gradient problem is less severe, since the hidden layers have already been trained and thus can be expected to be close to their optimal parameterization. Only the added output layer needs full training.

Multi-layer perceptrons with several hidden layers built such a way have been applied very successfully for handwritten digit recognition (see, for example, [36]). In such an application it is often assumed, though, that the handwriting has already been preprocessed in order to separate the digits (or, in other cases, the letters) from each other, so that each input tuple represents a single digit (or letter). However, one would like to use similar networks also for more general applications, like, for example, recognizing whole lines of handwriting or analyzing photos in order to identify their parts as sky, landscape, house, pavement, tree, human being, etc. For such applications, where it is advantageous that the features constructed in hidden layers are not localized to a specific part of the image (as it is appropriate for already preprocessed and centered images), a special form of such a network, known as a **convolutional neural network**, has been developed.

This network type is inspired by the structure of the human retina, in which sensory neurons have a so-called receptive field, that is, a limited region in which

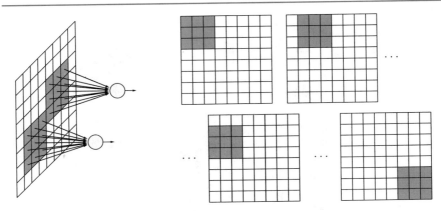

Fig. 5.52 In a convolutional neural network each neuron of the (first) hidden layer is connected to a small number of input neurons that refer to a contiguous region of the input image (left). Connection weights are shared, so that the same (partial) network is evaluated at different locations in the image. This computes a convolution of the connection weight matrix with the input image, as shown on the right. The input field is moved step by step over the whole image (right)

they respond to a (visual) stimulus. This is mimicked in a convolutional neural network by connecting each neuron in the (first) hidden layer to only a small number of contiguous (in the image) input neurons. This is illustrated in Fig. 5.52 on the left, which shows two neurons together with their receptive field in an input image. There are usually multiple neurons with the same receptive field that respond to different stimuli and thus compute different features. Weights are shared for a set of neurons with pairwise different locations of their receptive fields, so that such a set of neurons effectively evaluates the same neuron at all different locations in the image. As a result, a convolution of the weight matrix of the neuron with the input image is computed. As there are usually multiple neurons with the same receptive field, multiple such convolutions, for different weight matrices, are computed. Each weight matrix, which is often also referred to as **kernel (matrix)** or **filter (matrix)**, is supposed to detect (the occurrence locations of) a certain feature in the image.

A single such convolution computation is shown in Fig. 5.53 for one part of the so-called Sobel operator (named after Irwin Sobel), which is represented by the white weight matrix in the middle of the top row. This operator is meant to detect vertical edges in a picture by turning a (strong) brightness or color gradient into a line. (The second part of the Sobel operator consists of the same weight matrix rotated by 90 degrees and serves the purpose to detect horizontal edges.) In Fig. 5.53, this operator is convolved (written as ∗) with a simple 5×5 gray scale image shown at the top left, which has a horizontal brightness gradient (expressed by the numbers in the pixels as well as their shading). As the bottom three rows show, the operator is applied at every possible location in the image (shown by copies of the operator/weight matrix indicated in red). The convolution value is computed by multiplying the weights with the pixel values and summing these products. In Fig. 5.53 the convolution results are indicated below the corresponding copies of the input image, as well as in the output

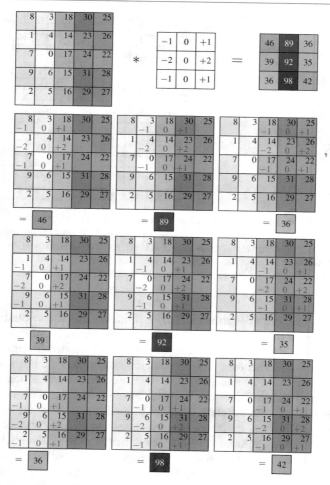

Fig. 5.53 Example of a convolution of a 5×5 image (shown in the top row on the left) with a 3×3 kernel (actually the Sobel operator for detecting vertical edges, shown in the top row in the middle). The red boxes in the bottom rows show the different locations at which the kernel is applied. Each location yields one pixel of the output image, which is shown under each copy of the input image as well as in the top row on the right. Note how the gray scale gradient in the input image is detected as a vertical edge, indicated by high values in the middle column

image at the top right. Note that the gradient has indeed been turned into a vertical line in the middle of the output image. By combining this result with the result of the other Sobel operator (same matrix rotated by 90 degrees), edges of arbitrary direction can be detected. This is illustrated in Fig. 5.54, which shows the result for a simple image with three gray circles.

Note that a convolution with a weight matrix/kernel of size $k \times k$ (where k is usually an odd integer), yields an output image with $(k - 1)$ pixels less in each

Fig. 5.54 Application of both Sobel operators (horizontal and vertical) to the image on the left, which, after combining the results pixel-wise via $s = \sqrt{h^2 + v^2}$, produces the image on the right

dimension. In Fig. 5.53, for example, a 5×5 image, by convolving it with a 3×3 weight matrix/kernel, leads to a 3×3 image. This is often avoided by adding so-called **padding** around the input image, which consists of $(k - 1)$ pixel rows and columns with pixels having some user-defined value (most commonly zero). With such padding, the size of the input image is preserved, with is desirable in many applications, even though padding causes some distortions at the image borders.

In a successor layer, in order to reduce the complexity, a maximum pooling over small regions is commonly computed (that is, only the maximum activation of the neurons in this region is passed on). The idea underlying such pooling is that it maintains knowledge that a feature, as represented by a set of neurons with shared weights, has been detected and only coarsens the information where exactly it was found. This is usually justified by the assumption that the rough relative position of features is more important than their exact location in an image.

In a convolutional neural network multiple stages of convolution, maximum pooling, and rectified linear units can be present, which build a hierarchy of image features. The resulting features may then be used, in a final, fully connected layer, to perform, for instance, image segmentation or classification. A tutorial that discusses such networks, which we described only very briefly here, in more detail can be found at https://towardsdatascience.com/simple-introduction-to-convolutional-neural-networks-cdf8d3077bac (accessed 2021-08-19).

Another type of deep neural network that has become hugely popular since 2016, when the ResNet neural network won the ImageNet Large Scale Visual Recognition Challenge (ILSVRC) [12], is the so-called residual neural network. It was inspired by the surprising observation of performance degradation: from a certain point on adding more layers to a neural network does no longer improve performance, but rather degrades it. This is surprising, because one would expect that an additonal layer, if it is not needed, could simply compute the identity and hence should not worsen the performance compared to a network in which this layer is simply missing. This led to the hypothesis that the training procedure may have trouble to train layers to compute the identity. If this is true (which is not really confirmed, but plausible), then using a network structure that can more easily learn the identity should avoid the degradation problem. The architecture called **residual neural network** achieves this by not letting a layer compute a transformation into a completely new space, but by letting it learn the residuals (hence the name of the architecture) between the input and a (desired) output. For this, the input and the output of the layer are summed, which is imagined as the input propagating along two paths: through the trainable layer as well as through a so-called skip or shortcut connection, which propagates the inputs unchanged. The outputs of both paths are then simply summed, and errors signals are backpropagated along both paths. Such an architecture makes it possible to train much deeper networks (ResNet had up to 152 layers [12]). A tutorial about residual neural networks and the ResNet architecture can be found at https://towardsdatascience.com/introduction-to-resnets-c0a830a288a4 (accessed 2012-08-19).

If a reader wants to dive deeper into deep learning (pun intended), we highly recommend [10] as a comprehensive compendium.

5.13 Sensitivity Analysis

(Artificial) neural networks have the serious disadvantage that the knowledge that they learned from training examples is often difficult to understand, because it is encoded in the connection weights, that is, stored in a matrix of real-valued numbers. In the preceding sections we tried to visualize the operation of neural networks with the help of geometric interpretations, but such an approach encounters severe problems for the complex networks that we meet in practice. Especially if the input space is high-dimensional, human imagination is bound to fail completely. A complex neural network thus easily appears to be a "black box", which produces its output from its input in somewhat mysterious ways.

However, we can improve this situation to some degree if we carry out a so-called **sensitivity analysis**, which determines what influence the different inputs have on the output of the network. To execute a sensitivity analysis, we sum the derivatives of the output w.r.t. the external inputs over all output neurons and all training patterns. This sum is divided by the number of training patterns, to make the result independent of the size of the training data set. That is, we compute

$$\forall u \in U_{\text{in}} : \quad s(u) = \frac{1}{|L_{\text{fixed}}|} \sum_{l \in L_{\text{fixed}}} \sum_{v \in U_{\text{out}}} \frac{\partial \text{out}_v^{(l)}}{\partial \text{ext}_u^{(l)}}.$$

The resulting value $s(u)$ indicates how "important" the input that is assigned to the neuron u is for the computations and eventually the output of the multi-layer perceptron. On this basis we may then, for example, simplify the neural network by removing the least important inputs, that is, those with the lowest values $s(u)$.

In order to obtain the exact sensitivity computation rule, we start by applying the chain rule—just as we did to derive the gradient descent rules:

$$\frac{\partial \text{out}_v}{\partial \text{ext}_u} = \frac{\partial \text{out}_v}{\partial \text{out}_u} \frac{\partial \text{out}_u}{\partial \text{ext}_u} = \frac{\partial \text{out}_v}{\partial \text{net}_v} \frac{\partial \text{net}_v}{\partial \text{out}_u} \frac{\partial \text{out}_u}{\partial \text{ext}_u}.$$

If the output function of the input neurons is the identity, as we assume here, the last factor can be neglected, because then

$$\frac{\partial \text{out}_u}{\partial \text{ext}_u} = 1.$$

For the second factor, we obtain in the general case

$$\frac{\partial \text{net}_v}{\partial \text{out}_u} = \frac{\partial}{\partial \text{out}_u} \sum_{p \in \text{pred}(v)} w_{vp} \text{out}_p = \sum_{p \in \text{pred}(v)} w_{vp} \frac{\partial \text{out}_p}{\partial \text{out}_u}.$$

On the right-hand side, we see a derivative of the output of a neuron p w.r.t. the output of the input neuron u, so that we arrive at the layer-wise recursion formula

$$\frac{\partial \text{out}_v}{\partial \text{out}_u} = \frac{\partial \text{out}_v}{\partial \text{net}_v} \frac{\partial \text{net}_v}{\partial \text{out}_u} = \frac{\partial \text{out}_v}{\partial \text{net}_v} \sum_{p \in \text{pred}(v)} w_{vp} \frac{\partial \text{out}_p}{\partial \text{out}_u}.$$

However, in the first hidden layer (or for a two-layer perceptron) we obtain

$$\frac{\partial \text{net}_v}{\partial \text{out}_u} = w_{vu}, \qquad \text{and thus} \qquad \frac{\partial \text{out}_v}{\partial \text{out}_u} = \frac{\partial \text{out}_v}{\partial \text{net}_v} w_{vu},$$

since all terms vanish except the one having $p = u$. This formula defines the starting point of the recursion. Then we apply the recursion formula to it until we reach the output layer, where we can finally compute the term of the value $s(u)$ that corresponds to the training pattern l by summing over the output neurons.

Like when we derived error backpropagation we also consider here the special case of a logistic activation function and the identity as the output function. In this case, we obtain a particularly simple recursion formula

$$\frac{\partial \text{out}_v}{\partial \text{out}_u} = \text{out}_v(1 - \text{out}_v) \sum_{p \in \text{pred}(v)} w_{vp} \frac{\partial \text{out}_p}{\partial \text{out}_u}$$

and the recursion start (v is a neuron in the first hidden layer)

$$\frac{\partial \text{out}_v}{\partial \text{out}_u} = \text{out}_v(1 - \text{out}_v) w_{vu}.$$

The command line programs mentioned at the end of the preceding section allow to carry out a sensitivity analysis of a multi-layer perceptron based on these formulas.

It should be noted, though, that a sensitivity analysis produces reliable and reproducible results only if training is carried out with weight decay (see Sect. 5.8.12). Otherwise the initial conditions (initial weights and bias values) can have a strong effect on the assessment of the relative importance of the different inputs. We illustrate this with the iris data (cf. page 102). Table 5.3 shows the result of sensitivity analyses of eight 3-layer perceptrons that were trained on the iris data (using the same setup as on page 102ff). Four networks were trained without weight decay ($\xi = 0$) and four with weight decay ($\xi = 0.0001$). As can be seen from the results, which state the "importance" of the measured quantities, using weight decay leads to fairly stable results, while the importance scores that are assigned without weight decay

Table 5.3 Sensitivity analysis of trained 3-layer perceptrons for the iris data.

Attribute	$\xi = 0$				$\xi = 0.0001$			
Sepal length	0.0216	0.0399	0.0232	0.0515	0.0367	0.0325	0.0351	0.0395
Sepal width	0.0423	0.0341	0.0460	0.0447	0.0385	0.0376	0.0421	0.0425
Petal length	0.1789	0.2569	0.1974	0.2805	0.2048	0.1928	0.1838	0.1861
Petal width	0.2017	0.1356	0.2198	0.1325	0.2020	0.1962	0.1750	0.1743

vary much more, which is a result of the random initialization. For more complex networks, the variations can even be much larger.

References

1. Anderson ES (1935) The Irises of the Gaspé Peninsula. Bull Am Iris Soc 59:2–5. American Iris Society, Philadelphia, PA, USA
2. Bengio Y, Courville A, Vincent P (2013) Representation learning: a review and new perspectives. IEEE Trans Pattern Anal Mach Intell 35(8):1798–1828. IEEE Press, Piscataway, NJ, USA
3. Bengio Y, LeCun Y, Hinton G (2015) Deep learning. Nature 521:436–444. Nature Publishing Group, London, United Kingdom
4. Dozat T (2016) Incorporating nesterov momentum into adam. In: Workshop Proceedings of 4th international conference on learning representations (ICLR 2016, San Juan, Puerto Rico). openreview.net
5. Duchi J, Hazan E, Singer Y (2011) Adaptive subgradient methods for online learning and stochastic optimization. J Mach Learn Res 12:2121–2159. JMLR Publishers
6. Fredkin E, Toffoli T (1982) Conservative logic. Int J Theor Phys 21(3/4):219–253. Plenum Press, New York, NY, USA
7. Fahlman SE (1988) An empirical study of learning speed in backpropagation networks. In: [Touretzky et al. 1988]
8. Fisher RA (1936) The use of multiple measurements in taxonomic problems. Ann Eugenics 7(2):179–188. Wiley, Chichester, United Kingdom
9. Glorot X, Bengio Y (2010) Understanding the difficulty of training deep feedforward neural networks. In: Proceedings 13th international conference on artificial intelligence and statistics (AISTATS 2010, Chia Laguna Resort, Sardinia, Italy), pp 249–256. PMLR Press
10. Goodfellow I, Bengio Y, Courville A (2015) Deep learning. MIT Press, Cambridge, MA, USA
11. He K, Zhang X, Ren S, Sun J (2015) Delving deep into rectifiers: surpassing human-level performance on ImageNet classification. In: Proceedings IEEE international conference on computer vision (ICCV 2015, Las Condes, Chile), pp 1026–1034. IEEE Press, Piscataway, NJ, USA

12. He K, Zhang X, Ren S, Sun J (2016) Learning deep residual, for image recognition. In: Proceedings IEEE conference on computer vision and pattern recognition (CVPR, (2016) Las vegas, NV), pp 770–778. IEEE Press, Piscataway, NJ, USA

13. Hochreiter S (1991) Untersuchungen zu dynamischen neuronalen Netzen. Diploma thesis, Institut für Informatik, Technische Universit"at M"unchen, Germany

14. Hochreiter S, Bengio Y, Frasconi P, Schmidhuber J (2001) gradient flow in recurrent nets: the difficulty of learning long-term dependencies. In: [Kremer and Kolen 2001]

15. Hornik K (1991) Approximation capabilities of multilayer feedforward networks. Neural Netw 4(2):251–257. Elsevier Science, Amsterdam, Netherlands

16. Ioffe S, Szegedy C (2015) Batch normalization: accelerating deep network training by reducing internal covariate shift. In: Proceedings of 32nd international conference machine learning (ICML 2015, Lille, France), pp 448–456. PMLR Press

17. Jakobs RA (1988) Increased rates of convergence through learning rate adaption. Neural Netw 1:295–307. Pergamon Press, Oxford, United Kingdom

18. Jang H, Park A, Jung K (2008) Using neural network implementation, CUDA and OpenMP. In: Proceedings of digital image computing: techniques and applications (DICTA, (2008) Canberra, Australia), pp 155–161. IEEE Press, Piscataway, NJ, USA

19. Kingma DP, Ba JL (2015) Adam: a method for stochastic optimization. In: Proceedings of 3rd international conference on learning representations (ICLR 2015, San Diego, CA). openreview.net

20. Kremer SC, Kolen JF (eds) (2001) A field guide to dynamical recurrent neural networks. IEEE Press, Piscataway, NJ, USA

21. McCluskey EJ, Jr (1956) Minimization of boolean functions. Bell Syst Techl J 35(6):1417–1444. American Telephone and Telegraph Company, New York, NY, USA

22. Nesterov YE (1983) A method of solving a convex programming problem with convergence rate $O(1/k^2)$. Soviet Math Doklady 27(2):372–376. Akademia Nauk USSR, Moscow, USSR

23. Pinkus A (1999) Approximation theory of the mlp model in neural networks. Acta Numerica 8:143-196. Cambridge University Press, Cambridge, United Kingdom

24. Polyak BT (1964) Some methods of speeding up the convergence of iteration methods. USSR Comput Math Math Phys 4(5):1–17. Dorodnitsyn Computing Centre, Moscow, USSR

25. Press WH, Teukolsky SA, Vetterling WT, Flannery BP (1992) Numerical recipes in C—the art of scientific computing, 2nd edn. Cambridge University Press, Cambridge, United Kingdom

26. Quine WV (1952) The problem of simplifying truth functions. Am Math Mon 59(8):521–531. Mathematical Association of America, Washington, DC, USA

27. Quine WV (1955) A way to simplify truth functions. Am Math Mon 62(9):627–631. Mathematical Association of America, Washington, DC, USA

28. Riedmiller M, Braun H (1992) Rprop—A Fast Adaptive Learning Algorithm. Technical Report, University of Karlsruhe, Karlsruhe, Germany

29. Riedmiller M, Braun H (1993) A direct adaptive method for faster backpropagation learning: the RPROP algorithm. In: International conference on neural networks (ICNN-93, San Francisco, CA). IEEE Press, Piscataway, NJ, USA, pp 586–591

30. Rosenbrock HH (1960) An automatic method for finding the greatest or least value of a function. Comput J 3(3):175–184. Oxford University Press, Oxford, United Kingdom

31. Rumelhart DE, Hinton GE, Williams RJ (1986) Learning representations by back-propagating errors. Nature 323:533–536

32. Srivastava N, Hinton G, Krizhevsky A, Sutskever I, Salakhutdinov R (2014) Dropout: a simple way to prevent neural networks from overfitting. J Mach Learn Res 15:1929–1958. MIT Press, Cambridge, MA, USA

33. Tieleman T, Hinton G (2012) Lecture 6.5-rmsprop: divide the gradient by a running average of its recent magnitude. Coursera: Neural Netw Mach Learn 4:26–31

34. Tollenaere T (1990) SuperSAB: fast adaptive backpropagation with good scaling properties. Neural Netw 3:561–573

35. Touretzky D, Hinton G, Sejnowski T (eds) Proceedings of the connectionist models summer school (Carnegie Mellon University). Morgan Kaufman, San Mateo, CA, USA
36. Vincent P, Larochelle H, Lajoie I, Bengio Y, Manzagol P-A (2010) Stacked denoising autoencoders: learning useful representations in a deep network with a local denoising criterion. J Mach Learn Res 11:3371–3408. MIT Press, Cambridge, MA, USA
37. Werbos PJ (1974) Beyond Regression: New Tools for Prediction and Analysis in the Behavioral Sciences. PhD Thesis, Harvard University, Cambridge, MA, USA
38. Zeiler MD (2012) AdaDelta: an adaptive learning rate method. Comput Res Repository (CoRR)

Radial Basis Function Networks

6

Like multi-layer perceptrons, radial basis function networks are feed-forward neural networks with a strictly layered structure. However, the number of layers is always three, that is, there is exactly one hidden layer. In addition, radial basis function networks differ from multi-layer perceptrons in the network input and activation functions, especially in the hidden layer. In this, hidden layer **radial basis functions** are employed, which are responsible for the name of this type of neural network. With these functions, a kind of "catchment region" is assigned to each neuron, which mainly influences the output of the neural network.

6.1 Definition and Examples

Definition 6.1 A **radial basis function network** (RBF network) is a neural network with a graph $G = (U, C)$ that satisfies the following (additional) conditions:

1. $U_{\text{in}} \cap U_{\text{out}} = \emptyset$,
2. $C = (U_{\text{in}} \times U_{\text{hidden}}) \cup C'$, $\quad C' \subseteq (U_{\text{hidden}} \times U_{\text{out}})$

The network input function of each hidden neuron is a **distance function** of the input vector and the weight vector, that is,

$$\forall u \in U_{\text{hidden}} : \qquad f_{\text{net}}^{(u)}(\text{in}_u, \mathbf{w}_u) = d(\text{in}_u, \mathbf{w}_u),$$

where $d : \mathbb{R}^n \times \mathbb{R}^n \to \mathbb{R}_0^+$ is a function that satisfies $\forall \mathbf{x}, \mathbf{y}, \mathbf{z} \in \mathbb{R}^n$:

$$
\begin{aligned}
&(i) \ d(\mathbf{x}, \mathbf{y}) = 0 \ \Leftrightarrow \ \mathbf{x} = \mathbf{y}, && \text{(identity of indiscernables)} \\
&(ii) \ d(\mathbf{x}, \mathbf{y}) = d(\mathbf{y}, \mathbf{x}) && \text{(symmetry),} \\
&(iii) \ d(\mathbf{x}, \mathbf{z}) \leq d(\mathbf{x}, \mathbf{y}) + d(\mathbf{y}, \mathbf{z}) && \text{(triangle inequality),}
\end{aligned}
$$

and thus fulfills the definition of a *distance* or a *metric*.

© Springer Nature Switzerland AG 2022
R. Kruse et al., *Computational Intelligence*, Texts in Computer Science,
https://doi.org/10.1007/978-3-030-42227-1_6

The network input function of the output neurons is the weighted sum of the inputs (weighted with the connection weights), that is,

$$\forall u \in U_{\text{out}}: \quad f_{\text{net}}^{(u)}(\text{in}_u, \mathbf{w}_u) = \mathbf{w}_u^{\top} \text{in}_u = \sum_{v \in \text{pred}(u)} w_{uv} \, \text{out}_v.$$

The activation function of each hidden neuron is a **radial function** (as we call it here), that is, a monotonically non-increasing function

$$f : \mathbb{R}_0^+ \to [0, 1] \quad \text{with} \quad f(0) = 1 \quad \text{and} \quad \lim_{x \to \infty} f(x) = 0.$$

The activation function of each output neuron is a linear function, namely

$$f_{\text{act}}^{(u)}(\text{net}_u, \theta_u) = \text{net}_u - \theta_u.$$

Note that a radial basis function network always has exactly three layers and that the input layer and the hidden layer are always fully connected because of the distance computation (that is, all coordinates are used to determine the distance).

The network input function and the activation functions of a hidden neuron describes a kind of "catchment region" of this neuron. The weights of the connections from the input layer to a neuron of the hidden layer state the **center** of this region, since the distance (network input function) is measured between the weight vector and the input vector. The type of distance function determines the shape of the catchment region. To illustrate this fact, we consider the well-known family of distance functions (the so-called *Minkowski family*) that is defined as

$$d_k(\mathbf{x}, \mathbf{y}) = \left(\sum_{i=1}^{n} (x_i - y_i)^k \right)^{\frac{1}{k}}.$$

Well-known special members of this family are:

$k = 1:$ Manhattan or city block distance,
$k = 2:$ Euclidean distance,
$k \to \infty:$ Maximum distance, that is, $d_\infty(\mathbf{x}, \mathbf{y}) = \max_{i=1}^{n} |x_i - y_i|$.

Distance functions like these can easily be illustrated by considering how a circle looks with them. The reason is that a circle is defined as the set of points that have the same fixed distance from a given point. This fixed distance is called the *radius* of the circle. For the three special cases listed above, circles are shown in Fig. 6.1. All three circles have the same radius. With these examples, we have an immediate impression of the possible shapes of the catchment region of a hidden neuron.

Intuitively, the activation function of a hidden neuron and its parameters determine the "size" of the catchment region of the neuron by specifying how strong the influence of an input vector is depending on its distance from the weight vector. We call this activation function a **radial function**, because it is defined along a ray (lat. *radius*: ray) from a center, which is described by the weight vector and thus assigns to each radius (that is, to each distance from the center) an activation. Examples of radial activation functions, all of which possess a parameter, namely a **(reference) radius** σ, are shown in Fig. 6.2 (cf. also Fig. 5.2 on p. 55).

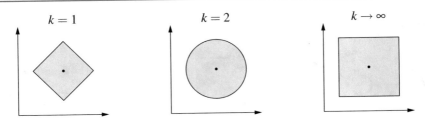

Fig. 6.1 Circles for different distance functions. All circles have the same radius

rectangular function:

$$f_{\mathrm{act}}(\mathrm{net}, \sigma) = \begin{cases} 0 & \text{if net} > \sigma, \\ 1 & \text{otherwise.} \end{cases}$$

triangular function:

$$f_{\mathrm{act}}(\mathrm{net}, \sigma) = \begin{cases} 0 & \text{if net} > \sigma, \\ 1 - \frac{\mathrm{net}}{\sigma} & \text{otherwise.} \end{cases}$$

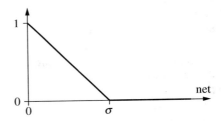

cosine down to zero:

$$f_{\mathrm{act}}(\mathrm{net}, \sigma) = \begin{cases} 0 & \text{if net} > 2\sigma, \\ \frac{\cos(\frac{\pi}{2\sigma}\,\mathrm{net})+1}{2} & \text{otherwise.} \end{cases}$$

Gaussian function:

$$f_{\mathrm{act}}(\mathrm{net}, \sigma) = e^{-\frac{\mathrm{net}^2}{2\sigma^2}}$$

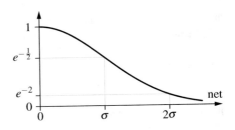

Fig. 6.2 Different radial activation functions

Note that not all of these radial activation functions limit the catchment region crisply. That is, not for all of these functions there exists a radius, beyond which the activation is 0. For instance, the Gaussian function yields a positive activation regardless of how far an input vector is from the center, even though this activation may be, due to the exponential decay of the Gaussian function, extremely small.

The output layer of a radial basis function network serves the purpose to combine the activations of the hidden neurons into the output of the network (weighted sum as the network input function), similar to the operation of a multi-layer perceptron. Note, however, that the activation function of the output neurons in a radial basis

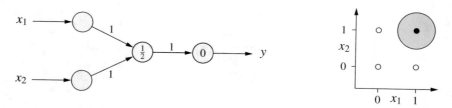

Fig. 6.3 A radial basis function network for the conjunction with Euclidean distance and rectangular activation function

function network is a linear function. The reason for this choice, which is important for initializing the parameters, is explained in Sect. 6.3.

As a first example, we consider, in analogy to Sect. 3.1, how the conjunction of two Boolean variables x_1 and x_2 is computed. A radial basis function network that solves this task is shown in Fig. 6.3 on the left. It possesses only a single hidden neuron, whose weight vector (center of the radial basis function) is the input vector for which an output of 1 is desired, that is, the point $(1, 1)$. The (reference) radius of the activation function is $\frac{1}{2}$. Like the parameter θ (bias value) of a neuron in a multi-layer perceptron, it is written into the circle that represents the hidden neuron. The diagram does not express explicitly that we use a Euclidean distance and a rectangular activation function. Due to the fact that the connection to the output neuron has weight 1 and the fact that the output neuron has the bias value 0, the output of the network coincides with the output of the hidden neuron.

In the same manner as the computations of threshold logic units (cf. Sect. 3.2), the computations of radial basis function networks can be interpreted geometrically, especially, if rectangular activation functions are used, see Fig. 6.3 on the right. The radial function describes a circle with radius $\frac{1}{2}$ around the point $(1, 1)$. Inside this circle, the activation of the hidden neuron (and thus the output of the network) is 1, outside it is 0. In this way, it is easy to see that the network shown in Fig. 6.3 on the left indeed computes the conjunction of its inputs.

Of course, the network shown in Fig. 6.3 is not the only possible one for computing the conjunction. For example, we may use a different radius, as long as it is less than 1, or we may shift the center a little, as long as the point $(1, 1)$ stays inside the circle and none of the other points enters the circle. We may also change the distance function or the activation function. However, we may also find a solution that exploits an entirely different principle, as it is shown, for instance, in Fig. 6.4. By using a bias value of -1 in the output neuron, a base output of 1 is produced, which is reduced to zero inside a circle with radius $\frac{6}{5}$ around the point $(0, 0)$ (note the negative weight of the connection to the output neuron). Intuitively, we may say that we punched a circular disk from a "carpet" of thickness 1 in such a way that all points, for which an output of 0 is desired, lie inside this disk.

As another example, we consider a radial basis function network that computes the biimplication, as shown in Fig. 6.5 on the left. It contains two hidden neurons, which are assigned to the two points for which an output of 1 is desired (namely $(0, 0)$ and

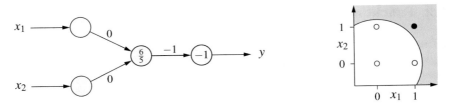

Fig. 6.4 Another radial basis function network for the conjunction with Euclidean distance and rectangular activation functions

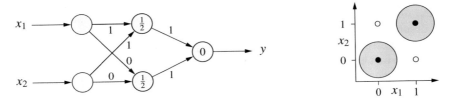

Fig. 6.5 A radial basis function network for the biimplication with Euclidean distance and rectangular activation functions

$(1, 1)$). Inside circles of radius $\frac{1}{2}$ around these two points the corresponding hidden neurons are activated (that is, outputs a 1), see Fig. 6.5 on the right. The output neuron merely combines these outputs: the output of the network is 1 if the input vector lies inside one of the two circles.

From a logical point of view, the top hidden neuron computes the conjunction of the inputs, while the bottom hidden neuron computes their negated disjunction. The output neuron combines the outputs of the hidden neurons disjunctively (where at most one of the two hidden neurons can be active). That is, the biimplication is represented by exploiting the logical equivalence

$$x_1 \leftrightarrow x_2 \quad \equiv \quad (x_1 \wedge x_2) \vee \neg(x_1 \vee x_2).$$

(Compare the related decomposition used in Sect. 3.4.)

Note that here we may as well draw on the possibility to create a base output of 1 (by using a bias value of -1 in the output neuron), which is reduced to zero by circles of radius $\frac{1}{2}$ (or some other radius less than 1) around the points $(1, 0)$ and $(0, 1)$. Note also that, like for threshold logic units, there is no way to compute the biimplication with only a single (hidden) neuron, unless we employ the *Mahalanobis distance* as the distance function. However, this extension of the computing power of radial basis function networks is discussed only later, namely in Sect. 6.6.

6.2 Function Approximation

After the examples of the preceding section, in which we examined only simple logical functions, we consider now, in analogy to Sect. 5.3, how we can approximate real-valued functions with the help of radial basis function networks. The principle is the same as in Sect. 5.3: the function to represent is approximated by a step function, which can easily be computed by a radial basis function network if we model it as a weighted sum of rectangular functions. We illustrate this principle with the help of the same example function as in Sect. 5.3, see Fig. 6.6.

For each step, a radial basis function is employed, whose center lies in the middle of the step and whose radius is half the step width. In this way, rectangular pulses are described (see Fig. 6.6 on the bottom right), which are weighted with the corresponding step height and then summed. Thus, we obtain the step function shown in Fig. 6.6 on the top right. The corresponding radial basis function network, which possesses one neuron for each rectangular pulse, is shown in Fig. 6.7.

Note that we actually compute the sum of the step heights at the borders (not shown in Fig. 6.6), since the rectangular pulses overlap at these points. This does not affect the approximation quality, though, because the error is measured, as in Sect. 5.3, as the area between the step function and the function to approximate. The deviations occur at finitely many points and thus do not contribute to the error.

Since we used the same principle as in Sect. 5.3, we can immediately transfer the approximation theorem derived there:

Theorem 6.1 *Every Riemann-integrable function can be approximated with arbitrary accuracy by a radial basis function network.*

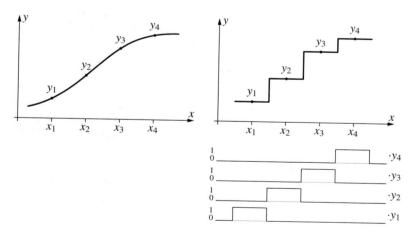

Fig. 6.6 Representing a step function by a weighted sum of rectangular functions (with centers x_i). Naturally, the step heights y_i may also be negative. However, at the step borders incorrect function values are computed (sum of the step heights)

Fig. 6.7 A radial basis
function network that
computes the step function
shown in Fig. 6.6 or the
piecewise linear function
shown in Fig. 6.8 (depending
in the activation functions of
the hidden neurons). It is
$\sigma = \frac{1}{2}\Delta x = \frac{1}{2}(x_{i+1} - x_i)$ or
$\sigma = \Delta x = x_{i+1} - x_i$,
respectively

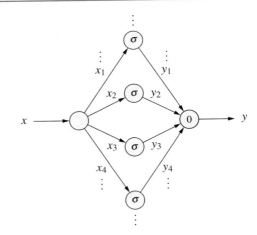

Note here again that, like for the approximation theorem about multi-layer perceptrons, this theorem only requires that the function to represent is Riemann integrable.
It need *not* be continuous. That is, the function to represent may have "jumps." However, it may have only finitely many "jumps" of finite height in the region in which it
is to be approximated by a radial basis function network. In other words, the function
must be continuous "almost everywhere."

Even though it does not impede the validity of the above theorem, the anomalies,
which occur at the step borders and make it differ from a pure step function, are at
least unattractive. However, they vanish automatically if we replace the rectangular activation function by a triangular function—in analogy to Sect. 5.3, where we
replaced the (Heaviside or unit) step function with a semi-linear function. With this
modification, the step function is turned into a piecewise linear function, which is
computed by a radial basis function network as the weighted sum of overlapping
triangular functions, see Fig. 6.8. Thus, the approximation is considerably improved
(or the same approximation quality can be achieved with fewer neurons).

The approximation quality can be improved even further if one increases the
number of support points, especially in regions where the function is strongly curved
(as already mentioned in the analogous discussion in Sect. 5.3). In addition, we
can eliminate the kinks of the piecewise linear function if we employ an activation
function like the Gaussian function, which can produce "smooth" transitions.

To illustrate how a function can be approximated by Gaussian functions, we consider how the function shown in Fig. 6.9 on the left can be approximated by a weighted
sum of three Gaussian bell curves (see Fig. 6.9 on the right). The corresponding radial
basis function network is shown in Fig. 6.10.

It should be clear that the principle of approximating a real-valued function,
which we exploited in this section, is not limited to unary functions, but may as
well be transferred to functions with multiple arguments. In contrast to multi-layer
perceptrons, we see immediately here that three layers are always sufficient, because
the basis functions influence the output of the network only locally.

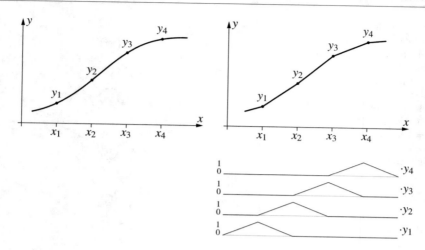

Fig. 6.8 Representing a piecewise linear function by a weighted sum of triangular functions (with centers x_i)

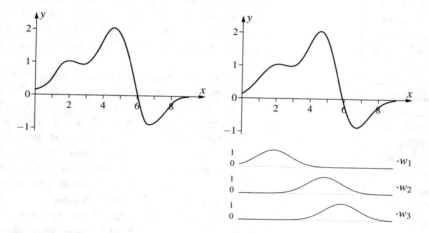

Fig. 6.9 Approximating a function by a sum of Gaussian functions with radius $\sigma = 1$. It is $w_1 = 1$, $w_2 = 3$ and $w_3 = -2$

Fig. 6.10 A radial basis function network that computes the weighted sum of Gaussian functions shown in Fig. 6.9

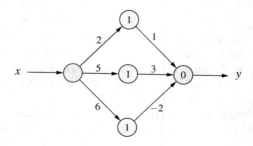

6.3   Initializing the Parameters

When we discussed multi-layer perceptrons, we treated initializing the parameters (that is, the connection weights and bias values) only in passing, since it is (mostly) trivial: simply choose random values (see Sect. 5.10). In principle, we can apply the same method to radial basis function networks, but it usually leads to fairly poor results. In addition, the hidden layer and the output layer of a radial basis function network differ considerably, since they possess different network input and activation functions—in contrast to a multi-layer perceptron, in which the hidden layer and the output layer is homogeneous. As a consequence, these two layers should be treated separately, and thus we devote a separate, somewhat longer section to how to initialize the parameters of a radial basis function network.

In order to simplify the presentation as much as possible, we start with the special case of a so-called **simple radial basis function network**, in which each training example is covered by its own radial basis function. That is, the hidden layer contains exactly as many neurons as there are training examples. The weights of the connections from the input neurons to the neurons of the hidden layer are determined by the training examples: each hidden neuron receives a training example and the weights of the connections to the hidden neurons are simply initialized with the elements of the input vector of this training example.

Formally: let $L_{\text{fixed}} = \{l_1, \ldots, l_m\}$ be a fixed learning task with m training patterns $l = (\mathbf{i}^{(l)}, \mathbf{o}^{(l)})$. Since each pattern is used as the center of its own radial function, there are m neurons in the hidden layer. Let these neurons be $v_1, \ldots, v_m$. We set

$$\forall k \in \{1, \ldots, m\}: \qquad \mathbf{w}_{v_k} = \mathbf{i}^{(l_k)}.$$

If the most commonly employed Gaussian activation function is chosen, the radii σ_k are often initialized to equal values according to the heuristics

$$\forall k \in \{1, \ldots, m\}: \qquad \sigma_k = \frac{d_{\max}}{\sqrt{2m}},$$

where $d_{\max}$ is the maximal distance between the input vectors of two training patterns (computed with the network input function chosen for the hidden neurons, which is a distance function d), that is,

$$d_{\max} = \max_{l_j, l_k \in L_{\text{fixed}}} d\left(\mathbf{i}^{(l_j)}, \mathbf{i}^{(l_k)}\right).$$

This choice has the advantages that the Gaussian bell curves are not too narrow (that is, they are not isolated peaks in the input space), but also not too wide (that is, they are also not overlapping too much, at least if the data set is "benevolent" or "well behaved", that is, there are no individual training examples that are located far away from all other training examples—no (extreme) outliers).

The weights from the hidden layer to the output layer and the bias values of the output neurons are computed with the following idea: since the parameters of the hidden layer (centers and radii) are known, we can compute the output of the hidden neurons for every training example. We now have to determine the connection weights and the bias values in such a way that, from these outputs, the desired outputs of the network are computed. Since the network input function of the output neurons is a weighted sum of its inputs and its activation and output functions are both linear, each training example l yields for each output neuron u one linear equation

$$\sum_{k=1}^{m} w_{uv_m} \, \text{out}_{v_m}^{(l)} - \theta_u = o_u^{(l)}.$$

(This is the main reason for choosing linear activation and output functions for the output neurons.) Thus, we obtain for each output neuron a linear equation system with m equations (one equation for each training example) and $m+1$ unknowns (m weights and one bias value). That this equation system is under-determined (more unknowns than equations), we can easily fix by simply setting the surplus parameter $\theta_u = 0$. In matrix and vector notation, the equation system to solve reads

$$\mathbf{A} \cdot \mathbf{w}_u = \mathbf{o}_u,$$

where $\mathbf{w}_u = (w_{uv_1}, \ldots, w_{uv_m})^\top$ is the weight vector of the output neuron u and $\mathbf{o}_u = \left(o_u^{(l_1)}, \ldots, o_u^{(l_m)}\right)^\top$ is the vector of the desired outputs of the output neuron u for the different training examples. $\mathbf{A}$ is an $m \times m$ matrix with the outputs of the neurons of the hidden layer for the different training examples, namely

$$\mathbf{A} = \begin{pmatrix} \text{out}_{v_1}^{(l_1)} & \text{out}_{v_2}^{(l_1)} & \ldots & \text{out}_{v_m}^{(l_1)} \\ \text{out}_{v_1}^{(l_2)} & \text{out}_{v_2}^{(l_2)} & \ldots & \text{out}_{v_m}^{(l_2)} \\ \vdots & \vdots & & \vdots \\ \text{out}_{v_1}^{(l_m)} & \text{out}_{v_2}^{(l_m)} & \ldots & \text{out}_{v_m}^{(l_m)} \end{pmatrix}.$$

That is, each matrix row contains the outputs of the different neurons for one training example, each column contains the outputs of one hidden neuron for the different training examples. Since the elements of this matrix can be computed from the training examples and the desired outputs are known as well, the weights can be found by simply solving this equation system with the usual methods of linear algebra.

For the subsequent considerations, it is advantageous to compute the solution of the linear equation system by inverting the matrix $\mathbf{A}$, that is, by

$$\mathbf{w}_u = \mathbf{A}^{-1} \cdot \mathbf{o}_u,$$

even though this method requires that the matrix $\mathbf{A}$ has full rank. In practice, this is usually, but not always the case. If $\mathbf{A}$ does not have full rank, weights have to be chosen randomly until the remaining equation system is uniquely solvable.

Note that this initialization method already guarantees that the error of a simple radial basis function network vanishes on the training data. Since the equation system, we have to solve is at most under-determined, we can always find connection weights with which the desired outputs are computed exactly. Hence, it is not necessary to train a simple radial basis function network.

To illustrate the procedure, we consider a radial basis function network for the biimplication $x_1 \leftrightarrow x_2$, in which the neurons of the hidden layer possess a Gaussian

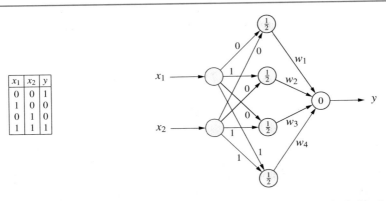

x_1	x_2	y
0	0	1
1	0	0
0	1	0
1	1	1

Fig. 6.11 Training examples for the biimplication and the simple radial radial basis function network as it is already partially determined by them

activation function. The training examples as well as the radial basis function network as it is already determined by them is shown in Fig. 6.11. The radii σ were chosen according to the heuristics mentioned above: clearly, we have $d_{\max} = \sqrt{2}$ (diagonal of the unit square) and $m = 4$, therefore $\sigma = \frac{\sqrt{2}}{\sqrt{2 \cdot 4}} = \frac{1}{2}$.

We are left with determining the four weights $w_1, \ldots, w_4$. (Note that the bias value of the output neuron is set to 0, since otherwise the equation system to solve is under-determined.) In order to compute these weights, we set up the matrix

$$\mathbf{A} = (a_{jk}) \qquad \text{with} \qquad a_{jk} = e^{-\frac{|\mathbf{i}_j - \mathbf{i}_k|^2}{2\sigma^2}} = e^{-2|\mathbf{i}_j - \mathbf{i}_k|^2},$$

where $\mathbf{i}_j$ and $\mathbf{i}_k$ are the input vectors of the jth and kth training example (numbered according to the table shown in Fig. 6.11). Therefore, it is

$$\mathbf{A} = \begin{pmatrix} 1 & e^{-2} & e^{-2} & e^{-4} \\ e^{-2} & 1 & e^{-4} & e^{-2} \\ e^{-2} & e^{-4} & 1 & e^{-2} \\ e^{-4} & e^{-2} & e^{-2} & 1 \end{pmatrix}.$$

The inverse of this matrix is the matrix

$$\mathbf{A}^{-1} = \begin{pmatrix} \frac{a}{D} & \frac{b}{D} & \frac{b}{D} & \frac{c}{D} \\ \frac{b}{D} & \frac{a}{D} & \frac{c}{D} & \frac{b}{D} \\ \frac{b}{D} & \frac{c}{D} & \frac{a}{D} & \frac{b}{D} \\ \frac{c}{D} & \frac{b}{D} & \frac{b}{D} & \frac{a}{D} \end{pmatrix},$$

where

$$D = 1 - 4e^{-4} + 6e^{-8} - 4e^{-12} + e^{-16} \approx 0.9287$$

is the determinant of the matrix $\mathbf{A}$ and

$$\begin{aligned} a &= 1 & - 2e^{-4} + e^{-8} &\approx 0.9637, \\ b &= -e^{-2} + 2e^{-6} - e^{-10} &\approx -0.1304, \\ c &= e^{-4} - 2e^{-8} + e^{-12} &\approx 0.0177. \end{aligned}$$

From this matrix and the output vector $\mathbf{o}_u = (1, 0, 0, 1)^\top$, we can now easily compute the connection weights. We obtain

$$\mathbf{w}_u = \mathbf{A}^{-1} \cdot \mathbf{o}_u = \frac{1}{D} \begin{pmatrix} a+c \\ 2b \\ 2b \\ a+c \end{pmatrix} \approx \begin{pmatrix} 1.0567 \\ -0.2809 \\ -0.2809 \\ 1.0567 \end{pmatrix}.$$

The computations of the radial basis function network that is initialized in this way are shown in Fig. 6.12. The left diagram shows a single basis function, namely the one with the center $(0, 0)$. The middle diagram shows all four basis functions (overlain, no sum). The output of the whole network is shown in the right diagram. It is clearly visible how the radial basis functions of the two centers, for which an output of 1 is desired, are weighted positively, while the two others are weighted negatively, so that in total exactly the desired outputs are computed.

Obviously, simple radial basis function networks are very easy to initialize, because the training examples already fix the parameters of the hidden layer. The weights from the hidden layer to the output layer can be computed, as we saw, by solving a simple linear equation system. In practice, however, simple radial basis function networks are of little use. In the first place, the number of training examples is generally too large to create a separate neuron for each of them: the resulting network would become too large to be feasible. In addition, one desires, for obvious reasons, that several training examples are covered by the same radial basis function.

Therefore, **radial basis function networks** (without the qualifier "simple") possess fewer neurons in the hidden layer than there are training examples. To initialize them, one often selects a random (though hopefully representative) subset of the training examples as the centers of the radial basis functions, namely one training example for each hidden neuron. (However, this is only one possible method for choosing the center coordinates; another is discussed below.) The weights from the input layer to the hidden layer are again fixed by these selected training examples: the coordinates of the input vectors are simply copied into the weight vectors. Likewise, the radii are chosen heuristically, using the same values as above (although we should rather refer to the selected subset of training examples instead of the whole set) in order to avoid having to deal with square roots.

single basis function   all basis functions   output

Fig. 6.12 Radial basis functions and output of a radial basis function network with four hidden neurons for the biimplication

In the step in which the weights of the connections from the hidden layer to the output neurons are determined, we now face the problem that the equation system to solve is over-determined. Since we selected a subset of the training examples, say k examples, we have m equations for each output neuron (one for each training example in the full set), but only $k + 1$ unknowns (k weights and one bias value) with $k < m$. Thus, we do not choose the bias value to be simply zero, but treat it by converting it into a weight (cf. Fig. 3.20 on p. 29).

In analogy to simple radial basis function networks, we set up a $m \times (k+1)$ matrix

$$\mathbf{A} = \begin{pmatrix} 1 & \text{out}_{v_1}^{(l_1)} & \text{out}_{v_2}^{(l_1)} & \dots & \text{out}_{v_k}^{(l_1)} \\ 1 & \text{out}_{v_1}^{(l_2)} & \text{out}_{v_2}^{(l_2)} & \dots & \text{out}_{v_k}^{(l_2)} \\ \vdots & \vdots & \vdots & & \vdots \\ 1 & \text{out}_{v_1}^{(l_m)} & \text{out}_{v_2}^{(l_m)} & \dots & \text{out}_{v_k}^{(l_m)} \end{pmatrix}$$

of the activations of the hidden neurons. (Note the ones in the first column, which refer to the fixed input 1 for the bias value.) We now have to determine for each output neuron u an (extended) weight vector $\mathbf{w}_u = (-\theta_u, w_{uv_1}, \dots, w_{uv_k})^\top$, so that

$$\mathbf{A} \cdot \mathbf{w}_u = \mathbf{o}_u,$$

where again $\mathbf{o}_u = \left(o_u^{(l_1)}, \dots, o_u^{(l_m)} \right)^\top$. However, since the equation system is over-determined (at least if the matrix has a rank $> k+1$), this equation does not generally possess a solution. In other words, the matrix $\mathbf{A}$ is not square and thus not invertible. Fortunately, a good approximate solution (so-called minimum norm solution) can be found with the help of the so-called (Moore–Penrose) **pseudo inverse $\mathbf{A}^+$** of the matrix $\mathbf{A}$ [1]. This pseudo inverse is computed as

$$\mathbf{A}^+ = (\mathbf{A}^\top \mathbf{A})^{-1} \mathbf{A}^\top.$$

The weights can finally be determined with the equation

$$\mathbf{w}_u = \mathbf{A}^+ \cdot \mathbf{o}_u = (\mathbf{A}^\top \mathbf{A})^{-1} \mathbf{A}^\top \cdot \mathbf{o}_u$$

(cf. the computations on p. 136).

We illustrate this procedure again with the help of the biimplication. From the training examples, we choose the first, i.e. $l_1 = (\mathbf{i}^{(l_1)}, \mathbf{o}^{(l_1)}) = ((0, 0), (1))$, and the last, i.e. $l_4 = (\mathbf{i}^{(l_4)}, \mathbf{o}^{(l_4)}) = ((1, 1), (1))$. That is, we start from the partially determined radial basis function network shown in Fig. 6.13 and now have to compute the weights w_1 and w_2 and the bias value θ. To do so, we set up the 4×3 matrix

$$\mathbf{A} = \begin{pmatrix} 1 & 1 & e^{-4} \\ 1 & e^{-2} & e^{-2} \\ 1 & e^{-2} & e^{-2} \\ 1 & e^{-4} & 1 \end{pmatrix} \cdot \cdot$$

The pseudo inverse of this matrix is the matrix

$$\mathbf{A}^+ = (\mathbf{A}^\top \mathbf{A})^{-1} \mathbf{A}^\top = \begin{pmatrix} a & b & b & a \\ c & d & d & e \\ e & d & d & c \end{pmatrix},$$

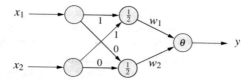

Fig. 6.13 Radial basis function network with only two hidden neurons that correspond to two selected training examples

where
$$a \approx -0.1810, \quad b \approx 0.6810,$$
$$c \approx 1.1781, \quad d \approx -0.6688, \quad e \approx 0.1594.$$

From this matrix and the output vector $\mathbf{o}_u = (1, 0, 0, 1)^\top$, we can compute the connection weights easily. We obtain

$$\mathbf{w}_u = \begin{pmatrix} -\theta \\ w_1 \\ w_2 \end{pmatrix} = \mathbf{A}^+ \cdot \mathbf{o}_u \approx \begin{pmatrix} -0.3620 \\ 1.3375 \\ 1.3375 \end{pmatrix}.$$

The computations of the network that is initialized in this way are shown in Fig. 6.14. The left and the middle diagram show the two radial basis functions, the right diagram the output of the network. Note the non-vanishing bias value and that (somewhat surprisingly) exactly the desired outputs are computed. In practice, this is usually not the case, because the equation system to solve is over-determined. Thus, in general, one has to be satisfied with an approximate solution. However, since the equation system is not really over-determined here (due to reasons of symmetry: from the matrix $\mathbf{A}$ and the output vector $\mathbf{o}_u$, it is easy to see that the second and the third equation are actually identical), we obtain an exact solution nevertheless.

Up to now we chose the centers of the radial basis functions (randomly) from the training examples. It would be better, though, if we could determine proper centers with a different method, because a random selection does not guarantee that the centers fit the training examples sufficiently well to allow for a good approximation of the desired outputs. In order to find the centers, basically any prototype-based

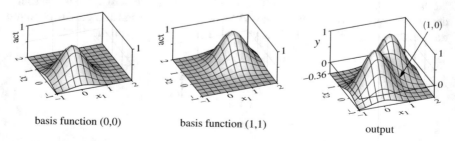

basis function (0,0)   basis function (1,1)

output

Fig. 6.14 Radial basis functions and output of a radial basis function network with two hidden neurons for the biimplication clustering method can be applied. Among these is, for instance, learning vector quantization, which is treated in the next chapter. Here, we consider a method of classical statistics instead, which is known as **c-means clustering** (also known as k-means clustering) [2]. The letter c (or k) in the name of this method stands for a parameter, namely the number of clusters to be found.

This method is very simple and straightforward (see Sect. 22.2.3 for more details). At the beginning, c cluster centers are chosen randomly, usually from the training examples. Then the training examples are divided into c groups (clusters) by assigning to each cluster center all training examples that are closer to it than to any other cluster center. In a second step, new cluster centers are computed by finding the "centers of gravity" of the formed groups of data points. That is, one computes the vector sum of the training examples of a group and divides by the number of these training example. The result is the new center. Next, the first step (that is, forming the groups) is executed again and so forth, until the cluster centers do not change anymore. The cluster centers found in this way can immediately be used to initialize the centers of a radial basis function network. Even the radii can be chosen with this method from the data: for example, one chooses the average distance of the training examples that are assigned to a cluster center from this cluster center.

6.4   Training the Parameters

Simple radial basis function networks cannot be improved (on the training data): due to the large number of neurons in the hidden layer we always obtain exactly the desired output if we initialize them with the method described in the preceding section. However, if we choose fewer hidden neurons than there are training examples, the quality of a radial basis function network can usually be improved by training.

Like the parameters of a multi-layer perceptron, the parameters of a radial basis function network are trained by **gradient descent**. Therefore, in order to find the adaptation rules for the weights, we proceed in basically the same manner as in Sect. 5.5. For the parameters of the output neurons, that is, for the weights of the connections from the hidden neurons and the bias values of the output neurons, we even obtain the same result as for a multi-layer perceptron: the gradient for a single output neuron u and a single training pattern l is (see p. 83)

$$\nabla_{\mathbf{w}_u} e_u^{(l)} = \frac{\partial e_u^{(l)}}{\partial \mathbf{w}_u} = -2\left(o_u^{(l)} - \mathrm{out}_u^{(l)}\right) \frac{\partial\, \mathrm{out}_u^{(l)}}{\partial\, \mathrm{net}_u^{(l)}}\, \mathrm{in}_u^{(l)},$$

where in_u is the vector of outputs of the predecessors of the neuron u, $o_u^{(l)}$ is the desired output of the neuron u, $\mathrm{net}_u^{(l)}$ is its network input and $\mathrm{out}_u^{(l)}$ its actual output if the input vector $\mathbf{i}^{(l)}$ of the training pattern l is fed into the network. Recall that the actual output $\mathrm{out}_u^{(l)}$ of the neuron u for the training pattern l is computed from its network input via its output function f_{out} and its activation function f_{act}, that is,

$$\text{out}_u^{(l)} = f_{\text{out}}\big(f_{\text{act}}\big(\text{net}_u^{(l)}\big)\big).$$

Like in Sect. 5.5, we assume, for reasons of simplicity, that the output function is the identity. In addition, we insert the linear activation function of the output neuron of a radial basis function network. Thus, we obtain

$$\frac{\partial \, \text{out}_u^{(l)}}{\partial \, \text{net}_u^{(l)}} = \frac{\partial \, \text{net}_u^{(l)}}{\partial \, \text{net}_u^{(l)}} = 1.$$

Note that the bias value θ_u of the output neuron u is already contained in the network input $\text{net}_u^{(l)}$, since we work with extended input and weight vectors in order to avoid inconvenient distinctions (just as we did in Sect. 5.5). Therefore, we have

$$\nabla_{\mathbf{w}_u} e_u^{(l)} = \frac{\partial e_u^{(l)}}{\partial \mathbf{w}_u} = -2\big(o_u^{(l)} - \text{out}_u^{(l)}\big)\,\text{in}_u^{(l)},$$

from which we derive the online adaptation rule

$$\Delta \mathbf{w}_u^{(l)} = -\frac{\eta_3}{2}\nabla_{\mathbf{w}_u} e_u^{(l)} = \eta_3\big(o_u^{(l)} - \text{out}_u^{(l)}\big)\,\text{in}_u^{(l)}$$

for the weights (and thus implicitly also the bias value θ_u). Note that the negative sign of the gradient vanishes, because we have to "descend in the error landscape" and thus have to move against the direction of the gradient. The factor 2 is incorporated into the learning rate η_3. (The index 3 of this learning rate already indicates that we will meet two more learning rates later.) As usual, for batch training the weight changes $\Delta \mathbf{w}_u$ have to be summed over all training patterns and are applied to the weights only after all training patterns have been processed.

The adaptation rules for the weights of the connections from the input neurons to the hidden neurons and for the radii of the radial basis functions are derived in a similar way as error backpropagation was derived in Sect. 5.6. We merely have to take the special network input and activation functions into account. This implies, though, that we can no longer work with extended weight and input vectors, but have to treat the weights (the centers of the radial basis functions) and the radius separately. Hence, for reasons of clarity, we go through the whole derivation here.

We start from the total error of a radial basis function network with output neurons U_{out} w.r.t. a fixed learning task L_{fixed}:

$$e = \sum_{l \in L_{\text{fixed}}} e^{(l)} = \sum_{u \in U_{\text{out}}} e_u = \sum_{l \in L_{\text{fixed}}} \sum_{u \in U_{\text{out}}} e_u^{(l)}.$$

Let v be a neuron in the hidden layer. Let its predecessors (input neurons) be the neurons $\text{pred}(v) = \{p \in U_{\text{in}} \mid (p, v) \in C\} = \{p_1, \ldots, p_n\}$ (which is actually the whole set U_{in}, see Definition 6.1). Furthermore, let $\mathbf{w}_v = (w_{vp_1}, \ldots, w_{vp_n})$ be the corresponding weight vector and σ_v the corresponding radius. We start by computing the gradient of the total error w.r.t. the connection weights (center coordinates):

$$\nabla_{\mathbf{w}_v} e = \frac{\partial e}{\partial \mathbf{w}_v} = \left(\frac{\partial e}{\partial w_{vp_1}}, \ldots, \frac{\partial e}{\partial w_{vp_n}}\right).$$

Since the total error is a sum over all training patterns, we have

$$\frac{\partial e}{\partial \mathbf{w}_v} = \frac{\partial}{\partial \mathbf{w}_v} \sum_{l \in L_{\text{fixed}}} e^{(l)} = \sum_{l \in L_{\text{fixed}}} \frac{\partial e^{(l)}}{\partial \mathbf{w}_v}.$$

Hence, we can confine ourselves, in analogy to Sect. 5.6, to the error $e^{(l)}$ for a single training pattern l. This error depends on the weights in $\mathbf{w}_v$ only via the network input $\text{net}_v^{(l)} = d(\mathbf{w}_v, \text{in}_v^{(l)})$ with the network input vector $\text{in}_v^{(l)} = (\text{out}_{p_1}^{(l)}, \ldots, \text{out}_{p_n}^{(l)})$. Therefore, we can apply the chain rule and obtain in analogy to Sect. 5.6:

$$\nabla_{\mathbf{w}_v} e^{(l)} = \frac{\partial e^{(l)}}{\partial \mathbf{w}_v} = \frac{\partial e^{(l)}}{\partial \text{net}_v^{(l)}} \frac{\partial \text{net}_v^{(l)}}{\partial \mathbf{w}_v}.$$

To compute the first factor, we examine the error $e^{(l)}$ for the pattern $l = (\mathbf{i}^{(l)}, \mathbf{o}^{(l)})$:

$$e^{(l)} = \sum_{u \in U_{\text{out}}} e_u^{(l)} = \sum_{u \in U_{\text{out}}} \left(o_u^{(l)} - \text{out}_u^{(l)}\right)^2,$$

that is, the error sum over all output neurons. Thus, we obtain

$$\frac{\partial e^{(l)}}{\partial \text{net}_v^{(l)}} = \frac{\partial \sum_{u \in U_{\text{out}}} \left(o_u^{(l)} - \text{out}_u^{(l)}\right)^2}{\partial \text{net}_v^{(l)}} = \sum_{u \in U_{\text{out}}} \frac{\partial \left(o_u^{(l)} - \text{out}_u^{(l)}\right)^2}{\partial \text{net}_v^{(l)}}.$$

Since only the actual output $\text{out}_u^{(l)}$ of an output neuron u depends on the network input $\text{net}_v^{(l)}$ of the neuron v under consideration, it is

$$\frac{\partial e^{(l)}}{\partial \text{net}_v^{(l)}} = -2 \sum_{u \in U_{\text{out}}} (o_u^{(l)} - \text{out}_u^{(l)}) \frac{\partial \text{out}_u^{(l)}}{\partial \text{net}_v^{(l)}}.$$

Let the neurons $\text{succ}(v) = \{s \in U_{\text{out}} \mid (v, s) \in C\}$ be the successors (output neurons) of the neuron v we consider. The output $\text{out}_u^{(l)}$ of an output neuron u is affected by the network input $\text{net}_v^{(l)}$ of the considered neuron v only if there exists a connection from v to u, that is, if u is among the successors $\text{succ}(v)$ of v. Therefore, we can confine the sum over the output neurons to the successors of v. Furthermore, the output $\text{out}_s^{(l)}$ of a successor s of neuron v depends on the network input $\text{net}_v^{(l)}$ of the considered successor only via the network input $\text{net}_s^{(l)}$ of the considered neuron v. Thus, we obtain with the chain rule

$$\frac{\partial e^{(l)}}{\partial \text{net}_v^{(l)}} = -2 \sum_{s \in \text{succ}(v)} (o_s^{(l)} - \text{out}_s^{(l)}) \frac{\partial \text{out}_s^{(l)}}{\partial \text{net}_s^{(l)}} \frac{\partial \text{net}_s^{(l)}}{\partial \text{net}_v^{(l)}}.$$

Since the successors $s \in \text{succ}(v)$ are output neurons, we can insert (as when we looked at the output neurons)

$$\frac{\partial \text{out}_s^{(l)}}{\partial \text{net}_s^{(l)}} = 1.$$

We are left with computing the partial derivative of the network input. Since the neurons s are output neurons, it is

$$\mathrm{net}_s^{(l)} = \mathbf{w}_s^{\top}\,\mathrm{in}_s^{(l)} - \theta_s = \left(\sum_{p\in\,\mathrm{pred}(s)} w_{sp}\,\mathrm{out}_p^{(l)} \right) - \theta_s,$$

where one element of $\mathrm{in}_s^{(l)}$ is the output $\mathrm{out}_v^{(l)}$ of the neuron v under consideration. Clearly, $\mathrm{net}_s^{(l)}$ depends on $\mathrm{net}_v^{(l)}$ only via this element $\mathrm{out}_v^{(l)}$. Therefore, it is

$$\frac{\partial\,\mathrm{net}_s^{(l)}}{\partial\,\mathrm{net}_v^{(l)}} = \left(\sum_{p\in\,\mathrm{pred}(s)} w_{sp}\,\frac{\partial\,\mathrm{out}_p^{(l)}}{\partial\,\mathrm{net}_v^{(l)}} \right) - \frac{\partial\theta_s}{\partial\,\mathrm{net}_v^{(l)}} = w_{sv}\,\frac{\partial\,\mathrm{out}_v^{(l)}}{\partial\,\mathrm{net}_v^{(l)}},$$

since all terms vanish except the one having $p = v$. In total, we have for the gradient

$$\nabla_{\mathbf{w}_v} e^{(l)} = \frac{\partial e^{(l)}}{\partial \mathbf{w}_v} = -2 \sum_{s\in\,\mathrm{succ}(v)} \left(o_s^{(l)} - \mathrm{out}_s^{(l)}\right) w_{su} \frac{\partial\,\mathrm{out}_v^{(l)}}{\partial\,\mathrm{net}_v^{(l)}} \frac{\partial\,\mathrm{net}_v^{(l)}}{\partial \mathbf{w}_v},$$

from which we obtain the online adaptation rule

$$\Delta\mathbf{w}_v^{(l)} = -\frac{\eta_1}{2}\nabla_{\mathbf{w}_v} e^{(l)} = \eta_1 \sum_{s\in\,\mathrm{succ}(v)} \left(o_s^{(l)} - \mathrm{out}_s^{(l)}\right) w_{sv} \frac{\partial\,\mathrm{out}_v^{(l)}}{\partial\,\mathrm{net}_v^{(l)}} \frac{\partial\,\mathrm{net}_v^{(l)}}{\partial \mathbf{w}_v}.$$

Note again that the minus sign vanishes, since we have to move against the direction of the gradient and that the factor 2 is incorporated into the learning rate η_1. Note also the index 1 of the learning rate, which indicates that it differs from the learning rate η_3 for the connection weights from the hidden to the output layer. For batch training, these weight changes have to be summed over all training patterns and are applied to the weights only after all training patterns have been processed.

Unfortunately, it is not possible to generally compute the derivative of the output w.r.t. the network input or the derivative of the network input w.r.t. the weights, which are still contained in the weight adaptation rule, since radial basis function networks can employ different distance and different radial functions. We consider here the special case of a Euclidean distance and a Gaussian activation function, which are the most common choices. In this case (Euclidean distance), we have

$$\mathrm{net}_v^{(l)} = d\!\left(\mathrm{in}_v^{(l)}, \mathbf{w}_v\right) = d\!\left((\mathrm{out}_{p_1}^{(l)}, \ldots, \mathrm{out}_{p_n}^{(l)}), \mathbf{w}_v\right) = \sqrt{\sum_{i=1}^{n}\left(\mathrm{out}_{p_i}^{(l)} - w_{vp_i}\right)^2}.$$

Therefore, the second factor becomes

$$\frac{\partial\,\mathrm{net}_v^{(l)}}{\partial\mathbf{w}_v} = \frac{1}{2}\left(\sum_{i=1}^{n}\left(\mathrm{out}_{p_i}^{(l)} - w_{vp_i}\right)^2\right)^{-\frac{1}{2}} 2\!\left(\mathrm{in}_v^{(l)} - \mathbf{w}_v\right)(-1) = -\frac{\mathrm{in}_v^{(l)} - \mathbf{w}_v}{\mathrm{net}_v^{(l)}}.$$

For the first factor (Gaussian function), we obtain (under the simplifying assumption that the output function is the identity)

$$\frac{\partial\,\mathrm{out}_v^{(l)}}{\partial\,\mathrm{net}_v^{(l)}} = \frac{\partial}{\partial\,\mathrm{net}_v^{(l)}} \exp\!\left(-\frac{(\mathrm{net}_v^{(l)})^2}{2\sigma_v^2}\right) = -\frac{\mathrm{net}_v^{(l)}}{\sigma_v^2} \exp\!\left(-\frac{(\mathrm{net}_v^{(l)})^2}{2\sigma_v^2}\right).$$

Collecting all the pieces derived above finally yields

$$\Delta \mathbf{w}_v^{(l)} = \eta_1 \left(\sum_{s \in \mathrm{succ}(v)} \left(o_s^{(l)} - \mathrm{out}_s^{(l)} \right) w_{sv} \right) \exp\left(-\frac{(\mathrm{net}_v^{(l)})^2}{2\sigma_v^2} \right) \frac{\mathrm{in}_v^{(l)} - \mathbf{w}_v}{\sigma_v^2}.$$

Finally, we have to compute the gradient for the radius parameter σ_v of the hidden neurons. In principle, computing this gradient follows the same paths as computing the gradient for the weights. Indeed, it is even somewhat simpler, since we need not take the network input function into account. Hence we only state the result here:

$$\frac{\partial e^{(l)}}{\partial \sigma_v} = -2 \sum_{s \in \mathrm{succ}(v)} \left(o_s^{(l)} - \mathrm{out}_s^{(l)} \right) w_{su} \frac{\partial \, \mathrm{out}_v^{(l)}}{\partial \sigma_v}.$$

As the online adaptation rule, we thus obtain

$$\Delta \sigma_v^{(l)} = -\frac{\eta_2}{2} \frac{\partial e^{(l)}}{\partial \sigma_v} = \eta_2 \sum_{s \in \mathrm{succ}(v)} \left(o_s^{(l)} - \mathrm{out}_s^{(l)} \right) w_{sv} \frac{\partial \, \mathrm{out}_v^{(l)}}{\partial \sigma_v}.$$

As usual, the negative sign disappears because we have to move against the gradient direction and the factor 2 is incorporated into the learning rate. Naturally, for (mini-)batch training the radius changes have to be summed over all (used) training patterns and are applied to the radius σ_v only after these have been processed.

The derivative of the output of the neuron v w.r.t. the radius σ_v cannot be determined in a general form, since the neurons of the hidden layer may employ different radial functions (see Fig. 6.2 on p. 127). Therefore, we consider again, as an example, the special case of a Gaussian activation function (and, for reasons of simplicity, the identity as the output function). Then it is

$$\frac{\partial \, \mathrm{out}_v^{(l)}}{\partial \sigma_v} = \frac{\partial}{\partial \sigma_v} \exp\left(-\frac{(\mathrm{net}_v^{(l)})^2}{2\sigma_v^2} \right) = \frac{\left(\mathrm{net}_v^{(l)} \right)^2}{\sigma_v^3} \exp\left(-\frac{(\mathrm{net}_v^{(l)})^2}{2\sigma_v^2} \right).$$

However, for a Gaussian activation function it may actually be more convenient to use σ_v^2 instead of σ_v as the parameter to adapt. In this case, we get

$$\frac{\partial \, \mathrm{out}_v^{(l)}}{\partial \sigma_v^2} = \frac{\partial}{\partial \sigma_v^2} \exp\left(-\frac{(\mathrm{net}_v^{(l)})^2}{2\sigma_v^2} \right) = \frac{\left(\mathrm{net}_v^{(l)} \right)^2}{2\sigma_v^4} \exp\left(-\frac{(\mathrm{net}_v^{(l)})^2}{2\sigma_v^2} \right),$$

leading to the adaptation of the alternative parameters σ_v^2 by

$$\Delta \sigma_v^{2(l)} = -\frac{\eta_2}{2} \frac{\partial e^{(l)}}{\partial \sigma_v^2} = \eta_2 \left(\sum_{s \in \mathrm{succ}(v)} \left(o_s^{(l)} - \mathrm{out}_s^{(l)} \right) w_{sv} \right) \frac{\left(\mathrm{net}_v^{(l)} \right)^2}{2\sigma_v^4} \exp\left(-\frac{(\mathrm{net}_v^{(l)})^2}{2\sigma_v^2} \right).$$

Note that in the computations carried out above, we do not have a single learning rate for all neurons as in a multi-layer perceptron. Rather we have a total of three: one learning rate for the weights of the connections to the hidden neurons (η_1), which represent the coordinates of the centers of the radial basis functions, one for the radii σ of the radial basis functions (η_2, which may alternatively apply to a variance σ^2), and one for the weights of the connections to the output neurons and the bias values of the output neurons (η_3). According to recommendations in [3], these learning rates should be chosen considerably smaller than the (one) learning rate for training a multi-layer perceptron. Especially the third learning rate η_3 should be small, because the weights of the connections to the output neurons and the bias values of the output neurons have a strong influence on the function that is computed by a radial basis function network. In addition, using online training is often discouraged, since it is a lot less stable than for multi-layer perceptrons. Full or at least mini-batch training should be preferred.

6.5 Example of Training

To illustrate training radial basis function networks on real-world data we use again the example of the iris data (cf. p. 102). We confine ourselves here to petal length and width as these are the most informative attributes w.r.t. the class. To represent the class attribute, 1-in-n encoding (see Sect. 4.3) is used, which means that the radial basis function network has three output neurons: one per class. It also has three hidden neurons, since the structure of the data is simple enough to achieve acceptable results with such a small number of radial basis functions.

The weights from the input layer to the hidden layer are initialized by the result of simple c-means clustering (see above and Sect. 22.2.3) with $c = 3$. The initial radii could be determined by the rule of thumb given above, namely $\sigma = d_{max}/\sqrt{2m}$. In the space spanned by petal length and width, we have $d_{max} \approx 6.26$ and $m = 150$ (as there are 150 sample cases of iris flowers) and hence $d_{max} \approx 0.36$. However, even though using it works fine, this fairly small radius does not lend itself well to visualization, and hence we rather use $\sigma = 1$ as a starting point.

The weights of the connections from the hidden layer to the output layer are initialized with the help of the pseudo-inverse method as it was explained above (compute the activations of the hidden neurons and solve the resulting over-determined linear equation system approximately). The result is shown in Fig. 6.16: the top row shows the radial basis functions (activations of the hidden neurons), the bottom row the network outputs for the three types of iris flowers. All diagrams show the input space as shown in Fig. 6.15 on the right. As in this diagram, the data points are relatively small, the left diagram repeats the data at a larger scale.

A radial basis function network initialized in this way already produces very satisfactory output. If always the class of the output neuron is assigned that yields the highest activation, only six sample cases of the training data are misclassified. However, with the help of gradient descent training, an even better adaptation of the network parameters to the training data can be achieved.

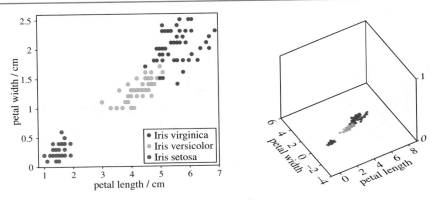

Fig. 6.15 The iris data in the space spanned by petal length and petal width, diagram on the right as used in Figs. 6.16 and 6.17 to depict the radial basis functions and the network output

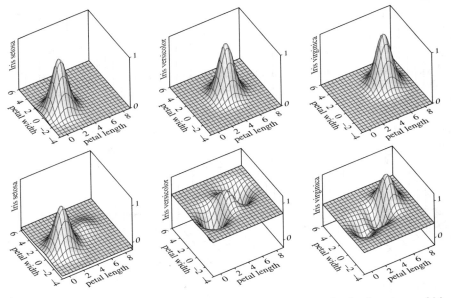

Fig. 6.16 Radial basis functions (top) and network output (bottom) for the three types of iris flowers directly after initialization (centers determined by c-means clustering, initial $\sigma \approx 3.1$)

For training, we rely on the sum of squared deviations, applied to a 1-in-n encoding of the classes. As learning rates, we choose $\eta_1 = 0.02$, $\eta_2 = 0.01$ and $\eta_3 = 0.001$, which are generally recommendable values. After 10,000 epochs of full batch training, the result shown in Fig. 6.17 is reached. Again the top row shows the radial basis functions (activations of the hidden neurons), the bottom row the network outputs for the three types of iris flowers. Note how the locations and the radii of the radial basis functions have changed due to the training, and what effect this has

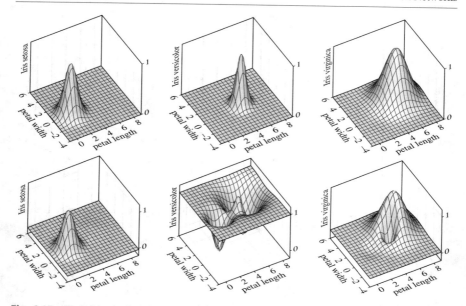

Fig. 6.17 Radial basis functions (top) and network output (bottom) for the three types of iris flowers after training (all three radii have been adapted to $\sigma \approx 1$)

on the network output. The trained network achieves only four misclassifications on the training data (which is not likely to exhibit substantial overfitting here, because the model is so simple as it employs only three radial basis functions).

6.6 Generalized Form

Up to now, we always used distance functions that are either isotropic (direction independent), like the Euclidean distance, or with which deviations from isotropy are fixed by the coordinate axes, like the city block distance or the maximum distance (see Fig. 6.1 on p. 127). However, if the training examples form point clouds with an "oblique" orientation in the input space, they cannot be captured well with such a distance function. In this case, one either needs a larger number of radial basis functions, which are stringed together along the data point cloud and increase the complexity of the network, or one has to accept that larger regions outside of the point clouds are covered as well, which can harm the performance on new data.

In such a case, one desires distance function that can describe ellipses (or generally hyper-ellipsoids) with arbitrary orientation. Such a distance function is the **Mahalanobis distance**, which is defined as

$$d(\mathbf{x}, \mathbf{y}) = \sqrt{(\mathbf{x} - \mathbf{y})^\top \Sigma^{-1} (\mathbf{x} - \mathbf{y})}.$$

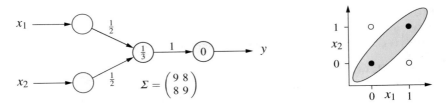

Fig. 6.18 A radial basis function network for the biimplication with Mahalanobis distance and rectangular activation function

Here Σ is a matrix which is called *covariance matrix* (this name is due to certain connections to statistics, which, however, we do not explain in detail here) and which describes the (direction dependence) of the distance. Note that the Mahalanobis distance is identical to the Euclidean distance if Σ is the unit matrix.

To illustrate the possibilities that result from using the Mahalanobis distance, we reconsider representing the biimplication with a radial basis function network: the Mahalanobis distance enables us to solve this task with only one neuron in the hidden layer. The corresponding network and the covariance matrix that is now necessary as an additional parameter of the network input function is shown in Fig. 6.18 on the left. As the activation function, we assume (like in the examples shown in Figs. 6.3 and 6.4 on p. 128) a rectangular function. The computations of this network are illustrated in Fig. 6.18 on the right. Inside the gray ellipse, the output is 1, while outside it is 0. This computes exactly the biimplication, as desired.

For radial basis function networks that employ the Mahalanobis distance, gradients can be derived for the shape parameter (i.e., the elements of the covariance matrix) as well. The computations follow essentially the same paths as those executed in Sect. 6.4. An explicit treatment is beyond the scope of this book and hence we confine ourselves to merely stating the results for a Gaussian function:

$$\Delta \mathbf{w}_v^{(l)} = \eta_1 \left(\sum_{s \in \mathrm{succ}(v)} \left(o_s^{(l)} - \mathrm{out}_s^{(l)}\right) w_{sv} \right) \exp\left(-\frac{1}{2}(\mathrm{net}_v^{(l)})^2\right) \mathbf{z},$$

$$\Delta \Sigma_v^{(l)} = \eta_2 \left(\sum_{s \in \mathrm{succ}(v)} \left(o_s^{(l)} - \mathrm{out}_s^{(l)}\right) w_{sv} \right) \frac{1}{2} \exp\left(-\frac{1}{2}(\mathrm{net}_v^{(l)})^2\right) \mathbf{z}\mathbf{z}^\top$$

where $\mathbf{z} = \Sigma^{-1}\left(\mathrm{in}_v^{(l)} - \mathbf{w}_v\right)$ and $(\mathrm{net}_v^{(l)})^2 = (\mathrm{in}_v^{(l)} - \mathbf{w}_v)^\top \Sigma^{-1}(\mathrm{in}_v^{(l)} - \mathbf{w}_v)$ is the squared Mahalanobis distance between the weight vector $\mathbf{w}_v$ (radial basis function center) and an input point $\mathrm{in}_v^{(l)}$. Note that the radius of the radial function is captured in the covariance matrix Σ and hence does not appear explicitly in the argument of the exponential function, but is implicit in $(\mathrm{net}_v^{(l)})^2$.

References

1. Albert A (1972) Regression and the Moore-Penrose pseudoinverse. Academic, New York, NY, USA
2. Hartigan JA, Wong MA (1979) A k-means clustering algorithm. Appl Stat 28:100–108. Blackwell, Oxford, United Kingdom
3. Zell A (1996) Simulation Neuronaler Netze. Addison-Wesley, Stuttgart, Germany

Self-organizing Maps

Self-organizing maps are closely related to radial basis function networks. They can be seen as radial basis function networks without an output layer, or, rather, the hidden layer of a radial basis function network is already the output layer of a self-organizing map. This output layer also has an internal structure, since the neurons are arranged in a grid. The neighborhood relationships resulting from this grid are exploited in the training process in order to determine a topology preserving map.

7.1 Definition and Examples

Definition 7.1 A **self-organizing map** (SOM) or **Kohonen feature map** is a neural network with a graph $G = (U, C)$ that satisfies the following conditions:

1. $U_{\text{hidden}} = \emptyset$, $U_{\text{in}} \cap U_{\text{out}} = \emptyset$,
2. $C = U_{\text{in}} \times U_{\text{out}}$.

The network input function of each output neuron is a **distance function** of the input and the weight vector (cf. Definition 6.1 on p. 125). The activation function of each output neuron is a **radial function** (cf. also Definition 6.1 on p. 125), that is, a monotone non-increasing function

$$f : \mathbb{R}_0^+ \to [0, 1] \quad \text{with} \quad f(0) = 1 \quad \text{and} \quad \lim_{x \to \infty} f(x) = 0.$$

The output function of each output neuron is the identity. The output may be discretized according to the "winner takes all" principle: the neuron with the highest activation produces the output 1, all other neurons produce the output 0.

© Springer Nature Switzerland AG 2022
R. Kruse et al., *Computational Intelligence*, Texts in Computer Science,
https://doi.org/10.1007/978-3-030-42227-1_7

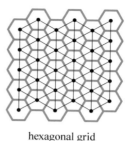

quadratic grid   hexagonal grid

Fig. 7.1 Examples of arrangements of the output neurons of a self-organizing map. Each black dot corresponds to an output neuron. The thin lines connect each neuron to its nearest neighbors, the thick gray lines indicate cells that may be used for visualization (see below for details)

On the neurons of the output layer, a **neighborhood relationship** is defined, which is described by a distance function

$$d_{\text{neurons}} : U_{\text{out}} \times U_{\text{out}} \to \mathbb{R}_0^+ .$$

This function assigns a non-negative real number to each pair of output neurons.

As a consequence, a self-organizing map is a two-layered neural network without hidden neurons (that is, there is only an input and an output layer). Its structure corresponds essentially to the input and the hidden layer of the radial basis function networks as they were discussed in the preceding chapter. Their alternative name **Kohonen feature map** refers to their inventor [2,6].

In analogy to radial basis function networks, the weights of the connections from the input to the output neurons state the coordinates of a **center**, from which the distance of an input pattern is measured. In connection to self-organizing maps, this center is often called a **reference vector**. The closer an input pattern is to a reference vector, the higher the activation of the corresponding neuron. Usually, all output neurons have the same network input function (distance function) and the same activation function (radial function) with the same **(reference) radius** σ.

The neighborhood relationship of the output neurons is usually defined by arranging these neurons into a (usually two-dimensional) grid. Examples of such grids are shown in Fig. 7.1. Each dot stands for one output neuron. The lines connecting the dots are meant to make the neighborhood structure more easily visible by indicating the closest neighbors. The thick gray lines indicate a possible visualization of a self-organizing map, which we study in more detail below.

However, the neighborhood relationship may also be missing, which is formally represented by an extreme distance measure for the neurons: each neuron has the distance 0 to itself and an infinite distance to all other neurons. By choosing this distance measure, the neurons become effectively independent.

If a neighborhood relationship is missing and the output is discretized (that is, the output neuron with the highest activation produces output 1, all other neurons produce output 0), a self-organizing map describes a so-called **vector quantization** of the input space: the input space is divided into as many regions as there are output

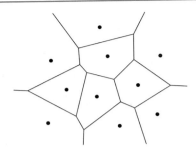

Fig. 7.2 Voronoi diagram of a vector quantization of a two-dimensional region with ten reference vectors. It depicts how the input space is divided by the reference vectors

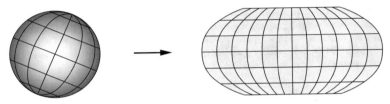

Fig. 7.3 Example of a topology preserving map: Robinson projection of the surface of a sphere onto a plane, as it is commonly used for world maps

neurons. This is achieved by assigning to an output neuron all points of the input space for which the neuron yields the highest activation. If the distance and activation functions are identical for all output neurons, we may also say: to an output neuron all points of the input space are assigned that are closer to the neuron's reference vector than to the reference vector of any other output neuron. This "tessellation" into regions can be represented by a so-called **Voronoi diagram** [1], as it is shown in Fig. 7.2 for two-dimensional inputs. The dots indicate the position of the reference vectors, the lines the division into regions.

The neighborhood relationship of the output neurons constrains the vector quantization. The objective is that reference vectors that are close to each other in the input space belong to output neurons that have a small distance from each other. That is, the neighborhood relationship of the output neurons is meant to reflect the relative position of the corresponding reference vectors in the input space (at least approximately). If this is the case, the self-organizing map describes a (quantized) **topology preserving map**, that is, a map that (approximately) preserves the position relationships between points (Greek $\tau o \pi o \varsigma$: position, location).

A well-known example of a topology preserving map is shown in Fig. 7.3, namely the so-called **Robinson projection** of the surface of a sphere onto a plane, as it is commonly used for world maps. Each point on the surface of a sphere (shown on the left) is mapped to a point of an approximately oval shape (shown on the right). With this map, the position relationships between points are approximately preserved, even though the ratio of the distances between two points in the projection to the

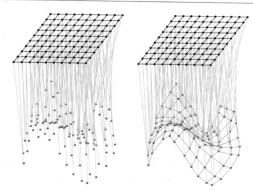

Fig. 7.4 Illustration of a topology preserving map from a three-dimensional input space (bottom) to a two-dimensional neuron grid (top) represented as a self-organizing map. The diagram on the left shows only the points in the input space that are the reference vectors associated with the neurons on the grid (association indicated by gray lines). The diagram on the right also connects the points in the input space that are associated with neighboring neurons

distance of its originals on the sphere is the larger, the farther these two points are from the equator of the sphere. Therefore, this projection, if it is used for a world map, does not always accurately reflect the actual distances between points on the surface of the earth. Nevertheless, it conveys a reasonably good impression of the relative location of cities, countries, and continents.

Transferred to self-organizing maps, the intersection points of the grid lines on the sphere could indicate the position of the reference vectors in the input space, while the intersection points of the grid lines in the projection indicates the position of the output neurons and their neighborhood relationship. However, in this case, the map is quantized, since points inside the grid cells are mapped only discretely through the reference vectors (although one may think about adding some interpolation).

A second, more concrete example of a topology preserving map, which also maps a three-dimensional input space to a two-dimensional (square) neuron grid is shown in Fig. 7.4. Although in practice, the input space would usually have many more dimensions, using only three makes, it possible to depict the mapping. The (tilted) square grid at the top shows the neuron grid, with lines connecting each neuron to its nearest neighbors. Each neuron is associated with a point in the input space, which is encoded as the weight vector of the neuron (reference vector). This association is indicated by gray lines in the diagrams in Fig. 7.4. While the left diagram shows only the reference vectors as blue points, the right diagram also connects the points in the input space that are associated with neurons that are connected in the grid, thus making the topology preservation of the mapping better visible.

The advantage of topology preserving maps is that they allow us to map high-dimensional structures onto low-dimensional spaces. In particular, a map to a space with only two or three dimensions is interesting, since then we can display the image of the high-dimensional structure graphically. For this one exploits the cell structure that is indicated in Fig. 7.1 by the gray lines. Obviously, this cell structure corresponds to a Voronoi diagram in the space of the output neurons. To each of these (two-dimensional) neuron Voronoi cells, a (generally higher-dimensional) cell of the input space is mapped by the reference vector belonging to the corresponding output neuron. Hence, we can visualize the relative position of points in the input space by finding the reference vector Voronoi cells in which they lie and then, for instance, coloring the corresponding neuron Voronoi cells. An even better impression is obtained if one chooses a different color for each represented point and not only colors the single neuron Voronoi cell, in whose corresponding reference vector Voronoi cell the point lies, but colors all neuron Voronoi cells in such a way that the color saturation represents the activation of the corresponding neuron. An example of such visualization can be found in Sect. 7.3.

7.2 Learning Vector Quantization

In order to explain the training of self-organizing maps, we first neglect the neighborhood relationship of the output neurons and thus confine ourselves to the so-called **learning vector quantization** [3]. The objective of this method is to organize the data into clusters, similar to what we have discussed for initializing radial basis function networks in Sect. 6.3 on p. 139: with the help of c-means clustering we tried to find good starting points for the centers of the radial basis functions. We also mentioned that learning vector quantization is an alternative.

In both c-means clustering and learning vector quantization, the individual clusters are represented by a center (or a reference vector, but that is merely a different name for center). This center is to be positioned in such a way that it lies roughly in the middle of the data point cloud that constitutes the cluster. An example is shown in Fig. 7.5: each group of data points (depicted as o) is assigned to a reference vector (depicted as ●). Thus, the input space is divided—as indicated by the lines—in such a way that each point lies in its own Voronoi cell.

The two methods mainly differ in how the cluster centers or the reference vectors are adapted. While in c-means clustering, the two steps of assigning the data points to the clusters and recomputing the cluster centers as the center of gravity of the assigned

Fig. 7.5 Clustering of data by learning vector quantization: each group of data points (○) is assigned to a reference vector (●)

Fig. 7.6 Adaptation of reference vectors (•) with a training pattern (○), $\eta = 0.4$. Left: attraction rule, right: repulsion rule

data points are alternatingly executed, learning vector quantization processes the data points one by one and adapts only one reference vector per data point. The procedure is known as **competitive learning**: the training patterns (data points) are traversed one by one. For each training pattern, a "competition" is carried out, which is won by the output neuron that yields the highest activation for this training pattern (if distance and activation functions are the same for all output neurons, we may say equivalently: whose reference vector is closest to the training vector). Only this "winner neuron" is adapted, namely in such a way that its reference vector is moved closer to the training pattern. Hence, the rule for adapting the reference vectors is

$$\mathbf{r}^{(\text{new})} = \mathbf{r}^{(\text{old})} + \eta\big(\mathbf{p} - \mathbf{r}^{(\text{old})}\big),$$

where $\mathbf{p}$ is the training pattern, $\mathbf{r}$ is the reference vector of the winner neuron for $\mathbf{p}$, and η is a learning rate with $0 < \eta < 1$. This rule is illustrated in Fig. 7.6 on the left: the learning rate η determines by what fraction of the distance $d = |\mathbf{p} - \mathbf{r}|$ between reference vector and training pattern the reference vector is shifted.

As for threshold logic units, multi-layer perceptrons and radial basis function networks, we distinguish again between **online training** and **batch training**. In the former, the reference vector is adapted immediately and thus in the next step, that is, when processing the next training pattern, the reference vector already has its new position. In the latter (batch training), the changes are aggregated and the reference vectors are adapted only at the end of an epoch, that is, after all training patterns have been processed. Note that the batch mode of learning vector quantization is very similar to c-means clustering: the assignment of the data points is clearly identical, since in the batch procedure, the position of the reference vectors is constant during an epoch. However, due to the learning the new position of the reference vector is not necessarily the center of gravity of the assigned data points, but generally a point between the old position and this center of gravity (depending on the learning rate).

As an illustration, Fig. 7.7 shows the process of learning vector quantization for the data points shown in Fig. 7.5 with online training on the left and batch training on the right. Since only a few epochs have been computed, the reference vectors have not yet reached their final positions, which are shown in Fig. 7.5. However, it is already clear that the desired clustering will finally be reached.

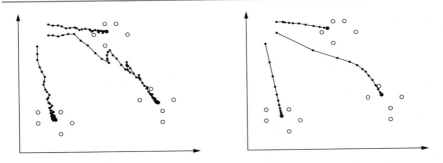

Fig. 7.7 Learning vector quantization for the data points shown in Fig. 7.5 with three reference vectors that start in the upper left corner. Left: online training with learning rate $\eta = 0.1$, right: batch training with learning rate $\eta = 0.05$ (training not conducted until convergence)

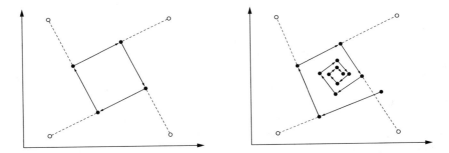

Fig. 7.8 Adaptation of a reference vector with four training patterns. Left: constant learning rate $\eta(t) = 0.5$, right: decreasing learning rate $\eta(t) = 0.6 \cdot 0.85^t$. In the first step, it is $t = 0$

Up to now, we used a constant learning rate η and merely required $0 < \eta < 1$. However, especially if online training is applied, a fixed learning rate can lead to problems, as shown in Fig. 7.8 on the left. In this diagram, a reference vector is repeatedly adapted with four data points, which causes the reference vector to move on a cyclical trajectory. The center of the four data points, which would be the desired result, is never reached. In order to solve this problem, a **time-dependent learning rate** is introduced, for instance,

$$\eta(t) = \eta_0 \alpha^t, \quad 0 < \alpha < 1, \quad \text{or} \quad \eta(t) = \eta_0 t^\kappa, \quad \kappa < 0.$$

This continuously decreasing learning rate turns the cyclical movement into a spiral that leads into the center (as desired), see Fig. 7.8 on the right. As a consequence, the desired destination of the reference vector is finally reached.

Although a time-dependent learning rate guarantees that the procedure converges, one should keep in mind that the learning rate must also not decrease too quickly, because otherwise the procedure may end in what is often called "starvation." That is, the adaptation steps become very small very quickly, so that the reference vectors never reach their natural destinations or may not even get close to them. On the other hand, the learning rate should not decrease too slowly, because then the learning process may converge only very slowly. As we already saw with other network types, choosing a proper learning rate is a difficult problem.

Although their main purpose is to find groups of data points, learning vector quantization can be applied not only for clustering, that is, to solve a free learning task. It can be extended in such a way that it takes classes into account that are assigned to the data points. In this way, fixed learning tasks may be solved, at least such learning tasks in which the desired output comes from a finite set of values (classes). To this end, the output neurons—and thus the reference vectors—are endowed with class labels and the adaptation rule is split into two types. If the class of the data point and the class of (the reference vector of) the winner neurons coincide, the **attraction rule** is applied, which is identical to the rule considered above:

$$\mathbf{r}^{(\text{new})} = \mathbf{r}^{(\text{old})} + \eta\left(\mathbf{p} - \mathbf{r}^{(\text{old})}\right).$$

That is, the reference vector is moved toward the training pattern (in other words: it is "attracted" by the training pattern). However, if the classes of the data point and the reference vector differ, the **repulsion rule** is applied:

$$\mathbf{r}^{(\text{new})} = \mathbf{r}^{(\text{old})} - \eta\left(\mathbf{p} - \mathbf{r}^{(\text{old})}\right).$$

That is, the reference vector is moved away from the training pattern (in other words: it is "repelled" by the training pattern), see Fig. 7.6 on the right. In this way, the reference vectors move toward groups of data points that carry the same class label as they do themselves. A trained vector quantization yields, for a new input vector to be classified, the class that is assigned to the output neuron with the highest activation (*nearest prototype classifier* if all distance and radial functions are identical).

Improved versions of learning vector quantization for fixed learning tasks do not only adapt the one reference vector that is closest to the current data point, but the two closest reference vectors [4,6]. Let $\mathbf{r}_j$ and $\mathbf{r}_k$ be these two closest reference vectors. They are adapted if the classes c_j and c_k assigned to them differ, but one of them coincides with the class z of the data point $\mathbf{p}$. Without loss of generality, we assume that $c_k = z$. Then the adaptation rules read

$$\mathbf{r}_j^{(\text{new})} = \mathbf{r}_j^{(\text{old})} + \eta\left(\mathbf{p} - \mathbf{r}_j^{(\text{old})}\right) \quad \text{and} \quad \mathbf{r}_k^{(\text{new})} = \mathbf{r}_k^{(\text{old})} - \eta\left(\mathbf{p} - \mathbf{r}_k^{(\text{old})}\right).$$

All other reference vectors remain unchanged. However, if the classes c_j and c_k of the two closest reference vectors coincide (independent of whether they also coincide with the class z of the data point or not), no reference vector is adapted. These rules often yield good nearest prototype classifiers [4].

Unfortunately, though, it was observed in practical tests that this version of learning vector quantization tends, in certain situations, to drive the reference vectors further and further apart instead of leading to stable convergence. In order to counteract this clearly undesirable behavior, [4] introduced a so-called **window rule** into the adaptation: the reference vectors are adapted only if the data point $\mathbf{p}$ lies close to the classification border, that is, close to the (hyper-)surface, which separates regions in which different classes are predicted. The somewhat vague notion "close" is made formally precise by requiring

$$\min\left(\frac{d(\mathbf{p}, \mathbf{r}_j)}{d(\mathbf{p}, \mathbf{r}_k)}, \frac{d(\mathbf{p}, \mathbf{r}_j)}{d(\mathbf{p}, \mathbf{r}_k)}\right) > \theta, \quad \text{with} \quad \theta = \frac{1 - \xi}{1 + \xi}.$$

Here, ξ is a parameter that has to be chosen by a user. Intuitively, ξ describes the "width" of the window around the classification border in which a data point $\mathbf{p}$ has to lie in order to cause an adaptation of reference vectors. This rule prevents a divergence of the reference vectors, because the adaptations caused by a data point stop as soon as the classification border is far enough away.

One has to concede, though, that this window rule is not particularly intuitive. Therefore, it would be desirable if one could do without it. This is indeed possible if one derives the adaptation rule for the reference vectors with a gradient descent approach for a specific objective function [7]. This approach starts from the assumption that the probability distribution of the data points for each class can be described sufficiently well by a mixture of (multi-dimensional) normal distributions. That is, each class consists of multiple clusters, each of which is covered by one normal distribution. Furthermore, it is assumed that all of these normal distributions have fixed and equal standard deviations σ. Intuitively, this means that all clusters have the same size and a (hyper-)spherical shape. Finally, it is assumed that all clusters are equally likely, that is, that all clusters comprise (roughly) the same number of data points. With these restrictions, only the distance of a data point from a reference vector decides how it is classified.

The adaptation procedure is derived from a maximization of an objective function that describes the probability that a data point is correctly classified. That is, one tries to maximize the so-called **likelihood ratio**. As we show below, the attraction rule considered above thus results from maximizing the posterior probability of the correct class (that is, the true class that is assigned to a data point), while the repulsion rule is a consequence of minimizing the posterior probability of a wrong class [7]. Formally, we execute a gradient descent on the maximum likelihood ratio. Starting from the assumptions made above, we obtain for the likelihood ratio (or rather its natural logarithm, which simplifies handling the normal distributions)

$$
\ln L_{\text{ratio}} = \sum_{j=1}^{n} \ln \sum_{i \in I(z_j)} \exp\left(-\frac{(\mathbf{p}_j - \mathbf{r}_i)^\top (\mathbf{p}_j - \mathbf{r}_i)}{2\sigma^2} \right)
$$
$$
- \sum_{j=1}^{n} \ln \sum_{i \notin I(z_j))} \exp\left(-\frac{(\mathbf{p}_j - \mathbf{r}_i)^\top (\mathbf{p}_j - \mathbf{r}_i)}{2\sigma^2} \right),
$$

where $I(z)$ contains the indices of the reference vectors that are labeled as belonging to class z. Note that the normalization factors, which appear in the well-known formula for a normal distribution, cancel, since all clusters have the same standard variation (or variance, respectively). In the same way, the prior probabilities of the different clusters cancel, since we assumed that they are all the same.

From this objective function, we obtain almost immediately as the online adaptation rule for a gradient descent

$$
\mathbf{r}_i^{(\text{new})} = \mathbf{r}_i^{(\text{old})} + \eta \cdot \nabla_{\mathbf{r}_i} \ln L_{\text{ratio}}\big|_{\mathbf{r}_i^{(\text{alt})}}
$$
$$
= \mathbf{r}_i^{(\text{old})} + \eta \cdot
\begin{cases}
u_{ij}^{\oplus\,(\text{old})} \cdot \left(\mathbf{p}_j - \mathbf{r}_i^{(\text{old})}\right) & \text{if } z_j = c_i, \\
-u_{ij}^{\ominus\,(\text{old})} \cdot \left(\mathbf{p}_j - \mathbf{r}_i^{(\text{old})}\right) & \text{if } z_j \neq c_i,
\end{cases}
$$

where c_i is again the class that is assigned to the ith reference vector and z_j the class of the data point $\mathbf{p}_j$. The "membership degrees" $u_{ij}^{\oplus}$ and $u_{ij}^{\ominus}$, with which a data point $\mathbf{p}_j$ belongs to the cluster of the reference vector $\mathbf{r}_i$, are given by

$$u_{ij}^{\oplus\,(\text{old})} = \frac{\exp\left(-\frac{1}{2\sigma^2}\left(\mathbf{p}_j - \mathbf{r}_i^{(\text{old})}\right)^\top \left(\mathbf{p}_j - \mathbf{r}_i^{(\text{old})}\right)\right)}{\sum\limits_{k \in I(z_j)} \exp\left(-\frac{1}{2\sigma^2}\left(\mathbf{p}_j - \mathbf{r}_k^{(\text{old})}\right)^\top \left(\mathbf{p}_j - \mathbf{r}_k^{(\text{old})}\right)\right)} \quad \text{and}$$

$$u_{ij}^{\ominus\,(\text{old})} = \frac{\exp\left(-\frac{1}{2\sigma^2}\left(\mathbf{p}_j - \mathbf{r}_i^{(\text{old})}\right)^\top \left(\mathbf{p}_j - \mathbf{r}_i^{(\text{old})}\right)\right)}{\sum\limits_{k \notin I(z_j)} \exp\left(-\frac{1}{2\sigma^2}\left(\mathbf{p}_j - \mathbf{r}_k^{(\text{old})}\right)^\top \left(\mathbf{p}_j - \mathbf{r}_k^{(\text{old})}\right)\right)}.$$

The split into the two cases $z_j = c_i$ (the class of the data point coincides with the class of the reference vector) and $z_j \neq c_i$ (the reference vector and the data point belong to different classes) results from the fact that each reference vector r_i appears in only one of the two sums: either its index i is in $I(z_j)$, and then only the first sum contributes, or its index is not in $I(z_j)$, and then only the second sum contributes. The denominators of the fractions result from the derivative of the natural logarithm.

Thus, we obtained a scheme for a "soft" learning vector quantization, where "soft" means that all reference vectors are adapted, but with different strength: all reference vectors with the same class as the data point are "attracted", while all reference vectors with a different class are "repelled" [7].

A "hard" learning vector quantization can easily be derived from this scheme by letting the standard deviations (or variances, respectively) that are assigned to the reference vectors go to zero. In the limit, we obtain a hard assignment

$$u_{ij}^{\oplus} = \delta_{i,k^{\oplus}(j)}, \text{ where } k^{\oplus}(j) = \operatorname*{argmin}_{l \in I(z_j)} d(\mathbf{p}_j, \mathbf{r}_l), \quad \text{and}$$
$$u_{ij}^{\ominus} = \delta_{i,k^{\ominus}(j)}, \text{ where } k^{\ominus}(j) = \operatorname*{argmin}_{l \notin I(z_j)} d(\mathbf{p}_j, \mathbf{r}_l),$$

and $\delta_{i,k}$ is the *Kronecker symbol* ($\delta_{i,k} = 1$ if $i = k$ and $\delta_{i,k} = 0$ otherwise).

Note, however, that this scheme is *not* identical to the one discussed above, which was proposed in [4,6]. While in Kohonen's scheme the two closest reference vectors are determined and adapted only if they are labeled with different classes, this scheme *always* adapts two reference vectors, namely the one that is closest among those that carry the same class label (this vector is attracted) and the closest among those that carry a different class label (this vector is repelled). Note that these two need not be the two closest ones among all reference vectors: although one of them must be the closest overall, the second may be much farther away than several other reference vectors.

An advantage of this approach is that it explains why sometimes diverging behavior can be observed. (Details are beyond the scope of this book—an interested reader should consult [7].) Furthermore, it suggests a method how the divergence can be avoided without having to introduce a window rule. The idea consists in a minor modification of the objective function [7] (cf. p. 157 for the objective function originally proposed):

$$\ln L_{\text{ratio}} = \sum_{j=1}^{n} \ln \sum_{i \in I(z_j)} \exp\left(-\frac{(\mathbf{x}_j - \mathbf{r}_i)^\top (\mathbf{x}_j - \mathbf{r}_i)}{2\sigma^2}\right)$$
$$- \sum_{j=1}^{n} \ln \sum_{i} \exp\left(-\frac{(\mathbf{x}_j - \mathbf{r}_i)^\top (\mathbf{x}_j - \mathbf{r}_i)}{2\sigma^2}\right).$$

Obviously, the difference consists only of the fact that the second sum now runs over all reference vectors (and not only over those with a different class label than the data point). That is, we no longer compare to the likelihood of seeing a data point with a different class, but to the likelihood of seeing a data point at all. Again we obtain an adaptation rule for a "soft" learning vector quantization:

$$\mathbf{r}_i^{(\text{new})} = \mathbf{r}_i^{(\text{old})} + \eta_{\mathbf{r}} \cdot \nabla_{\mathbf{r}_i} \ln L_{\text{ratio}}$$
$$= \mathbf{r}_i^{(\text{old})} + \eta_{\mathbf{r}} \cdot \begin{cases} \left(u_{ij}^{\oplus \, (\text{old})} - u_{ij}^{(\text{old})}\right) \cdot \left(\mathbf{p}_j - \mathbf{r}_i^{(\text{old})}\right) & \text{if } z_j = c_i, \\ \qquad\quad - u_{ij}^{(\text{old})} \cdot \left(\mathbf{p}_j - \mathbf{r}_i^{(\text{old})}\right) & \text{if } z_j \neq c_i. \end{cases}$$

(Note that this time the reference vectors that carry the correct class label occur in both terms of the likelihood ratio, which explains the sum $u_{ij}^{\oplus} - u_{ij}$ in the first case.) The "membership degrees" $u_{ij}^{\oplus}$ and u_{ij} are given as

$$u_{ij}^{\oplus \, (\text{old})} = \frac{\exp\left(-\frac{1}{2\sigma^2}\left(\mathbf{x}_j - \mathbf{r}_i^{(\text{old})}\right)^\top \left(\mathbf{x}_j - \mathbf{r}_i^{(\text{old})}\right)\right)}{\sum\limits_{k \in I(z_j)} \exp\left(-\frac{1}{2\sigma^2}\left(\mathbf{x}_j - \mathbf{r}_k^{(\text{old})}\right)^\top \left(\mathbf{x}_j - \mathbf{r}_k^{(\text{old})}\right)\right)} \quad \text{and}$$

$$u_{ij}^{(\text{old})} = \frac{\exp\left(-\frac{1}{2\sigma^2}\left(\mathbf{x}_j - \mathbf{r}_i^{(\text{old})}\right)^\top \left(\mathbf{x}_j - \mathbf{r}_i^{(\text{old})}\right)\right)}{\sum\limits_{k} \exp\left(-\frac{1}{2\sigma^2}\left(\mathbf{x}_j - \mathbf{r}_k^{(\text{old})}\right)^\top \left(\mathbf{x}_j - \mathbf{r}_k^{(\text{old})}\right)\right)}.$$

A "hard" variant can again be obtained by letting the standard variations σ go to zero. This leads to

$$u_{ij}^{\oplus} = \delta_{i,k^{\oplus}(j)}, \quad \text{where } k^{\oplus}(j) = \operatorname*{argmin}_{l \in I(z_j)} d(\mathbf{x}_j, \mathbf{r}_l), \quad \text{and}$$
$$u_{ij} = \delta_{i,k(j)}, \quad \text{where } \quad k(j) = \operatorname*{argmin}_{l} d(\mathbf{x}_j, \mathbf{r}_l).$$

This adaptation rule is very similar to the one of [4,6]. Intuitively, we can interpret it as follows: if the closest reference vector carries the same class label as the data point, no adaptation is carried out. However, if the class of the closest reference vector differs from the class of the data point, then this reference vector is repelled, while the closest reference vector among those with the same class label is attracted. In other words, the reference vectors are adapted only if a data point would be classified wrongly by the closest reference vector. Otherwise, the current positions of the reference vectors are maintained.

Fig. 7.9 Screenshots of the demonstration program `xlvq`, without and with class information

For learning vector quantization with a time-dependent learning rate for classified and unclassified training patterns (though only with batch training and without any window rule or any of the improved adaptation rules), the web page

http://www.borgelt.net/lvqd.html

offers the programs `wlvq` (for Microsoft Windows™) and `xlvq` (for Linux). With them clusters can be found for two-dimensional data (selectable from a larger number of dimensions) and the movement of the reference vectors can be followed. An extensive explanation of these programs can be found on the web page named above, screenshots are shown in Fig. 7.9 (for the iris data, cf. Fig. 5.42 on p. 103).

7.3 Neighborhood of the Output Neurons

Up to now, we neglected the neighborhood relationship of the output neurons: the reference vectors could move independently of each other. As a consequence, learning vector quantization does not generally allow us to read anything from the (relative) position of the output neurons about the (relative) position of the reference vectors. In order to learn a topology preserving map in which the (relative) position of the output neurons reflects the (relative) position of the reference vectors, the neighborhood relationship has to be respected in the training process. Only in this case, one calls the network a **self-organizing map** [2,6].

Self-organizing maps are trained—like vector quantization—with **competitive learning**. That is, the training patterns are visited one after the other and for each training pattern, the neuron is determined that yields the highest activation. In self-organizing maps, it is mandatory that all output neurons have the same distance function and the same activation function. Thus, we can definitely say here equivalently: we find the output neuron, whose reference vector is closest to the training pattern. This neuron is the "winner" of the competition.

However, in contrast to learning vector quantization, not only the reference vector of the winner neuron is adapted. At the end of training, the reference vectors of its neighbor neurons are supposed to be close to its reference vector. Hence, the neighboring reference vectors are adapted as well, though possibly less severe than that of the winner neuron. In this way, it is achieved that the reference vectors of neighboring neurons cannot move arbitrarily far apart, since they are adapted analogously. Thus, it can be expected that in the learning result neighboring output neurons possess reference vectors that are close to each other in the input space.

Another important difference to learning vector quantization is that self-organizing maps are almost exclusively used for free learning tasks. Since, prior to training, nothing is known about the relative position of different classes, it is difficult to assign meaningful classes to the reference vectors. One may assign classes to the output neurons *after* training is completed, though, namely by simply assigning the class that is most frequent among the training patterns for which the output neuron yields the highest activation. However, in this case, the class information does not influence the training of the map (and thus the position of the reference vectors). Hence, classifying data with the help of a self-organizing map that has been extended in this way is not necessarily recommended. Such a class assignment may, however, provide a good impression of the distribution and the relative position of different classes in the input space and thus may be a useful analysis tool.

Since only free learning tasks can reasonably be handled, there is only one adaptation rule for the reference vectors, which is analogous to the attraction rule discussed in the preceding section. This rule reads

$$\mathbf{r}_u^{(\text{new})} = \mathbf{r}_u^{(\text{old})} + \eta(t) \cdot f_{\text{nb}}\big(d_{\text{neurons}}(u, u_*), \varrho(t)\big) \cdot \big(\mathbf{p} - \mathbf{r}_u^{(\text{old})}\big),$$

where p is the considered training pattern, $\mathbf{r}_u$ is the reference vector of neuron u, u_* is the winner neuron, $\eta(t)$ is a time-dependent learning rate and $\varrho(t)$ is a time-dependent neighborhood radius. The variable d_{neurons} measures the distance of output neurons (cf. Definition 7.1 on p. 149), here specifically the distance of the neuron to adapt from the winner neuron. (Since in a self-organizing map the neighbors of the winner neuron are adapted as well, we can no longer restrict the adaptation rule to the winner neuron.) How severely the reference vectors of other output neurons can be adapted depends, according to this rule, via a function f_{nb} (nb for neighbor) on the distance of the neuron from the winner neuron and the size of the radius $\varrho(t)$ that determines the neighborhood.

The function f_{nb} is a radial function, that is, it is of the same type as the functions that we used to compute the activation of a neuron depending on the distance of training patterns from a reference vector (cf. Fig. 6.2 on p. 127). It assigns a number between 0 and 1 to each output neuron, which depends on its distance to the winner neuron[1] and describes the strength of the adaptation of its reference vector relative to the strength of the adaptation of the reference vector of the winner neuron. If the function f_{nb} is, for example, a rectangular function, then all output neurons

---

[1] Note that the distance is computed from the grid, in which the output neurons are arranged, and *not* from the position of the reference vectors or the distance measure in the input space.

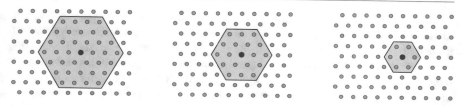

Fig. 7.10 Illustration of a time-dependent neighborhood for a rectangular neighborhood function

in a certain radius around the winner neuron are adapted with full strength, while all other output neurons are left unchanged. Most commonly, however, a Gaussian neighborhood function is used, so that the strength of the adaptation of the reference vectors decays exponentially with the distance from the winner neuron.

A **time-dependent learning rate** is used for the same reasons as discussed for learning vector quantization, namely to avoid update cycles. Therefore, it can be defined in the same way, for example, as

$$\eta(t) = \eta_0 \alpha_\eta^t, \quad 0 < \alpha_\eta < 1, \quad \text{or} \quad \eta(t) = \eta_0 t^{\kappa_\eta}, \quad \kappa_\eta < 0.$$

Analogously, a **time-dependent neighborhood radius** is defined, for instance, as

$$\varrho(t) = \varrho_0 \alpha_\varrho^t, \quad 0 < \alpha_\varrho < 1, \quad \text{or} \quad \varrho(t) = \varrho_0 t^{\kappa_\varrho}, \quad \kappa_\varrho < 0.$$

That the neighborhood radius decreases with time is reasonable, because then the self-organizing map can properly "unfold" in the first training steps (larger neighborhood), while in later training steps (smaller neighborhood) the position of the reference vectors are fitted more closely to the positions of the training patterns. An illustration of a reducing neighborhood radius for a rectangular neighborhood function at certain time steps (not necessarily consecutive) is shown in Fig. 7.10.

As an example of a training process with the stated adaptation rule, we consider a self-organizing map with 100 output neurons, which are arranged in a square 10×10 grid. This map is trained with points that are chosen randomly (uniform distribution) from the square $[-1, 1] \times [-1, 1]$. The training procedure is depicted in Fig. 7.11. All diagrams show the input space, with the frame indicating the square $[-1, 1] \times [-1, 1]$. The grid of the output neurons is projected into this space by connecting, with a straight line, the reference vector of an output neuron to the reference vectors of its immediate neighbors. The top left diagram shows the situation directly after the reference vectors have been initialized with random weights from the interval $[-0.5, 0.5]$. Due to the randomness of the initialization, the (relative) position of the reference vectors is completely independent of the (relative) position of the output neurons, so that no grid structure can be discerned.

The following diagrams (first the top row from left to right, then the bottom row from left to right) show the state of the self-organizing map after 10, 20, 40, 80, and 160 training steps (in each training step one training pattern is processed; learning rate $\eta(t) = 0.6 \cdot t^{-0.1}$, Gaussian neighborhood function f_{nb}, neighborhood radius $\varrho(t) = 2.5 \cdot t^{-0.1}$). These diagrams show nicely how the self-organizing map slowly "unfolds" and thus adjusts itself to the input space. That the grid structure becomes

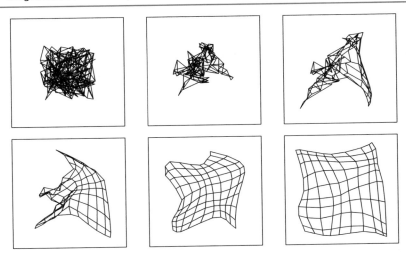

Fig. 7.11 Unfolding of a self-organizing map trained with random patterns from the square $[-1, 1] \times [-1, 1]$ (indicated by the frames). The lines connect the reference vectors of neighboring neurons

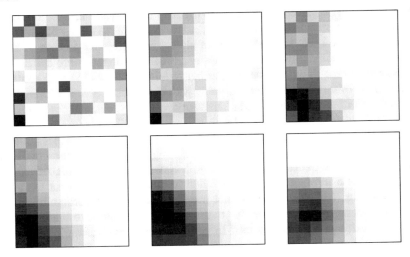

Fig. 7.12 Coloring of the training steps of the self-organizing map shown in Fig. 7.11 for the input pattern $(-0.5, -0.5)$ using a Gaussian activation function

visible demonstrates how the arrangement of the output neurons is transferred to the arrangement of the reference vectors in the input space.

For the same example, Fig. 7.12 shows a way to visualize a self-organizing map, which we mentioned in Sect. 7.1. All diagrams show the grid structure of the output neurons (*not* the input space as in Fig. 7.11). Each neuron is represented by a small square (cf. Fig. 7.1 on p. 150). The shades of gray encode the activation of the out-

Fig. 7.13 If the initialization is unfavorable or the learning rate or the neighborhood is too small, the result can be a twisted map

put neurons if the pattern $(-0.5, -0.5)$ is fed into the network (Gaussian activation function): the darker a square, the higher the activation. With this representation, the training can also be followed well. After the initialization, the heavily activated neurons are still randomly distributed on the map. However, the further the training proceeds, the better they group together. Note the activation structures after 20 training patterns (3rd diagram) and after 40 training patterns (4th diagram) and compare them to the corresponding diagrams of the self-organizing map in the input space shown in Fig. 7.11: since in these phases the map is not well unfolded on the left side, many neurons on the left are strongly activated.

The example we just studied shows an almost ideal training process of a self-organizing map. After only a few training steps, the map is already unfolded and very well adapted to the training patterns. Further training stretches the map somewhat more, until it covers the region of the training patterns almost uniformly (even though the projection of the neuron grids, as can easily be worked out, can never reach the edges of the square $[-1, 1] \times [-1, 1]$).

However, training is not always so successful. If, by accident, the map is initialized in an unfavorable way, and particularly if the learning rate or the neighborhood radius are chosen too small or decrease too quickly, "twisted" maps can result. An example of the result of a training process that failed in this way is shown for the square example in Fig. 7.13. The map did not properly unfold. The corners of the unit square have been assigned "wrongly" to the corners of the grid, so that the middle of the map exhibits a kind of "knot." In general, such a twist cannot be amended, regardless of how long training is continued. However, most of the time, it can be avoided by starting with a large learning rate and particularly a large neighborhood radius (in the order of the grid side length of the self-organizing map).

In order to illustrate the dimension reduction that can be achieved with a (quantized) topology preserving map, as it is implemented by a self-organizing map, Fig. 7.14 shows the projection of the neuron grid of a self-organizing map with 10×10 output neurons into a three-dimensional input space. The map shown on the left was trained with random points from a rotated parabola, the map in the middle with random points from a binary cubic function (cf. also Fig. 7.4), and the map on the right with random points from the surface of a sphere. Since, in these cases, the

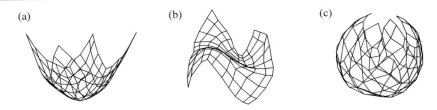

(a)   (b)   (c)

Fig. 7.14 Self-organizing maps that were trained with random points on **a** a rotated parabola, **b** a cubic function, **c** the surface of a sphere

input space is actually two-dimensional (all training patterns lie on, though curved, surfaces), a self-organizing map can adjust very well to the training patterns.

As a further illustration of the training of self-organizing maps, the web page

http://www.borgelt.net/somd.html

provides the programs wsom (for Microsoft Windows™) and xsom (for Linux). With these programs, a self-organizing map with a square grid can be trained with points that are chosen randomly from certain two-dimensional regions (a square, a circle, or a triangle) or three-dimensional surfaces (e.g., surface of a sphere or a rotated parabola). Figures 7.11, 7.13 and 7.14 show training processes and training results that were obtained with these programs.

With these programs, it can also be studied what happens if the training patterns truly have a higher-dimensional structure, so that it is not possible to map them onto a two-dimensional grid with only little loss. As an example, one may train with one of these programs a self-organizing map with at least 30×30 output neurons for training patterns that are randomly selected from a cube (volume, not surface). The self-organizing map will be folded in several places in order to fill the space as uniformly as possible. Although self-organizing maps are still useful in such cases, one has to pay attention to the fact that due to the folds it may happen that an input pattern activates output neurons that are fairly far away from each other in the grid structure of the map, simply because they lie on the two sides of a fold.

References

1. Aurenhammer F (1991) Voronoi diagrams— survey of a fundamental geometric data structure. ACM Comput Surv 23(3):345–405. ACM Press, New York, NY, USA
2. Kohonen T (1982) Self-organized formation of topologically correct feature maps. Biol Cybern. Springer, Heidelberg, Germany
3. Kohonen T (1986) Learning vector quantization for pattern recognition. Technical Report TKK-F-A601. Helsinki University of Technology, Finland

4. Kohonen T (1990) Improved versions of learning vector quantization. In: Proceedings of international joint conference on neural networks (IJCNN 1990, San Diego, CA), vol 1, pp 545–550. IEEE Press, Piscataway, NJ, USA
5. Kohonen T, Kangas J, Laaksonen J, Torkkola T (1992) The learning vector quantization program package, version 2.1. LVQ Programming Team, Helsinki University of Technology, Finland
6. Kohonen T (1995) Self-organizing Maps, 3rd ext. edn 2011. Springer, Heidelberg, Germany
7. Seo S, Obermayer K (2003) Soft learning vector quantization. Neural Comput 15(7):1589–1604. MIT Press, Cambridge, MA, USA

Hopfield Networks

8

In the preceding Chaps. 5 to 7, we studied so-called *feed forward networks*. That is, networks with an acyclic graph (no directed cycles). In this and the next chapter, however, we turn to so-called *recurrent networks*, that is, networks, the graph of which may contain (directed) cycles. We start with one of the simplest forms, the so-called **Hopfield networks** [12, 13], which originated as physical models to describe magnetism. Hopfield networks are indeed closely related to the Ising model of magnetism [15] (see below).

8.1 Definition and Examples

Definition 8.1 A **Hopfield network** is a neural network with a graph $G = (U, C)$ that satisfies the following conditions:

1. $U_{\text{hidden}} = \emptyset$, $\quad U_{\text{in}} = U_{\text{out}} = U$,
2. $C = U \times U - \{(u, u) \mid u \in U\}$.

The connection weights are symmetric, that is, we have $\forall u, v \in U, u \neq v \colon w_{uv} = w_{vu}$. The network input function of each neuron u is the weighted sum of the outputs of all other neurons, that is,

$$\forall u \in U : \quad f_{\text{net}}^{(u)}(\mathbf{w}_u, \mathbf{in}_u) = \mathbf{w}_u \mathbf{in}_u = \sum_{v \in U - \{u\}} w_{uv} \, \text{out}_v \,.$$

The activation function of each neuron u is a threshold function

$$\forall u \in U : \quad f_{\text{act}}^{(u)}(\text{net}_u, \theta_u) = \begin{cases} 1 & \text{if } \text{net}_u \geq \theta_u, \\ -1 & \text{otherwise.} \end{cases}$$

The output function of each neurons is the identity, that is,

$$\forall u \in U : \quad f_{\text{out}}^{(u)}(\text{act}_u) = \text{act}_u \,.$$

© Springer Nature Switzerland AG 2022
R. Kruse et al., *Computational Intelligence*, Texts in Computer Science,
https://doi.org/10.1007/978-3-030-42227-1_8

Fig. 8.1 A simple Hopfield
network that can oscillate if
the activations of the two
neurons are updated in
parallel, but reaches a stable
state if the neurons are
updated alternatingly

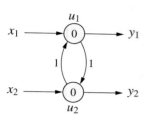

Note that there are no loops in a Hopfield network, that is, no neuron receives its own output as input. All feedback loops run through other neurons: a neuron u receives the outputs of all other neurons as its input and all other neurons receive the output of the neuron u as input.

The neurons of a Hopfield network work exactly like the threshold logic units that we studied in Chap. 3: depending on whether the weighted sum of the inputs exceeds a threshold θ_u or not, their activation is set to 1 or -1. Although in Chap. 3 the inputs and activations usually had the values 0 and 1, we mentioned in Sect. 3.6 the ADALINE variant that uses the values -1 and 1 instead. Section 10.3 shows how the two versions can be transformed into each other.

Sometimes the activation function of the neurons of a Hopfield network is defined, drawing on the old activation act_u, as follows:

$$\forall u \in U: \quad f_{\text{act}}^{(u)}(\text{net}_u, \theta_u, \text{act}_u) = \begin{cases} 1 & \text{if net}_u > \theta_u, \\ -1 & \text{if net}_u < \theta_u, \\ \text{act}_u & \text{if net}_u = \theta_u. \end{cases}$$

This is certainly advantageous for the physical interpretation of a Hopfield network (see below) and also simplifies a proof which we present in the next section. Nevertheless we stick to the definition given above, because it has other advantages.

To carry out the computations in this section, it is beneficial to represent the connection weights in a weight matrix (cf. also Chaps. 4 and 5). In order to do so, we set the missing weights $w_{uu} = 0$ (feedback loops of a neuron to itself), as this is, due to the special network input function of Hopfield neurons, equivalent to a missing connection. Because of the symmetric weights, the weight matrix is obviously symmetric (that is, it is equal to its transpose) and because of the missing feedback loops of neurons to themselves, its diagonal is 0. That is, a Hopfield network with n neurons $u_1, \ldots, u_n$ can be described by the $n \times n$ matrix

$$\mathbf{W} = \begin{pmatrix} 0 & w_{u_1 u_2} & \cdots & w_{u_1 u_n} \\ w_{u_1 u_2} & 0 & \cdots & w_{u_2 u_n} \\ \vdots & \vdots & & \vdots \\ w_{u_1 u_n} & w_{u_2 u_n} & \cdots & 0 \end{pmatrix}.$$

As a first example for a Hopfield network, we consider the network with two neurons that is shown in Fig. 8.1. The weight matrix of this network is

$$\mathbf{W} = \begin{pmatrix} 0 & 1 \\ 1 & 0 \end{pmatrix}.$$

Fig. 8.2 A simple Hopfield network with three neurons u_1, u_2 and u_3 (from top to bottom)

Fig. 8.3 Simplified representation of the Hopfield network shown in Fig. 8.2 that exploits the symmetry of the weights

Both neurons have the threshold 0. Like we did for the threshold logic units studied in Chap. 3, we write this threshold into the circle that represents the corresponding neuron. A second example of a simple Hopfield network is shown in Fig. 8.2. The weight matrix of this network is

$$\mathbf{W} = \begin{pmatrix} 0 & 1 & 2 \\ 1 & 0 & 1 \\ 2 & 1 & 0 \end{pmatrix}.$$

Again all neurons have the threshold 0. This example shows that the large number of connections can make a graphical representation difficult to read. In order to obtain a simpler representation, we exploit that every neuron is both input as well as output neuron. Therefore we need not draw input and output arrows explicitly, since these only serve the purpose to point out the input and output neurons. Furthermore, we know that the weights must be symmetric. Hence we can combine the connections of two neurons into a double arrow, which is labeled only once with the corresponding weight. Thus, we reach the representation shown in Fig. 8.3.

Let us now turn to the computations of a Hopfield network. We start with the Hopfield network with two neurons shown in Fig. 8.1. We assume that this network is fed with the values $x_1 = -1$ and $x_2 = 1$. This means that in the input phase the activation of the neuron u_1 is set to -1 (i.e., $\text{act}_{u_1} = -1$) and the activation of the neuron u_2 is set to 1 (i.e., $\text{act}_{u_2} = 1$). This is in no way different from how we always proceeded up to now (according to the general description of the operation of a neuron, as it was given in Sect. 4.2). However, due to the cycle in the graph that underlies this network, we face the question how the activations of the neurons are to be recomputed in the work phase. Up to now there was no need to consider this question, because in a feed forward network the update order is irrelevant: regardless of the order in which the neurons recompute their activation and output, we always obtain the same result. However, as we know from the example discussed in Sect. 4.2

Table 8.1 Computations of
the Hopfield network shown
in Fig. 8.1 for the inputs
$x_1 = -1$ and $x_2 = 1$ if the
activations are updated in
parallel

	u_1	u_2
input phase	-1	1
work phase	1	-1
	-1	1
	1	-1
	-1	1

(see p. 44), in a network with cycles the result can depend on the order in which the activations of the neurons are updated.

We try first to update the activations **synchronously** (at the same time, in parallel). That is, we compute the new activations and new outputs of both neurons from their old outputs. This leads to the computations shown in Table 8.1. Obviously, no stable activation state is reached, but the network oscillates between the states $(-1, 1)$ and $(1, -1)$. However, if we update the activations **asynchronously**, that is, if we update the activation and output only for one neuron at a time and already use the new output in subsequent updates (of other neurons), the network always reaches a stable state. As an illustration Table 8.2 shows the two possible update sequences, in which both neurons are updated alternatingly. In both cases, a stable state is reached. However, which state is reached depends on which neuron is updated first. Arguments of symmetry easily show that for any other inputs one of these two stable states is reached as well.

A similar observation can be made for the Hopfield network with three neurons that is shown in Fig. 8.2. If the input pattern $(-1, 1, 1)$ is fed into the network, a synchronous update of the neurons lets the network oscillate between the states $(-1, 1, 1)$ and $(1, 1, -1)$, while an asynchronous update leads either to the stable state $(1, 1, 1)$ or to the stable state $(-1, -1, -1)$.

For any other inputs one of the same two stable states is finally reached as well, regardless of the order we choose to update the neurons. This is most easily seen with the help of a **state graph** as it is shown in Fig. 8.4. Each state (that is, each combination of activations of the neurons) is represented by an ellipse, into which the signs of the activations of the three neurons u_1, u_2 and u_3 are written (from left to right). The arrows between the states are labeled with the neuron, an update of which causes the corresponding state transition. Since for each state the caused transitions are stated for an update of each of the three neurons, we can easily read from this graph which state sequences are traversed for any order in which the activations of the neurons are updated. It is easy to check that finally one of the two stable states $(+1, +1, +1)$ or $(-1, -1, -1)$ will be reached.

8.2   Convergence of the Computations

As we saw from the examples of the preceding sections, the network may oscillate between different activation states if the activations of the different neurons are updated synchronously. However, if we update them asynchronously, a stable state

Table 8.2 If the activations of the neurons of the Hopfield network shown in Fig. 8.1 are updated alternatingly, a stable state is reached. The result depends, though, on which neuron is updated first

	u_1	u_2
input phase	-1	1
work phase	1	1
	1	1
	1	1
	1	1

	u_1	u_2
input phase	-1	1
work phase	-1	-1
	-1	-1
	-1	-1
	-1	-1

Fig. 8.4 State graph of the Hopfield network shown in Fig. 8.2. The arrows are labeled with the neurons, an update of which causes the corresponding state transition. The two stable states are drawn in gray

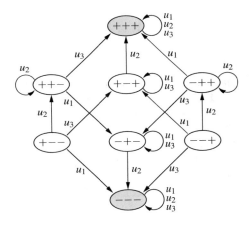

is always reached in these examples. Indeed, one can show that an asynchronous update of the activations cannot lead to oscillations and must lead to a stable state.

Theorem 8.1 (Convergence Theorem for Hopfield Networks) *If the activations of the neurons of a Hopfield network are updated asynchronously, a stable state is reached in a finite number of steps. If the neurons are traversed in an arbitrary, but fixed cyclic fashion, at most $n \cdot 2^n$ steps (updates of individual neurons) are needed, where n is the number of neurons of the network.*

Proof This theorem is proven with a method that one may call (in analogy to Fermat's method of infinite descent) the *method of finite descent*. We define a function that maps every state of a Hopfield network to a real-valued number, which is reduced with every state transition or at least stays the same. This function is commonly called the **energy function** of the Hopfield network. The number this function maps an activation state to is called the **energy** of this state. (The reason for these names stems from the physical interpretation of a Hopfield network, since the energy function corresponds to the Hamilton operator that describes the energy of the magnetic field; see below). By drawing on an additional insight in case of a transition that leaves the state energy unchanged, we can easily show that a state, once it has been left, can never be reached again. Since a Hopfield network possesses only a finite number of possible states, we must reach a situation in which further transitions cannot descend any further and thus a stable state must be reached.

The energy function of a Hopfield network with n neurons $u_1, \ldots, u_n$ is

$$E = -\frac{1}{2} \text{act}^\top \mathbf{W} \text{act} + \boldsymbol{\theta}^\top \text{act},$$

where the vector $\text{act} = (\text{act}_{u_1}, \ldots, \text{act}_{u_n})^\top$ describes the activation state of the network, $\mathbf{W}$ is the weight matrix of the Hopfield network and $\boldsymbol{\theta} = (\theta_{u_1}, \ldots, \theta_{u_n})^\top$ collects the thresholds of the neurons in a vector. If it is spelled out with individual connection weights, this energy function reads

$$E = -\frac{1}{2} \sum_{u,v \in U, u \neq v} w_{uv} \, \text{act}_u \, \text{act}_v + \sum_{u \in U} \theta_u \, \text{act}_u .$$

In this form, we can easily understand the reason for the factor $\frac{1}{2}$ in front of the first sum. Due to the symmetry of the weights, every term in the first sum occurs twice, which is compensated by the factor $\frac{1}{2}$.

We now show that the energy cannot increase in a state transition caused by updating a neuron. Since the neurons are updated asynchronously, a state transition means updating the activation of only one neuron u. We assume that due to this update it changes its activation from $\text{act}_u^{(\text{old})}$ to $\text{act}_u^{(\text{new})}$. The difference of the energy of the old and the new activation state consists of all terms that contain the activation act_u. All other terms cancel, because they are contained in both the old and the new energy. Hence, we have

$$\Delta E = E^{(\text{new})} - E^{(\text{old})} = \left(- \sum_{v \in U - \{u\}} w_{uv} \, \text{act}_u^{(\text{new})} \, \text{act}_v + \theta_u \, \text{act}_u^{(\text{new})} \right)$$
$$- \left(- \sum_{v \in U - \{u\}} w_{uv} \, \text{act}_u^{(\text{old})} \, \text{act}_v + \theta_u \, \text{act}_u^{(\text{old})} \right).$$

The factor $\frac{1}{2}$ vanishes because of the symmetry of the weights, due to which every term of the sum occurs twice. From the above sums we can extract the new and the old activation of the neuron u and thus reach

$$\Delta E = \left(\text{act}_u^{(\text{old})} - \text{act}_u^{(\text{new})} \right) \underbrace{\left(\sum_{v \in U - \{u\}} w_{uv} \, \text{act}_v - \theta_u \right)}_{= \, \text{net}_u}.$$

We now have to distinguish two cases. If $\text{net}_u < \theta_u$, then the second factor is less than 0. In addition, it is $\text{act}_u^{(\text{new})} = -1$, and since we assumed that the activation changed due to the update, we know $\text{act}_u^{(\text{old})} = +1$. Therefore the first factor is greater than 0 and hence $\Delta E < 0$. However, if $\text{net}_u \geq \theta_u$, then the second factor is no less than 0. In addition, we have $\text{act}_u^{(\text{new})} = +1$, which implies $\text{act}_u^{(\text{old})} = -1$. It follows that the first factor is less than 0 and hence $\Delta E \leq 0$.

If a state transition reduced the energy, the original state cannot be reached anymore, because this would require an (impossible) energy increase. However, the second case allows for transitions that keep the energy constant. Hence, we have to exclude the possibility of cycles of states having the same energy. Fortunately, such a transition increases the number of +1 activations (because it can occur only if the activation of the updated neuron changed from −1 to +1, see above). Therefore, the original state cannot be reached again in this situation either. As a consequence, every state transition reduces the number of reachable states, and since there are only finitely many states, we must finally reach a stable state.

It should be noted that the additional criterion (number of +1 activations) is not needed if the activation function is defined as stated on p. 168 as an alternative, that is, if the old activation is maintained if the network input coincides with the threshold. In this case the activation only changes from −1 to +1 if $\text{net}_u > \theta_u$. As a consequence, we obtain for the second case $\Delta E < 0$ as well and therefore it is sufficient to study the behavior of the energy of the Hopfield network.

We remark here also that it is only guaranteed that the computations converge to a state of (locally) minimal energy if no neurons are excluded from updates of their activation after a certain point in time. Otherwise we could update, for instance, the same neuron over and over again, which does not cause the current activation state to be left. That no neuron is excluded from an activation update is certain if the neurons are updated in an arbitrary, but fixed cyclic order. In this case, we can derive the following corollary: either a traversal of all neurons did not change any activation. Then we obviously reached a stable state. Or at least one activation changed. Then (at least) one of the 2^n possible activation states (n neurons, each with two possible activations) has been excluded, since the old state cannot be reached again as we proved above. Therefore, a stable state must be reached after at most 2^n traversals of the neurons, that is, after at most $n \cdot 2^n$ updates of neuron activations.

The energy function we introduced in the proof of the above theorem plays an important role in the following. Therefore, but also to illustrate the above theorem, we consider it for the example of the simple Hopfield network shown in Fig. 8.2. The energy function of this network is

$$E = -\operatorname{act}_{u_1} \operatorname{act}_{u_2} - 2\operatorname{act}_{u_1} \operatorname{act}_{u_3} - \operatorname{act}_{u_2} \operatorname{act}_{u_3} .$$

If we arrange the activation states of this network (cf. Fig. 8.4) according to their energy, discarding the (self-)loops and the edge labels for reasons of clarity, we obtain Fig. 8.5, in which the two stable states are clearly visible as the two states with the lowest energy. Note that there are no transitions from a state lower in the diagram to a state higher up in the diagram, which would correspond to an increase in energy. Note also that all transitions between states having the same energy increase the number of +1 activations. This illustrates the arguments of the above proof.

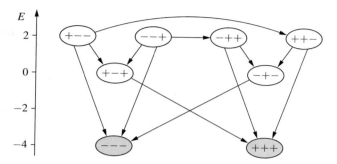

Fig. 8.5 (Simplified) state graph of the Hopfield network shown in Fig. 8.2, in which the states are arranged according to their energy. The two stable states are shown in gray

Fig. 8.6 A Hopfield network with three neurons and non-vanishing thresholds

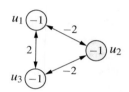

It is, of course, not necessary that the state graph is as highly symmetric as this, even if the network exhibits a lot of strong symmetries. As an example, we consider the Hopfield network shown in Fig. 8.6. Even though this network has essentially the same symmetry structure as the one shown in Fig. 8.3, it possesses, due to the non-vanishing thresholds, a very different state graph. Here, we only study the form in which the states are arranged according to the energy function

$$E = 2\,\mathrm{act}_{u_1}\,\mathrm{act}_{u_2} - 2\,\mathrm{act}_{u_1}\,\mathrm{act}_{u_3} + 2\,\mathrm{act}_{u_2}\,\mathrm{act}_{u_3} - \mathrm{act}_{u_1} - \mathrm{act}_{u_2} - \mathrm{act}_{u_3}$$

of this network. This state graph is shown in Fig. 8.7. Note that the asymmetries of this graph are basically the effect of the non-vanishing thresholds.

To conclude this section we remark that the energy function of a Hopfield network establishes the connection to physics, which we already mentioned at the beginning of this chapter. In physics, Hopfield networks are used as (microscopic) models of magnetism, based on the relations between physical and neural notions that are shown in Table 8.3. More specifically, a Hopfield network corresponds to the so-called Ising model of Magnetism [15]. This physical analogy also provides (another) reason why the activation function of the neurons of a Hopfield network is sometimes defined in such a way that the neuron does not change its activation if its network input equals its threshold (see p. 168): if the effects of the external magnetic field and the magnetic coupling of the atoms cancel, the atom should maintain its magnetic moment.

Fig. 8.7 (Simplified) state graph of the Hopfield network shown in Fig. 8.6, in which the states are arranged according to their energy. The two stable states are drawn in gray. Note that the energy scale is broken between −1 and −7 and thus the bottom state lies actually much further down

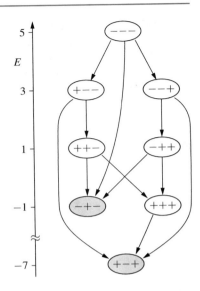

Table 8.3 Physical interpretation of a Hopfield network as a (microscopic) model of magnetism (Ising model, [15])

Physical	Neural
Atom	Neuron
Magnetic moment (spin)	Activation state
Strength of the external magnetic field	Threshold
Magnetic coupling of the atoms	Connection weights
Hamilton operator of the magnetic field	Energy function

8.3 Associative Memory

Hopfield networks are very well suited to implement so-called **associative memory**, that is, a kind of memory that is addressed by its contents. If a pattern is presented to an associative memory, the memory returns whether this pattern coincides with one of the stored patterns. This coincidence need not be exact. An associative memory may also return a stored pattern that is as similar as possible to the presented pattern. In this way "noisy" input patterns may be recognized as well.

Hopfield networks are employed as associative memory by exploiting their stable states, of which one is eventually reached in the work phase. If we determine the weights and the thresholds of a Hopfield network in such a way that the patterns to store are exactly the stable states, the normal update procedure of a Hopfield network finds for any input pattern a similar stored pattern. In this way, "noisy" patterns may be corrected or disturbed patterns may still be recognized.

In order to simplify the following computations, we start by considering how a single pattern $\mathbf{p} = (\text{act}_{u_1}, \ldots, \text{act}_{u_n})^\top \in \{-1, 1\}^n, n \geq 2$, can be stored in a Hopfield network. To this end we have to determine the weights and the thresholds in such a way that the pattern becomes a stable state (also: an attractor) of the Hopfield network. Therefore we need to ensure

$$S(\mathbf{W}\mathbf{p} - \boldsymbol{\theta}) = \mathbf{p},$$

where $\mathbf{W}$ is the weight matrix of the Hopfield network, $\boldsymbol{\theta} = (\theta_{u_1}, \ldots, \theta_{u_n})^\top$ is the threshold vector and S is a function

$$S : \mathbb{R}^n \rightarrow \{-1, 1\}^n,$$
$$\mathbf{x} \mapsto \mathbf{y},$$

where the vector $\mathbf{y}$ is determined by

$$\forall i \in \{1, \ldots, n\} : \quad y_i = \begin{cases} 1 & \text{if } x_i \geq 0, \\ -1 & \text{otherwise.} \end{cases}$$

That is, the function S is a kind of element-wise threshold function.

If we set $\boldsymbol{\theta} = \mathbf{0}$, that is, if we choose all thresholds to be 0, a suitable matrix $\mathbf{W}$ can easily be found, since it obviously suffices if

$$\mathbf{W}\mathbf{p} = c\mathbf{p} \quad \text{with } c \in \mathbb{R}^+$$

holds. Algebraically: we have to find a matrix $\mathbf{W}$ that possesses w.r.t. $\mathbf{p}$ a positive eigenvalue c.[1] We choose now

$$\mathbf{W} = \mathbf{p}\mathbf{p}^\top - \mathbf{1}$$

with the $n \times n$ unit matrix $\mathbf{1}$. The term $\mathbf{p}\mathbf{p}^\top$ is the so-called **outer product** (or **matrix product**) of the vector $\mathbf{p}$ with itself. It yields a symmetric $n \times n$ matrix. The unit matrix $\mathbf{1}$ has to be subtracted from this matrix in order to ensure that the diagonal of the weight matrix becomes 0, since there are no (self-)loops in a Hopfield network. With this matrix $\mathbf{W}$ we have for the pattern $\mathbf{p}$:

$$\mathbf{W}\mathbf{p} = (\mathbf{p}\mathbf{p}^\top)\mathbf{p} - \underbrace{\mathbf{1}\mathbf{p}}_{=\mathbf{p}} \overset{(*)}{=} \mathbf{p} \underbrace{(\mathbf{p}^\top\mathbf{p})}_{=|\mathbf{p}|^2=n} - \mathbf{p}$$
$$= n\mathbf{p} - \mathbf{p} = (n-1)\mathbf{p}.$$

The equality $(*)$ holds, because matrix and vector multiplications are associative and thus we may change the order of the operations (in other words: may place the parentheses in the expressions differently). In this form, we first have to compute the **inner product** (or **scalar product**) of the vector $\mathbf{p}$ with itself. This yields its squared length. Since we know that $\mathbf{p} \in \{-1, 1\}^n$, we derive that $\mathbf{p}^\top\mathbf{p} = |\mathbf{p}|^2 = n$. Because we assumed $n \geq 2$, it is $c = (n - 1) > 0$, as required. Therefore, the pattern $\mathbf{p}$ is a stable state of the Hopfield network.

---

[1] In linear algebra one usually studies the inverse problem, that is, given a matrix, one tries to find the eigenvalues and eigenvectors.

If we write the computations with individual weights, we obtain

$$w_{uv} = \begin{cases} 0 \text{ if } u = v, \\ 1 \text{ if } u \neq v, \text{act}_u^{(p)} = \text{act}_v^{(p)}, \\ -1 \text{ otherwise.} \end{cases}$$

This rule is also known as the **Hebbian learning rule** [7]. Originally, this rule was derived from a biological analogy: the connection between two neurons that are synchronously active is strengthened ("cells that fire together, wire together").

Note, however, that with this method the pattern $-\mathbf{p}$, which is complementary to the pattern $\mathbf{p}$, becomes a stable state as well. The reason is that with

$$\mathbf{Wp} = (n - 1)\mathbf{p} \quad \text{we also have} \quad \mathbf{W}(-\mathbf{p}) = (n - 1)(-\mathbf{p}).$$

Unfortunately, it is impossible to avoid that the complementary pattern is also stored.

If multiple patterns $\mathbf{p}_1, \ldots, \mathbf{p}_m, m < n$, are to be stored, we determine a matrix $\mathbf{W}_i$ for each pattern $\mathbf{p}_i$ in the way described above. The weight matrix $\mathbf{W}$ is then computed as the sum of these matrices, that is,

$$\mathbf{W} = \sum_{i=1}^m \mathbf{W}_i = \left(\sum_{i=1}^m \mathbf{p}_i \mathbf{p}_i^\top \right) - m\mathbf{1}.$$

If the patterns to be stored are pairwise orthogonal (that is, if the corresponding vectors are perpendicular to each other, or formally, if their scalar product vanishes), we have with this matrix $\mathbf{W}$ for an arbitrary pattern $\mathbf{p}_j, j \in \{1, \ldots, m\}$:

$$\mathbf{Wp}_j = \sum_{i=1}^m \mathbf{W}_i \mathbf{p}_j = \left(\sum_{i=1}^m (\mathbf{p}_i \mathbf{p}_i^\top)\mathbf{p}_j \right) - m \underbrace{\mathbf{1}\mathbf{p}_j}_{=\mathbf{p}_j}$$

$$= \left(\sum_{i=1}^m \mathbf{p}_i (\mathbf{p}_i^\top \mathbf{p}_j) \right) - m\mathbf{p}_j$$

Since we assumed that the patterns are pairwise orthogonal, it is

$$\mathbf{p}_i^\top \mathbf{p}_j = \begin{cases} 0 \text{ if } i \neq j, \\ n \text{ if } i = j, \end{cases}$$

because the scalar product of orthogonal vectors vanishes, while the scalar product of a vector with itself yields the squared length of the vector. This length equals n, because $\mathbf{p}_j \in \{-1, 1\}^n$ (see above). Therefore, we obtain

$$\mathbf{Wp}_j = (n - m)\mathbf{p}_j,$$

and hence $\mathbf{p}_j$ is a stable state of the Hopfield network if $m < n$. Note that in this case the pattern $-\mathbf{p}_j$, which is complementary to the pattern $\mathbf{p}_j$ is again a stable state as well, since due to

$$\mathbf{Wp}_j = (n - m)\mathbf{p}_j \quad \text{we also have} \quad \mathbf{W}(-\mathbf{p}_j) = (n - m)(-\mathbf{p}_j).$$

Although we can choose n pairwise orthogonal vectors in an n-dimensional space, this method allows to store only $n - 1$ patterns (and their complements), because we have to satisfy $n - m > 0$ (see above). Compared to the number of possible states (2^n, since we have n neurons with two possible states each) the storage capacity of a Hopfield network is fairly small.

If the patterns are not pairwise orthogonal, as it is often the case in practice, we have for an arbitrary pattern $\mathbf{p}_j$, $j \in \{1, \dots, m\}$:

$$\mathbf{W}\mathbf{p}_j = (n - m)\mathbf{p}_j + \underbrace{\sum_{\substack{i=1 \\ i \neq j}}^{m} \mathbf{p}_i(\mathbf{p}_i^\top \mathbf{p}_j)}_{\text{"disturbance term"}} \quad .$$

Now the state $\mathbf{p}_j$ may be stable nevertheless, namely if $n - m > 0$ and the "disturbance term" is sufficiently small. This is the case if the patterns $\mathbf{p}_i$ are "approximately" orthogonal, because then the scalar products $\mathbf{p}_i^\top \mathbf{p}_j$ are small. Clearly, the larger the number of patterns to store, the smaller this disturbance term must be, because a growing m means a reducing $n - m$, which makes the state more "susceptible" to disturbances. Hence, in practice, the theoretical limit for the storage capacity of a Hopfield network is never reached.

To illustrate the discussed method, we determine the weight matrix of a Hopfield network with four neurons that stores the two patterns $\mathbf{p}_1 = (+1, +1, -1, -1)^\top$ and $\mathbf{p}_2 = (-1, +1, -1, +1)^\top$. It is

$$\mathbf{W} = \mathbf{W}_1 + \mathbf{W}_2 = \mathbf{p}_1\mathbf{p}_1^\top + \mathbf{p}_2\mathbf{p}_2^\top - 2 \cdot \mathbf{1}$$

with the individual matrices

$$\mathbf{W}_1 = \begin{pmatrix} 0 & 1 & -1 & -1 \\ 1 & 0 & -1 & -1 \\ -1 & -1 & 0 & 1 \\ -1 & -1 & 1 & 0 \end{pmatrix}, \quad \mathbf{W}_2 = \begin{pmatrix} 0 & -1 & 1 & -1 \\ -1 & 0 & -1 & 1 \\ 1 & -1 & 0 & -1 \\ -1 & 1 & -1 & 0 \end{pmatrix}.$$

The weight matrix of the Hopfield network thus reads

$$\mathbf{W} = \begin{pmatrix} 0 & 0 & 0 & -2 \\ 0 & 0 & -2 & 0 \\ 0 & -2 & 0 & 0 \\ -2 & 0 & 0 & 0 \end{pmatrix}.$$

It is easy to check that with this matrix we have

$$\mathbf{W}\mathbf{p}_1 = (+2, +2, -2, -2)^\top \quad \text{and} \quad \mathbf{W}\mathbf{p}_1 = (-2, +2, -2, +2)^\top.$$

Therefore, both patterns are indeed stable states. However, their complements—that is, the patterns $-\mathbf{p}_1 = (-1, -1, +1, +1)$ and $-\mathbf{p}_2 = (+1, -1, +1, -1)$—are stable states as well, as an analogous computation shows.

Another possibility to determine the parameters of a Hopfield network consists in mapping the network to a (single) threshold logic unit, which is then trained with the delta rule [19]. This approach works as follows: if a pattern $\mathbf{p} = (\text{act}_{u_1}^{(p)}, \dots, \text{act}_{u_n}^{(p)}) \in$

$\{-1, 1\}^n$ is to be a stable state of a Hopfield network, the following n equations must hold

$$s(0 \qquad\qquad + w_{u_1 u_2}\, \text{act}_{u_2}^{(p)} + \ldots + w_{u_1 u_n}\, \text{act}_{u_n}^{(p)} - \theta_{u_1}) = \text{act}_{u_1}^{(p)},$$
$$s(w_{u_2 u_1}\, \text{act}_{u_1}^{(p)} + 0 \qquad\qquad + \ldots + w_{u_2 u_n}\, \text{act}_{u_n}^{(p)} - \theta_{u_2}) = \text{act}_{u_2}^{(p)},$$
$$\vdots \qquad\qquad \vdots \qquad\qquad \vdots \qquad\qquad \vdots \qquad\quad \vdots$$
$$s(w_{u_n u_1}\, \text{act}_{u_1}^{(p)} + w_{u_n u_2}\, \text{act}_{u_2}^{(p)} + \ldots + 0 \qquad\quad - \theta_{u_n}) = \text{act}_{u_n}^{(p)}.$$

Here, s is the usual threshold function

$$s(x) = \begin{cases} 1 & \text{if } x \geq 0, \\ -1 & \text{otherwise.} \end{cases}$$

For training we transform the weight matrix into a weight vector, namely by traversing the rows of the upper triangle of the matrix (excluding the diagonal; the lower triangle is neglected, because the weights are symmetric). This weight vector is then extended by appending the thresholds:

$$\mathbf{w} = (\ w_{u_1 u_2},\ w_{u_1 u_3},\ \ldots,\ w_{u_1 u_n},$$
$$w_{u_2 u_3},\ \ldots,\ w_{u_2 u_n},$$
$$\ddots \quad \vdots$$
$$w_{u_{n-1} u_n},$$
$$-\theta_{u_1},\ -\theta_{u_2},\ \ldots,\ -\theta_{u_n}\quad).$$

For this weight vector, we can find input vectors $\mathbf{z}_1, \ldots, \mathbf{z}_n$ such that the arguments of the threshold function appearing in the above equations can be written as scalar products $\mathbf{w}^\top \mathbf{z}_i$. For example, we may choose

$$\mathbf{z}_2 = (\text{act}_{u_1}^{(p)}, \underbrace{0, \ldots, 0}_{n-2\text{zeros}}, \text{act}_{u_3}^{(p)}, \ldots, \text{act}_{u_n}^{(p)}, \ldots 0, 1, \underbrace{0, \ldots, 0}_{n-2\text{zeros}}).$$

In this way, we can map training a Hopfield network to training a threshold logic unit with the threshold 0 and the weight vector $\mathbf{w}$ for the training patterns $l_i = (\mathbf{z}_i, \text{act}_{u_i}^{(p)})$. This threshold logic unit may now be trained, for instance, with the delta rule (cf. Sect. 3.5). If multiple patterns are to be stored, we obtain more input patterns $\mathbf{z}_i$. Note, though, that this option is more of theoretical interest than of practical value.

To illustrate how a Hopfield network can be used to recognize patterns, we consider a simple number recognition example (derived from [6]). The patterns (two-dimensional black and white pictures) shown in Fig. 8.8 are stored in a Hopfield network with $10 \times 12 = 120$ neurons, encoding a dark field as $+1$ and a white one as -1. The resulting patterns are not exactly, but sufficiently orthogonal, so that they can be turned into stable states of a Hopfield network with the method described above. If a pattern is presented to the Hopfield network determined in this way, the computations of the network reconstruct one of these stored patterns, as shown in Fig. 8.8 on the right. Note, however, that there are multiple update steps between two consecutive diagrams in this figure.

In order to better understand this example, the web page

http://www.borgelt.net/hfnd.html

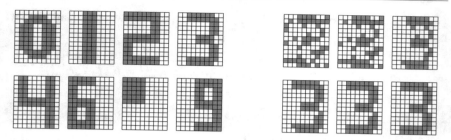

Fig. 8.8 Example patterns that are stored in a Hopfield network (left) and the reconstruction of a pattern from disturbed input (right)

provides the programs whopf (for Microsoft Windows™) and xhopf (for Linux). With them one can store two-dimensional patterns in a Hopfield network and then retrieve them again. The patterns shown in Fig. 8.8 are available in a file.

However, with these programs some of the problems of the discussed method become obvious as well. We already know that this method also stores the complementary patterns. As a consequence, they may be produced as the result of the computations. Besides these patterns there are additional stable states, which differ only marginally from those stored. Among other things, these problems result from the fact that the patterns are not exactly orthogonal.

8.4   Solving Optimization Problems

By exploiting their energy function, Hopfield networks can be used to solve optimization problems. The core idea is as follows: by updating a Hopfield network a (local) minimum of its energy function is reached. If we can rewrite a given function to optimize in such a way that it can be interpreted as an energy function (to minimize) of a Hopfield network, we can construct a Hopfield network by reading the weights and thresholds from its terms. Then this Hopfield network is initialized randomly—that is, it is placed into a random activation state—and the update computations are carried out as usual. The network eventually reaches a stable state, which corresponds to a minimum of the energy function and thus an optimum of the function to optimize. Note, however, that this optimum may only be a local one.

This principle is obviously very simple. The only difficulty we face is that, if we try to solve an optimization problem, we often have to respect certain additional constraints. For example, it may be that the arguments of the function to optimize must not leave certain ranges of values. In such a case it is not sufficient to simply turn the function to optimize into an energy function of a Hopfield network, but we also have to take precautions that the additional constraints are respected, so that the solution found with the help of a Hopfield network is actually valid.

In order to incorporate the additional constraints, we proceed in essentially the same way as for optimizing the objective function. For each additional constraint we construct a function, which is minimized if the constraint is respected and then transform this function into an energy function of a Hopfield network. Finally, we combine the energy function that describes the objective function and the energy functions that result from the additional constraints, exploiting the following lemma:

Lemma 8.1 *Suppose we have two Hopfield networks defined on the same set U of neurons with weights $w_{uv}^{(i)}$, thresholds $\theta_u^{(i)}$ and energy functions*

$$E_i = -\frac{1}{2} \sum_{u \in U} \sum_{v \in U - \{u\}} w_{uv}^{(i)}\, act_u\, act_v + \sum_{u \in U} \theta_u^{(i)}\, act_u$$

for $i = 1, 2$. (The index i states which of the two networks the quantities refer to.) Furthermore, let $a, b \in \mathbb{R}$. Then $E = aE_1 + bE_2$ is the energy function of the Hopfield network with neurons in U that has the weights $w_{uv} = aw_{uv}^{(1)} + bw_{uv}^{(2)}$ and the thresholds $\theta_u = a\theta_u^{(1)} + b\theta_u^{(2)}$.

This lemma allows us to construct the energy function that is to be optimized by a Hopfield network as a linear combination of several energy function. Proving this lemma is trivial (simply calculate $E = aE_1 + bE_2$ and compare the terms on both sides of the equation). Hence, we do not spell out the proof in detail here.

As an example for the described procedure, we consider how the well-known traveling salesman problem (TSP) could be solved (approximately) with the help of a Hopfield network. This problem consists in the task to find for a traveling salesman the best tour through a given set of n cities, so that each city is visited exactly once. In order to solve this problem with the help of a Hopfield network, we use the activations 0 and 1 for the neurons, because this simplifies setting up the energy functions. That we may deviate from the original definition in this way, because we can always transform the weights and thresholds into those needed for a Hopfield network with activations 1 and -1, is shown in Sect. 10.3, in which the needed transformation formulas are derived.

A tour through n given cities is encoded as follows: we set up a binary $n \times n$ matrix $\mathbf{M} = (m_{ij})$, the columns of which correspond to the cities and the rows of which correspond to the steps of the tour. We write a 1 into row i and column j of this matrix ($m_{ij} = 1$) if the city j is visited in the ith step of the tour. Otherwise we write a 0 into this element ($m_{ij} = 0$). For example, the matrix shown in Fig. 8.9 on the right describes the tour through the cities 1 to 4 shown in the same figure on the left. Note that a cyclic permutation of the steps (rows) describes the same tour, since we did not define any city as the one to start from.

The Hopfield network we have to construct possesses one neuron for each element of this $n \times n$ matrix, which we denote by the coordinates (i, j) of the corresponding matrix element and whose activation corresponds to the value of this matrix element. After the update process has reached a stable state we can thus read the found tour

Fig. 8.9 A tour through four cities and a 4×4 matrix representing it

from the activations of the neurons. Note that in the following we always use index i to refer to steps and index j to refer to cities.

With the help of the matrix $\mathbf{M}$ we can describe the objective function that has to be minimized in order to solve the traveling salesman problem as

$$E_1 = \sum_{j_1=1}^{n} \sum_{j_2=1}^{n} \sum_{i=1}^{n} d_{j_1 j_2} \cdot m_{i j_1} \cdot m_{(i \bmod n)+1, j_2}.$$

Here $d_{j_1 j_2}$ is the distance between city j_1 and city j_2. With the two factors that refer to the matrix $\mathbf{M}$ we ensure that only distances between cities are summed that are visited consecutively on the tour, that is, if the city j_1 is visited in the ith step and the city j_2 is visited in the $((i \bmod n) + 1)$th step of the tour. Only in this case both matrix elements are 1. On the other hand, if the cities are not visited consecutively in the tour, at least one of the matrix elements and thus the term is 0.

Following the plan laid down above, we now have to transform the function E_1 in such a way that it takes the form of the energy function of a Hopfield network with the neurons (i, j), where the matrix elements m_{ij} play the role of the activations of the neurons. In order to do so, we have to introduce a second sum over the steps of the tour (index i). We achieve this by using two indices i_1 and i_2 for the steps in which the cities j_1 and j_2 are visited. In addition, we ensure by an additional factor that only such terms are formed in which these two indices are related in the desired way (that is, i_2 follows i_1). Thus we obtain

$$E_1 = \sum_{(i_1, j_1) \in \{1, \ldots, n\}^2} \sum_{(i_2, j_2) \in \{1, \ldots, n\}^2} d_{j_1 j_2} \cdot \delta_{(i_1 \bmod n)+1, i_2} \cdot m_{i_1 j_1} \cdot m_{i_2 j_2},$$

where δ_{ab} is the so-called *Kronecker symbol* ($\delta_{ab} = 1$ if $a = b$ and $\delta_{ab} = 0$ otherwise).

All that is still missing now to reach the form of an energy function is the factor $-\frac{1}{2}$ in front of the sums. This factor can easily be introduced, for example, by moving a factor -2 into the sums. However, it seems to be more appropriate to move only a factor of -1 into the sums and to obtain the factor 2 by making the factor with the Kronecker symbol symmetric. Clearly, it is irrelevant whether i_2 follows i_1 or vice versa: both cases describe the same relation between the cities (only the tour is reversed). If we allow for both orders, every distance between two cities is automatically considered twice. Thus we finally arrive at

$$E_1 = -\frac{1}{2} \sum_{\substack{(i_1,j_1)\in\{1,\dots,n\}^2 \\ (i_2,j_2)\in\{1,\dots,n\}^2}} -d_{j_1 j_2} \cdot (\delta_{(i_1 \bmod n)+1,i_2} + \delta_{i_1,(i_2 \bmod n)+1}) \cdot m_{i_1 j_1} \cdot m_{i_2 j_2}.$$

This function has the form of the energy function of a Hopfield network. However, we cannot use it directly, because it is obviously minimized if and only if all $m_{ij} = 0$, regardless of the distances between the cities. Indeed, we have to respect two constraints when minimizing the above function, namely

- Each city is visited in exactly one step of the tour:

$$\forall j \in \{1,\dots,n\}: \quad \sum_{i=1}^{n} m_{ij} = 1,$$

  that is, every column of the matrix contains exactly one 1.
- In each step of the tour exactly one city is visited:

$$\forall i \in \{1,\dots,n\}: \quad \sum_{j=1}^{n} m_{ij} = 1,$$

  that is, every row of the matrix contains exactly one 1.

These two constraints exclude the trivial solution (all $m_{ij} = 0$). Since these constraints have the same structure, we demonstrate in detail only how the first is turned into an energy function. Clearly, the first constraint is satisfied if and only if

$$E_2^* = \sum_{j=1}^{n} \left(\sum_{i=1}^{n} m_{ij} - 1 \right)^2 = 0.$$

Since E_2^* cannot be negative (due to the squared terms), the first constraint is satisfied if and only if E_2^* is minimized. Simply computing the square yields

$$E_2^* = \sum_{j=1}^{n} \left(\left(\sum_{i=1}^{n} m_{ij} \right)^2 - 2 \sum_{i=1}^{n} m_{ij} + 1 \right)$$

$$= \sum_{j=1}^{n} \left(\left(\sum_{i_1=1}^{n} m_{i_1 j} \right) \left(\sum_{i_2=1}^{n} m_{i_2 j} \right) - 2 \sum_{i=1}^{n} m_{ij} + 1 \right)$$

$$= \sum_{j=1}^{n} \sum_{i_1=1}^{n} \sum_{i_2=1}^{n} m_{i_1 j} m_{i_2 j} - 2 \sum_{j=1}^{n} \sum_{i=1}^{n} m_{ij} + n.$$

The constant term n can be neglected, because it does not change the location of the minimum. In order to obtain the form of an energy function, we merely have to duplicate the sum over the cities (index j), using the same principle that we already applied to derive the objective function E_1. This leads to

$$E_2 = \sum_{(i_1,j_1)\in\{1,\dots,n\}^2} \sum_{(i_2,j_2)\in\{1,\dots,n\}^2} \delta_{j_1 j_2} \cdot m_{i_1 j_1} \cdot m_{i_2 j_2} - 2 \sum_{(i,j)\in\{1,\dots,n\}^2} m_{ij}.$$

By moving the factors -2 into both sums we finally arrive at

$$E_2 = -\frac{1}{2} \sum_{\substack{(i_1,j_1)\in\{1,\dots,n\}^2 \\ (i_2,j_2)\in\{1,\dots,n\}^2}} -2\delta_{j_1 j_2} \cdot m_{i_1 j_1} \cdot m_{i_2 j_2} + \sum_{(i,j)\in\{1,\dots,n\}^2} -2m_{ij}$$

and thus the form of the energy function of a Hopfield network. In a completely analogous fashion we obtain for the second constraint

$$E_3 = -\frac{1}{2} \sum_{\substack{(i_1,j_1)\in\{1,\dots,n\}^2 \\ (i_2,j_2)\in\{1,\dots,n\}^2}} -2\delta_{i_1 i_2} \cdot m_{i_1 j_1} \cdot m_{i_2 j_2} + \sum_{(i,j)\in\{1,\dots,n\}^2} -2m_{ij}.$$

Finally we combine the three energy functions E_1 (objective function), E_2 (first constraint) and E_3 (second constraint) to obtain the total energy function

$$E = aE_1 + bE_2 + cE_3.$$

The factors $a, b, c \in \mathbb{R}^+$ have to be chosen in such a way that it is not possible to reduce the value of the energy function by violating the constraints. This is certainly the case if

$$\frac{b}{a} = \frac{c}{a} > 2 \max_{(j_1,j_2)\in\{1,\dots,n\}^2} d_{j_1 j_2},$$

that is, if the maximum improvement achievable by a (local) modification of the tour is less than the minimum degradation that results from a violation of a constraint.

Since the matrix elements m_{ij} correspond to the activations $\text{act}_{(i,j)}$ of the neurons (i, j) of the Hopfield network, we read from the total energy function E the following weights and thresholds:

$$w_{(i_1,j_1)(i_2,j_2)} = \underbrace{-ad_{j_1 j_2} \cdot (\delta_{(i_1 \bmod n)+1,i_2} + \delta_{i_1,(i_2 \bmod n)+1})}_{\text{from } E_1} \underbrace{-2b\delta_{j_1 j_2}}_{\text{from } E_2} \underbrace{-2c\delta_{i_1 i_2}}_{\text{from } E_3},$$

$$\theta_{(i,j)} = \underbrace{0a}_{\text{from } E_1} \underbrace{-2b}_{\text{from } E_2} \underbrace{-2c}_{\text{from } E_3} = -2(b+c).$$

The resulting Hopfield network is now initialized randomly (that is, the activations of the neurons are set to randomly chosen values from $\{0, 1\}$) and then these activations are repeatedly updated until a stable state is reached. The solution—that is, the found tour—can then be read from this state.

Note, however, that the presented approach to solve the traveling salesman problem is, despite its plausibility, of very limited use for practical purposes. One of the main problems is that the Hopfield network is unable to switch from a found tour to another with a lower total length. The reason is that transforming a matrix that represents a tour into another matrix that represents a different tour requires that at least four neurons (matrix elements) change their activations. (For example, if the steps, in which two cities are visited, are exchanged, two neurons must change their activation from 1 to 0 and two others must change their activations from 0 to 1.) However, each of these changes, executed individually, violates at least one of the two constraints and thus leads to an increase in energy. Only all four changes together can result in

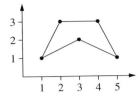

Fig. 8.10   A very simple traveling sales man problem with 5 cities and its solution

a smaller energy, but cannot be executed together due to the asynchronous update. Therefore the normal activation updates can never change an already found tour into another, even if this requires only a marginal change of the tour. As a consequence, it is highly likely that the Hopfield network gets stuck in a local minimum of the energy function. Although it can never be made absolutely sure that the process does not get stuck in a local minimum, this problem is particularly annoying here, since it occurs even in situations in which the modifications, which are necessary to improve the tour, are, so to say, "obvious" (for example, exchanging one city with another or reversing a sub-tour).

Unfortunately, the situation is actually even worse. Although we introduced the energy functions E_1 and E_2 to represent the constraints that describe a valid tour, it is not guaranteed that the resulting stable state actually represents a valid tour. The reason is that there are also situations, in which a matrix that does *not* describe a valid tour can be changed into a matrix that *does* describe a valid tour only through intermediate states that possess a higher energy. For example, if a column of the matrix contains two ones (and thus the first constraint is violated), but these two ones are the only ones in their respective columns, the violation of the first constraint can only be resolved by, at least temporarily, violating the second constraint. Since both constraints are equivalent, the activations remain unchanged.

The above considerations can be checked with the program tsp.c, which is available on the web page for this book. This program tries to solve the very simple traveling salesman problem with 5 cities that is shown in Fig. 8.10 with the help of a Hopfield network. The found solution is not always a valid tour and even if it is valid, it appears to be chosen entirely at random.

Therefore, it cannot really be recommended to use a Hopfield network to solve the traveling salesman problem. We considered this problem nevertheless, because it allowed us to explain the procedure of setting up the energy functions in a very clear and straightforward way. Note, however, that the problems encountered here have to be taken into account also if Hopfield networks are applied to other optimization problems, which may render Hopfield networks a sub-optimal choice.

A certain improvement is possible if one does not use **discrete Hopfield networks** with only two possible activations per neuron as we studied them up to now, but rather **continuous Hopfield networks**, in which the activation of a neuron can be any number from $[-1, 1]$ (or $[0, 1]$, respectively). This transition corresponds roughly to the generalization of the activation function that led us from threshold logic units to multi-layer perceptrons (see Fig. 5.2 on p. 55). With continuous Hopfield networks, which also have the advantage that they are better suited for being implemented in hardware with the help of an (analog) electrical circuit, solving the traveling salesman problem is more successful [14].

8.5 Simulated Annealing

The problems discussed in the preceding section, which are encountered if we try to use Hopfield networks to solve optimization problems, are mainly caused by the possibility that the process gets stuck in a local minimum of the energy function. Since the same problem is—not surprisingly—also encountered with other optimization methods, the idea suggests itself to transfer approaches to handle this problem, which have been developed for other optimization methods, to Hopfield networks. One such approach is so-called **simulated annealing**.

The idea of simulated annealing [16,18] is to start with a randomly generated candidate solution of the optimization problem and to evaluate it. In every later step, the current candidate solution is modified (slightly) and re-evaluated. If the new solution is better than the old, it is accepted and replaces the old solution. However, if it is worse, it is accepted only with a certain probability that depends on how much worse the new solution is. In addition, this probability is reduced over time, so that in the limit new candidate solutions are only accepted if they are actually better than the old solution. Furthermore, the best solution found so far is usually recorded in parallel.

The reason why a worse candidate solution is sometimes accepted is that the method would otherwise be very similar to a gradient descent. The only difference would be that the descent direction is not determined by computing a gradient, but is found by trial and error. However, we saw in Chap. 5 that a gradient descent can easily get stuck in a local minimum (see Fig. 5.24 on p. 75). However, if solutions that are actually worse than the current candidate are accepted at least sometimes, this undesired behavior can be counteracted to some degree. Intuitively, it allows to overcome "barriers" (regions of the search space in which candidate solutions have low quality) that separate local minima from the global minimum. Only later, when the probability for accepting candidate solutions that are worse has been decreased, the objective function is locally optimized.

The name "simulated annealing" for this approach originates from the fact that it is very similar to the physical minimization of the lattice energy of the atoms if a heated piece of metal is cooled slowly. This process is usually called "annealing" and serves the purpose to make a metal easier to work or to machine by relieving tensions and correcting lattice malformations. Seen from a physical perspective, the thermal energy of the atoms prevents them from settling in a configuration that is only a local minimum of the lattice energy. They rather "jump out" of this configuration. However, the "deeper" the (local) energy minimum is, the harder it is for the atoms to abandon the configuration. Therefore it is likely that they finally settle in a configuration of

Fig. 8.11 Simulated
annealing can be illustrated
by a ball rolling around on a
curved surface. Here Ω
denotes the search space
(parameter space) and f is
the function to minimize

very low lattice energy, the optimum of which, in the case of a metal, is a mono-crystalline structure.

It should be clear, though, that it cannot be guaranteed that the global minimum of the lattice energy is reached. Especially, if the metal is not heated long enough or is cooled down too quickly, it is likely that the atoms settle in a configuration that is only a local minimum of the energy (which in the case of the metal is a polycrystalline structure). Hence it is important that the temperature is reduced slowly, so that the probability that local minima are abandoned again is sufficiently large.

This energy minimization can also be illustrated by a ball that rolls around on a curved surface (see Fig. 8.11 for a sketch). In this case, the function to minimize is the potential energy of the ball. In the beginning the ball is endowed with a certain kinetic energy, which enables it to roll up some slopes of the surface. In the course of time, however, friction reduces the kinetic energy of the ball more and more, so that it finally comes to a rest in a valley of the surface (a minimum of the potential energy). Since a larger kinetic energy is needed to escape from a deep valley than to escape from a shallow one, it is likely that the point at which the ball finally stops is located in a fairly deep valley, possibly even in the deepest one available (the global minimum). Although this particular process contains little randomness (as simulated annealing usually does), it is still a useful way of envisioning the method.

The reduction of the thermal energy of the atoms in the simulated annealing process or the reduction of the kinetic energy of the ball in the illustration is modeled by the reducing probability for accepting a worse solution. Often an explicit temperature parameter is introduced, with the help of which the probability is computed. Since the probability distribution over the velocities of atoms is often an exponential distribution (for example, the Maxwell-Boltzmann distribution, which describes the velocity distribution for an ideal gas [5]), a function like

$$P(\text{accept the solution}) = \frac{1}{c} e^{-\frac{\Delta Q}{kT}}$$

is a common choice to compute the probability of accepting a solution that is worse than the current one. ΔQ is the quality difference between the current and the new candidate solution, T the temperature parameter that is reduced in the course of time, c is a normalization constant and k Boltzmann's constant (see also the next section).

Applying simulated annealing to Hopfield networks is very simple: after the activations have been initialized randomly, the neurons of the Hopfield network are traversed (for example, in some random order) and it is determined whether an activation change leads to a reduction of the network energy or not. An activation change that reduces the network energy is always accepted (in the normal update process, only such changes occur, see above). However, if an activation change increases the network energy, it is accepted with a probability that is computed with the formula stated above. Note that in this case we have simply

$$\Delta Q = \Delta E = |\, net_u - \theta_u|$$

(cf. the proof of Theorem 8.1 on p. 171).

8.6 Boltzmann Machines

A network model that is closely related to Hopfield networks is the so-called **Boltzmann machine** [1,8]. Special variants of this model, which we will consider below, gained considerable interest in recent years, especially in the field of deep learning (cf. Sect. 5.12), for learning probability distributions. A standard Boltzmann machine differs from a Hopfield network mainly in how the states of the neurons are updated (details are discussed below). Like Hopfield networks for solving optimization problems (cf. Sect. 8.4) it relies on the fact that one can define an **energy function** that assigns a numeric value (an *energy*) to each state of the network. With the help of this energy function a probability distribution over the states of the network is defined based on the **Boltzmann distribution** (also known as **Gibbs distribution**) of statistical mechanics, namely

$$P(\mathbf{s}) = \frac{1}{c}\, e^{-\frac{E(\mathbf{s})}{kT}},$$

where the vector $\mathbf{s}$ describes the (discrete) state of the system, c is a normalization constant (ensuring that the sum of the probabilities of all possible states equals 1), E is the function that yields the energy of a state $\mathbf{s}$, T is the thermodynamic temperature of the system and k is Boltzmann's constant ($k \approx 1.38 \cdot 10^{-23}$ J/K, that is, the unit is Joule (energy) divided by Kelvin (temperature)).

For Boltzmann machines (which only employ a physical analogy, but usually do not have an actual physical interpretation) the product kT is often replaced by merely T, thus combining the temperature and Boltzmann's constant into a single formal parameter. Furthermore, the state $\mathbf{s}$ consists of the vector act of the activations of the neurons and the energy function is (cf. p. 172)

$$E(\text{act}) = -\frac{1}{2}\text{act}^\top \mathbf{W}\text{act} + \boldsymbol{\theta}^\top \text{act},$$

where $\mathbf{W}$ is the matrix of connection weights and θ the vector of threshold values.

In analogy to the discussion on p. 172f, we now look at a single neuron u and work out the change in energy that is brought about by this neuron changing its state. That is, we consider the (absolute) difference in energy between $\text{act}_u = 0$ and $\text{act}_u = 1$, while all other neurons keep their activations. This difference is

$$\Delta E_u = E_{\text{act}_u=1} - E_{\text{act}_u=0} = \sum_{v \in U-\{u\}} w_{uv}\, \text{act}_v - \theta_u$$

(cf. p. 172). Writing the energies in terms of the Boltzmann distribution yields

$$\Delta E_u = -kT \ln(P(\text{act}_u = 1)) - (-kT \ln(P(\text{act}_u = 0))),$$

which can easily be rewritten as

$$\frac{\Delta E_u}{kT} = \ln(P(\text{act}_u = 1)) - \ln(P(\text{act}_u = 0))$$
$$= \ln(P(\text{act}_u = 1)) - \ln(1 - P(\text{act}_u = 1)),$$

since the probability of the unit u having activation 0 and the probability of it having activation 1 must sum to 1 (as there are only these two possible states). Solving this equation for $P(\text{act}_u = 1)$ finally yields

$$P(\text{act}_u = 1) = \frac{1}{1 + e^{-\frac{\Delta E_u}{kT}}},$$

that is, the probability of a neuron being active is a logistic function (cf. Fig. 5.2 on p. 55) of the (scaled) energy difference between its active and inactive state. Since the energy difference is closely related to the network input of the neuron, namely as (see also above)

$$\Delta E_u = \sum_{v \in U - \{u\}} w_{uv}\, \text{act}_v - \theta_u = \text{net}_u - \theta_u,$$

this formula suggests a stochastic update procedure for the network, which derives additional plausibility from the fact that for multi-layer perceptrons we used the logistic function as a generalization of the step function used in threshold logic units.

The update proceeds as follows: a neuron u of a given Boltzmann machine is chosen. Its network input, from it the energy difference ΔE_u and finally (via the logistic function) the probability of it having activation 1 is computed. The neuron is set to activation 1 with this probability and to activation 0 with the complement probability. Technically, this can easily be achieved by sampling a random number x from a uniform distribution on the unit interval $[0, 1]$ (for which a function is available in practically all programming languages) and comparing it to the computed probability $P(\text{act}_u = 1)$. If $x \leq P(\text{act}_u = 1)$, then $\text{act}_u^{(\text{new})} = 1$, otherwise $\text{act}_u^{(\text{new})} = 0$.

This update is repeated many times for randomly chosen neurons. In addition, simulated annealing (cf. the preceding section) is carried out by slowly lowering the temperature T over time. After this update process, which can also be described as a **Markov Chain Monte Carlo** (**MCMC** or $(\text{MC})^2$) procedure (see, for example, [3] for an introduction), has been carried out for sufficiently many steps, the probability that the network is in a specific activation state depends only on the energy of that state and has become independent of the specific initial activation state the update process was started with. This final state is also referred to as **thermal equilibrium**. As a consequence, the Boltzmann machine can be seen as a representation of and be used as a sampling mechanism for the Boltzmann distribution that is defined by its connection weights and threshold values.

Based on these considerations the idea suggests itself to develop a training procedure with which the probability distribution that a Boltzmann machine represents via its energy function can be adapted to a given sample of data points, in order to obtain a probabilistic model of the data. Of course, this can only work reasonably well if the data points are actually a sample from a Boltzmann distribution. Otherwise the model cannot, in principle, be made to fit the sample data well.

In order to mitigate this restriction, a deviation from the structure of Hopfield networks, as we studied them at the beginning of this chapter, is introduced. While in Hopfield networks all neurons are input as well as output neurons (and thus *visible* neurons, as their activations are all visible to the environment), the neurons of a Boltzmann machine are divided into *visible neurons*, which receive the data points as input, and *hidden neurons*, the activations of which are not fixed by the data points and thus allow for a more flexible adaptation to the sample data.

The objective of training a Boltzmann machine is to adapt the connection weights and threshold values in such a way that the true distribution underlying a given data sample is approximated, as well as possible, by the probability distribution represented by the Boltzmann machine. A natural approach to achieve this objective is to start from a measure that describes the difference between two probability distributions and then to carry out a gradient descent in order to minimize this difference measure. Among the best known such measures is the so-called **Kullback–Leibler information divergence** [17], which for two probability distributions p_1 and p_2 that are defined over the same sample space Ω reads (cf. also Definition 4.5 on p. 50, where it was considered in a different context)

$$I_{\text{KLdiv}}(p_1, p_2) = \sum_{\omega \in \Omega} p_1(\omega) \ln \frac{p_1(\omega)}{p_2(\omega)}.$$

Applied to Boltzmann machines, the probability distributions refer to the data sample (p_1) and to the visible neurons of the Boltzmann machine (p_2). On the resulting function a gradient descent is then carried out in order to minimize it.

Without formal derivation, we only state here the final training procedure: In each training step the Boltzmann machine is ran twice. In the first run, which is also referred to as the "positive" phase, the visible neurons are fixed to a data point that is randomly chosen from the data sample to fit and only the hidden neurons are updated according to the procedure described above. The update is carried out until thermal equilibrium is reached. In the second run, which is also referred to as the "negative phase", all units, both visible and hidden, are updated until thermal equilibrium. In the two phases statistics about individual neurons and pairs of neurons (both visible and hidden) being activated (simultaneously) are collected: let p_u^+ be the probability that neuron u is active in the positive phase, p_u^- the probability that neuron u is active in the negative phase, p_{uv}^+ the probability that the neurons u and v are both active simultaneously in the positive phase, and p_{uv}^- be the probability that the neurons u and v are both active simultaneously in the negative phase. (All of these probabilities are estimated from the relative frequency with which the corresponding situation was observed in the respective phase.) Then, the update can be performed according to the following two equations [1]:

$$\Delta w_{uv} = \frac{1}{\eta}(p_{uv}^+ - p_{uv}^-) \quad \text{and} \quad \Delta \theta_u = -\frac{1}{\eta}(p_u^+ - p_u^-).$$

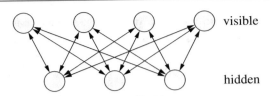

visible

hidden

Fig. 8.12 A restricted Boltzmann machine has a bipartite graph structure, in which the two groups of visible and hidden units are fully connected to each other, but there are no connections within each group

Intuitively: if a neuron is more often active when a training example is presented (i.e., visible units are fixed) than when the network is allowed to run freely (i.e., visible units are updated), the probability of the neuron being active is too low, so the threshold should be reduced. Similarly, if two neurons are more often active together when a training example is presented than when the network is allowed to run freely, the connection weight between them should be increased, so that they become more likely to be active together. Note that this training method is very similar to the **Hebbian learning rule** [7] as we mentioned it in Sect. 8.3. Derived from a biological analogy it says: the connection between two neurons that are synchronously active is strengthened ("cells that fire together, wire together").

Unfortunately, training general Boltzmann machines like we described them up to now is impractical unless the networks are very small. The main reason for this is the fact that the larger the network, the more update steps need to be carried out in order to obtain sufficiently reliable statistics for the neuron activation (pairs) needed in the update formulas, which is due to the larger number of parameters and the resulting slower convergence of the network. As a consequence, general Boltzmann machines, though theoretically very powerful devices, are rarely used in practice.

However, a special variant of the Boltzmann machine, which was originally introduced under the name **Harmonium** [21], but nowadays is generally referred to as **restricted Boltzmann machine** [4,9], has achieved considerable attention in recent years, after efficient learning procedures based on **contrastive divergence** had been developed [9–11,22]. The restriction referred to in the name of this network model consists in revoking the condition that the graph underlying the network must be a fully connected graph (as we introduced it originally for Hopfield networks, cf. Definition 8.1 on p. 167). Rather, one uses a bipartite graph, in which the vertices are split into two groups, namely the visible and the hidden neurons. Connections only exist between neurons from different groups, but not between neurons from the same group (see Fig. 8.12).

The advantage of a restricted Boltzmann machine is that, due to the lack of connections within the visible units and within the hidden units, training can proceed by repeating the following three steps (a detailed discussion can be found in [9] and particularly in the practical guide [11]): in the first step the visible units are fixed to a randomly chosen training example from the given data set and the hidden units are updated once and in parallel according to the sampling procedure described above (that is, activate a unit according to the value of the logistic function of the energy difference). Let $\mathbf{x}$ be the training sample to which the visible neurons are fixed and let $\mathbf{y}$ be the vector of activations of the hidden neurons that have been sampled with the visible neurons fixed to $\mathbf{x}$. The outer product $\mathbf{x}\mathbf{y}^\top$ is called the *positive gradient* for the weight matrix.

In the second step, fix the hidden neurons to the computed vector $\mathbf{y}$ and update the visible units once and in parallel according to the sampling procedure described above (that is, activate a unit according to the value of the logistic function of the energy difference). This produces a "reconstruction" $\mathbf{x}^*$ of the training example $\mathbf{x}$. Fix the visible neurons to this "reconstruction" and update the hidden neurons once more, which produces a second activation vector $\mathbf{y}^*$ for the hidden neurons. The outer product $\mathbf{x}^*\mathbf{y}^{*\top}$ is called the *negative gradient* for the weight matrix.

In the third step, update the connection weights with the difference of the positive and the negative gradient, that is, according to (η is a learning rate)

$$\Delta w_{uv} = \eta(x_u y_v - x_u^* y_v^*),$$

where the indices u and v indicate the elements of the activation vectors to be used by the neurons they refer to. The threshold values are updated with the simple differences of the original and the "reconstructed" activation vectors.

While this is the basic training procedure, many improvements have been suggested, including techniques like a momentum term (cf. Sect. 5.8) and using, for the reconstruction of the training example, actual probabilities instead of binary samples. Furthermore, online-like training based on small batches of the given data is advantageous. These and many other recommendations for training restricted Boltzmann machines in practice can be found in [11].

Finally, it is worth mentioning that restricted Boltzmann machines have been used to build deep networks in a fashion similar to stacked auto-encoders as we discussed them in Sect. 5.12 for multi-layer perceptrons. The idea is to train a restricted Boltzmann machine, then to create a data set of hidden neuron activations by sampling from the trained Boltzmann machine, and to build another restricted Boltzmann machine from the obtained data set. This procedure can be repeated several times and the resulting Boltzmann machines can then easily be stacked (like we stacked the encoder parts of the auto-encoders trained in Sect. 5.12). Details about this approach and how the final stack can be fine-tuned with a procedure similar to back-propagation can be found in [10].

References

1. Ackley DH, Hinton GE, Sejnowski TJ (1985) A learning algorithm for boltzmann machines. Cognitive Sci 9(1):147–169. Elsevier Science, Amsterdam, Netherlands
2. Anderson JA, Rosenfeld E (1988) Neurocomputing: foundations of research. MIT Press, Cambridge, MA, USA

3. Andrieu C, De Freitas N, Doucet A, Jordan MI (2003) An introduction to MCMC for machine learning. Mach Learn 50:5–43. Kluwer, Dordrecht, Netherlands
4. Freund Y, Haussler D (1992) Unsupervised learning of distributions on binary vectors using two layer networks. Adv Neural Inf Process Syst 4:912–919. Morgan Kaufmann, San Mateo, CA, USA
5. Greiner W, Neise L, Stöcker H (2000) Thermodynamik und Statistische Mechanik (Series: Theoretische Physik). Verlag Harri Deutsch, Thun, Frankfurt am Main, Germany, (1987) English edition: Thermodynamics and Statistical Physics. Springer-Verlag, Berlin, Germany
6. Haykin S (2008) Neural networks and learning machines. Prentice Hall, Englewood Cliffs, NJ, USA
7. Hebb DO (1949) The Organization of Behaviour. Wiley, New York, NY, USA. Chap. 4: "The First Stage of Perception: Growth of an Assembly" reprinted in [Anderson und Rosenfeld 1988], pp 45–56
8. Hinton G, Sejnowski TJ (1986) Learning and relearning in boltzmann machines. In: [Rumelhart and McClelland 1986], pp 282–317
9. Hinton GE (2002) Training products of experts by minimizing contrastive divergence. Neural Comput 14(8):1771–1800. MIT Press, Cambridge, MA
10. Hinton GE, Osindero S, Teh YW (2006) A fast learning algorithm for deep belief nets. Neural Comput 18(7):1527–1554. MIT Press, Cambridge, MA
11. Hinton GE (2010) A practical guide to training restricted boltzmann machines. Technical Report 2010–003, Department of Computer Science, University of Toronto, Canada
12. Hopfield JJ (1982) Neural networks and physical systems with emergent collective computational abilities. Proc Natl Acad Sci 79:2554–2558. USA
13. Hopfield JJ (1984) Neurons with graded response have collective computational properties like those of two-state neurons. Proc Natl Acad Sci 81:3088–3092. USA
14. Hopfield J, Tank D (1985) "Neural" computation of decisions in optimization problems. Biol Cybern 52:141–152. Springer, Heidelberg, Germany
15. Ising E (1925) Beitrag zur Theorie des Ferromagnetismus. Zeitschrift für Physik 31(253)
16. Kirkpatrick S, Gelatt CD, Vercchi MP (1983) Optimization by simulated annealing. Science 220:671–680. High Wire Press, Stanford, CA, USA
17. Kullback S., Leibler RA (1951) On information and sufficiency. Ann Math Stat 22:79–86. Institute of Mathematical Statistics, Hayward, CA, USA
18. Metropolis N, Rosenblut N, Teller A, Teller E (1953) Equation of state calculations for fast computing machines. J Chem Phys 21:1087–1092. American Institute of Physics, Melville, NY, USA
19. Rojas R (1996) Theorie der neuronalen Netze—Eine systematische Einführung. Springer, Berlin, Germany
20. Rumelhart DE, McClelland JL (eds) (1986) Parallel distributed processing: explorations in the microstructure of cognition. volume 1: foundations. MIT Press, Cambridge, MA, USA
21. Smolensky P (1986) Information processing in dynamical systems: foundations of harmony theory. In: [Rumelhart and McClelland 1986], pp 194–281
22. Tieleman T (2008) Training restricted boltzmann machines using approximations to the likelihood gradient. Proceedings 21st international conference on machine learning (ICML 2008, Helsinki, Finland). ACM Press, New York, NY, USA

Recurrent Networks

9

The Hopfield networks that we discussed in the preceding chapter are special recurrent networks, which have a very constrained structure. In this chapter, however, we lift all restrictions and consider recurrent networks without any constraints. Such general recurrent networks are well suited to represent **differential equations** and to solve them (approximately) in a numerical fashion. If the type of the differential equation is known that describes a given system, but the values of the parameters appearing in it are unknown, one may also try to train a suitable recurrent network with the help of example patterns in order to determine the system parameters.

9.1 Simple Examples

In contrast to all preceding chapters, we do not start with a definition. The reason is that all special types of neural networks we studied before were obtained from the general definition in Chap. 4 by introducing specific restrictions. In this chapter, however, all constraints are lifted. Hence, we turn immediately to examples.

As a first example, we consider the cooling (or warming) of a body with the temperature ϑ_0 that is placed into a medium with the temperature ϑ_a (ambient temperature), which is held constant. Depending on whether the initial temperature of the body is higher or lower than the ambient temperature, the body will dissipate heat to the medium or absorb heat from it until its temperature reaches the ambient temperature ϑ_a. It is plausible that the amount of heat dissipated or absorbed per unit of time—and thus the temperature change—is proportional to the difference between the current temperature $\vartheta(t)$ of the body and the ambient temperature ϑ_a. That is, we assume that

$$\frac{d\vartheta}{dt} = \dot{\vartheta} = -k(\vartheta - \vartheta_a).$$

© Springer Nature Switzerland AG 2022
R. Kruse et al., *Computational Intelligence*, Texts in Computer Science,
https://doi.org/10.1007/978-3-030-42227-1_9

This equation is called **Newton's Cooling Law** [5]. The minus sign in front of the (positive) **cooling constant** k, which depends on the considered body, indicates that the temperature change reduces the temperature difference.

Of course, a differential equation as simple as this can be solved analytically. With standard methods we obtain (see, for example, [5])

$$\vartheta(t) = \vartheta_a + (\vartheta_0 - \vartheta_a)e^{-k(t-t_0)}$$

with the initial temperature $\vartheta_0 = \vartheta(t_0)$ of the body. Here, however, we consider a numerical approximation of the solution, namely with the help of the so-called **Euler–Cauchy polygonal course** [5]. The idea of this method consists of the insight that with the differential equation, we can compute the derivative $\dot{\vartheta}(t)$ of the function $\vartheta(t)$ for arbitrary points in time t, that is, we know the course of the function $\vartheta(t)$ locally. With a given initial value $\vartheta_0 = \vartheta(t_0)$, we can thus compute any value $\vartheta(t)$ *approximately* as follows: we divide the interval $[t_0, t]$ into n parts of equal length $\Delta t = \frac{t-t_0}{n}$. The split points are given as

$$\forall i \in \{0, 1, \ldots, n\}: \qquad t_i = t_0 + i\Delta t.$$

We proceed from the starting point $P_0 = (t_0, \vartheta_0)$ along a *straight line* with the slope $\dot{\vartheta}(t_0)$ given by the differential equation until we reach the time t_1 and thus the point $P_1 = (t_1, \vartheta_1)$. It is

$$\vartheta_1 = \vartheta(t_1) = \vartheta(t_0) + \dot{\vartheta}(t_0)\Delta t = \vartheta_0 - k(\vartheta_0 - \vartheta_a)\Delta t.$$

In this point P_1, the slope of the function is described by the differential equation as $\dot{\vartheta}(t_1)$. Again we proceed on a straight line with this slope until we reach the point $P_2 = (t_2, \vartheta_2)$ for the time t_2. Then we have

$$\vartheta_2 = \vartheta(t_2) = \vartheta(t_1) + \dot{\vartheta}(t_1)\Delta t = \vartheta_1 - k(\vartheta_1 - \vartheta_a)\Delta t.$$

By repeating this procedure, we compute step by step the points $P_k = (t_k, \vartheta_k)$, $k = 1, \ldots, n$, the second coordinate ϑ_k of which can always be found with the recursion

$$\vartheta_i = \vartheta(t_i) = \vartheta(t_{i-1}) + \dot{\vartheta}(t_{i-1})\Delta t = \vartheta_{i-1} - k(\vartheta_{i-1} - \vartheta_a)\Delta t.$$

Finally, we reach the point $P_n = (t_n, \vartheta_n)$ and the desired approximation $\vartheta_n = \vartheta(t_n)$.

Intuitively, the described method approximates the function $\vartheta(t)$ by a *polygonal course*, since we always proceed on a *straight line* from one point to the next (hence the name *Euler-Cauchy polygonal course*). More formally, we can derive the above recursion formula for the values ϑ_i by approximating the differential quotient by a difference quotient, that is, by using

$$\frac{d\vartheta(t)}{dt} \approx \frac{\Delta\vartheta(t)}{\Delta t} = \frac{\vartheta(t + \Delta t) - \vartheta(t)}{\Delta t}$$

with a sufficiently small Δt. In this case, we clearly have

$$\vartheta(t + \Delta t) - \vartheta(t) = \Delta\vartheta(t) \approx -k(\vartheta(t) - \vartheta_a)\Delta t,$$

from which the recursion formula follows directly.

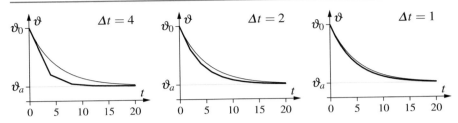

Fig. 9.1 Euler–Cauchy polygonal courses as approximate solutions of Newton's cooling law for different step widths Δt. The thin curve is the exact solution

It should be clear that the accuracy of the computed approximation is the better, the smaller the **step width** Δt is chosen, because then the computed polygonal course will differ less from the actual course of the function $\vartheta(t)$ under consideration. To illustrate this, Fig. 9.1 shows, for an ambient temperature $\vartheta_a = 20$, a cooling constant $k = 0.2$ and the initial values $t_0 = 0$ and $\vartheta_0 = 100$ the exact solution $\vartheta(t) = \vartheta_a + (\vartheta_0 - \vartheta_a)e^{-k(t-t_0)}$ (thin line) as well as its approximation by Euler–Cauchy polygonal courses with step widths $\Delta t = 4, 2, 1$ in the interval $[0, 20]$. Compare the deviation from the exact solution, for example, for $t = 8$ or $t = 12$.

To represent the recursion formula derived above by a recurrent neural network, we merely have to expand the right hand side. Thus, we obtain

$$\vartheta(t + \Delta t) - \vartheta(t) = \Delta\vartheta(t) \approx -k\Delta t\vartheta(t) + k\vartheta_a\Delta t$$

and therefore

$$\vartheta_i \approx \vartheta_{i-1} - k\Delta t\vartheta_{i-1} + k\vartheta_a\Delta t.$$

The form of this equation corresponds exactly to the computations of a single neuron with a feedback loop. As a consequence, we can approximate the function $\vartheta(t)$ with the help of a neural network with a single neuron u with the network input function

$$f_{\mathrm{net}}^{(u)}(w, x) = \underbrace{-k\Delta t}_{=w}\, x$$

and the activation function (assuming an internal feedback of the activation)

$$f_{\mathrm{act}}^{(u)}(\mathrm{net}_u, \mathrm{act}_u, \theta_u) = \mathrm{act}_u + \mathrm{net}_u - \theta_u$$

with $\theta_u = -k\vartheta_a\Delta t$. This network is shown in Fig. 9.2, where—as usual—the bias value θ_u is written into the neuron.

Note that this network actually contains two feedback loops, namely (1) the feedback that is shown explicitly, which describes the temperature change as a function of the current temperature, and (2) the implicit feedback which results from the fact that the current activation of the neuron u is a parameter of its activation function. Because of this second feedback loop, the network input is not used to recompute

Fig. 9.2 A recurrent neural network with a single neuron for Newton's cooling law

$$\vartheta(t_0) \longrightarrow \boxed{-k\vartheta_a\Delta t} \overset{-k\Delta t}{\circlearrowright} \longrightarrow \vartheta(t)$$

Fig. 9.3 A mass on a spring.
Its movement can be
described by a simple
differential equation

the activation of the neuron u from scratch, but only to compute the change of its activation (cf. Fig. 4.2 on page 42 and the accompanying explanations about the general structure of a generalized neuron).

Alternatively, we could have used the network input function

$$f_{\text{net}}^{(u)}(x, w) = (1 - k\Delta t)x$$

(that is, the connection weight $w = 1 - k\Delta t$) and the activation function

$$f_{\text{act}}^{(u)}(\text{net}_u, \theta_u) = \text{net}_u - \theta_u$$

(again with $\theta_u = -k\vartheta_a \Delta t$). In this way, we could have avoided the implicit feedback. However, the first form corresponds better to the structure of the differential equation and thus we prefer it here.

As a second example, we consider a mass on a spring as it is shown in Fig. 9.3. The height $x = 0$ denotes the equilibrium position of the mass m. We assume now that the mass m is lifted by a certain distance $x(t_0) = x_0$ and then dropped (that is, it has the initial velocity $v(t_0) = v_0 = 0$). Since the gravitational force acting on the mass m is the same at all heights x, we can ignore its influence. The spring force is governed by **Hooke's Law** [2,5], according to which the exerted force F is proportional to the length change Δl of the spring and directed opposite to the direction of this change. That is, we have

$$F = c\Delta l = -cx,$$

where c is a constant that depends on the spring. According to **Newton's Second Law** $F = ma = m\ddot{x}$, this force causes an acceleration $a = \ddot{x}$ of the mass m. Therefore, we arrive at the differential equation

$$m\ddot{x} = -cx, \qquad \text{or} \qquad \ddot{x} = -\frac{c}{m}x.$$

Of course, this differential equation can easily be solved analytically. The general solution, which can be obtained with standard methods, is

$$x(t) = a\sin(\omega t) + b\cos(\omega t)$$

with the parameters

$$\omega = \sqrt{\frac{c}{m}}, \qquad \begin{aligned} a &= x(t_0)\sin(\omega t_0) + v(t_0)\cos(\omega t_0), \\ b &= x(t_0)\cos(\omega t_0) - v(t_0)\sin(\omega t_0). \end{aligned}$$

With the given initial values $x(t_0) = x_0$ and $v(t_0) = v_0 = 0$ and the additional definition $t_0 = 0$, we obtain the simple expression

$$x(t) = x_0 \cos\left(\sqrt{\frac{c}{m}}\, t\right).$$

In order to construct a recurrent neural network that approximates this solution numerically, we rewrite the differential equation, which is of second order, as a system of two coupled differential equations of first order. We achieve this by introducing the velocity v of the mass as an intermediary quantity and obtain

$$\dot{x} = v \qquad \text{and} \qquad \dot{v} = -\frac{c}{m}x.$$

Next we approximate, as for Newton's cooling law, the differential quotient by a difference quotient, which leads to

$$\frac{\Delta x}{\Delta t} = \frac{x(t + \Delta t) - x(t)}{\Delta t} = v \quad \text{and} \quad \frac{\Delta v}{\Delta t} = \frac{v(t + \Delta t) - v(t)}{\Delta t} = -\frac{c}{m}x.$$

From these equations, we obtain the recursion formulas

$$x(t_i) = x(t_{i-1}) + \Delta x(t_{i-1}) = x(t_{i-1}) + \Delta t \cdot v(t_{i-1}) \qquad \text{and}$$

$$v(t_i) = v(t_{i-1}) + \Delta v(t_{i-1}) = v(t_{i-1}) - \frac{c}{m}\Delta t \cdot x(t_{i-1}).$$

Now we only have to create a neuron for each of these formulas and to read the connection weights from the formulas. This yields the neural network that is shown in Fig. 9.4. The network input function and the activation function of the top neuron u_1 are

$$f_{\text{net}}^{(u_1)}(v, w_{u_1 u_2}) = w_{u_1 u_2} v = \Delta t\, v \qquad \text{and}$$
$$f_{\text{act}}^{(u_1)}(\text{act}_{u_1}, \text{net}_{u_1}, \theta_{u_1}) = \text{act}_{u_1} + \text{net}_{u_1} - \theta_{u_1}.$$

The corresponding functions of the bottom neuron u_2 are

$$f_{\text{net}}^{(u_2)}(x, w_{u_2 u_1}) = w_{u_2 u_1} x = -\frac{c}{m}\Delta t\, x \qquad \text{and}$$
$$f_{\text{act}}^{(u_2)}(\text{act}_{u_2}, \text{net}_{u_2}, \theta_{u_2}) = \text{act}_{u_2} + \text{net}_{u_2} - \theta_{u_2}.$$

The output function of both neurons is the identity. Obviously, these choices implement exactly the recursion formulas stated above.

Note that this network not only produces approximations of the location $x(t)$ (output of the neuron u_1), but also approximations of the velocity $v(t)$ (output of the neuron u_2). Note also that we may draw on the following consideration [2]: the approximation becomes more accurate if we do not compute the velocity $v(t)$ for the times $t_i = t_0 + i\Delta t$, but for the midpoints of the intervals, that is for the times

Fig. 9.4 Recurrent neural network that computes the movement of a mass on a spring (governed by Hooke's law)

t	v	x
0.0	0.0000	1.0000
0.1	−0.5000	0.9500
0.2	−0.9750	0.8525
0.3	−1.4012	0.7124
0.4	−1.7574	0.5366
0.5	−2.0258	0.3341
0.6	−2.1928	0.1148

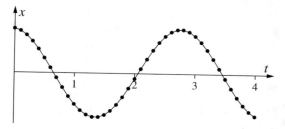

Fig. 9.5 The first computation steps of the neural network shown in Fig. 9.4 and the computed movement of a mass on a spring (dots mark vertices of the polygonal course)

$t_i' = t_0 + i\,\Delta_t + \frac{\Delta t}{2}$. In this case the bottom neuron does not receive $v(t_0)$ as input, but rather $v(t_0 + \frac{\Delta t}{2}) \approx v_0 - \frac{c}{m}\frac{\Delta t}{2}x_0$.

Example computations of the neural network shown in Fig. 9.4 for the parameters $\frac{c}{m} = 5$ and $\Delta t = 0.1$ are shown in the table and the diagram in Fig. 9.5. The table contains in the columns labeled t and x the coordinates of the first seven points of the diagram. The update starts with the neuron u_2, which receives $v(t_0)$ as input.

9.2 Representing Differential Equations

From the examples we discussed in the preceding section, we can derive a simple principle how arbitrary explicit differential equations[1] can be represented by recurrent neural networks: A given explicit differential equation of nth order

$$x^{(n)} = f(t, x, \dot{x}, \ddot{x}, \ldots, x^{(n-1)})$$

($\dot{x}$ denotes the first, $\ddot{x}$ the second and $x^{(i)}$ the ith derivative of x w.r.t. t) is transformed into a system of n coupled differential equation of first order by introducing $n - 1$ intermediary quantities

$$y_1 = \dot{x}, \qquad y_2 = \ddot{x}, \qquad \ldots \qquad y_{n-1} = x^{(n-1)}.$$

This yields the system of differential equations

$$\dot{x} = y_1,$$
$$\dot{y}_1 = y_2,$$
$$\vdots$$
$$\dot{y}_{n-2} = y_{n-1},$$
$$\dot{y}_{n-1} = f(t, x, y_1, y_2, \ldots, y_{n-1}).$$

---

[1] Due to the special operation scheme of neural networks, it is not possible to solve arbitrary differential equations numerically with the help of recurrent networks. It suffices, however, if the differential equation can be solved for the independent variable or for one of the occurring derivatives of the dependent variable. Here we consider, as an example, the special case in which the differential equation can be written with the highest occurring derivative isolated on one side.

In analogy to the examples studied in the preceding section, each differential quotient that occurs in these equations is replaced by a difference quotient. This yields the n recursion formulas

$$
\begin{aligned}
x(t_i) &= x(t_{i-1}) + \Delta t \cdot y_1(t_{i-1}), \\
y_1(t_i) &= y_1(t_{i-1}) + \Delta t \cdot y_2(t_{i-1}), \\
&\ \ \vdots \\
y_{n-2}(t_i) &= y_{n-2}(t_{i-1}) + \Delta t \cdot y_{n-3}(t_{i-1}), \\
y_{n-1}(t_i) &= y_{n-1}(t_{i-1}) + f(t_{i-1}, x(t_{i-1}), y_1(t_{i-1}), \ldots, y_{n-1}(t_{i-1})).
\end{aligned}
$$

For each of these equations, we create a neuron, which extrapolates the quantity on the left-hand side with the help of the right-hand side. If the differential equation depends explicitly on t (and not only indirectly through the quantities x, $\dot{x}$, etc., which depend on t), an additional neuron is necessary, which updates the value of t with the help of the simple formula

$$
t_i = t_{i-1} + \Delta t.
$$

This produces the recurrent neural network that is shown in Fig. 9.6. The bottom neuron advances only the time t by subtracting in each step the bias value $-\Delta t$ from the current activation. The top $n - 1$ neurons have the network input function

$$
f_{\text{net}}^{(u)}(z, w) = wz = \Delta t\, z,
$$

the activation function

$$
f_{\text{act}}^{(u)}(\text{act}_u, \text{net}_u, \theta_u) = \text{act}_u + \text{net}_u - \theta_u
$$

and the identity as their output function. The weights of the connections to the second neuron from the bottom, its bias value as well as its network input, activation, and output function depend on the form of the differential equation. For example, if the

Fig. 9.6 General structure of a recurrent neural network representing an explicit differential equation of nth order. The weights of the feedback loops and the input function of the second neuron from the bottom depend on the form of the differential equation. Of course, this network not only produces approximations of $x(t)$, but also of $\dot{x}(t)$, $\ddot{x}(t)$ etc

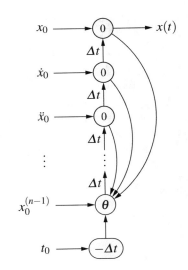

differential equation is linear and has constant coefficients, then the network input function is the weighted sum (as for the neurons of a multi-layer perceptron), the activation function is a linear function, and the output function is the identity.

9.3 Vectorial Neural Networks

Up to now we considered only differential equations of a function $x(t)$. However, in practice, one often meets systems of differential equations with more than one function. A simple example is the differential equations of a two-dimensional movement, for example, of an oblique throw: a (punctiform) body is thrown at time t_0 from the point (x_0, y_0) of a coordinate system with horizontal x-axis and vertical y-axis, namely with the initial velocity $v_0 = v(t_0)$ and with the (upward) angle φ, $0 \leq \varphi \leq \frac{\pi}{2}$, w.r.t. the x-axis (see Fig. 9.7). In this case, we have to compute the functions $x(t)$ and $y(t)$, which describe the position of the body at time t. If we ignore air friction, we have the two equations

$$\ddot{x} = 0 \qquad \text{and} \qquad \ddot{y} = -g,$$

where $g = 9.81\,\mathrm{ms}^{-2}$ is the gravitational acceleration on the surface of the earth. That is, the body moves uniformly in horizontal direction (no acceleration) and in vertical direction accelerated by the gravitational attraction of the earth. In addition, we have the initial conditions $x(t_0) = x_0$, $y(t_0) = y_0$, $\dot{x}(t_0) = v_0 \cos \varphi$ and $\dot{y}(t_0) = v_0 \sin \varphi$. By introducing—according to the general principle described in the preceding section—the intermediary quantities $v_x = \dot{x}$ and $v_y = \dot{y}$, we arrive at the system of differential equations

$$\dot{x} = v_x, \qquad \dot{v}_x = 0,$$
$$\dot{y} = v_y, \qquad \dot{v}_y = -g,$$

from which we can derive the recursion formulas

$$x(t_i) = x(t_{i-1}) + \Delta t\, v_x(t_{i-1}), \qquad v_x(t_i) = v_x(t_{i-1}),$$
$$y(t_i) = y(t_{i-1}) + \Delta t\, v_y(t_{i-1}), \qquad v_y(t_i) = v_y(t_{i-1}) - \Delta t\, g.$$

The result is a recurrent neural network with two independent sub-networks consisting of two neurons each, one of which updates the position coordinate, while the other updates the corresponding velocity.

However, it appears to be more natural to combine the two coordinates x and y into a position vector r of the body. Since the differentiation rules transfer directly

Fig. 9.7 Oblique throw of a body

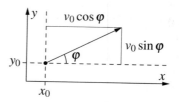

from scalar functions to vector functions (see, for instance, [4]), we may treat the derivatives of this position vector in the same way as those of a scalar quantity. The differential equation, from which we start in this case, is

$$\ddot{r} = -g\,e_y.$$

Here, $e_y = (0, 1)$ is the unit vector in y-direction, with which we describe the direction of the gravitational force. The initial conditions are $r(t_0) = r_0 = (x_0, y_0)$ and $\dot{r}(t_0) = v_0 = (v_0 \cos\varphi, v_0 \sin\varphi)$. As before, we introduce a (now vectorial) intermediary quantity $v = \dot{r}$ to obtain the system

$$\dot{r} = v, \qquad\qquad \dot{v} = -g\,e_y$$

of coupled differential equations. From this system, we read the recursion formulas

$$r(t_i) = r(t_{i-1}) + \Delta t\, v(t_{i-1}),$$
$$v(t_i) = v(t_{i-1}) - \Delta t\, g\,e_y,$$

which can be represented by two vectorial neurons.

The advantages of such a vectorial representation, which may appear fairly small at this point, become obvious if we refine our model by taking air friction into account. If a body moves in a medium (for example, in air), one distinguishes two types of friction: Stokesian friction, which is proportional to the velocity of the body, and Newtonian friction, which is proportional to the squared velocity of the body [4]. If the occurring velocities are low, one usually neglects the Newtonian, if they are high, one neglects the Stokesian friction. As an example, we consider here only Stokesian friction. In this case, the equation

$$a = -\beta v = -\beta\dot{r}$$

describes the deceleration of the body that is caused by air friction, where β is a constant that depends on the shape and the volume of the body. In total, we thus have the differential equation

$$\ddot{r} = -\beta\dot{r} - g\,e_y.$$

With the help of the intermediary quantity $v = \dot{r}$ we obtain

$$\dot{r} = v, \qquad\qquad \dot{v} = -\beta v - g\,e_y,$$

from which we derive the recursion formulas

$$r(t_i) = r(t_{i-1}) + \Delta t\, v(t_{i-1}) \qquad \text{and}$$
$$v(t_i) = v(t_{i-1}) - \Delta t\, \beta\, v(t_{i-1}) - \Delta t\, g\,e_y.$$

The corresponding network is shown in Fig. 9.8. Stokesian friction is taken into account by the feedback loop of the bottom neuron.

An example computation with $v_0 = 8$, $\varphi = 45^o$, $\beta = 1.8$, and $\Delta t = 0.05$ is shown in Fig. 9.9. Note the steeper right branch of the trajectory, which demonstrates the decelerating effect of Stokesian friction. Without this friction, the trajectory would be a parabola (equal slopes on the left and right branch at the same height).

Fig. 9.8 A vectorial recurrent neural network to compute an oblique throw, taking Stokesian friction into account

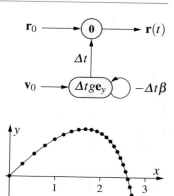

Fig. 9.9 Trajectory of an oblique throw of a body that is computed by the recurrent neural network shown in Fig. 9.8

As a second example, we consider the orbit of a planet [2]. The movement of a planet around a central body (sun) of mass m at the origin of the coordinate system can be described by the vectorial differential equation

$$\ddot{r} = -\gamma\, m\, r\, |r|^{-3},$$

where $\gamma = 6.672 \cdot 10^{-11}\,\mathrm{m^3 kg^{-1} s^{-2}}$ is the gravitational constant. This equation describes the acceleration of the planet that is caused by the mass attraction between the sun and the planet. As before, we introduce the vectorial velocity $v = \dot{r}$ as an intermediary quantity and thus arrive at the system of differential equations

$$\dot{r} = v, \qquad \dot{v} = -\gamma\, m\, r\, |r|^{-3}.$$

From this system, we derive the vectorial recursion formulas

$$r(t_i) = r(t_{i-1}) + \Delta t\, v(t_{i-1}) \quad \text{and}$$
$$v(t_i) = v(t_{i-1}) - \Delta t\, \gamma\, m\, r(t_{i-1}) |r(t_{i-1})|^{-3},$$

which can be represented by two vectorial neurons as shown in Fig. 9.10. Note that, however, in this case the bottom neurons need a somewhat unusual network input function (compared to the previous examples): simply multiplying the output of the top neuron with the connection weight is no longer enough.

An example computation with $\gamma m = 1$, $r_0 = (0.5, 0)$ and $v_0 = (0, 1.63)$ (following an example of [2]) is shown in Fig. 9.11. The diagram nicely shows the elliptical course on which the planet moves faster if it close to the sun (perihelion) than farther

Fig. 9.10 A vectorial recurrent neural network to compute the orbit of a planet

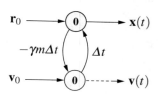

Fig. 9.11 Orbit of a planet that is computed by the recurrent neural network that is shown in Fig. 9.10. The sun is located at the origin of the coordinate system

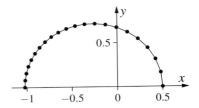

away from the sun (aphelion). This illustrates the content of the first two of **Kepler's Laws**, according to which the orbit of a planet is an ellipse (first law) and a radius from the sun to the planet sweeps out equal areas in equal intervals of time (second law).

9.4   Error Backpropagation Through Time

Computations like those executed in the preceding sections are, of course, only possible if one knows both the differential equation that describes the considered (physical) system as well as the values of the parameters appearing in it. However, in practice we often face the problem that we know the form of the differential equation, but not the values of the parameters appearing in it. If measurement data about the considered system are available, one may try in such a case to find the system parameters by training a recurrent neural network, which represents the differential equation. Since the weights and bias values of the neural network are functions of the system parameters, the actual parameter values can be read from them.

   In principle, recurrent neural networks are trained in the same way as multi-layer perceptrons, namely by error backpropagation (see Figs. 5.5 and 5.6). However, due to the feedback loops, this method cannot be applied directly, since these loops would propagate the error signals in a cyclic fashion. This problem is solved by **unfolding** the network **in time** between two training patterns. This special form of error backpropagation is called **error backpropagation through time**.

   We illustrate here only the basic principle with the help of Newton's cooling law, which we studied in Sect. 9.1 (see page 196). We assume that we have measurement values from the cooling (or warming) of a body at our disposal, which state the temperature of the body at different points in time. In addition, we assume that the temperature ϑ_a of the medium is known, into which the body is placed (ambient temperature). From these measurement values we desire to determine the value of the (unknown) cooling constant k of the body.

   Like for the training of multi-layer perceptrons, the weight and the bias value are initialized randomly. The time between two consecutive measurement values is divided—in analogy to Sect. 9.1—into a certain number of intervals. According to the chosen number of intervals, the feedback loop of the network is then "unfolded." For example, if there are four intervals between one measurement value and the next, that is, if $t_{j+1} = t_j + 4\Delta t$, we obtain the network shown in Fig. 9.12, which

$$\vartheta(t_0) \longrightarrow \bigcirc \xrightarrow{1-k\Delta t} \theta \xrightarrow{1-k\Delta t} \theta \xrightarrow{1-k\Delta t} \theta \xrightarrow{1-k\Delta t} \theta \longrightarrow \vartheta(t)$$

Fig. 9.12 Unfolding in time of the recurrent neural network shown in Fig. 9.2 (four steps)

possesses five neurons. Note that the neurons of this network do not possess feedback loops, neither explicit nor implicit ones. As a consequence, the connection weights are $1 - k\Delta t$: the 1 represents the implicit feedback loop of the network in Fig. 9.2 (cf. the explanations given on page 198).

If a measurement value ϑ_j (temperature of the body at time t_j) is fed into this network, it computes—with the current values of the weight and the bias values—an approximation for the next measurement value ϑ_{j+1} (temperature of the body at time $t_{j+1} = t_j + 4\Delta t$). By comparing this value with the actual value ϑ_{j+1}, we obtain an error signal, which is propagated with the known formulas of error backpropagation and thus causes adaptations of the weights and the bias values.

However, we have to pay attention to the fact that the network shown in Fig. 9.12 actually possesses only one weight and one bias value, since all weights refer to the same feedback loop and all bias values refer to the same neuron. Hence, the derived adaptations have to be aggregated and must only be applied at the end of the procedure to change the one connection weight and the one bias value that actually exists. Furthermore, note that both the weight as well as the bias value contain the unknown parameter k, but only known constants apart from it. Therefore, it is advisable to transform the weight and bias changes computed by error backpropagation into changes of this single free parameter k. Thus, only one quantity needs to be adapted, from which both weight and bias value are then computed.

It should be clear that, in practice, one does not proceed in the described manner for such a simple differential equation as Newton's cooling law. Since this equation can be solved analytically, better and more direct methods are available to determine the value of the unknown parameter k. Starting from the analytical solution of the differential equation, the problem may be solved, for instance, with the regression methods studied in Sect. 5.4: with a suitable transformation of the measurement data the problem is reduced to the task of finding the best fit line (regression line) through the origin, the only parameter of which is the cooling constant k.

However, there are many practical problems for which it makes sense to find unknown system parameters by training a recurrent neural network. In general, this is always the case if the differential equations describing the system under consideration cannot be solved analytically. As an example, we mention here the problem of finding tissue parameters for virtual surgery, especially virtual laparoscopy[2] [10]. The systems of coupled differential equations occurring in this application

---

[2] A laparoscope is a medical instrument with which a physician can examine the abdominal cavity through small incisions. In virtual laparoscopy, an examination of the abdominal cavity is simulated with a computer and a force-feedback device in the shape of a laparoscope in order to teach medical students how to use this instrument properly.

are too complex (mainly due to the high number of equations), so that they cannot be solved analytically. By training recurrent neural networks, however, remarkable results could be achieved.

9.5 Long Short-Term Memory

A very popular structure of a recurrent neural network is the so-called **Long Short-Term Memory (LSTM)** [3,7]. Its development was originally triggered by an observation that was discussed on page 112 in Sect. 5.12 (which deals with deep learning): If gradient descent is used to train the parameters of chained functions (as they appear in deep neural networks) and thus has to rely on the chain rule, parameters further away from the output are trained more slowly due to the often contracting behavior of the (activation) functions, which let the gradient "vanish" the more, the more functions it is backpropagated through [6] (see also [8]).

We meet this problem again, in a possibly even amplified fashion, in recurrent neural networks due to the fact that they are trained by backpropagation through time (see the preceding section), which can lead to very long backpropagation chains. This makes it difficult to train networks so that they capture long-term dependencies: If an event far back in time should have an influence on the current output, this influence may not be captured properly, because the gradient backpropagated to the time at which the event occurred may have become so small that it almost vanishes, making it impossible to adapt the corresponding parameters effectively.

In order to cope with this problem, [7] proposed a structure that allows for a constant (that is, non-decaying) flow of the error signal through an arbitrary number of time steps. The core idea is to view a neuron's activation as a **cell state**, which, in the absence of any network input to the neuron, is simply copied from the cell state in the preceding time step. Formally, the next cell state is computed (in the absence of any network input) by applying the identity function to the previous cell state. This may be seen as applying a special activation function that has a unit derivative. Because of this unit derivative, the error signal cannot decay merely due to the unfolding of the computations in time. [7] refer to this as the *constant error carousel*.

Note that this view is completely analogous to how we viewed, for example, the neuron in Fig. 9.2 on page 197: We assumed an internal feedback connection, so that the previous activation enters the activation function (cf. also the structure of a generalized neuron as shown in Fig. 4.2 on page 42). Hence, the neuron's activation is not recomputed from the network input (alone), but the network input rather describes how the activation should change. Therefore, if there is no network input, the activation (and hence the cell state) remains unchanged. In principle, this allows us to propagate an error signal arbitrarily far back in time.

The difference between an LSTM unit and a standard neuron as it is shown in Fig. 4.2 on page 42 is how the network input influences the activation (cell state). Up to now we merely assumed that there is some activation function that takes the

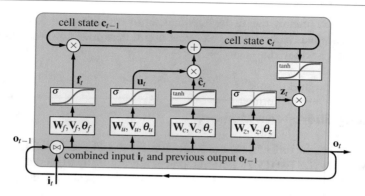

Fig. 9.13 Structure of a Long Short-Term Memory (LSTM) unit

network input and, here, also the previous activation and produces the new activation of the neuron. In an LSTM unit, this computation is described more explicitly, leading to a more complex internal structure, which is shown in Fig. 9.13. Like the units of the simple recurrent networks shown in Figs. 9.8 and 9.10, this unit is vectorial in nature. That is, all quantities, like the input i_t (where the index t refers to the time step) fed into this unit, are real-valued vectors.

The cycle at the top of the unit shows how the previous cell state is reproduced, unless it is modified by the input the unit receives. In addition to this internal feedback connection, there is an explicit external backward connection that feeds the unit's output o_{t-1} of the previous time step back into it, where it is combined with the actual input i_t coming from other neurons (depicted by the circle with a ⋈ in it at the bottom left of the LSTM unit). This combination of current input i_t and previous output o_{t-1} is then fed into multiple so-called **gates**: the **forget gate**, which computes the forget gate activation f_t (this element was not present in the original proposal by [7] and was introduced only later in [3]), the **update gate** (often also referred to as the **input gate**), which computes the update gate activation u_t and the **output gate**, which computes the output gate activation z_t. Furthermore, there is a gate-like structure (which, however, is usually not referred to as a gate) that computes a candidate cell state $\hat{c}_t$. Finally, the output o_t is computed from the cell state and the output gate activation. The output o_t of an LSTM unit is often also referred to (confusingly, as we believe) as the **hidden state** vector and then denoted by h_t.

Formally, the forward propagation of a long short-term memory (LSTM) unit is governed by the following formulae:

$$f_t = \sigma(\mathbf{W}_f i_t + \mathbf{V}_f o_{t-1} + \boldsymbol{\theta}_f) \quad \text{(forget gate activation)},$$
$$u_t = \sigma(\mathbf{W}_u i_t + \mathbf{V}_u o_{t-1} + \boldsymbol{\theta}_u) \quad \text{(update gate activation)},$$
$$\hat{c}_t = \tanh(\mathbf{W}_c i_t + \mathbf{V}_c o_{t-1} + \theta_c) \quad \text{(candidate cell state)},$$
$$z_t = \sigma(\mathbf{W}_z i_t + \mathbf{V}_z o_{t-1} + \boldsymbol{\theta}_z) \quad \text{(output gate activation)},$$
$$c_t = f_t \odot c_{t-1} + u_t \odot \hat{c}_t \quad \text{(new cell state)},$$
$$o_t = z_t \odot \tanh(c_t) \quad \text{(output)},$$

where the index t refers to the time step in which the different quantities are computed and the initial values of the cell state and the output are $c_0 = o_0 = \mathbf{0}$. The function σ denotes the logistic function (see Fig. 5.2 on page 55) and tanh the hyperbolic tangent (see Fig. 5.3 on page 55), both applied element-wise. The operator $\odot$ denotes the element-wise product (also known as the Hadamard product).

More intuitively, the formulas are interpreted like this: The candidate cell state $\hat{c}_t$ is computed like the activation of our standard neurons. The input i_t and the previous output vector o_{t-1} are multiplied with weight matrices $\mathbf{W}_c$ and $\mathbf{V}_c$, respectively, and a bias vector θ_c is added to the result. This yields what we called the network input of our standard neurons, and it is turned into an activation (a candidate cell state) by applying an activation function. The only difference is that here the activation function is a bipolar sigmoid function (hyperbolic tangent, which we also mentioned as a possible activation function in Sect. 5.1).

However, this candidate cell state $\hat{c}_t$ is not simply overwriting the cell state c_t, but is used only to update the cell state in an additive manner (indicated by the circle with a $+$ shown in the middle of the cell state cycle in Fig. 9.13). The severity of the update is controlled by the update gate activation u_t, which may prevent an update (almost) altogether, namely if we have $u_t \approx 0$, since the candidate cell state and the update gate activation are multiplied with each other (indicated by the circle with the $\times$ symbol combining the update gate activation and the candidate cell state) before the result is added to the cell state. Note that for the update gate activation u_t a kind of network input is computed from the input i_t and the previous output vector o_t by multiplying them with weight matrices $\mathbf{W}_u$ and $\mathbf{V}_u$, respectively, and adding a bias vector θ_u. The actual activation is then obtained by applying the logistic function as an activation function. Hence, we may say that the update gate functions roughly like a standard neuron, and so do the other gates.

The output of an LSTM unit is computed from the (new) cell state c_t transformed with a hyperbolic tangent (as a means to normalize it to the range $(-1, 1)$ in all vector elements). However, this does not directly yield the output. Rather, it is modified by multiplying it with the output gate activation z_t, so that the output does not only depend on the current cell state, but its magnitude also on the current input. The output gate activation z_t is again computed in a similar way as the activation of our standard neurons: A kind of network input to this gate is computed from the input i_t and the previous output vector o_{t-1} by multiplying them with weight matrices $\mathbf{W}_z$ and $\mathbf{V}_z$, respectively, and adding a bias vector θ_z. The actual activation is obtained by applying the logistic function as an activation function.

The forget gate functions in the same fashion: A kind of network input is computed from the input i_t and the previous output vector o_{t-1} by multiplying them with weight matrices $\mathbf{W}_f$ and $\mathbf{V}_f$, respectively, and adding a bias vector θ_f. The forget gate activation f_t is again obtained by applying the logistic function as an activation function. The forget gate, which—as already mentioned—was not present in the original proposal by [7], was introduced by [3] to allow a cell to "forget" its cell state. This is achieved by multiplying the forget gate activation f_t with the previous cell state c_{t-1} (indicated by the circle with the $\times$ symbol at the top left in Fig. 9.13), which can potentially turn the cell state into a zero vector $\mathbf{0}$, thus "forgetting" its contents.

Fig. 9.14 Unfolding in time of a Long Short-Term Memory (LSTM) unit

Fig. 9.15 Structure of a
Gated Recurrent Unit (GRU)

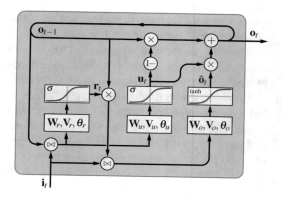

While Fig. 9.13 shows the structure of a single LSTM unit with an internal feed-back connection for the cell state, it is also common to depict such cells unfolded in time. Such a representation is shown in Fig. 9.14. Note that, in this case, there is no internal feedback loop (as it has been unfolded). Rather the different time instances of the cell are connected by passing the cell state and the output of the previous time step to the instance for the next time step. Clearly, this does not change the operation, but merely demonstrates how LSTM units can be trained with backpropagation through time, as it was discussed in the previous section.

Long short-term memory has been very successfully applied to process and forecast time series data, especially, if there are time lags of unknown duration between important events. Since LSTM units are, in principle, able to let an error signal flow arbitrarily far back in time, there is no fundamental reason why long-duration dependencies cannot be captured. In practice, one finds that they can be captured fairly well and significantly better than with standard recurrent neural networks.

A more recent alternative to long short-term memory units is the so-called **gated recurrent unit (GRU)** [1]. It is somewhat simpler than an LSTM unit, since it lacks an output gate and a forget gate (where the latter, however, was also not present in the original proposal by [7] and was only added by [3]) and since it uses the internal cell state directly as the output. The structure of a gated recurrent unit is shown in Fig. 9.15.

Formally, the forward propagation of a gated recurrent unit (GRU) is governed by the following formulae:

$$
\begin{aligned}
\boldsymbol{r}_t &= \sigma(\mathbf{W}_r \boldsymbol{i}_t + \mathbf{V}_r \boldsymbol{o}_{t-1} + \boldsymbol{\theta}_r) && \text{(reset gate activation),}\\
\boldsymbol{u}_t &= \sigma(\mathbf{W}_u \boldsymbol{i}_t + \mathbf{V}_u \boldsymbol{o}_{t-1} + \boldsymbol{\theta}_u) && \text{(update gate activation),}\\
\hat{\boldsymbol{o}}_t &= \tanh(\mathbf{W}_o \boldsymbol{i}_t + \mathbf{V}_o \boldsymbol{o}_{t-1} + \boldsymbol{\theta}_o) && \text{(candidate output vector),}\\
\boldsymbol{o}_t &= (1 - \boldsymbol{u}_t) \odot \boldsymbol{o}_{t-1} + \boldsymbol{u}_t \odot \hat{\boldsymbol{o}}_t && \text{(output),}
\end{aligned}
$$

where he index t refers to the time step in which the different quantities are computed and the initial value of the output is $\boldsymbol{o}_0 = \mathbf{0}$. The function σ denotes the logistic function (see Fig. 5.2 on page 55) and tanh the hyperbolic tangent (see Fig. 5.3 on page 55), both applied element-wise. The operator $\odot$ denotes the element-wise product (also known as the Hadamard product).

More intuitively, a gated recurrent unit computes, similar to an LSTM unit, a candidate cell state, which, since the cell state is directly the output of a gated recurrent unit equals a candidate output. The candidate output $\hat{\boldsymbol{o}}_t$ is computed like the activation of our standard neurons. The input and the previous output vector are multiplied with weight matrices and a bias vector is added to the result. This yields a kind of network input, which is turned into a candidate output by applying an activation function. This is, as in LSTM units, a bipolar sigmoid function (hyperbolic tangent), ensuring an output in $[-1, 1]$. The difference to an LSTM unit is that the input to the candidate output computation is not simply the previous cell state (here: $\boldsymbol{o}_{t-1}$), but this previous state is first multiplied with the reset gate activation $\boldsymbol{r}_t$ (as indicated by the circle with the $\times$ symbol in the middle of Fig. 9.15). In this way, there is, though not forgetting of previous cell states, control of how much the previous cell state should influence the candidate cell state/output.

Another difference is that the candidate output $\hat{\boldsymbol{o}}_t$ (modified by the update gate activation) is not simply added to the old cell state (or here: output) $\boldsymbol{o}_{t-1}$, but that a convex combination of the old output $\boldsymbol{o}_{t-1}$ and the candidate output $\hat{\boldsymbol{o}}_t$ is computed, with combination weights $(1 - \boldsymbol{u}_t)$ (for the previous output) and $\boldsymbol{u}_t$ (for the candidate output). Note that this ensures that the output is always in $[-1, 1]$. In contrast to this, the cell state $\boldsymbol{c}_t$ of LSTM units can be any value, which is then transformed to the interval $[-1, 1]$ only with the help of a hyperbolic tangent (cf. Fig. 9.13).

Gated recurrent units have proved to perform similarly to LSTM units on many tasks and appear to exhibit better performance on certain small data sets.

References

1. Cho K, van Merrienboer B, Gulcehre C, Caglar D, Bahdanau D, Bougares F, Schwenk H, Bengio Y (2014) Representations learning phrase, using RNN encoder-decoder for statistical machine translation. In: Proceedings of conference on empirical methods in natural language processing (EMNLP, (2014) Doha, Qatar). Association for Computational Linguistics, Stroudsberg, PA, USA, pp 1724–1734

2. Feynman RP, Leighton RB, Sands M (1963) The feynman lectures on physics, vol. 1: mechanics, radiation, and heat. Addison-Wesley, Reading, MA, USA
3. Gers FA, Schmidhuber J, Cummins F (2000) Learning to forget: continual prediction with LSTM. Neural Computation 12(10):2451–2471. MIT Press, Cambridge, MA
4. Greiner W, Teil M (1989) 1 (Series: Theoretische Physik) Verlag Harri Deutsch, Thun, Frankfurt am Main, Germany. English edition: Classical Mechanics. Springer, Berlin, Germany
5. Heuser H (1989) Gewöhnliche Differentialgleichungen. Teubner, Stuttgart, Germany
6. Hochreiter S (1991) Untersuchungen zu dynamischen neuronalen Netzen. Diploma thesis, Institut für Informatik, Technische Universit"at M"unchen, Germany
7. Hochreiter S, Schmidhuber J (1997) Long short-term memory. Neural Comput 9(8):1735–1780. MIT Press, Cambridge, MA
8. Hochreiter S, Bengio Y, Frasconi P, Schmidhuber J (2001) Gradient flow in recurrent nets: the difficulty of learning long-term dependencies. In: [Kremer and Kolen 2001]
9. Kremer SC, Kolen JF (eds) (2001) A Field Guide to Dynamical Recurrent Neural Networks. IEEE Press, Piscataway, NJ, USA
10. Radetzky A, Nürnberger A (2002) Visualization and simulation techniques for surgical simulators using actual patient's data. Artif Intell Med 26:3, 255–279. Elsevier Science, Amsterdam, Netherlands

Neural Networks: Mathematical Remarks

10

The following sections treat mathematical topics that were presupposed in the text (Sect. 10.1 on straight line equations and Sect. 10.2 on regression), or side remarks, which would have disturbed the flow of the exposition (Sect. 10.3 on activation transformation in a Hopfield network).

10.1 Equations for Straight Lines

In this section a few important facts about straight lines and their equations have been collected, which are used in Chap. 3 on threshold logic units. More extensive explanations can be found in any textbook on linear algebra.

Straight lines are commonly described in one of the following forms:

$$\begin{aligned}
\text{explicit form:} \quad & g \equiv x_2 = bx_1 + c \\
\text{implicit form:} \quad & g \equiv a_1x_1 + a_2x_2 + d = 0 \\
\text{point-direction form:} \quad & g \equiv \boldsymbol{x} = \boldsymbol{p} + k\boldsymbol{r} \\
\text{normal form:} \quad & g \equiv (\boldsymbol{x} - \boldsymbol{p})\boldsymbol{n} = 0
\end{aligned}$$

with the parameters

$b :$ slope of the line
$c :$ intercept (x_2-value at $x_1 = 0$)
$\boldsymbol{p} :$ position vector of a point of the line (support vector)
$\boldsymbol{r} :$ direction vector of the line
$\boldsymbol{n} :$ normal vector of the line.

It is a disadvantage of the explicit form that straight lines that are parallel to the x_2-axis cannot be represented. All other forms can represent arbitrary lines.

© Springer Nature Switzerland AG 2022
R. Kruse et al., *Computational Intelligence*, Texts in Computer Science,
https://doi.org/10.1007/978-3-030-42227-1_10

Fig. 10.1 A straight line and the parameters describing it

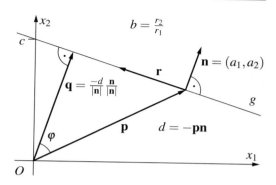

The implicit form and the normal form are closely related to each other, because the coefficients a_1 and a_2 of the variables x_1 and x_2, respectively, are the coordinates of a normal vector of the line. That is, we may use $\mathbf{n} = (a_1, a_2)$ in the normal form. Expanding the normal form also shows that $d = -\mathbf{pn}$.

The relations between the parameters of the different forms of stating a straight line are shown in Fig. 10.1. Particularly important is the vector $\mathbf{q}$, which provides an interpretation for the parameter d of the implicit form. The vector $\mathbf{q}$ is obtained by projecting the support vector $\mathbf{p}$ onto the direction orthogonal to the straight line. This is achieved with the scalar product. It is

$$\mathbf{pn} = |\mathbf{p}\,||\mathbf{n}| \cos \varphi.$$

From the diagram we see that $|\mathbf{q}\,| = |\mathbf{p}\,| \cos \varphi$. Therefore we have

$$|\mathbf{q}\,| = \frac{|\mathbf{pn}|}{|\mathbf{n}|} = \frac{|d|}{|\mathbf{n}|}.$$

Hence $|d|$ measures the distance of the straight line from the origin of the coordinate system relative to the length of the normal vector. If $\sqrt{a_1^2 + a_2^2} = 1$, that is, if the normal vector has unit length, then $|d|$ measures this distance directly. In this case the normal form is called the **Hessian normal form** of the line equation.

If one takes into account that $\mathbf{pn}$ becomes negative if $\mathbf{n}$ does not point away from the origin (as in the diagram), but towards it, one finally obtains

$$\mathbf{q} = \frac{\mathbf{pn}}{|\mathbf{n}|} \frac{\mathbf{n}}{|\mathbf{n}|} = \frac{-d}{|\mathbf{n}|} \frac{\mathbf{n}}{|\mathbf{n}|}.$$

Note that $\mathbf{q}$ always points from the origin to the straight line, regardless of whether $\mathbf{n}$ points towards the origin or away from it. Therefore we can read the location of the origin from the sign of d:

$d = 0$: The straight line contains the origin,
$d < 0$: $\mathbf{n} = (a_1, a_2)$ points away from the origin,
$d > 0$: $\mathbf{n} = (a_1, a_2)$ points towards the origin.

Fig. 10.2 Determining the side of straight line on which a point x lies

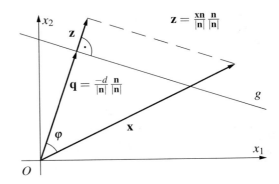

Of course, we can carry out these computations not only for a support vector p of the straight line, but for an arbitrary vector x (see Fig. 10.2). Thus we obtain a vector z that is the projection of the vector x onto the direction orthogonal to the line. By comparing this vector to the vector q considered above, we can determine on which side of the straight line the point lies that has the position vector x:

A point with position vector x lies on that side of the straight line to which the normal vector n points if $xn > -d$, and on the other side if $xn < -d$. If $xn = -d$, the point lies on the straight line.

It should be clear that these considerations are not restricted to straight lines, but can be transferred immediately to planes and hyperplanes. Thus we can easily determine for them as well on which side a point with given position vector lies.

10.2 Regression

This section recalls the **method of least squares**, also known as **regression**, which is well known in calculus and statistics. It is used to determine best fit straight lines (regression lines) and generally best fit polynomials (regression polynomials).

(Physical) measurement data rarely show the exact relationship of the measured quantities as it is described by physical laws, since they are inevitably afflicted by errors. If one wants to determine the relationship of the quantities nevertheless (at least approximately), one faces the task to find a function that fits the measurement points as well as possible, so that the measurement errors are somehow "balanced." Naturally, in order to achieve this, we should have a hypothesis about the type of relationship, so that we can choose a function class and thus reduce the problem to the selection of the parameters of a function of a specific type.

For example, if we expect two quantities x and y to exhibit a linear dependence (for instance, because a scatter plot of the measurement points suggests such a relationship), we have to determine the parameters a and b of the straight line $y = g(x) = a + bx$. However, due to the inevitable measurement errors it will generally not be possible to find a straight line in such a way that all n given measurement points (x_i, y_i), $1 \le i \le n$ lie exactly on this straight line. Rather we have to try to find a straight line that deviates from the measurement points as little as possible. Therefore it is plausible to determine the parameters a and b in such a way that the sum of the squared differences

$$F(a, b) = \sum_{i=1}^{n} (g(x_i) - y_i)^2 = \sum_{i=1}^{n} (a + bx_i - y_i)^2$$

is minimized. That is, the y-values that can be computed from the line equation should deviate (in total) as little as possible from the measured values. The reasons for choosing the squared deviations are basically the same as those given in Sect. 4.4: in the first place using squares makes the error functions continuously differentiable everywhere. In contrast to this, the derivative of the absolute value, which would be an obvious alternative, does not exist/is not continuous at 0. Secondly, squaring the deviations weights large deviations more heavily than small ones, so that there is a tendency to avoid individual large deviations from the measured data.[1]

A necessary condition for a minimum of the error function $F(a, b)$ defined above is that the partial derivatives of this function w.r.t. the parameters a and b vanish:

$$\frac{\partial F}{\partial a} = \sum_{i=1}^{n} 2(a + bx_i - y_i) = 0 \quad \text{and}$$

$$\frac{\partial F}{\partial b} = \sum_{i=1}^{n} 2(a + bx_i - y_i)x_i = 0.$$

From these equations we obtain, after a few simple transformations, the so-called **normal equations**

$$na + \left(\sum_{i=1}^{n} x_i \right) b = \sum_{i=1}^{n} y_i$$

$$\left(\sum_{i=1}^{n} x_i \right) a + \left(\sum_{i=1}^{n} x_i^2 \right) b = \sum_{i=1}^{n} x_i y_i,$$

that is, a linear equation system with two equations and two unknowns a and b. It can be shown that this equation system has a unique solution unless the x-values of all measurement points are identical (that is, unless $x_1 = x_2 = \ldots = x_n$), and that this solution indeed describes a minimum of the function F. The straight line $y = g(x) = a + bx$ determined in this way is called the **best fit (straight) line** or the **regression line** for the data set $(x_1, y_1), \ldots, (x_n, y_n)$.

---

[1] Note, however, that this can also be a disadvantage. If the data set contains "outliers" (that is, measurement values that due to random, disproportionally large measurement errors deviate strongly from the true value), the position of the regression line may be influenced heavily by very few measurement points (precisely the outliers), which can lead to an unusable result.

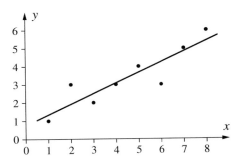

Fig. 10.3 An example data set and a regression line that was computed with the method of least squares

To illustrate the procedure, we consider a simple example. Let the data set consisting of eight data points $(x_1, y_1), \ldots, (x_8, y_8)$ be given that is shown in the following table (see also Fig. 10.3):

x	1	2	3	4	5	6	7	8
y	1	3	2	3	4	3	5	6

In order to set up the system of normal equations, we compute

$$\sum_{i=1}^{8} x_i = 36, \quad \sum_{i=1}^{8} x_i^2 = 204, \quad \sum_{i=1}^{8} y_i = 27, \quad \sum_{i=1}^{8} x_i y_i = 146.$$

Thus we obtain the equation system (normal equations)

$$8a + 36b = 27,$$
$$36a + 204b = 146,$$

which possesses the solution $a = \frac{3}{4}$ and $b = \frac{7}{12}$. Therefore the regression line is

$$y = \frac{3}{4} + \frac{7}{12}x.$$

This line is shown, together with the data points we started from, in Fig. 10.3.

The method we just considered is, of course, not limited to straight lines, but can be extended at least to polynomials. In this case one tries to find a polynomial

$$y = p(x) = a_0 + a_1 x + \ldots + a_m x^m$$

with a given, fixed degree m that approximates the n data points $(x_1, y_1), \ldots, (x_n, y_n)$ as well as possible. In this case we have to minimize

$$F(a_0, a_1, \ldots, a_m) = \sum_{i=1}^{n} (p(x_i) - y_i)^2 = \sum_{i=1}^{n} (a_0 + a_1 x_i + \ldots + a_m x_i^m - y_i)^2.$$

Necessary conditions for a minimum are again that the partial derivatives w.r.t. the parameters a_0 to a_m vanish, that is,

$$\frac{\partial F}{\partial a_0} = 0, \quad \frac{\partial F}{\partial a_1} = 0, \quad \ldots, \quad \frac{\partial F}{\partial a_m} = 0.$$

In this way we obtain the system of normal equations

$$
na_0 + \left(\sum_{i=1}^{n} x_i\right) a_1 + \ldots + \left(\sum_{i=1}^{n} x_i^m\right) a_m = \sum_{i=1}^{n} y_i
$$

$$
\left(\sum_{i=1}^{n} x_i\right) a_0 + \left(\sum_{i=1}^{n} x_i^2\right) a_1 + \ldots + \left(\sum_{i=1}^{n} x_i^{m+1}\right) a_m = \sum_{i=1}^{n} x_i y_i
$$

$$
\vdots \qquad\qquad \vdots \qquad\qquad \vdots \qquad\qquad \vdots
$$

$$
\left(\sum_{i=1}^{n} x_i^m\right) a_0 + \left(\sum_{i=1}^{n} x_i^{m+1}\right) a_1 + \ldots + \left(\sum_{i=1}^{n} x_i^{2m}\right) a_m = \sum_{i=1}^{n} x_i^m y_i,
$$

from which the parameters a_0 to a_m can be derived with the usual methods of linear algebra (Gaussian elimination, Cramer's rule, inverting the coefficient matrix, etc.). The resulting polynomial $p(x) = a_0 + a_1 x + a_2 x^2 + \ldots + a_m x^m$ is called **best fit polynomial** or **regression polynomial** of degree m for the data set $(x_1, y_1), \ldots, (x_n, y_n)$.

Furthermore the method of least squares cannot only be used, as considered up to now, to compute regression polynomials, but may as well be employed to fit functions with more than one argument. This case is called **multiple** or **multivariate regression**. We consider, as an example, only the special case of **multi-linear regression** and confine ourselves first to a function with two arguments. That is, we consider, how one can find a best fitting function of the form

$$
z = f(x, y) = a + bx + cy
$$

for a given data set $(x_1, y_1, z_1), \ldots, (x_n, y_n, z_n)$ in such a way that the sum of squared errors is minimized. In this case the normal equations are derived in a perfectly analogous way. We have to minimize

$$
F(a, b, c) = \sum_{i=1}^{n} (f(x_i, y_i) - z_i)^2 = \sum_{i=1}^{n} (a + bx_i + cy_i - z_i)^2.
$$

Necessary conditions for a minimum are

$$
\frac{\partial F}{\partial a} = \sum_{i=1}^{n} 2(a + bx_i + cy_i - z_i) = 0,
$$

$$
\frac{\partial F}{\partial b} = \sum_{i=1}^{n} 2(a + bx_i + cy_i - z_i) x_i = 0,
$$

$$
\frac{\partial F}{\partial c} = \sum_{i=1}^{n} 2(a + bx_i + cy_i - z_i) y_i = 0.
$$

Therefore we obtain the system of normal equations

$$na + \left(\sum_{i=1}^{n} x_i\right) b + \left(\sum_{i=1}^{n} y_i\right) c = \sum_{i=1}^{n} z_i$$

$$\left(\sum_{i=1}^{n} x_i\right) a + \left(\sum_{i=1}^{n} x_i^2\right) b + \left(\sum_{i=1}^{n} x_i y_i\right) c = \sum_{i=1}^{n} z_i x_i$$

$$\left(\sum_{i=1}^{n} y_i\right) a + \left(\sum_{i=1}^{n} x_i y_i\right) b + \left(\sum_{i=1}^{n} y_i^2\right) c = \sum_{i=1}^{n} z_i y_i$$

from which a, b and c can easily be computed.

For the general multivariate linear case we start with

$$y = f(x_1, \ldots, x_m) = a_0 + \sum_{k=1}^{m} a_k x_k.$$

As usual, we minimize the sum of squared errors, only that this time we use matrix-vector notation, which is more convenient in this case. That is, we consider

$$F(a) = (Xa - y)^\top (Xa - y), \qquad \text{where}$$

$$X = \begin{pmatrix} 1 & x_{11} & \ldots & x_{1m} \\ \vdots & \vdots & \ddots & \vdots \\ 1 & x_{n1} & \ldots & x_{nm} \end{pmatrix}, \qquad y = \begin{pmatrix} y_1 \\ \vdots \\ y_n \end{pmatrix}, \qquad \text{and} \qquad a = \begin{pmatrix} a_0 \\ a_1 \\ \vdots \\ a_m \end{pmatrix}.$$

Note the additional first column of all ones in the (data) matrix X, which captures the intercept term a_0. A necessary condition for a minimum is

$$\nabla_a F(a) = \nabla_a (Xa - y)^\top (Xa - y) = 0.$$

For this derivative we obtain

$$\nabla_a F(a) = \nabla_a \left((Xa - y)^\top (Xa - y)\right)$$

$$= (\nabla_a (Xa - y))^\top (Xa - y) + \left((Xa - y)^\top (\nabla_a (Xa - y))\right)^\top$$

$$= (\nabla_a (Xa - y))^\top (Xa - y) + (\nabla_a (Xa - y))^\top (Xa - y)$$

$$= 2X^\top (Xa - y)$$

$$= 2X^\top Xa - 2X^\top y \overset{!}{=} 0.$$

The necessary condition for a minimum can therefore be written as

$$\nabla_a F(a) = \nabla_a (Xa - y)^\top (Xa - y) = 2X^\top Xa - 2X^\top y \overset{!}{=} 0.$$

As a consequence we obtain the system of **normal equations**

$$X^\top Xa = X^\top y.$$

This system has a solution unless $\mathbf{X}^\top\mathbf{X}$ is singular. If it is regular, we have

$$a = (\mathbf{X}^\top\mathbf{X})^{-1}\mathbf{X}^\top y.$$

Here $(\mathbf{X}^\top\mathbf{X})^{-1}\mathbf{X}^\top$ is called the (Moore–Penrose-)**pseudo-inverse** of the matrix $\mathbf{X}$.

With the matrix-vector representation of the regression problem an extension to **multivariate polynomial regression** is straightforward: Simply add the desired products of powers (monomials) to the (data) matrix $\mathbf{X}$. How it may also be extended, under certain conditions, to other function classes is demonstrated in Sect. 5.4 with the help of the example of **logistic regression**.

A program for multivariate polynomial regression that uses ideas from dynamic programming to quickly compute the needed power products can be found at

http://www.borgelt.net/regress.html

10.3 Activation Transformation

In this section we demonstrate how the weights and thresholds of a Hopfield network that works with activations 0 and 1 can be transformed into the corresponding parameters of a Hopfield network that works with the activations -1 and $+1$ (and vice versa). This shows that the two network types are essentially equivalent, and thus that it was justified to choose in Chap. 8 whichever form was more suitable for the specific task under consideration.

In the following we indicate by an upper index of the considered quantities what the range of activation values of the neural network is, to which they refer

^{0} : quantity of a network with $\mathrm{act}_u \in \{\,0, 1\}$,

$^-$: quantity of a network with $\mathrm{act}_u \in \{-1, 1\}$.

Clearly we must always have

$$\mathrm{act}_u^0 = \frac{1}{2}(\mathrm{act}_u^- + 1) \quad \text{and} \quad \mathrm{act}_u^- = 2\,\mathrm{act}_u^0 - 1.$$

That is, the neuron u either has activation 1 in both networks or it has activation 0 in one network and activation -1 in the other. In order to achieve that both network types exhibit the same behavior, it must also hold that

$$s(\mathrm{net}_u^- - \theta_u^-) = s(\mathrm{net}_u^0 - \theta_u^0),$$

where

$$s(x) = \begin{cases} 1, & \text{if } x \geq 0, \\ -1, & \text{otherwise.} \end{cases}$$

Only if this is the case, the activation changes are the same in both networks. The above equation clearly holds if

$$\mathrm{net}_u^- - \theta_u^- = \mathrm{net}_u^0 - \theta_u^0.$$

(Note that this is a sufficient, but not a necessary condition.) Using the relations between the activations stated above, we obtain from this equation

$$\mathrm{net}_u^- -\theta_u^- = \sum_{v\in U-\{u\}} w_{uv}^- \,\mathrm{act}_u^- -\theta_u^-$$

$$= \sum_{v\in U-\{u\}} w_{uv}^-(2\,\mathrm{act}_u^0 -1) - \theta_u^-$$

$$= \sum_{v\in U-\{u\}} 2w_{uv}^- \,\mathrm{act}_u^0 - \sum_{v\in U-\{u\}} w_{uv}^- - \theta_u^-$$

$$\stackrel{!}{=} \mathrm{net}_u^0 -\theta_u^0$$

$$= \sum_{v\in U-\{u\}} w_{uv}^0 \,\mathrm{act}_u^0 -\theta_u^0.$$

This equation holds if we choose

$$w_{uv}^0 = 2w_{uv}^- \qquad \text{and}$$

$$\theta_u^0 = \theta_u^- + \sum_{v\in U-\{u\}} w_{uv}^-.$$

For the opposite direction we obtain

$$w_{uv}^- = \frac{1}{2}w_{uv}^0 \qquad \text{and}$$

$$\theta_u^- = \theta_u^0 - \sum_{v\in U-\{u\}} w_{uv}^- = \theta_u^0 - \frac{1}{2}\sum_{v\in U-\{u\}} w_{uv}^0.$$

Part II
Evolutionary Algorithms

Introduction to Evolutionary Algorithms

11

Evolutionary algorithms comprise a class of optimization techniques that imitate principles of biological evolution. They belong to the family of **metaheuristics**, which also includes, for example, particle swarm [14] and ant colony optimization [8], which are inspired by other biological structures and processes, as well as classical methods like simulated annealing [15, 18], which is inspired by a thermodynamical process. The core principle of evolutionary algorithms is to apply evolution principles like mutation and selection to populations of candidate solutions in order to find a (sufficiently good) solution for a given optimization problem.

11.1 Metaheuristics

Metaheuristics are fairly general computational techniques that are typically used to solve numerical and combinatorial optimization problems approximately in several iterations (as opposed to analytically and exactly in a single step). Metaheuristics are generally defined as an abstract sequence of operations on certain objects and are applicable to essentially arbitrary problems. However, the objects operated on and the steps to be carried out must be adapted to the specific problem at hand. Thus, the core task is usually to find a proper mapping of a given problem to the abstract structures and operations that constitute the metaheuristic.

Metaheuristics are usually applied to problems for which no efficient solution algorithm is known, that is, problems, for which all known algorithms have an (asymptotic) time complexity that is exponential in the problem size. In practice, such problems can rarely be solved exactly, due to the prohibitively high demands on computing time and/or computing power. As a consequence, approximate solutions have to be accepted, and this is what metaheuristics can provide. Although there is no guarantee that they will find the optimal solution or even a solution of

© Springer Nature Switzerland AG 2022
R. Kruse et al., *Computational Intelligence*, Texts in Computer Science,
https://doi.org/10.1007/978-3-030-42227-1_11

a given minimum quality (although this is not impossible either), they usually offer good chances of finding a "sufficiently good" solution.

The success and the execution time of metaheuristics depend critically on a proper mapping of the problem to the steps of the metaheuristic and the efficient implementation of every single step. Many metaheuristics work by iteratively improving a set of so-called candidate solutions. They differ in the methods they employ to vary solutions in order to possibly improve them, in the principles by which partial solutions are combined or elements of found solutions are exploited to find new solutions, as well as in the principles by which a new set of candidate solutions is selected from the previously created ones. However, they share that they usually carry out a **guided random search** in the space of solution candidates. That is, these algorithms carry out a search that contains certain random elements to explore the search space, but they are also guided by some measure of the solution quality, which governs which (parts of) solution candidates are focused on or at least kept for further exploration and which are discarded, because they are not promising.

An important advantage of metaheuristics is the fact that they can usually be terminated after any iteration step (so-called **anytime algorithms**), because they have, at any point in time, at least some solution candidates available. From these the best solution candidate found so far is then retrieved and returned, regardless of whether some other termination criterion is met or not. However, it should be clear that the solution quality is usually the better, the longer the search can run.

A wide range of metaheuristics based on various principles has been proposed, many of which are nature-inspired. While evolutionary algorithms rely on principles of biological evolution in various forms, (particle) swarm optimization [14] mimics the behavior of swarms of animals (like fish or birds) that search for food in schools or flocks. Ant colony optimization [7,8] mimics the path finding behavior of ants and termites. Other biological entities that inspired metaheuristics include, among others, honey bees [19] or the immune system of vertebrates [11]. Alternatives are algorithms that draw on physical rather than biological analogies, like simulated annealing [15,18], which mimics annealing processes, threshold accepting [10], or the deluge algorithm [9].

11.2 Biological Evolution

Evolutionary algorithms are among the oldest and most popular metaheuristics. They are essentially based on Charles Darwin's theory of biological evolution, which he proposed in his seminal book "The Origin of Species" [3].[1] This theory explains the diversity and complexity of all forms of living organisms and allows us to unify all

---

[1] The full title "The Origin of Species by Means of Natural Selection, or the Preservation of Favoured Races in the Struggle for Life" is usually shortened to merely "The Origin of Species."

biological disciplines. Non-technical modern introductions to the theory of biological evolution can be found in the popular and very readable books [4–6].
The core principle of biological evolution can be formulated as

**Beneficial traits resulting from random variation
are favored by natural selection.**

That is, individuals with beneficial traits have better chances to procreate and multiply, which may also be captured by the expression **differential reproduction**.

New or at least modified traits may be created by various processes. There is, in the first place, the both blind and purely random modification of genes, that is, **mutation**, which affects both sexually and asexually reproducing life forms. Mutations may occur due to exposure to radioactivity (e.g., caused by earth or cosmic radiation or nuclear reactor disasters) or to so-called *mutagens* (i.e., chemical compounds that disturb the copying process of the genetic information), but may also happen simply naturally, due to an unavoidable susceptibility of the complex genetic copying process to errors. In sexual reproduction equally blindly and purely randomly selected halves of the (diploid) chromosome sets of the parents are (re-) combined, thus creating new combinations of traits and physical characteristics. In addition, during *meiosis* (i.e., the cell division process that produces the germ cells or *gametes*), parts of (homologous) chromosomes, in a process called **crossover** (or *crossing over*), may cross each other, break, and rejoin in a modified fashion, thus exchanging genetic material between (homologous) chromosomes. As a result, offspring with new or at least modified genetic plans and thus physical traits are created.

The vast majority of these (genetic) modifications are unfavorable or even harmful, in the worst case rendering the resulting individual unable to live. However, there is a (small) chance that some of these modifications result in (small) improvements, endowing the individual with traits that help it to survive. For example, they may make it better able to find food, to defend itself against predators or at least to hide or run from them, to attract mates for reproduction, etc. Generally speaking, each individual is put to the test in its natural environment where it either proves to have a high fitness, making it more likely to procreate and multiply, or where it fails to survive or mate, thus causing it or its traits to disappear.

Note that the natural selection process is driven by both the natural environment and the individual traits, leading to different reproduction rates or probabilities. Life forms with traits that are better fitted to their environment usually have more offspring on average. Consequently, their traits become more frequent with each generation of individuals. On the other hand, life forms with traits less favorable in their environment usually have less offspring on average and thus might even become extinct after some generations (at least in this environment).

It is important to understand that a trait is not beneficial or harmful in itself, but only w.r.t. the environment. For example, while the dark skin color of many Africans protects their skin against the intense sun in regions close to the equator, their skin pigmentation can turn out to be a disadvantage in regions where sunlight is scarce, because it increases the risk of vitamin D deficiency [12], as vitamin D is produced in the skin under the influence of ultraviolet light. On the other hand, humans with little skin pigmentation may be less likely to suffer from vitamin D deficiency, but may have to take more care to protect their skin against sunlight to reduce the risk of premature skin aging and skin cancer [2]. Another example is sickle cell anemia, which is a usually harmful misshaping of hemoglobin (red blood cells), because it reduces the capability of hemoglobin to transport oxygen. However, it has certain protective effects against malaria and thus can be an advantage in regions in which malaria is common [1].

Note also that it is the interaction of random variation and natural selection that explains why there are so many complex life forms, even though these life forms are extremely unlikely to come into existence as the result of pure chance. However, evolution is *not* just pure chance. Although the variations are blind and random, their selection is not, but strictly driven by their benefit for the survival and procreation of the individuals endowed with them. As a consequence, small improvements can accumulate over many generations and finally lead to surprising complexity and strikingly fitting adaptations to an environment.

One of the reasons why we tend to make wrong estimates of the probability of complex and adapted life forms, seeing them as (much) less likely than they actually are, is that we tend to use the following model: consider a box full of scrap metal parts collected from a junkyard, all of which are car parts. If we shake this box long enough there is a certain (though extremely small) probability that after some time a drivable car is assembled in the box. This model represents creating something complex by mere chance. It clearly makes it effectively impossible that anything even remotely complex, let alone a living organisms, comes into existence.

In evolution, however, each small variation is immediately put to the test w.r.t. an environment and only the beneficial variations are kept and extended. Therefore a better model is the following thought experiment, suggested by Skinner [5]: Suppose we want to convince people that we can predict the outcome of horse races. We send a letter (or nowadays maybe rather an email) to 10,000 people, predicting the winning horse in the next race. There are ten horses and since we actually have no clue which one will win, we predict to the first 1000 people that horse number 1 will win, to the next 1000 people that horse number 2 will win, etc. Thus, after the race has taken place, we are sure to have 1000 people to whom we predicted the correct horse, while we forget about the 9000 others. We repeat the process with another horse race, predicting horse number 1 to the first 100 people (of the 1000 that remained after the first race), horse number 2 to the second 100, etc. After the race we have 100 people to whom we predicted the correct horse twice, while we forget about the 900 others, to whom we predicted the wrong horse. Doing another round in the same manner, we end up with 10 people to whom we predicted the winning horse in three consecutive races. In another letter (or email) to these 10 people, we propose to them to predict the winning horse in yet another race, only that this time we charge a fee. Consider the situation from the point of view of these 10 people: they know that we correctly predicted the winning horse in three consecutive races. The chance of achieving such accuracy by mere guessing is 1 in 1000 and thus highly unlikely. Therefore these 10 people may be inclined to think that we have some insider knowledge that allows us to predict the winning horse much better than by pure chance and thus may be willing to pay the fee.

However, from the process in which these 10 people were selected, we know that it was certain that we would end up with 10 people to which we predicted the winning horse three times in a row. We did not (and need not) know anything about the horses or the races. All we did was to select the successes, that is, the people to whom we made the correct prediction, and ignore the failures, that is, the people to whom we sent a wrong prediction. Evolution works in essentially the same manner: it focuses on the successes (the life forms that survive and procreate) and forgets about the failures (the life forms that go extinct). Since we tend to ignore the failures—simply because we do not see them as they do not exist anymore—we underestimate the probability of seeing an evolved complex individual.

In addition, we have trouble imagining the time—actually billions of years—that has passed since the first, extremely simple cells assembled. There was so much time for variation and selection, for tiny improvements to accumulate, that complex life forms may not even be unlikely, but may actually be almost unavoidable once the process started (at least according to some authors like [5]).

Up to now we focused on variation (mutation and recombination) and selection as the fundamental principles of (biological) evolution. These may indeed be the most important constituents, and often a description of evolutionary processes is reduced to these core elements. However, a more detailed analysis reveals many more principles, of which we list some of the more important in the following. A more detailed discussion can be found in [20] or [13].

- **Diversity**
  All forms of life—even organisms of the same species—*differ* from each other, and not just physically, but already in their genetic material (*biological diversity* or *diversity of species*). Nevertheless, the currently actually existing life forms are only a tiny fraction of the theoretically possible ones.
- **Variation**
  Mutation and genetic recombination (in sexual reproduction) continuously create *new variants*. These new variants may exhibit a new combination of already existing traits or may introduce a modified and thus new trait.
- **Inheritance**
  As long as variations enter the germ line, they are *heritable*, that is, they are genetically passed on to the next generation. However, there is generally no inheritance of acquired traits (so-called *Lamarckism* [17]).
- **Speciation**
  Individuals and populations *diverge genetically*. Thus new *species* are created once their members cannot crossbreed any longer. Speciation gives the phylogenetic "pedigree" its characteristic branching structure.
- **Birth Surplus/Overproduction**
  Nearly all life forms produce *more offspring* than can ever become mature enough to procreate themselves.

- **Adaptation/Natural Selection/Differential Reproduction**
  On average, the survivors of a population exhibit such hereditary variations which *increase* their *adaptation* to the local environment. Herbert Spencer's expression "survival of the fittest" is rather misleading, though. We prefer to speak of *differential reproduction due to different fitness*, because mere survival without offspring is obviously without effect in the long run, especially since the life span of most organisms is limited.
- **Randomness/Blind Variation**
  Variations are *random*. Although *triggered*, *initiated*, or *caused by* something, they do not favor certain traits or beneficial adaptions. In this sense they are *non-teleological* (from the Greek $\tau\varepsilon\lambda o\varsigma$: goal, purpose, objective).
- **Gradualism**
  Variations happen in comparatively *small steps* as measured by the complete information content (entropy) or the complexity of an organism. Thus phylogenetic changes are *gradual* and relatively slow. (In contrast to this *saltationism*—from Latin *saltare*: to jump—means fairly large and sudden changes in development).
- **Evolution/Transmutation/Inheritance with Modification**
  Due to the adaptation to the environment, species are not immutable. They rather *evolve* in the course of time. Hence the theory of evolution opposes *creationism*, which claims that species are immutable.
- **Discrete Genetic Units**
  The genetic information is stored, transferred, and changed in discrete ("atomic", from the Greek $\overset{\smile}{\alpha}\tau o\mu o\varsigma$: indivisible) units. There is no continuous blend of hereditary traits. Otherwise we might see the so-called *Jenkins Nightmare*, that is, a complete disappearance of any differences in a population due to averaging.
- **Opportunism**
  The processes of evolution are extremely opportunistic. They work exclusively with what is present and not with what once was or could be. Better or optimal solutions are not found if the intermediary stages (that are evolutionarily necessary to build these solutions) exhibit certain fitness handicaps.
- **Evolution-Strategic Principles**
  Not only organisms are optimized, but also the mechanisms of evolution. These include parameters like reproduction and mortality rates, life spans, vulnerability to mutations, mutation step sizes, evolutionary speed, etc.
- **Ecological Niches**
  Competitive species can tolerate each other if they occupy different ecological niches ("biospheres" in a wider sense) or even create them themselves. This is the only way the observable biological diversity of species is possible in spite of competition and natural selection.
- **Irreversibility**
  The course of evolution is irreversible and unrepeatable.
- **Unpredictability**
  The course of evolution is neither determined, nor programmed, nor purposeful and thus not predictable.

- **Increasing Complexity**
  Biological evolution has led to increasingly more complex systems. However, an open problem in evolutionary biology is the question how we can actually measure the complexity of life forms.

11.3 Simulated Evolution

Given that biological evolution has created complex life forms and solved difficult adaptation problems, it is reasonable to assume that the same optimization principles can be used to find good solutions for (complex) optimization problems. Hence, we start by formally defining optimization problems in Sect. 11.3.1 and consider some of their main properties. In Sect. 11.3.2 we then transfer some basic notions of biological evolution to simulated evolution, pointing out what requirements have to be fulfilled in order to make an evolutionary algorithm approach worthwhile. In Sect. 11.3.3 we then present the building blocks of an evolutionary algorithm in an abstract fashion, before we turn in Sect. 11.4 to a concrete illustrative example.

11.3.1 Optimization Problems

Definition 11.1 (*Optimization problem*). An *optimization problem* is a pair (Ω, f) consisting of a (search) space Ω of potential solutions and an evaluation function $f : \Omega \to \mathbb{R}$ that assigns a quality assessment $f(\omega)$ to each candidate solution $\omega \in \Omega$. An element $\omega \in \Omega$ is called an (exact) **solution** of the optimization problem (Ω, f) if and only if it is a **global maximum** of f, that is, if $\forall \omega' \in \Omega : f(\omega') \leq f(\omega)$.

Note that, even though the above definition requires a solution to maximize the value of the evaluation function f, this is no actual restriction. If we have a function that yields smaller values for better solutions, we may simply use $-f$ in the definition. In order to capture this, we will speak of a solution as a *global optimum*, thus covering maxima and minima. Note also that a solution need not be unique. There may be multiple elements of Ω for which f yields the same (optimal) value.

As a simple example of an optimization problem consider the task of finding the side lengths of a box with fixed surface area S such that its volume is maximized. Here the search space Ω is the set of all triples (x, y, z), that is, the three side lengths, with $x, y, z \in \mathbb{R}^+$ (i.e., the set of all positive real numbers) and $2xy + 2xz + 2yz = S$, while the evaluation function f is simply $f(x, y, z) = xyz$. In this case the solution is unique, namely $x = y = z = \sqrt{S/6}$, that is, the box is a cube.

Note that this example already exhibits an important feature that will be considered in more detail later, namely that the search space is **constrained**: we do not consider all triples of real numbers, but only those with positive elements that satisfy $2xy + 2xz + 2yz = S$. In this example this is even necessary to have a well-defined solution: if x, y and z could be arbitrary (positive) real numbers, there would be no (finite) optimum. The problem thus consists in making sure that in the search for a solution we never leave the search space, that is, that we never consider as solution candidates objects that do not satisfy the constraints.

Optimization problems occur in many applications areas of which the following (certainly incomplete) list gives only a limited impression:

- **Parameter Optimization**
  In many situations, a set of (suitable) parameters has to be found such that a given real-valued function is optimized. Such parameters can be, for example, the angle and curvature of the air intake and exhaust pipes for automobile motors to maximize power, the relative quantities of the ingredients of a rubber mixture for tires to maximize grip under different conditions, and the temperatures in different components of a thermal power plant to maximize energy efficiency.
- **Routing Problems**
  The most famous routing problem is clearly the *traveling salesman problem*, which occurs in practice, for instance, if holes have to be drilled in a printed circuit board and the distance the drill is to be moved (and thus the movement time) is to be minimized. Other examples include the optimization of delivery routes from a central storage to individual shops or the arrangement of printed circuit board tracks with the objective to minimize length and number of layers.
- **Packing and Cutting Problems**
  Classical examples of packing problems include the *knapsack* (or *backpack* or *rucksack*) *problem*, in which a knapsack of a given (maximum) capacity is to be filled with given goods of known value and size (or weight) such that the total value is maximized, the *bin packing problem*, in which given objects of known size and shape are to be packed into boxes of given size and shape such that the number of boxes is minimized, and the *cutting stock problem* in its various forms, in which geometrical shapes are to be arranged in such a way as to minimize waste (e.g., the wasted cloth after the parts of a garment have been cut out).
- **Arrangement and Location Problems**
  An well-known example of this problem type is the so-called facility location problem, which consists in finding the best placement of multiple facilities (e.g., the distribution nodes in a telephone network), usually under certain constraints. It is also known as Steiner's problem, because certain specific cases are equivalent to the introduction of so-called Steiner points to minimize the length of a spanning tree in a geometric planar graph.
- **Scheduling Problems**
  Job shop scheduling in its various forms, in which jobs have to be assigned to resources at certain times in order to minimize the time to complete all jobs, is a common optimization problem. A special case is the reordering of instructions by a compiler in order to maximize the execution speed of a program. This problem type also includes setting up timetables for schools (where constraints like the number of classrooms and the need to avoid skip hours at least for the lower grades complicate the problem) or trains (in which the number of tracks available on certain lines and the different speeds of the trains render the problem difficult).

- **Strategy Problems**
  Finding optimal strategies of how to behave in the (iterated) prisoner's dilemma and other models of game theory is a common problem in economics. A related goal is to simulate the behavior of actors in economic life, where not only strategies are optimized, but also their prevalence in a population. We discuss a specific (and fairly simple) behavioral simulation in more detail in Sect. 13.5.1.

As a side remark, we mention that (not surprisingly) evolutionary algorithms can also be used for **biological modeling**. An example is the "Netspinner" program [16], which describes the web building behavior of certain spiders by parametrized rules (e.g., number of spokes, angle of the spiral, etc.). With the help of an evolutionary algorithm the program optimizes the rule parameters based on an evaluation function that takes both the metabolic cost of building the web as well as the chances of catching prey with it into account. The obtained results mimic the behavior observed in real spiders very well and thus help to understand the forces that cause spiders to build their webs the way they do.

Optimization problems can be tackled in many different ways, but all possible approaches can essentially be categorized into four classes:

- **Analytical Solution**
  Some optimization problems can be solved analytically. For example, the problem of finding the side lengths of a box with given surface area S such that its volume is maximized (see above) can easily be solved with the method of Lagrange multipliers, setting the partial derivatives of the constructed Lagrange functional w.r.t. the parameters equal to zero and solving the resulting equation system. If there is an analytical approach, it is often the method of choice, because it usually guarantees that the solution is actually optimal and that it can be found in a fixed number of steps. However, for many practical problems there is no (efficient) analytical method, either because the problem is not yet understood well enough or because it is too difficult in a fundamental way (e.g., because it is NP-hard), or it is a black-box problem (e.g., a simulation-based problem).

- **Complete/Exhaustive Exploration**
  Since the definition of an optimization problem already contains all candidate solutions in the form of the (search) space Ω, one may consider simply enumerating and evaluating all of its elements. Although this approach certainly guarantees that the optimal solution will be found, it can be extremely inefficient and thus is usually applicable only to (very) small search spaces Ω. It is clearly infeasible for parameter optimization problems over real domains, since then Ω is infinite and thus can not possibly be explored exhaustively.

- **(Blind) Random Search**
  Instead of enumerating all elements of the search space Ω (which may not be efficiently possible anyway), we may consider picking and evaluating random elements, always keeping track of the best solution (candidate) found so far. This approach is efficient and has the advantage that it can be stopped at any time, but suffers from the severe drawback that it depends on pure luck whether we obtain

a reasonably good solution. This approach corresponds to the model of a box with car parts that is shook to obtain a drivable car, as we discussed it above. It usually offers only very low chances of obtaining a satisfactory solution.

- **Guided (Random) Search**

  Instead of blindly picking random elements from the search space Ω, we may try to exploit the structure of the search space and how the evaluation function f assesses similar elements to control the search. The fundamental idea is to exploit information that has been gained from evaluating certain solution candidates to guide the choice of the next solution candidates to examine. Of course, for this to be possible, the evaluation of similar elements of the search space must be similar. Otherwise there is no basis on which we may transfer gathered information. Note that the choice of the next solution candidates to examine may still contain a random element (non-deterministic choice), though, but that the choice is constrained by the evaluation of formerly examined solution candidates.

All metaheuristics, including evolutionary algorithms, fall into the last category. They differ, as already pointed out above, mainly in how the gathered information is represented and exploited for picking the next solution candidates to evaluate. Although metaheuristics thus provide fairly good chances of obtaining a satisfactory solution, it should always be kept in mind that they *cannot guarantee that the optimal solution is found*. That is, the solution candidate they return may have a high quality, and this quality may be high enough for many practical purposes, but there might still be room for improvement. If the problem at hand requires to find a truly (guaranteed) optimal solution, evolutionary algorithms are not suited for the task. In such a case, one has to opt for an analytical solution or an exhaustive exploration.

It is also important to keep in mind that metaheuristics require that the evaluation function allows for **gradual improvement** (similar solution candidates have similar quality). Although in evolutionary algorithms the evaluation function is motivated by the biological fitness or aptitude in an environment and thus must differentiate better and worse candidate solutions, it must not possess large jumps at random points in the search space. Consider, for example, an evaluation function that assigns a value of 1 to exactly one solution candidate and 0 to all others. In this case, any evolutionary algorithm (actually any metaheuristic) cannot perform better than (blind) random search, because the quality assessment of non-optimal solution candidates does not provide any information about the location of the actual optimum.

11.3.2 Basic Notions and Concepts

In this section, we introduce the basic notions and concepts of evolutionary algorithms by transferring them from their biological counterparts, see Table 11.1.

An **individual**, which is a living organism in biology, corresponds to a candidate solution in computer science. Individuals are the entities to which a fitness is assigned and which are subject to the (natural) selection process. In both domains, an individual is described by a **chromosome** (from the Greek $\chi\rho\omega\mu\alpha$: color and

$\sigma\omega\mu\alpha$: body, thus "colored body", because they are the colorable substance in a cell nucleus), which is the carrier of the genetic information. In biology, a chromosome consists of deoxyribonucleic acid (DNA) and many histone proteins, while in computer science the genetic information is encoded as a sequence of computational objects like bits, characters, numbers, etc. A chromosome represents the "genetic blueprint" and encodes (parts of) traits of an individual. Most living organisms have several chromosomes, for example, humans have 46 chromosomes, which come in 23 so-called *homologous* pairs. In computer science, however, this complication is ignored and all genetic information is combined in a single chromosome.

Table 11.1 Basic evolutionary notions in biology and computer science

Notion	Biology	Computer science
Individual	Living organism	Solution candidate
Chromosome	DNA histone protein strand	Sequence of computational objects
	Describes the "construction plan" and thus (some of the) traits of an individual in encoded form	
	Usually multiple chromosomes Per individual	Usually only one chromosome Per individual
Gene	Part of a chromosome	Computational object (e.g., a bit, character, number, etc.)
	Is the fundamental unit of inheritance, which determines a (partial) characteristic of an individual	
Allele (allelomorph)	Form or "value" of gene	Value of a computational object
	In each chromosome there is at most one form/value of a gene	
Locus	Position of a gene	Position of a computational object
	At each position in a chromosome there is exactly one gene	
Phenotype	Physical appearance of a living organism	Implementation/application of a solution candidate
Genotype	Genetic constitution of a living organism	Encoding of a solution candidate
Population	Set of living organisms	Bag/multiset of chromosomes
Generation	Population at a point in time	Population at a point in time
Reproduction	Creating offspring of one or multiple (usually two) (parent) organisms	Creating (child) chromosomes from one or multiple (parent) chromosomes
Fitness	Aptitude/conformity of a living organism	Aptitude/quality of a solution candidate
	Determines chances of survival and reproduction	

A **gene** is the fundamental unit of inheritance as it determines (a part of) a trait or characteristic of an individual. An **allele** (from the Greek $\alpha\lambda\lambda\eta\lambda\omega\nu$: "each other", "mutual", because initially mainly two-valued genes were considered) refers to a possible form of a gene in biology. For example, a gene may represent the color of the iris in the eye of a human. This gene has alleles that code for blue, brown, green, gray, etc., irises. In computer science, an allele is simply the value of a computational object, which selects one of several possible properties of a solution candidate that the gene stands for. Note that in a given chromosome there is exactly one allele per gene. That is, the iris color gene may code for blue or brown or green or gray eyes, but only one of these possibilities, as specified by the corresponding allele, is present in a concrete chromosome. The **locus** is the position of a gene in its chromosome. At any locus in a chromosome there is exactly one gene. Usually, a gene can be identified by its locus. That is, a specific position (or a specific section) of chromosome codes for a specific trait of the individual.

In biology, **phenotype** refers to the physical appearance of an organism, that is, the shape, structure and organization of its body. Note that the phenotype is what interacts with the environment and hence that it is the phenotype that actually determines the fitness of the individual. Likewise, in computer science, the phenotype is the implementation or application of a candidate solution, from which the fitness of the corresponding individual can be read. In contrast to this, the **genotype** is the genetic configuration of an organism or the encoding of a candidate solution, respectively. Note that the genotype determines the fitness of an individual only indirectly through the phenotype it encodes and that, at least in biology, the phenotype also comprises acquired traits that are not represented in the genotype (for example, learned behavior and bodily changes like a limb lost due to an accident).

A **population** is a simple set of organisms, usually of the same species. Due to the complexity of biological genomes it is usually safe to assume that no two individuals from a population share exactly the same genetic configuration—homozygous twins being the only exception. In addition, even genetically identical individuals differ due to acquired traits, which are never perfectly identical (even for homozygous twins) and thus lead to different phenotypes. In computer science, however, due to the usually much more limited variability of a chromosome as they are used in evolutionary algorithms and the lack of acquired traits, we must allow for the possibility of identical individuals. As a consequence, a population of an evolutionary algorithm is a *bag* or a *multiset* of individuals. In both biological and simulated evolution **generation** refers to the population at a certain point in time.

A new generation is created by **reproduction**, that is, by the generation of offspring from one or more (in biology: if more than one, then usually two) organisms, in which genetic material of the parent individuals may be recombined. The same holds for computer science, only that the child creation process works directly on the chromosomes and that the number of parents may exceed two.

Finally, the **fitness** of an individual measures how high its chances of survival and reproduction are due to its adaptation to its environment. Since the quality of a biological organism w.r.t. its environment is difficult to assess objectively and simply defining fitness as the ability to survive can lead to a tautological "survival of the survivor," a formally more precise notion defines the fitness of an organism as the number of its offspring organisms that procreate themselves, thus linking (biological) fitness directly to the concept of differential reproduction. In computer science, the situation is simpler, because we are given an optimization problem that directly provides a fitness function with which solution candidates are to be evaluated.

It should be noted that, even though there are many parallels, simulated evolution is (usually) much simpler than biological evolution. For example, there are principles of biological evolution, e.g., speciation, that are usually not implemented in an evolutionary algorithm. On the genetic level, we already pointed out that in most life forms the genetic information is spread over multiple chromosomes, which often even come in so-called *homologous* pairs. These are pairs of chromosomes comprising the same genes, but possibly with different alleles, of which both or only one determine the corresponding phenotypical trait. Although such complications have their purpose in biological evolution, they are usually not simulated in a computer.

11.3.3   Building Blocks of an Evolutionary Algorithm

The general idea of an evolutionary algorithm is to employ evolution principles to generate increasingly better solution candidates for the optimization problem to solve. Essentially, this is achieved by evolving a population of solution candidates with the help of random variation and fitness-based selection of the next generation.

An evolutionary algorithm requires the following ingredients:

- an **encoding** for the solution candidates,
- a method to create an **initial population**,
- an **fitness function** to evaluate the individuals,
- a **selection method** on the basis of the fitness function,
- a set of **genetic operators** to modify chromosomes,
- a **termination criterion** for the search, and
- values for various **parameters**.

Since we want to evolve a population of solution candidates, we need a way of representing them as chromosomes, that is, we have to encode them, essentially as sequences of computational objects (like bits, characters, numbers, etc.). Such an encoding may be so direct that the distinction between the genotype, as it is represented by the chromosome, and the phenotype, which is the actual solution candidate, becomes blurred. For example, for the problem of finding the side lengths of a box with given surface area that has maximum volume (which we considered above), we may use the triples (x, y, z) of the side lengths, which are the solution candidates, directly as the chromosomes. In other cases there is a clear distinction between the solution candidate and its encoding, for example, if we have to turn the chromosome into some other structure (the phenotype) before we can evaluate its fitness. We will see several examples of such cases in later chapters.

Generally, the encoding of the solution candidates is highly problem-specific and there are no general rules. However, in Sect. 12.1 we discuss several aspects that attention should be paid to when choosing an encoding for a given problem. An inappropriate choice can severely reduce the effectiveness of the evolutionary algorithm or may even make it impossible to find a sufficiently good solution. Depending on the problem to solve, it is therefore highly recommended to spent considerable effort on finding a good encoding of the solution candidates.

Once we have decided on an encoding, we can create an initial population of solution candidates in the form of chromosomes representing them. Since chromosomes are simple sequences of computational objects, an initial population is commonly created by simply generating random sequences. However, depending on the problem to solve and the chosen encoding, more complex methods may be needed, especially, if the solution candidates have to satisfy certain constraints.

In order to mimic the influence of the environment in biological evolution, we need a fitness function with which we can evaluate the individuals of the created population. In many cases this fitness function is simply identical to the function to optimize, which is given by the optimization problem to solve. However, the fitness function may also contain additional elements that represent constraints that have to be satisfied in order for a solution candidate to be acceptable or that introduce a tendency towards certain additionally desired properties of a solution.

The (natural) selection process of biological evolution is simulated by a method to select candidate solutions according to their fitness. This method is used to choose the parents of offspring we want to create or to select those individuals that are transferred to the next generation without change. Such a selection method may simply transform the fitness values into a selection probability, such that better individuals have higher chances of getting chosen for the next generation.

The random variation of chromosomes is simulated by so-called genetic operators that modify and recombine chromosomes, for example, **mutation**, which randomly changes individual genes, and **crossover**, which exchanges parts of the chromosomes of parent individuals to produce offspring. Depending on the problem and the chosen encoding, the genetic operators can be very generic or highly problem-specific. The choice of the genetic operators is another element that effort should be spent on, especially in connection with the chosen encoding.

The ingredients described up to now allow us to generate a sequence of populations of (hopefully) increasingly better quality. However, while biological evolution is unbounded, we need a criterion when to stop the process in order to retrieve a final solution. Such a criterion may be, for example, that the algorithm is terminated (1) after a user-specified number of generations have been created, (2) there has been no improvement (of the best solution candidate) for a user-specified number of generations, or (3) a user-specified minimum solution quality has been obtained.

To complete the specification of an evolutionary algorithm, we have to choose the values of several parameters, which include, for example, the size of the population to evolve, the fraction of individuals that is chosen from each population to produce offspring, the probability of a mutation occurring in an individual, etc.

More formally, the procedure of an evolutionary algorithm is shown in Algorithm 11.1:

Algorithm 11.1 (General Scheme of an Evolutionary Algorithm).

procedure evoalg;
begin
   $t \leftarrow 0$;   (* initialize the generation counter *)
   initialize pop(t);   (* create the initial population *)
   evaluate pop(t);   (* and evaluate it (compute fitness) *)
   while not termination criterion **do**   (* loop until termination *)
      $t \leftarrow t + 1$;   (* count the created generation *)
      select pop(t) from pop($t - 1$);   (* select individuals based on fitness *)
      alter pop(t);   (* apply genetic operators *)
      evaluate pop(t);   (* evaluate the new population *)
      environmental selection (pop(t), pop($t - 1$));
   (* select individuals for the next population *)
   end end

That is, after having created and evaluated an initial population of solution candidates (in the form of chromosomes), a sequence of generations of solution candidates is computed. Each new generation is created by selecting individuals based on their fitness (with a higher fitness meaning a higher chance of getting selected). Then genetic operators (like mutation and crossover) are applied to the selected individuals. Next, the modified population (or at least the new individuals in it, which have been created by the genetic operators) is evaluated and the cycle starts over. This process continues until the chosen termination criterion is fulfilled.

11.4 The n-Queens Problem

The n-queens problem consists in the task to place n queens (a piece in the game of chess) of the same color onto an $n \times n$ chessboard in such a way that no rank (chess term for row), no file (chess term for column) and no diagonal contains more than one queen. Drawing on the rules of how a queen may move in the game of chess (namely horizontally, vertically or diagonally by any number of squares, but not onto a square that is occupied by a piece of the same color or beyond a square that is occupied by a piece of either color, see Fig. 11.1), we may say that the queens must be placed in such a way that none of them obstructs the possible moves of any other. As an example, Fig. 11.1 shows a solution of the 8-queens problem.

A well-known approach to solve the n-queens problem is a backtracking algorithm, which can be seen as an essentially exhaustive exploration of the space of candidate solutions with a depth-first search. Such an algorithm exploits the obvious fact that each rank (row) of the chessboard must contain exactly one queen. Hence

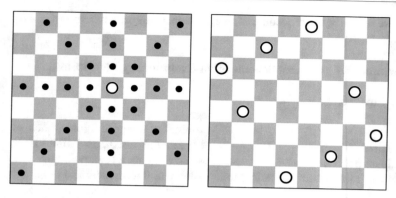

Fig. 11.1 Possible moves of a chess queen (left) and a solution of the 8-queens problem (right)

it proceeds by placing the queens rank by rank. For each rank, it is checked for each possible square whether a queen placed on it obstructs the moves of any queens placed earlier (that is, whether there is already a queen on the same file (column) or the same diagonal). If this is not the case, the algorithm proceeds recursively to the next rank. However, if the newly placed queen obstructs any of the queens placed earlier or if the recursion to the next rank returns with the result that no solution could be found because obstructions could not be avoided, the queen is removed again and the algorithm continues with the next square. More formally, this backtracking algorithm can be described, for example, by the following function:

Algorithm 11.2 (Backtracking Solution of the n -Queens Problem).

function queens (n: int, k: int, $board$: array of array of boolean) : boolean;
begin   (∗ recursively solve n-queens problem ∗)
  if $k \geq n$ **then return true**;   (∗ if all ranks filled, abort with success ∗)
  for $i = 0$ **up to** $n - 1$ **do begin**   (∗ traverse the squares of rank k ∗)
    $board[i][k] \leftarrow$ **true**;   (∗ place a queen on square (i, k) ∗)
    if   **not** $board[i][j] \, \forall j : 0 < j < k$(∗ if no other queen is obstructed ∗)
    and not $board[i - j][k - j] \, \forall j : 0 < j \leq \min(k, i)$
    and not $board[i + j][k - j] \, \forall j : 0 < j \leq \min(k, n - i - 1)$
    and queens $(n, k + 1, board)$ (∗ and the recursion succeeds, ∗)
    then return true;   (∗ a solution has been found ∗)
    $board[i][k] \leftarrow$ **false**;   (∗ remove queen from square (i, k) ∗)
  end   (∗ for $i = 0$ ...∗)
  return false;   (∗ if no queen could be placed, ∗)
end   (∗ abort with failure ∗)

This function is called with the number n of queens that defines the problem size, $k = 0$ indicating that the board should be filled starting from rank 0, and $board$ being an $n \times n$ Boolean matrix that is initialized to *false* in all elements. If the function returns *true*, the problem can be solved. In this case, one possible placement of the queens is indicated by the *true* entries in *board*. If the function returns *false*, the problem cannot be solved (the 3-queens problem, for example, has no solution). In this case the variable *board* is in its initial state of all *false* entries.

Note that the above algorithm can easily be modified to yield all possible solutions of an n-queens problem. In this case, the first if-statement, which checks whether all ranks have been filled, must be extended by a procedure that reports the found solution. In addition, the recursion must not be terminated if the recursion succeeds (and thus a solution has been found), but the loop over the squares of the current rank must be continued to find possibly existing other solutions.

Although a backtracking approach is very effective for sufficiently small n (up to, say, $n \approx 30$), it can take a long time to find a solution if n is larger. If we are interested in only one solution (i.e., one placement of the queens), there exists a better method, namely an analytical solution (which is slightly less well known than the backtracking approach). We compute the positions of the queens as follows:

Algorithm 11.3 (Analytical Solution of the n -Queens Problem).

- If $n = 2$ or $n = 3$, the n-queens problem does not have a solution.
- If n is odd (that is, if $n \bmod 2 = 1$),
  then we place a queen onto the square $(n - 1, n - 1)$ and decrement n by 1.
- If $n \bmod 6 \neq 2$, then we place
  the queens in the rows $y = 0, \ldots, \frac{n}{2} - 1$ in the columns $x = 2y + 1$ and
  the queens in the rows $y = \frac{n}{2}, \ldots, n - 1$ in the columns $x = 2y - n$.
- If $n \bmod 6 = 2$, then we place
  the queens in the rows $y = 0, \ldots, \frac{n}{2} - 1$ in the columns $x = (2y + \frac{n}{2}) \bmod n$ and
  the queens in the rows $y = \frac{n}{2}, \ldots, n - 1$ in the columns $x = (2y - \frac{n}{2} + 2) \bmod n$.

Due to this analytical solution, it is not quite appropriate to approach the n-queens problem with an evolutionary algorithm. Here we do so nevertheless, because this problem allows us to illustrate certain aspects of evolutionary algorithms very well.

In order to solve the n-queens problem with an evolutionary algorithm, we first need an encoding of the solution candidates. For this we draw on the same obvious fact that was already exploited for the backtracking algorithm, namely that each rank (row) of the chessboard must contain exactly one queen. Therefore we describe a candidate solution by a chromosome with n genes, each of which refers to one rank of the chessboard and has n possible alleles, namely the possible file (column) numbers 0 to $n - 1$. Such a chromosome is interpreted as demonstrated in Fig. 11.2 (for $n = 5$): the allele of each gene indicates the file in which the queen is placed in the rank to which the gene refers. Note that with this encoding we can clearly distinguish between the genotype, which is an array of numbers, and the phenotype, which is the actual placement of the queens on the chessboard.

Note that this way of encoding the solution candidates has the advantage that we already exclude candidate solutions with more than one queen per rank. As a

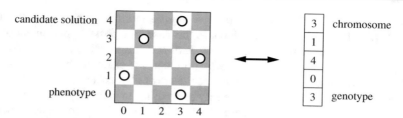

Fig. 11.2 Encoding of candidate solutions in the n-queens problem (here: $n = 5$)

Fig. 11.3 A solution candidate for the 5-queen problem with four obstructions and thus fitness value -2

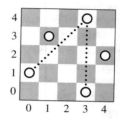

consequence, the search space becomes much smaller and thus can be explored more quickly and more effectively by an evolutionary algorithm. In order to reduce the search space even further, we may even consider restricting the chromosomes to permutations of the file (column) numbers. That is, each file number must occur for exactly one gene. However, although this clearly shrinks the search space even further, it introduces complications w.r.t. the genetic operators and thus we refrain from introducing this requirement here (however, cf. the discussion in Sect. 12.3).

In order to create an initial population we simple generate a random sequence of n numbers in $\{0, 1, \ldots, n - 1\}$ for each individual, because there are no special conditions that such a sequence has to satisfy to be an element of the search space.

The fitness function is derived directly from the defining characteristics of a solution: we compute for each queen the number of obstructions, that is, the number of other queens that obstruct its moves. Then we sum these numbers over the queens, divide by 2 and negate the result (see Fig. 11.3 for an example). Clearly, for an actual solution the fitness computed in this manner is zero, whereas it is negative for all other candidates. Note that we divide by two, because each obstruction is counted twice with the above procedure as obstruction is symmetric: if queen 1 obstructs queen 2, then queen 2 also obstructs queen 1. Note also that we negate the result, because we want a fitness function that has to be maximized. For the example shown in Fig. 11.3 we have four (pairwise symmetric) obstructions (we may also say: two collisions between queens) and thus the fitness value is -2.

The fitness function immediately fixes the termination criterion: since a solution has the (maximally possible) fitness value of 0, we stop the algorithm as soon as a solution candidate with fitness 0 has been generated. However, to be on the safe side, we should also introduce a limit for the number of generations, so that the algorithm

Fig. 11.4 One-point
crossover of two
chromosomes

is guaranteed to terminate. Note, though, that with such an additional criterion the evolutionary algorithm may stop without having found a solution.

For the selection operation we choose a simple, but often very effective form of so-called **tournament selection**. That is, from the individuals of the current population a (small) sample of individuals is drawn that carry out a tournament with each other. This tournament is won by the individual with the highest fitness (ties are broken randomly). A copy of the winning individual is then added to the next generation and the participants of the tournament are replaced into the current population. The process is repeated until the next generation is complete, which usually means that it has reached the same size as the current population. Alternative selection methods as well as variants of tournament selection are discussed in Sect. 12.2.

In order to alter the selected individuals, we need genetic operators for recombination and variation. For the former we use so-called **one-point crossover**, which chooses a random cut point on the chromosomes of two parent individuals and exchanges the part on one side of the cut point between these individuals to create two children. An example for $n = 5$ is shown in Fig. 11.4: the genes below the randomly chosen cut point (the second out of the four possible ones) are exchanged. This example demonstrates what one hopes a genetic recombination may achieve: by combining partial solutions that are present in two deficient individuals (that is, both parent individuals have a negative fitness) a complete solution is obtained (the left child has a fitness of 0 and thus is a solution of the 5-queens problem). Alternative crossover operators are discussed in Sect. 12.3.

As a variation operation, we use a random replacement of the alleles of randomly selected genes (so-called **standard mutation**). An example is shown in Fig. 11.5, in which two genes of a chromosome representing a candidate solution for the 5-queens problem receive new values. This example is fairly typical in the respect that most mutations reduce the fitness of the individual affected by them. However, mutation is nevertheless important, because alleles that are not present in the initial population cannot be created by recombination, which only reorganizes existing alleles. Generally, mutation can more easily introduce new alleles into chromosomes. Alternative mutation operators are discussed in Sect. 12.3.

Finally, we have to choose the values of several parameters. These include the size of the population to evolve (for example, $\mu = 1000$ individuals), the maximum number of generations to compute (for example, $g = 100$ generations), the size of the tournaments to carry out (for example, $\mu_t = 5$ individuals), the fraction of individuals that are subject to crossover (for example, $p_c = 0.5$), and the probability that a gene is subject to a mutation (for example, $p_m = 0.01$). Once all parameters

Fig. 11.5 Mutation of two genes of a chromosome

have been chosen, the evolutionary algorithm is fully specified and can be executed according to the general scheme presented in Algorithm 11.1 on page 242.

An implementation of this evolutionary algorithm, which can be found (as a command line program) on the web site for this book, shows that one can find solutions to the n-queens problem even for somewhat larger n than backtracking allows (also available as a command line program on the website for this book), at least if a sufficiently large population is used and a sufficient number of generations is computed. However, as it is not guaranteed that a solution can be found, the program sometimes ends with a candidate solution that has a high fitness (like -1 or -2), but does not really solve the problem since obstructions remain.

By experimenting with the parameters, especially the fraction of individuals that are subject to crossover or the probability that a gene gets mutated, one can discover some interesting properties. For example, it turns out that mutations seem to be more important than crossover, since the speed with which a solution is found or the quality of solutions that are found in a given number of generations is not reduced if the fraction of individuals that are subject to crossover is reduced to zero. On the other hand, disallowing mutations causes the (average) solution quality to degrade significantly. Note, however, that these are not general characteristics of evolutionary algorithms, but are exhibited only for this particular problem and chosen encoding, selection and genetic operators, etc. It cannot be transferred directly to other applications, where the crossover operation may contribute more to a solution being found and mutation, if its probability is chosen too large, rather degrades performance.

11.5   Related Optimization Techniques

In classical optimization (for example, in operations research) many techniques and algorithms have been developed that are fairly closely related to evolutionary algorithms. Among these are some of the best-known optimization techniques, for instance, gradient ascent or descent (Sect. 11.5.1), hill climbing (random ascent or descent, Sect. 11.5.2), simulated annealing (Sect. 11.5.3), threshold accepting (Sect. 11.5.4) and the great deluge algorithm (Sect. 11.5.5). All of these methods are sometimes called **local search methods**, because they make only small steps in the search space and thus carry out a local search for better solutions.

Like evolutionary algorithms, these techniques are based on the assumption that for similar solution candidates s_1 and s_2 the values of the function to optimize—that

Fig. 11.6 Intuitive
interpretation of the gradient
of a real-valued function
$z = f(x, y)$ at a
point $p = (x_0, y_0)$. It is
$\nabla z|_{(x_0, y_0)} =$
$\left(\frac{\partial z}{\partial x}|_{(x_0, y_0)}, \frac{\partial z}{\partial y}|_{(x_0, y_0)} \right)$

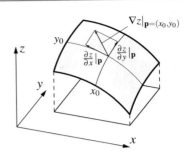

is, the values $f(s_1)$ and $f(s_2)$—are also similar. The main difference to evolutionary algorithms consists in the fact that the mentioned approaches focus on a single solution candidate rather than a whole population of solution candidates. These methods are relevant in the context of evolutionary algorithms, because one may see them, to some degree, as evolutionary algorithms with a population of size 1. In addition, they are often employed to improve solutions candidates locally or as a final optimization step for the output of an evolutionary algorithm. Finally, they illustrate some of the fundamental ideas underlying evolutionary algorithm techniques.

11.5.1 Gradient Ascent or Descent

While all methods discussed in the following sections only assume similarity of the quality of similar solution canditates, gradient ascent (or descent), requires in addition that the following two conditions hold:

- The search space is a subset of the n-dimensional space of real numbers: $\Omega \subseteq \mathbb{R}^n$.
- The function $f : \Omega \to \mathbb{R}$ to optimize is differentiable (everywhere).

Technically, the **gradient** is a differential operation that creates a vector field. That is, the gradient assigns to each point in the domain of the function of which the gradient is computed as a vector that points into the direction of the steepest ascent of the function in that point. An illustration of the gradient of a two-dimensional function $z = f(x, y)$ at a point $p = (x_0, y_0)$ is shown in Fig. 11.6 (see also Sect. 5.5 in Part I). Formally, the gradient at a point (x_0, y_0) is defined as

$$\nabla z|_{(x_0, y_0)} = \left(\frac{\partial z}{\partial x}\bigg|_{(x_0, y_0)}, \frac{\partial z}{\partial y}\bigg|_{(x_0, y_0)} \right).$$

The basic idea of gradient ascent (or descent) is to start at a randomly chosen point and then to make small steps in the search space Ω in (or against) the direction of the steepest slope of the function f until a (local) optimum is reached. More formally, gradient ascent or descent works according to the following general scheme:

Algorithm 11.4 (Gradient Ascent or Descent).

1. Choose a (random) starting point $x^{(0)} = \left(x_1^{(0)}, \ldots, x_n^{(0)}\right)$.
2. Compute the gradient at the current point $x^{(t)}$

$$\nabla_x f\left(x^{(t)}\right) = \left(\frac{\partial}{\partial x_1} f\left(x^{(t)}\right), \ldots, \frac{\partial}{\partial x_n} f\left(x^{(t)}\right)\right).$$

3. Make a small step in the direction of the gradient (for gradient ascent, positive sign) or against the direction of the gradient (for gradient descent, negative sign):

$$x^{(t+1)} = x^{(t)} \pm \eta \, \nabla_x f\left(x^{(t)}\right)$$

where η is a parameter that controls the step width (also called "learning rate" in the training of (artificial) neural networks, see Sect. 5.5).
4. Repeat steps 2 and 3 until some termination criterion is fulfilled (for example, until a user-specified number of steps has been executed or until the gradient is smaller than a user-specified threshold).

Although simple and often very effective, this optimization technique is not without drawbacks. As discussed in more detail in Sect. 5.7 in Part I, the choice of the step width η is critical. If its value is too small, it may take very long until a (local) optimum is reached, because only tiny steps are executed, especially if the gradient is small. On the other hand, if the step width is too large, the optimization process may oscillate (jump back and forth in the search space), never converging to a (local) optimum. Some approaches to mitigate this problem (like a using a momentum term or an adaptive step width parameter) are discussed in Sect. 5.8 in Part I.

A fundamental problem of gradient ascent or descent is the choice of the starting point, since a bad choice can make it essentially impossible to find the global or even just a good local optimum. Unfortunately, there is little that can be done to improve this situation. Thus the best option is to execute the procedure multiple times with different starting points and finally to choose the best result obtained. This provides an argument for looking at *populations* of solution candidates (as in evolutionary algorithms). Such an approach offers, in addition, the possibility to exchange information between solution candidates to improve the optimization process. In evolutionary algorithm, this is achieved by recombination operators.

11.5.2 Hill Climbing

If the function f is not differentiable, gradient ascent or descent is not an option. However, we may try to determine a direction in which f increases by evaluating random points in the vicinity of the current point. The result is known as *hill climbing*, which is probably the simplest local search method. It works as follows:

Algorithm 11.5 (Hill Climbing).

1. Choose a (random) starting point $s_0 \in \Omega$.
2. Choose a point $s \in \Omega$ "in the vicinity" of s_t
   (for example, by a small random variation of s_t).

Fig. 11.7 Getting stuck in local maxima is a core problem of gradient ascent as well as of hill climbing. In order to reach the global maximum (or at least a better local optimum) a temporary worsening of the solution quality has to be accepted

3. Set $\quad s_{t+1} = \begin{cases} s & \text{if } f(s) > f(s_t), \\ s_t & \text{otherwise.} \end{cases}$

4. Repeat steps 2 and 3 until a termination criterion is fulfilled.

As for gradient ascent or descent, the biggest problem of this fairly naïve approach is that it has a strong tendency to get stuck in local optima. This is illustrated in Fig. 11.7, where a solution canditate (indicated by the gray ball) cannot rise any further by merely local modifications and thus gets stuck in a local maximum. All subsequent methods try to mitigate this fundamental problem by accepting, under certain conditions, solution candidates s that are worse than the current solution s_t. The idea is to enable solutions candidates to cross regions of lower solution quality between the local maximum and a global one (or at least a better local one), see Fig. 11.7. The different methods we study below differ mainly in the exact conditions under which they accept solution candidates that are worse.

11.5.3   Simulated Annealing

Simulated annealing [15, 18] can be seen as extension of both hill climbing and gradient descent. The fundamental idea of this method, which is also discussed in another context in Sect. 8.5 in Part I, is that transitions from lower to higher (local) maxima should be more probable than transitions in the opposite direction. This is achieved as follows: we randomly create variants of the current solution—in exactly the same way as for hill climbing. Better solution candidates are always accepted. Solution candidates that are worse are accepted with a certain probability which depends on both the quality difference between the current and the new solution candidate and a temperature parameter that decreases over time. The guiding principles are that small reductions of the solution quality are more readily accepted than large ones, and that reductions of the solution quality are more easily accepted in early steps of the algorithm than in later ones.

Algorithm 11.6 (Simulated Annealing).

1. Choose a (random) starting point $s_0 \in \Omega$.
2. Choose a point $s \in \Omega$ "in the vicinity" of s_t
   (for example, by a small random variation of s_t).
3. Set $s_{t+1} = \begin{cases} s & \text{if } f(s) \geq f(s_t), \\ s & \text{with probability } p = e^{-\frac{\Delta f}{kT}} \text{ and} \\ s_t & \text{with probability } 1 - p \text{ otherwise.} \end{cases}$

   where $\Delta f = f(s_t) - f(s)$ is the quality reduction of the solution, $k = \Delta f_{\max}$ an estimate of the range of quality values and T is a temperature parameter that is (slowly) decreased over time.
4. Repeat steps 2 and 3 until a termination criterion is fulfilled.

For (very) small T this method is almost identical to hill climbing, because the probability of accepting a worse solution candidate is (very) small. For larger T, quality reductions are accepted with a non-negligible probability, thus allowing the solution candidates to cross regions of reduced quality. Nevertheless the algorithm cannot guarantee, of course, that the global optimum is reached. However, the risk of getting stuck in a local optimum is reduced and thus one obtains better chances of reaching the global optimum or at least a good local one.

11.5.4 Threshold Accepting

The idea of threshold accepting [10] is very similar to that of simulated annealing. Again worse solutions are sometimes accepted, however, with an upper bound for the quality degradation. Threshold accepting works as follows:

Algorithm 11.7 (Threshold Accepting).

1. Choose a (random) starting point $s_0 \in \Omega$.
2. Choose a point $s \in \Omega$ "in the vicinity" of s_t
   (for example, by a small random variation of s_t).
3. Set $s_{t+1} = \begin{cases} s & \text{if } f(s) \geq f(s_t) - \theta, \\ s_t & \text{otherwise,} \end{cases}$

   where $\theta > 0$ is a threshold for accepting worse solution candidates that is (slowly) decreased over time. ($\theta = 0$ is equivalent to standard hill climbing.)
4. Repeat steps 2 and 3 until a termination criterion is fulfilled.

11.5.5 Great Deluge Algorithm

Similar to simulated annealing and threshold accepting, the great deluge algorithm [9] also accepts worse solutions. The difference is that an absolute lower bound is used, which is slowly increased over time. One may imagine the procedure as if the "landscape" formed by the quality function to optimize is slowly "flooded" (in a great deluge, hence the name of the algorithm) and only solution candidates that "sit on dry land" are acceptable. The higher the water level rises, the more strongly the tendency becomes to accept only better solution candidates.

Algorithm 11.8 (Great Deluge Algorithm).

1. Choose a (random) starting point $s_0 \in \Omega$.
2. Choose a point $s \in \Omega$ "in the vicinity" of s_t
   (e.g., by a small random variation of s_t).
3. Set $s_{t+1} = \begin{cases} s & \text{if } f(s) \geq \theta_0 + t \cdot \eta, \\ s_t & \text{otherwise,} \end{cases}$
   where θ_0 is a lower bound for the quality of the candidate solutions at $t = 0$ (that is, an initial "water level") and η is a step width parameter that can be seen as corresponding to the "speed of the rain" causing the flood.
4. Repeat steps 2 and 3 until a termination criterion is fulfilled.

11.5.6 Record-to-Record Travel

Similar to the great deluge algorithm, record-to-record travel uses a rising water level [9]. However, it is linked to the fitness of the best individual that has been found so far. Thus, a possible degradation is always seen in relation to the best found individual. Only if there is an improvement, then the current individual is important. Similar to threshold accepting, a monotonously increasing sequence θ of real numbers controls the selection of poor individuals. More formally:

Algorithm 11.9 (Record-to-Record Travel).

1. Choose a (random) starting point $s_0 \in \Omega$ and set $s_{\text{best}} = s_0$.
2. Choose a point $s \in \Omega$ "in the vicinity" of s_t
   (for example, by a small random variation of s_t).
3. Set $s_{t+1} = \begin{cases} s & \text{if } f(s) \geq f(s_{\text{best}}) - \theta, \\ s_t & \text{otherwise,} \end{cases}$
   and $s_{\text{best}} = \begin{cases} s & \text{if } f(s) > f(s_{\text{best}}), \\ s_{\text{best}} & \text{otherwise,} \end{cases}$
   where $\theta > 0$ is a threshold for accepting solution candidates worse than the currently best solution that is (slowly) decreased over time.
4. Repeat steps 2 and 3 until a termination criterion is fulfilled.

11.6 The Traveling Salesman Problem

To illustrate the application of the local search methods discussed above to an optimization problem, we take a look at the famous **traveling salesman problem** (TSP): we are given both a set of n cities (idealized as points on a plane) and the distances or costs of the routes between the cities. A traveling salesman has business to conduct in each of the cities, but, of course, wants to travel as cheaply (or as shortly) as possible. Hence we desire to find a minimum cost (or minimum distance) round trip through the cities, so that each city is visited exactly once.

Mathematically, the traveling salesman problem is defined as follows: we are given a graph with weighted edges and we desire to find a so-called **Hamiltonian cycle** in this weighted graph, that is, an ordering of the vertices of the graph in which neighbors as well as the last and the first vertex are connected by an edge and which minimizes the total weight of the connecting edges. In the form of our definition of an optimization problem (as we gave it in Definition 11.1 on page 235) a traveling salesman problem can be stated as follows:

Definition 11.2 (*Traveling Salesman Problem*) Let $G = (V, E, w)$ be a weighted graph with the vertex set $V = \{v_1, \ldots, v_n\}$ (each v_i represents a city), the edge set $E \subseteq V \times V - \{(v, v) \mid v \in V\}$ (each edge represents a connection between two cities) and the edge weight function $w : E \to \mathbb{R}_+$ (which represents the distances or costs of the connections). The traveling salesman problem is the optimization problem $(\Omega_{\text{TSP}}, f_{\text{TSP}})$ where Ω_{TSP} contains all permutations π of the numbers $\{1, \ldots, n\}$ that satisfy $\forall k; 1 \leq k \leq n : (v_{\pi(k)}, v_{(\pi(k) \bmod n)+1}) \in E$ and the function f_{TSP} is defined as

$$f_{\text{TSP}}(\pi) = -\sum_{k=1}^{n} w((v_{\pi(k)}, v_{\pi((k \bmod n)+1)})).$$

A traveling salesman problem is called **symmetric** if

$$\forall i, j \in \{1, \ldots, n\}, i \neq j : (v_i, v_j) \in E \Rightarrow (v_j, v_i) \in E \ \wedge \ w((v_i, v_j))$$
$$= w((v_j, v_i)),$$

that is, if all connections can be traversed in both directions and the directions have the same costs. Otherwise the traveling salesman problem is called **asymmetric**.

No algorithm is known that solves this problem in polynomial time unless a non-deterministic formalism (like a non-deterministic Turing machine) is used (which is essentially irrelevant for practical purposes). More technically, one says that this problem is non-deterministic polynomial-time-complete (or *NP-complete* for short). Intuitively, this means that in order to guarantee that one obtains the optimal solution one cannot do fundamentally better than trying all possibilities (exhaustive exploration of the search space). As a consequence, for large n only approximate solutions can be computed in reasonable time, because an exhaustive exploration is exponential in n. Of course, we may actually find the best solution with a guided (random) search like evolutionary algorithms, but this is not guaranteed. Here we consider, as an illustration of the methods discussed above, how the traveling salesman problem can be tackled with hill climbing and simulated annealing. We work with the following algorithm, which captures hill climbing for $T \equiv 0$:

Algorithm 11.10 (Simulated Annealing for the Traveling Salesman Problem).

1. Order the cities randomly (that is, create a random round trip).
2. Randomly choose two pairs of cities such that each pair consists of cities that are neighbors in the current round trip and such that all four cities are distinct. Split the round trip between the cities of each pair and reverse the interjacent part.
3. If this new round trip is better (that is, shorter or cheaper) than the old, then replace the old round trip with the new one. Otherwise replace the old round trip with probability $p = \exp\left(-\frac{\Delta Q}{kT}\right)$ where ΔQ is the quality difference between the old and the new round trip, k is an estimate of the range of round trip qualities and T is a temperature parameter that is reduced over time, e.g., $T = \frac{1}{t}$.
4. Repeat steps 2 and 3 until a termination criterion is met.

Since we cannot know the range k of quality values, we estimate it, for example, by $k_t = \frac{t+1}{t}(\max_{i=1}^{t} Q_i - \min_{i=1}^{t} Q_i)$ where Q_i is the quality of the ith solution candidate and t is the current time step. The employed variation operator is introduced more generally as a mutation operator in Sect. 12.3.1 under the name **inversion**.

As an illustration, we consider the simple traveling salesman problem with only five cities that is shown in Fig. 11.8 on the left. We simply use the Euclidean distance of the points as the cost function, that is, we search for a round trip of minimum length. A possible starting point (chosen in step 1) is the round trip that is shown in Fig. 11.8 on the right, which has a length of $2\sqrt{2} + 2\sqrt{5} + 4 \approx 11.30$.

All five possible splits of this round trip that could be chosen in step 2 of the above algorithm are shown in Fig. 11.9: the two edges that are cut by a dashed line identify the cities of each pair. The new round trips that result from inverting the interjacent part (that is, traversing it in the opposite direction) are shown in Fig. 11.10 on the left. The new edges of the round trip are shown in gray.

Figure 11.10 also demonstrates that all five splits lead to a round trip that is worse than the initial tour (see the length computations on the right). As a consequence, a hill climbing approach (which accepts only better modifications) is unable to improve

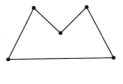

Fig. 11.8 Example of a traveling salesman problem (left) with initially created round trip (right)

Fig. 11.9 Possible separations of the exemplary round trip

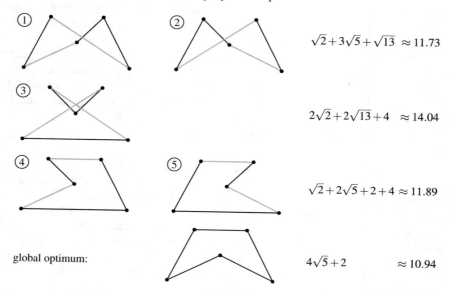

Fig. 11.10 Modifications of the initial round trip and the global optimum with the corresponding fitness values. Compared to the initial round trip, all its possible variations are worse and therefore hill climbing cannot reach the global optimum

the initial round trip and thus returns it as the solution. However, this round trip is only a local optimum, as can be seen from the tour that is shown at the bottom of Fig. 11.10, which is the global optimum (round trip of minimum length).

While hill climbing cannot find this optimum (from the chosen starting point—it can find it from a different starting point), simulated annealing offers at least certain chances to find it (although no guarantee). For instance, since the modifications 1, 2, 4 and 5 are only slightly worse than the initial tour, they may be accepted in the simulated annealing algorithm in the random choice in step 3. At least from the modifications 1 and 2 the optimal solution can be obtained with a single additional modification that reverses the right or the left edge, respectively.

It should be noted that it may depend on the operations with which a solution candidate "in the vicinity" of the current solution is created whether hill climbing can get stuck in a local optimum. If we permit other operations, the problem disappears for the example studied in Figs. 11.8, 11.9 and 11.10: by removing the city in the center and inserting it between the two cities at the bottom, the initial tour can be transformed directly into the optimal round trip and thus hill climbing is enabled to find the optimal solution. However, for this modified operation set another example can be found, in which hill climbing can get stuck in a local optimum.

References

1. Aidoo M, Terlouw DJ, Kolczak MS, McElroy PD, ter Kuile FO, Kariuki S, Nahlen BL, Lal AA, Udhayakumar V (2002) Protective effects of the sickle cell gene against malaria morbidity and mortality. Lancet 359:1311–1312. Elsevier, Amsterdam, Netherlands
2. Brenner M, Hearing VJ (2008) The protective role of melanin against UV damage in human skin. Photochem Photobiol 84(3):539–549. Wiley, New York, NY, USA
3. Darwin C (1859) On the origin of species by means of natural selection, or the preservation of favoured races in the struggle for life. John Murray, London, United Kingdom
4. Dawkins R (1976) The selfish gene, 2nd edn. Oxford University Press, Oxford, United Kingdom 1976; 2nd edn 1989
5. Dawkins R (1986) The blind watchmaker. W.W. Norton, New York, NY, USA
6. Dawkins R (2009) The greatest show on earth: the evidence for evolution. Free Press, New York, NY, USA
7. Dorigo M (1992) Optimization, Learning and Natural Algorithms. PhD Thesis, Politecnico di Milano, Milan, Italy 1992
8. Dorigo M, Stützle T (2004) Ant colony optimization. Bradford/MIT Press, Cambridge, MA, USA
9. Dueck G (1993) New optimization heuristics: the great deluge algorithm and the record-to-record travel. J Comput Phys 104(1):86–92. Elsevier, Amsterdam, Netherlands
10. Dueck G, Scheuer T (1990) Threshold accepting: a general purpose optimization algorithm appearing superior to simulated annealing. J Comput Phys 90(1):161–175. Elsevier, Amsterdam, Netherlands
11. Farmer JD, Packard N, Perelson A (1986) The immune system, adaptation and machine learning. Phys D: Nonlinear Phenom 2:187–204. Elsevier, Amsterdam, Netherlands
12. Harris SS (2006) Vitamin D and African Americans. J Nutr 136(4):1126–1129. American Society for Nutrition, Bethesda, MD, USA
13. Hartl DL, Clark AG (2007) Principles of population genetics, 4th edn. Sinauer Associates, Sunderland, MA, USA
14. Kennedy J, Eberhart R (1995) Particle swarm optimization. In: Proceedings of IEEE international conference on neural networks, vol 4. IEEE Press, Piscataway, NJ, USA, pp 1942–1948
15. Kirkpatrick S, Gelatt CD, Vercchi MP (1983) Optimization by simulated annealing. Science 220:671–680. High Wire Press, Stanford, CA, USA
16. Krink T, Vollrath F (1997) Analysing spider web-building behaviour with rule-based simulations and genetic algorithms. J Theor Biol 185(3):321–331. Elsevier, Amsterdam, Netherlands
17. Lamarck J-B (1809) Philosophie zoologique, ou. Exposition des considérations relative à l'histoire naturelle des animaux. Paris, France

18. Metropolis N, Rosenblut N, Teller A, Teller E (1953) Equation of state calculations for fast computing machines. J. Chem. Phys. 21:1087–1092. American Institute of Physics, Melville, NY, USA

19. Nakrani S, Tovey S (2004) On honey bees and dynamic server allocation in internet hosting centers. Adapt Behav 12:223–240. SAGE Publications, New York, NY, USA

20. Vollmer G (1995) Der wissenschaftstheoretische Status der Evolutionstheorie: Einwände und Gegenargumente. In: Vollmer G (ed) Biophilosophie, 92–106. Reclam, Stuttgart, Germany

Elements of Evolutionary Algorithms

12

Evolutionary algorithms are not fixed procedures, but contain several elements that must be adapted to the optimization problem to be solved. In particular, the encoding of the candidate solution needs to be chosen with care. Although there is no generally valid rule or recipe, we discuss in Sect. 12.1 some important properties a good encoding should have. In Sect. 12.2 we turn to the fitness function and review the most common selection techniques as well as how certain undesired effects can be avoided by adapting the fitness function or the selection method. Section 12.3 is devoted to genetic operators, which serve as tools to explore the search space, and covers sexual and asexual recombination and other variation techniques.

12.1 Encoding of Solution Candidates

The way in which candidate solutions to the given optimization problem are encoded can have a considerable impact on how easily an evolutionary algorithm finds a (good) solution. With a bad or unfavorable encoding it may not even find a useful solution at all. As a consequence, one should spend a lot of effort and care on designing a good encoding and the corresponding genetic operators.

For example, in the n-queens problem we discussed in Sect. 11.4, the encoding we chose, namely, using an array of n integer numbers with values in $\{0, \ldots, n-1\}$, is much better than representing the placement of the queens by a binary array with n^2 elements, in which each element refers to a square of the $n \times n$ chessboard and encodes whether this square is occupied by a queen (bit is 1) or not (bit is 0). The reason is that our encoding already rules out candidate solutions with more than one queen in the same rank (row) and thus considerably reduces the search space. In addition, it ensures that there are always exactly n queens on the board, while with a binary encoding the genetic operators we employed (standard mutation and one-point crossover) may produce a candidate solution with more or less queens.

© Springer Nature Switzerland AG 2022
R. Kruse et al., *Computational Intelligence*, Texts in Computer Science,
https://doi.org/10.1007/978-3-030-42227-1_12

An even better encoding than the one we chose is to restrict the candidate solutions to permutations of the numbers $\{0, \ldots, n-1\}$. Such an encoding not only guarantees that each rank contains exactly one queen, but at the same time that each file (column) contains exactly one queen. Hence it reduces the search space even further, making the task much easier for the evolutionary algorithm. We merely refrained from using this encoding in Sect. 11.4, because it would have caused problems with the genetic operators, since standard mutation and one-point crossover do not maintain that a candidate solution is a permutation. However, these problems can be solved with specialized operators as we demonstrate in Sect. 12.3.

Generally, it is important to pay attention to the interplay between the chosen encoding and the genetic operators. If the encoding reduces the search space, but it turns out to be difficult to find genetic operators that guarantee that the result of their application is in this reduced search space, additional efforts may be needed to handle such cases. These efforts may very well be so costly that it can turn out to be better to use an encoding that defines a larger search space (incorporating fewer constraints) but allows for simpler choices of genetic operators.

This brief discussion already shows that there are no general "cookbook recipes" with which one can find a (good) encoding. Nevertheless we can identify the following desirable properties that an encoding should have:

- Similar phenotypes should be represented by similar genotypes.
- Similarly encoded candidate solutions should have a similar fitness.
- If possible, the search space Ω (i.e., the set of all possible candidate solutions) should be closed under the used genetic operators.

Of course, these are not absolute rules, but rather guidelines. Depending on the problem at hand one may consider breaking them in order to gain other advantages. However, it is usually a good idea to refrain from using an encoding that does not have these properties unless there are very good reasons for doing so.

12.1.1 Hamming Cliffs

Similar phenotypes should be represented by similar genotypes

Two genotypes are clearly similar if they differ in few genes, because then few mutations, for example, are necessary to transform the one genotype into the other. That is, the similarity of genotypes is defined as how easy it is to transform one into the other with the available genetic operators. It may be measured, for example, as the needed minimum number of applications of genetic operators (like mutation).

However, what is subject to the fitness evaluation is the phenotype, that is, the actual solution candidate, and we only presuppose that similar solution candidates have similar fitness (so that fitness information can be exploited to guide the search for better solution candidates). As we only modify the chromosomes (and thus the genotype), this can clearly lead to problems if similar phenotypes and thus similar solution candidates are described by (very) dissimilar genotypes, because then it may be impossible to obtain a similar phenotype by (small) genetic modifications.

As an example we consider a parameter optimization problem: we are given an n-ary real-valued function $y = f(x_1, \ldots, x_n)$ and we desire to find an argument vector $(x_1, \ldots, x_n)$ that optimizes the value of y. We represent the real-valued arguments by encoding them as binary numbers, which are then concatenated to obtain a binary chromosome. Unfortunately, such an encoding can create a serious problem in an evolutionary algorithm, because it may introduce so-called **Hamming cliffs**.

To make the example precise, we briefly consider how such a binary encoding of real numbers is computed. Let a real interval $[a, b]$ and an encoding precision ε be given. We look for an encoding rule for any number $x \in [a, b]$ as a binary number z such that the encoded number z differs by less than ε from its actual value x. The basic idea is to divide the interval $[a, b]$ into equally sized segments of a length smaller than or equal to ε. That is, we create 2^k segments with $k = \lceil \log_2 \frac{b-a}{\varepsilon} \rceil$ which are mapped to the numbers $0, \ldots, 2^k - 1$. Thus we obtain the encoding

$$z = \left\lfloor \frac{x - a}{b - a} (2^k - 1) \right\rfloor .$$

Alternatively, we may use the encoding

$$z = \left\lfloor \frac{x - a}{b - a} (2^k - 1) + \frac{1}{2} \right\rfloor ,$$

which allows us to reduce the number of intervals to half, that is, we only need $k = \lceil \log_2 \frac{b-a}{2\varepsilon} \rceil$ segments. Decoding is performed by

$$x' = a + z \cdot \frac{b - a}{2^k - 1} .$$

Note that the decoded x' and the original x may differ, although by at most ε.

As an example, suppose that we want to encode the number $x = 0.637197$ in the interval $[-1, 2]$ with precision $\varepsilon = 10^{-6}$. Then k and z are computed as follows:

$$k = \left\lceil \log_2 \frac{2 - (-1)}{10^{-6}} \right\rceil = \lceil \log_2 3 \cdot 10^6 \rceil = 22 \quad \text{and}$$

$$z = \left\lfloor \frac{0.637197 - (-1)}{2 - (-1)} (2^{22} - 1) \right\rfloor = 2288966_{10}$$
$$= 1000101110110101000110_2 .$$

Studying such an encoding in more detail reveals that it may encode numbers from adjacent segments very differently. That is, although the numbers are close (they differ by no more than 2ε if they come from adjacent segments), their encodings may have a large Hamming distance, where the Hamming distance of two-bit strings is simply the number of different bits. Large Hamming distances can be overcome by mutation and crossover only with great difficulties, simply because they require many bits to be modified. As a consequence, they have been called "Hamming cliffs" to express that they are difficult to "climb" in a solution improvement process.

In order to make the problem perfectly clear, we consider the number range from 0 to 1 encoded by 4 bits. That is, we map the real numbers $\frac{k}{15}$ onto the integer numbers k, thus ensuring an ε of $\frac{1}{30}$. In this case, the encoding of $\frac{7}{15}$ is 0111 the encoding of $\frac{8}{15}$ is 1000. Therefore they have a Hamming distance of 4 because every bit is different. Although the phenotypical distance between the two numbers is comparatively small (only $2\varepsilon = \frac{1}{15}$), the genotypical distance is maximal. As a consequence, if the actual optimum of a parameter optimization problem is at $\frac{8}{15}$, it is of no help to the algorithm to discover that $\frac{7}{15}$ has a high fitness, because it cannot generate the encoding for $\frac{8}{15}$ from the encoding of $\frac{7}{15}$ with few genetic operations.

This problem can be solved by introducing so-called **Gray codes**, which are defined as any binary representation of integer numbers in which adjacent numbers differ in only one-bit. For 4-bit numbers, a possible Gray code is shown in Table 12.1. Note that a Gray code is not unique as can already be seen from the simple fact that any cyclic permutation of the codes in Table 12.1 is again a Gray code.

The most common form of Gray encoding and decoding, respectively, is

$$g = z \oplus \left\lfloor \frac{z}{2} \right\rfloor \quad \text{and} \quad z = \bigoplus_{i=0}^{k-1} \left\lfloor \frac{g}{2^i} \right\rfloor,$$

where $\oplus$ is the *exclusive or* (XOR gate) of the binary representation.

Table 12.1 Gray code of 4-bit numbers

Binary	Gray
0000	0000
0001	0001
0010	0011
0011	0010
0100	0110
0101	0111
0110	0101
0111	0100
1000	1100
1001	1101
1010	1111
1011	1110
1100	1010
1101	1011
1110	1001
1111	1000

As an illustration, we reconsider the example we discussed above, namely representing the real number $x = 0.637197$ in the interval $[-1, 2]$ with precision $\varepsilon = 10^{-6}$. Drawing on the binary number encoding derived above we obtain the Gray code

$$g = 1000101110110101000110_2$$
$$\oplus\, 0100010111011010100011_2$$
$$= 1100111001101111100101_2.$$

12.1.2 Epistasis

Similarly encoded candidate solutions should have a similar fitness

In biology **epistasis** means the phenomenon that one allele of a gene (the so-called epistatic gene) suppresses the effect of all possible alleles of another gene. It might even be that several other genes are suppressed by one epistatic gene. In evolutionary algorithms, epistasis describes the interaction between the genes of a chromosome. That is, how much the fitness of a solution candidate changes if a gene is modified, in extreme cases even which trait of a solution candidate is represented by some genes, strongly depends on the value(s) of (an)other gene(s). This is undesirable.

As a side remark, we mention that in biology epistasis explains certain deviations from the Mendelian laws. For instance, if homozygous black and white seed beans are crossed, then black, white and brown seed beans are obtained in the second offspring generation in a ratio of 12:1:3, which contradicts the Mendelian laws.

In order to illustrate the occurrence and effects of possible epistasis in evolutionary algorithms, we draw on the **traveling salesman problem** as an example (see Sect. 11.6). We consider two possible encodings of round trips through the cities (Hamiltonian cycles) in order to illustrate the epistasis problem:

1. A round trip is represented by a permutation of the cities (as in Definition 11.2 on page 250). This means that the city at the kth position is visited in the kth step. Such an encoding exhibits low epistasis. For instance, swapping two cities alters the fitness (that is, the cost of the round trip) by comparable amounts, regardless of which two cities are swapped. Such a swap also changes the tour only locally, because the part of the tour through the other cities is not affected.

2. A round trip is specified by a list of numbers that state the position of the next city to be visited in a (sorted) list from which all already visited cities have been deleted. An example of how a chromosome is interpreted in this encoding is shown in the top part of Table 12.2. In contrast, this encoding exhibits high epistasis. Modifying a single gene may alter a large part of the trip, and the more so, the closer to the front (top) the gene is located. The bottom part of Table 12.2 shows an extreme case in which the trip is changed entirely, even though the genetic modification is minimal: only a single gene is mutated.

Table 12.2 Impact of a mutation in the traveling salesman problem (second encoding option)

	chromosome	remaining cities	trip
before the mutation	5	1, 2, 3, 4, **5**, 6	5
	3	1, 2, **3**, 4, 6	3
	3	1, 2, **4**, 6	4
	2	1, **2**, 6	2
	2	1, **6**	6
	1	**1**	1

	chromosome	remaining cities	trip
after the mutation	**1**	**1**, 2, 3, 4, 5, 6	1
	3	2, 3, **4**, 5, 6	4
	3	2, 3, **5**, 6	5
	2	2, **3**, 6	3
	2	2, **6**	6
	1	**2**	2

Of course, the reason for the difference in the two encodings is that in the latter the interpretation of the values of any gene depends on the values of the preceding genes, thus introducing a strong dependence between genes, that is, high epistasis.

If the chosen encoding exhibits high epistasis, the optimization problem is often difficult to solve for an evolutionary algorithm. One reason is that epistasis in the extreme form we studied in the above example destroys the assumption that small changes to the chromosomes (which produce similar genotypes) also affect only small changes in the represented candidate solutions (the phenotypes). More generally, if small changes of the genotype can lead to large changes of the fitness, a fundamental assumption underlying evolutionary algorithms, namely that **gradual improvement** is possible (see Sect. 11.3.1), is no longer valid.

It has been tried to classify optimization problems as "easy or hard to solve by an evolutionary algorithm" based on the notion of epistasis [1]. However, this does not work, because epistasis is a property of the encoding and *not of the problem itself*. This can already be seen from the example of the traveling salesman problem, for which we presented an encoding with high and an encoding with low epistasis. In addition, there are problems that can be encoded with low epistasis and that are nevertheless difficult to solve with the help of an evolutionary algorithm.

12.1.3   Closedness of the Search Space

If possible, the search space Ω should be closed under the used genetic operators

In particular if the solution candidates have to satisfy certain constraints, it is not necessarily the case that the genetic operators, if applied to elements of the search space Ω, always yield other (valid) elements of this search space. Generally, we say that an (operator created) individual lies outside of the search space if

- its chromosome cannot be meaningfully interpreted or decoded,
- the represented candidate solution does not fulfill certain basic requirements,
- the represented candidate solution is evaluated incorrectly by the fitness function.

Clearly, such individuals are undesirable and require some action to treat them, so they should be avoided if possible at all. If one faces a situation nevertheless in which the chosen encoding of candidate solutions together with the preferred genetic operators can produce individuals outside of the search space, one has essentially the following options to tackle the problem:

- Choose or design a **different encoding**, which does not suffer from this problem. Note that this may require enlarging the search space.
- Choose or design **encoding-specific genetic operators** under which the search space is closed. That is, find genetic operators which ensure that only elements of the search space can be produced.
- Use **repair mechanisms**, with which an individual outside the search space is modified in such a way that it is brought back into the search space.
- Accept individuals outside the search space in the evolutionary algorithm, but introduce a **penalty term** that reduces the fitness of such individuals, so that the selection process is endowed with a clear tendency to eliminate them.

In order to illustrate the problem and its solutions, recall the n-queens problem as we discussed it in Sect. 11.4. We already mentioned at the beginning of Sect. 12.1 that a better encoding for this problem (than the one used in Sect. 11.4) are chromosomes that are permutations of the numbers $\{1, \ldots, n-1\}$, because this reduces the search space and thus makes it potentially easier for an evolutionary algorithm to find a solution. However, if we apply the same genetic operators, namely, one-point crossover and standard mutation (see Figs. 11.4 and 11.5 on page 247) with this encoding, we may produce chromosomes that are not permutations. This is immediately clear for mutation if it is applied to a single gene, because changing a single number in a permutation cannot possibly yield a permutation: for this at least two numbers have to be changed. A crossover example is shown in Fig. 12.1, in which two permutations are processed with one-point crossover, yielding two children, both of which are not permutations: the top one contains number 5 twice and lacks number 1 while it is the other way round for the bottom child. A closer inspection quickly reveals that it is actually fairly unlikely that a random choice of the parents and the cut point creates a situation in which both children are permutations.

According to the above list of options, we can solve this problem as follows:

- **Different encoding:** The problem vanishes if we allow any sequence of numbers in $\{1, \ldots, n-1\}$, not just permutations. We chose this solution in Sect. 11.4 in order to be able to employ the one-point crossover and standard mutation. However, this

Fig. 12.1 One-point crossover of a permutation

Fig. 12.2 Repair mechanism for permutations

has the disadvantage that the search space is enlarged, thus making it possibly more difficult for an evolutionary algorithm to find a solution.

- **Encoding-specific genetic operators:** Instead of standard mutation, one may use gene pair swaps as a mutation operation (cf. Sect. 12.3.1). Likewise, a permutation-preserving crossover operation can be designed (see Sect. 12.3.2). These choices ensure that the search space (i.e., the set of permutations of $\{1, \ldots, n-1\}$) becomes closed under the genetic operators.
- **Repair mechanisms:** If a genetic operator produced a chromosome that is not a permutation, repair it, that is, modify it in such a way that it becomes a permutation. For example, find and remove duplicate occurrences of the same file (i.e., column) numbers and append the missing numbers (see Fig. 12.2).
- **Penalty term:** Allow the population to contain chromosomes that are not permutations, but add a penalty term to the fitness function, which reduces the fitness of such non-permutations. For example, the fitness could be reduced by the number of missing file (i.e., column) numbers, possibly multiplied by a weighting factor.

For the n-queens problem the best solution is clearly to employ permutation-preserving genetic operators, since they are not much more costly than simple one-point crossover or standard mutation and the evolutionary algorithm certainly benefits from the reduced search space. This is also the solution one would choose for the traveling salesman problem, where essentially the same situation occurs: a solution candidate is best described by a permutation of the cities to visit.

Repair mechanisms are actually closely related to encoding-specific genetic operators, because the application of a genetic operator that may produce an individual outside of the search space followed by the application of a repair mechanism may be seen as one operation w.r.t. which the search space is closed.

However, in certain cases, encoding-specific genetic operators or repair mechanism may complicate the search. If the search space is disconnected, encoding-specific genetic operators never produce individuals in the "forbidden" regions and repair mechanisms immediately restore such individuals to a permissible area. The possibly resulting problem is illustrated in the sketch in Fig. 12.3: if the optimum is in a region of the search space in which no individuals of the initial population are located, it may be very difficult for an evolutionary algorithm to find the optimum, because the "forbidden" regions are difficult to cross. In such cases a penalty term may be preferable, which reduces the fitness of candidate solutions in "forbidden"

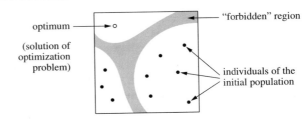

Fig. 12.3 Disconnected areas of a search space complicate the search for an evolutionary algorithm, especially when repair mechanisms are used

regions, but it does not eliminate them altogether. As a consequence, the individuals are left with at least some chance to transgress "forbidden" regions.

12.2 Fitness and Selection

The basic principle of **selection** is that better individuals (that is, individuals with a higher fitness) have better chances to procreate. We referred to this principle before as **differential reproduction** (see Sect. 11.2), because it states that individuals differ in their (expected) reproduction depending on their fitness. How strongly individuals with higher fitness are preferred for producing offspring is called **selective pressure**. High selective pressure means that even small fitness differences can cause considerably differing chances of procreation, while a low selective pressure means that reproduction chances depend only a little on fitness differences.

It should be clear that we need at least some selective pressure for evolutionary algorithms to work, because without any selective pressure the search for a solution is essentially random and thus unlikely to be successful (cf. Sect. 11.2, especially the car parts example). On the other hand, if the selective pressure is very high, the search may focus too quickly on individuals that happened to be the best in the initial population, trying to optimize them further. In such a case, other regions of the search space, which were under-represented in the initial population, may never be explored and thus good solutions contained in them may never be found. This effect is related to the principle of **opportunism** mentioned in Sect. 11.2, namely, that better and optimal solutions are not found if the intermediary stages exhibit certain fitness handicaps. (We consider this problem in more detail below).

More generally, when trying to adjust the selective pressure by choosing a selection mechanism or its parameters, we face the problem that there is a trade-off between **exploration of the search space** and **exploitation of good individuals**. With a low selective pressure we favor search space exploration, because the fitness has only little influence on the reproductive chances. Thus even individuals in regions of the search space with a generally lower fitness have good chances to multiply and hence (due to random modifications and recombinations) to spread over the corresponding region of the search space. On the other hand, with a high selective pressure we favor the exploitation of good individuals, because in this case only highly fit individuals have good reproduction chances and thus their vicinity in the search space will be focused on to find even better modifications of them.

The best strategy is usually a time-dependent selective pressure: in early generations the selective pressure is kept fairly low, so that the search space is well explored, with the hope that we obtain sub-populations in all promising regions of the search space. In later generations the selective pressure is then increased in order to find the best (local) modifications in the promising regions and thus the best solution candidates. The selective pressure can be controlled by either adapting (in particular, scaling) the fitness function (see Sect. 12.2.4) or by choosing a selection method and/or adapting its parameters (see Sects. 12.2.6 and 12.2.7).

12.2.1 Fitness Proportionate Selection

Roulette-wheel selection is certainly among the best-known selection methods. It takes the (absolute) fitness of each individual and computes their relative fitness as

$$f_{\text{rel}}(s) = \frac{f_{\text{abs}}(s)}{\sum_{s' \in \text{pop}(t)} f_{\text{abs}}(s')}.$$

This value is then interpreted as the probability with which the corresponding individual may be selected (to become a member of the next generation). Since with this method the selection probability of an individual is directly proportional to its fitness, this method is also called **fitness proportionate selection**.

Note that the absolute fitness $f_{\text{abs}}(s)$ must not be negative in order for this method to be applicable. This is no real constraint, though. If necessary, we may add a suitable (positive) number to all fitness values and/or we may set all (remaining) negative fitness values equal to zero (provided there are only few). Furthermore, the fitness function must be such that it is to be maximized. Otherwise this methods clearly selects bad individuals with high probability rather than good ones.

The name "roulette wheel selection" for this method stems from the fact that a roulette wheel is a good illustration of how this method works. Each individual is represented by a sector of a roulette wheel (see Fig. 12.4 on the left, which shows a roulette wheel for six individuals s_1 to s_6). The angle (or equivalently, the area) of the sector for an individual s represents its relative fitness value $f_{\text{rel}}(s)$. In addition, there is a marker at the top of the roulette wheel. An individual is selected by setting the roulette wheel into motion and waiting until it stops. Then the individual is chosen that corresponds to the sector the marker points to. For example, in Fig. 12.4 on the

Fig. 12.4 Fitness
proportionate selection and
the dominance problem

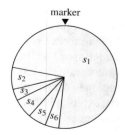

left, individual s_3 is selected. It should be clear that with this procedure, assuming that all angles are equally likely as stopping positions, the probability that a certain individual gets selected equals its relative fitness.

12.2.2 The Dominance Problem

A drawback of fitness proportionate selection is that an individual with a very high fitness may **dominate** the selection, (almost) suppressing all other individuals. This is illustrated in Fig. 12.4 on the right: the individual s_1 has a much higher fitness than all other individuals (even than all other individuals together). Thus it is extremely likely to get selected, while any other individual has only very low chances of getting into the next generation. In subsequent generations, this dominance may become even stronger, since many copies of the dominating individual and many individuals very similar to it will be present. As a consequence, we may observe what is often called **crowding**, that is, the population becomes very dense in one or only few regions of the search space (that is, these regions become "crowded"), while the rest of the search space becomes (almost) void of individuals.

Crowding usually results in very fast convergence to a (local) optimum in the crowded region(s). That is, only one or very few good individuals are exploited (i.e., optimized locally), while a wider exploration of the search space may (almost) cease. As already mentioned, this may be desirable in later generations to actually find the best solutions. In early generations, however, it should be avoided in favor of a thorough exploration of the search space in order to increase the chances that all promising regions are sufficiently covered. Otherwise, **premature convergence** may occur, that is, the search focuses too quickly on those regions of the search space in which the best individuals of the initial population happened to be located.

Figure 12.5 illustrates the drawbacks of such a behavior. Since the region Ω' is much larger than the region Ω'', it may easily happen that the (randomly created) initial population contains only individuals from Ω'. With crowding, the best individuals in this region may be found, but it is highly unlikely that, by random modification and recombination, individuals can reach the region Ω''. The reason is the opportunism of evolution processes: since the transition region between Ω' and Ω'' exhibits severe fitness handicaps, individuals in this region have low chances of surviving the selection process, let alone multiplying themselves. As a consequence, the better solutions in region Ω'' may never be found.

Fig. 12.5 The problem of premature convergence: individuals close to the transition area between Ω' and Ω'' have very low reproduction chances

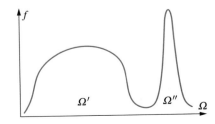

12.2.3 Vanishing Selective Pressure

While the dominance problem demonstrates the disadvantages of large fitness differences between individuals, (very) small differences are also undesirable as they can lead to **vanishing selective pressure**. That is, if the (relative) fitness differences between individuals are too small, fitness proportionate selection does not prefer good individuals enough to push solution candidates towards better and better fitness values. As an illustration, consider the fitness function g in Fig. 12.6. Since the (relative) differences between the fitness values are fairly small, the reproductive chances of all individuals are roughly the same, regardless of where they are located, thus rendering the search almost equivalent to a random search.

More generally, we have to consider how the absolute fitness values relate to their range over the population or even the whole search space. For example, compare the functions f and g depicted in Fig. 12.6. Both have the same shape and thus the same maxima. More precisely, it is $g \equiv f + c$ for some constant $c \in \mathbb{R}$. An evolutionary algorithm may work well for f, because the relative difference of the fitness values provided by f allows for sufficient selective pressure to find the maxima of f. However, since $c \gg \sup_{s \in \Omega} f(s)$, we have $\forall s \in \mathrm{pop}(t) : g_{\mathrm{rel}}(s) \approx \frac{1}{\mu}$ where μ is the population size and thus almost no selective pressure to find the maxima of g.

Since an evolutionary algorithm has a tendency to increase the (average) fitness of individuals from one generation to the next (as better individuals are selected with higher probability), it may even create the problem of vanishing selective pressure itself. Due to a rather random distribution of the initial fitness values, there may be a comparatively high selective pressure in early generations. In later generations, however, the selection of better individuals may have increased the average fitness. As a consequence, the range of fitness values and thus the selective pressure as it results from fitness proportionate selection may be reduced. This is illustrated in Fig. 12.7, which shows a possible situation in an early (or the initial) generation on the left and a possible situation in a later generation on the right. The gray curve is the fitness function, the black dots represent individuals (candidate solutions).

Note that the desirable behavior is exactly opposite: low selective pressure in early generations (to favor exploration) and higher selective pressure in later generations

Fig. 12.6 Vanishing selective pressure: although the maxima are located at the same points, they are differently easy to find with an evolutionary algorithm. With the function g there are only (too) small differences in the relative fitness of individuals

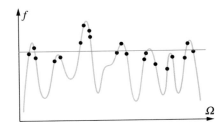

Fig. 12.7 Example of vanishing selective pressure. The dots represent individuals in a very early stage of evolution (left) and in a later generation (right). The average fitness is shown as a line

(to favor exploitation)—see the beginning of this section. As a consequence, this is a particularly unfortunate behavior, which should be counteracted with one of the methods that are presented in the following sections.

12.2.4 Adapting the Fitness Function

A popular possibility to overcome both the dominance problem and the problem of vanishing selective pressure is to **scale the fitness function**. A very simple approach is **linear dynamic scaling**, which scales the fitness values according to

$$f_{\text{lds}}(s) = \alpha \cdot f(s) - \min\left\{ f(s') \mid s' \in \text{pop}(t) \right\}, \qquad \alpha > 0.$$

The factor α determines the strength of the scaling. Instead of the minimum of $\text{pop}(t)$, one may also use the minimum of the last k generations.

An alternative to linear dynamic scaling is the popular σ-**scaling**, which transforms the fitness of all individuals according to

$$f_{\sigma}(s) = f(s) - (\mu_f(t) - \beta \cdot \sigma_f(t)), \qquad \beta > 0,$$

where $\mu_f(t) = \frac{1}{\mu} \sum_{s \in \text{pop}(t)} f(s)$ and $\sigma_f(t) = \sqrt{\frac{1}{(\mu-1)} \sum_{s \in \text{pop}(t)} (f(s) - \mu_f(t))^2}$ are the mean value and the standard deviation, respectively, of the fitness values of the individuals in the current population and β is a parameter.

Obviously, a problem of both approaches is how to choose the parameters α and β. To solve this problem one considers the so-called **coefficient of variation** v of the fitness function (in the current population), which is defined as

$$v = \frac{\sigma_f}{\mu_f} = \frac{\sqrt{\frac{1}{|\Omega|-1} \sum_{s' \in \Omega} \left(f(s') - \frac{1}{|\Omega|} \sum_{s \in \Omega} f(s) \right)^2}}{\frac{1}{|\Omega|} \sum_{s \in \Omega} f(s)}, \quad \text{or} \quad v(t) = \frac{\sigma_f(t)}{\mu_f(t)}.$$

Note that v is defined on the whole search space Ω (left formula). However, in practice (where we certainly cannot compute it for the complete search space Ω, as then we could also carry out a complete enumeration search), it is estimated from its value for the current population $\text{pop}(t)$ (right formula, $\text{pop}(t)$ instead of Ω).

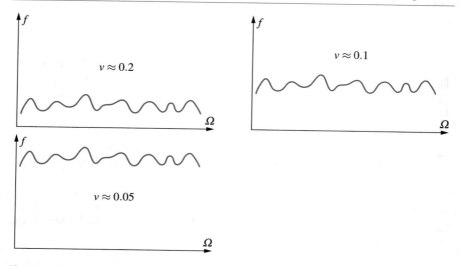

Fig. 12.8 Illustrations of the coefficient of variation. High coefficients of variation can cause premature convergence, low values cause vanishing selective pressure. Empirically it was found that $v \approx 0.1$ yields a good compromise

In order to illustrate the coefficient of variation, Fig. 12.8 shows the fitness function of Fig. 12.6 lifted to different levels, so that the coefficient of variation becomes roughly 0.2 (top left), 0.1 (top right) and 0.05 (bottom left).

Empirically it was found that a value of $v \approx 0.1$ yields a good rapport of exploration and exploitation. As a consequence, if v deviates (significantly) from this (optimal) value, one should try to adapt the fitness function f, for example, by scaling or exponentiation, so that $v \approx 0.1$ is obtained. That is, one computes $v(t)$ for the current populations and then adapts the fitness values in such a way that $v(t)$ becomes 0.1. This is particularly easy for σ-scaling, since β and v are directly related. To be precise, one should choose $\beta = \frac{1}{v^*}$ with $v^* = 0.1$.

Another advisable adaptation of the fitness function is to introduce a **dependence on time**. That is, we do not compute the relative fitness values directly from the function $f(s)$ to be optimized, but from $g(s) \equiv (f(s))^{k(t)}$. The time-dependent exponent $k(t)$ controls the selective pressure and should be chosen in such a way that the coefficient of variation v stays close to $v^* \approx 0.1$. [2] proposes

$$k(t) = \left(\frac{v^*}{v}\right)^{\beta_1} \left(\tan\left(\frac{t}{T+1} \cdot \frac{\pi}{2}\right)\right)^{\beta_2\left(\frac{v}{v^*}\right)^\alpha},$$

where v^*, β_1, β_2, α are parameters, v is the coefficient of variation (estimated, for instance, from the initial population), T is the maximum number of generations to be computed and t is the current point in time (that is, the generation index). For this function [2] recommends $v^* = 0.1$, $\beta_1 = 0.05$, $\beta_2 = 0.1$, $\alpha = 0.1$.

An alternative time-dependent fitness function is **Boltzmann selection**. It determines the relative fitness from $g(s) \equiv \exp\left(\frac{f(s)}{kT}\right)$. The time-dependent temperature parameter T controls the selective pressure and k is a normalization constant that is analogous to the Boltzmann constant. The temperature may decrease (usually linearly) until a predefined maximum number of generations is reached. The idea of this selection method resembles the idea underlying **simulated annealing** (see Sect. 11.5.3): in early generations, the temperature parameter is high and the relative differences between the fitness values are therefore small. In later generations, the temperature parameter is decreased, which causes increasing fitness differences. As a consequence, the selective pressure increases in the course of the generations.

12.2.5 The Variance Problem

Although roulette wheel selection strives to select individuals in proportion to their fitness, deviations from an exactly proportional behavior have to be expected, since the selection process is random. As a consequence, there is no guarantee that good individuals (say, individuals that are better than average) enter the next generation. Even the best individual of the population, although it certainly has the best *chances* to be selected and thus to procreate, may not get selected. More generally, the number of offspring individuals that are produced per individual of the current generation may differ considerably from the expected value. This phenomenon is also known as the **variance problem** of roulette wheel or fitness proportionate selection.

A very simple, though not really advisable solution to this problem is to discretize the range of fitness values. Based on the mean $\mu_f(t)$ and the standard deviation $\sigma_f(t)$ of the fitness values in the population, offspring is created according to the following rule: If $f(s) < \mu_f(t) - \sigma_f(t)$, then s is discarded and does not create a descendant. If $\mu_f(t) - \sigma_f(t) \le f(s) \le \mu_f(t) + \sigma_f(t)$, then s is allowed to have one descendant. If $f(s) > \mu_f(t) + \sigma_f(t)$, then two descendants of s are created.

An alternative way to solve the variance problem is to employ the **expected value model**, with which one strives to give each individual a number of children that is close to the expected value. More precisely, in this model $\lfloor f_{rel}(s) \cdot |pop(t)| \rfloor$ offspring individuals are created for every individual of the current population (where $|pop(t)|$ is the size of the population). However, this usually produces too few individuals, since generally $\sum_{s \in pop(t)} \lfloor f_{rel}(s) \cdot |pop(t)| \rfloor < |pop(t)|$. In order to fill the remaining spots in the next generation, different methods may be drawn on, with roulette wheel selection being among the most straightforward choices. Alternatives include techniques that are known from voting evaluation for the apportionment of political mandates or seats in a parliament. Among these are the *largest remainder method*, the *Haré-Niemeyer method*, the *D'Hondt method* etc.

A very elegant way of implementing the expected value model is so-called **stochastic universal sampling**, which can be seen as a variant of roulette wheel selection. As shown in Fig. 12.9, it uses a roulette wheel with as many markers as there are individuals in the population. These markers are located at equal distances around the roulette wheel. Instead of turning the roulette wheel once for each individual to be selected (as in standard roulette wheel selection), the roulette wheel is turned only once and each marker gives rise to one selected individual. For example,

Fig. 12.9 Stochastic universal sampling. This method implements the expected value model, according to which each individual is selected at least as often as the expected number of times rounded to the next lower integer

in Fig. 12.9, s_1 and s_2 get one child each, s_3 and s_5 get two children, whereas s_4 and s_6 do not get any children at all. It should be clear that this selection method guarantees that individuals with better than average fitness (and thus an expected value of children greater than 1) get at least one child and they may receive more than one. Individuals with a lower than average fitness (and thus an expected value of children less than 1) may not procreate at all or may get at most one child.

An alternative approach (which, however, does not guarantee each individual as many children as the integer part of the expected value indicates) is the following procedure: individuals are chosen by roulette wheel selection. After each selection of an individual, its fitness is reduced by a certain amount Δf (and the relative fitness of all individuals is recomputed). If an individual's fitness becomes negative, it is discarded and cannot produce offspring anymore. Methods for choosing Δf include

$$\Delta f = \frac{\sum_{s \in \mathrm{pop}(t)} f(s)}{|\,\mathrm{pop}(t)\,|},$$

which renders this model very similar to the expected value model, and

$$\Delta f = \frac{1}{k} \max\{f(s) \mid s \in \mathrm{pop}(t)\},$$

which restricts the best individual to at most k children. The latter specifically targets the dominance problem, which can be limited with this choice.

12.2.6   Rank-Based Selection

In **rank-based selection** the individuals are sorted w.r.t. their fitness and thus a rank is assigned to each individual of the population. The idea underlying this method stems from statistics, especially from distribution-free techniques like rank correlation. Each rank is assigned a probability, with higher ranks (and thus better individuals) receiving higher probabilities. The actual selection is then be carried out with roulette wheel selection or any of its variants (like the expected value model), with the rank probability taking the place of the relative fitness.

The advantage of rank-based selection is that it decouples the fitness value and the selection probability (which are proportional in standard roulette wheel selection). Only the order of the fitness values, but not their absolute value determines the selection probability. As a consequence, one can easily avoid the dominance problem, namely, by choosing the rank probabilities in such a way that the higher ranks, although endowed with higher probabilities, do not completely dominate the lower ones. In addition, the progress of an evolutionary algorithm no longer produces a vanishing selection pressure (cf. Sect. 12.2.3), since the distribution of the rank probabilities does not depend on the distribution of the fitness values. By adapting the rank probabilities over time, one even has a convenient way of controlling how the selective pressure develops. By moving probability mass from the lower to the higher ranks, one can easily install a drift from exploration to exploitation.

The only disadvantage of rank-based selection is the fact that the individuals need to be sorted which causes an computational effort of $O(|\,\mathrm{pop}\,| \cdot \log_2 |\,\mathrm{pop}\,|)$. In contrast, fitness proportionate selection, at least in the form of stochastic universal sampling, is linear in the number of individuals, because it only has to sum the fitness values and then to select $|\,\mathrm{pop}\,|$ individuals.

12.2.7 Tournament Selection

In **tournament selection** a certain number k of individuals are drawn at random from the current population. These individuals carry out a tournament, which is won by the individual with the highest fitness (ties are broken arbitrarily). As a prize, the winning individual receives a descendant in the next generation. After the tournament, all participants are returned to the current population (including the winner). Since each tournament selects one individual, $|\,\mathrm{pop}\,|$ tournaments have to be carried out to select all individuals of the next generation.

Note that all individuals have the same probability to be chosen to participate in a tournament. Their fitness does not influence how likely it is that they *participate* in a tournament, but only how likely it is that they *win* a tournament in which they participate. Clearly, individuals with a high fitness are more likely to win the tournaments in which they participate. However, individuals with a low fitness may still produce offspring, namely, if they happen to participate in a tournament in which all other participants have an even lower fitness.

The individuals that are to participate in a tournament may be drawn with or without replacement, which usually does not make much of a difference given the fact that typical population sizes in an evolutionary algorithm are in the thousands. The number k, the **tournament size**, is a parameter of this selection method that has to be chosen by a user, $k \in \{2, 3, \ldots, |\,\mathrm{pop}\,|\}$. With this parameter the selective pressure can be controlled: the larger the tournament, the higher the selective pressure. If tournaments are small, individuals with a low fitness have a higher chance of finding themselves in a tournament in which all of their opponents have even lower fitness. Only the $k - 1$ worst individuals do not have any chance of reproducing at all (assuming that all fitness values are different). On the other hand, the larger the tournaments, the higher the chance that at least one participant has a high fitness, wins the tournament, and thus deprives less fit participants of potential offspring.

Tournament selection is an excellent method to tackle the dominance problem, since the fitness value does not directly influence the selection probability. For example, even for the best individual the expected number of offspring is only the expected number of tournaments in which this individual participates. This number is controlled by the tournament size k, but not by the individual's fitness.

A modification of tournament selection is to replace the deterministic rule that the best participating individual wins the tournament by a fitness proportionate selection. That is, for each individual the tournament-specific relative fitness is computed and the winner is determined by roulette wheel selection. This modification allows worse individuals to produce offspring even if they participate only in tournaments in which at least one other participant has a better fitness.

An important advantage of tournament selection is that it lends itself perfectly to parallelization. While fitness proportionate selection requires a central agent that collects and normalizes the fitness values (computes the relative fitness values), arbitrarily many tournaments can be carried out in parallel without any need for centralized computations or communication between the tournament hosts. As a consequence, parallelized implementations often use the tournament selection.

12.2.8 Elitism

Due to the randomness of the selection procedures we discussed up to now, only the expected value model (and some of its variants) ensures that the best individual enters the next generation. However, even if this is the case, the best individual is not protected from modifications by genetic operators (mutation or recombination with another individual). As a consequence, it is not guaranteed that the quality of the best solution candidate never worsens from one generation to the next.

As this possibility is clearly undesirable, evolutionary algorithms often employ a technique known as **elitism**. That is, the best individual (or, alternatively, the k best individuals with k to be chosen by a user) are transferred without modification to the next generation. This ensures that the best solution(s) that have been found up to now—that is, the *elite* of the population—never gets lost or destroyed. Note, though, that (other copies of) these elite individuals still enter the normal selection and modification process in the hope that they may be improved by genetic operators.

A closely related technique is **local elitism**, which refers to the treatment of individuals that undergo modification by genetic operators. In a standard evolutionary algorithm the products of an application of a genetic operator (mutation or crossover) enter the new generation, while the originals are discarded. We may also say that the products (children) replace the originals (parents). With local elitism, however, the fitness of the involved individuals decides which actually enter the next generation. For example, a mutated individual replaces its original only if it has a better fitness. Of the four individuals involved in a crossover (two parents, two children), the best two are determined and passed on to the next generation (which may mean that the two parents are maintained and the children are discarded).

Evolutionary algorithms employing (global or local) elitism usually exhibit better convergence characteristics, since local optima are approached more consistently. However, especially local elitism bears a certain danger of converging prematurely and getting stuck in local optima, because as no (local) degradations are possible.

12.2.9 Environmental Selection

The environmental selection is used to select the individuals to be included in the new population in the next generation. The simplest environmental selection scheme is the so-called generational selection mechanism. In this case, the newly generated individuals take over the new population and in this way completely replace the old population. Using this mechanism, we ignore the individuals in the old population, even if some of them are better than the newly produced individuals. In order to avoid this, it is usually combined with elitism. The two other environmental selection mechanisms are (μ, λ) and $(\mu + \lambda)$ selection mechanisms. In both of these mechanisms, we assume that the population has μ individuals from which we have produced λ offspring and $\mu \leq \lambda$. In the (μ, λ) scheme, we take the best μ individuals from the offspring population for the new population in the next generation. In the $(\mu + \lambda)$ scheme, we take the best μ individuals from both the population and the offspring population. In this way, the good individuals can survive in the population for many generations. Another environmental selection scheme is the so-called steady-state approach. Steady-state mechanism usually produces one offspring from the population. This offspring replaces the worst solution of the population which builds the new population for the next generation. While this approach seems to be very slow (producing one individual in one generation), it removes the worst solutions from the population and these solutions do not have any survival chance. This can help to improve the convergence over several generations for certain problems.

12.2.10 Niche Techniques

The objective of **niche techniques** is to prevent crowding as it was discussed in Sect. 12.2.1. Here we consider *deterministic crowding* and *sharing*.

The idea of **deterministic crowding** is that generated offspring should always replace those individuals in the population that are most similar. As a consequence, the local density of individuals in the search space cannot grow so easily. Of course, in order to be able to apply this idea, we need a similarity or distance measure for the individuals. If chromosomes are binary coded, the *Hamming distance* may be a viable choice. In other cases specialized similarity or distance measures may be needed that take the concrete encoding of solutions candidates into account. As a consequence, it is not possible to provide generally applicable measures.

A variant of deterministic crowding, which includes ideas of elitism (see the preceding section), is the following approach: in a crossover, the two parents and two children are grouped into two pairs, each consisting of one parent and one child.

The guiding principle is that a child is assigned to the parent to which it is more similar. If both children happen to be assigned to the same parent, the child that is less similar is re-assigned to the other parent. Ties are broken arbitrarily. From each pair the better individual is selected and passed on into the next generation. The advantage of this variant is that much fewer similarity computations are needed than in a global approach that finds the most similar individuals in the population as a whole.

The idea of **sharing** is to reduce the fitness of an individual if there are other individuals in its neighborhood in the search space (and hence we we need again a similarity or distance measure for individuals). Intuitively, the individuals share the resources of a niche, that is, a region in the search space, which has a negative effect on their fitness. A possible choice for the fitness reduction is

$$f_{\text{share}}(s) = \frac{f(s)}{\sum_{s' \in \text{pop}(t)} g(d(s, s'))},$$

where d is a distance measure for individuals and g defines both shape and size of the niche. A concrete example is so-called **power law sharing**, which employs

$$g(x) = \begin{cases} 1 - \left(\frac{x}{\varrho}\right)^{\alpha} & \text{if } x < \varrho, \\ 0, & \text{otherwise,} \end{cases}$$

where ϱ is the radius of the niche and α controls the strength of the influence that individuals in the niche have on each other.

12.2.11 Characterization of Selection Methods

Selection methods are often characterized by certain terms, which describe their properties. Some of the most important ones are collected in Table 12.3.

The distinction of "static" and "dynamic" mainly refers to whether the selective pressure changes over time (preferably from exploration—low selective pressure—to exploitation—high selective pressure), which is governed by the (relative) selection probabilities for individuals with different fitness. Characterizing selection methods as "extinguishing" versus "preservative" based on whether selection probabilities may be zero or not may be somewhat misleading, since even preservative methods allow for solution candidates to go extinct. The reason is simply that due to the randomness of the selection procedures a positive probability does not guarantee survival. The purpose of allowing individuals to have offspring in only one generation ("pure-bred") rather than in multiple generations ("under-bred") is to reduce the danger of crowding, since offspring tends to be similar to its parents. "Left" selection methods, in which the best individuals are not allowed to reproduce, are meant to prevent premature convergence by forcing offspring to result from worse individuals, thus explicitly favoring exploration. In contrast to this, "right" selection methods do not introduce such a guidance. Finally, "on the fly" methods are constantly modifying the population, as it is also the case in nature, which "generational" methods employ a strict discretization of time.

Table 12.3 Characterization of selection methods

Term	Meaning
Static	The probability of selection remains constant
Dynamic	The probability of selection changes
Extinguishing	Probabilities of selection may be 0
Preservative	All probabilities of selection must be greater than 0
Pure-bred	Individuals can only have offspring in one generation
Under-bred	Individuals are allowed to have offspring in more than one generation
Right	All individuals of a population may reproduce
Left	The best individuals of a population may *not* reproduce
Generational	The set of parents is fixed until all offspring are created
On the fly	Created offspring directly replace their parents

12.3 Genetic Operators

Genetic operators are applied to a certain part of the individuals in a generation to create modifications and recombinations of existing solution candidates. Although the majority of these modifications can be expected to be harmful, there is reasonable hope that a few of these modifications result in (slightly) better individuals.

Genetic operators are commonly categorized according to the number of parents into **mutation** or **variation** operators (only one parent, see Sect. 12.3.1), **crossover** operators (two parents, see Sect. 12.3.2), and multi-parent operators (see Sect. 12.3.3). The latter two categories (that is, with more than one parent) are sometimes generally called **recombination** operators.

An important aspect of genetic operators is whether the search space is closed under their application (see Sect. 12.1.3). For example, if candidate solutions are encoded as permutations (see, for instance, the traveling salesman problem, Sect. 11.6), then the genetic operators should preserve this property. That is, if they are applied to permutations, the results should also be permutations.

12.3.1 Mutation Operators

Genetic one-parent operators are generally referred to as **mutation** or **variation** operators. Such operators mainly serve the purpose to introduce an element of **local search**, that is, to produce a solution candidate that is (very) similar to its parent.

If solution candidates are encoded as bit strings (that is, the chromosomes are composed of zeroes and ones), then **bit mutation** (also called *bit flipping*) is commonly chosen. It consists in flipping randomly chosen alleles, that is, turning a 1 into a 0 and *vice versa*. The following algorithm formalizes this operation:

Algorithm 12.1 (Bit Mutation)

procedure mutate_bits (**var** s: array of bit, p_m: real);
begin   (∗ mutation rate p_m ∗)
  for $i \in \{1, \ldots, \text{length}(s)\}$ **do begin**
    $u \leftarrow$ randomly choose according to $U([0, 1))$;
    if $u \leq p_m$ **then** $s_i \leftarrow 1 - s_i$; **end**
  end
end

Empirically it was found that choosing $p_m = 1/\text{length}(s)$ is often close to optimal.
  While the number of bits that are flipped in standard bit mutation may vary (to be precise: it follows a binomial distribution with the parameter p_m), a variant fixes the number of randomly chosen bits to be flipped. In this case the mutation rate p_m is replaced by a number n, $1 \leq n < \text{length}(s)$, of bits to be flipped, or a fraction p_b, $1 < p_b < 1$, of bits, which translates into the number of bits by $n = \lfloor p_b \cdot \text{length}(s) \rfloor$. We refer to this variant as n-**bit mutation**. The special case $n = 1$, which flips exactly one randomly chosen bit of the chromosome, is called **one-bit mutation**.
  For chromosomes that are arrays of real-valued numbers so-called **Gaussian mutation** is most commonly used. It adds a random number that is sampled from a normal or Gaussian distribution to every gene, as shown in the following algorithm:

Algorithm 12.2 (Gaussian Mutation)

procedure mutate_gauss (**var** x: array of real, σ: real);
begin   (∗ standard deviation σ ∗)
  for $i \in \{1, \ldots, \text{length}(x)\}$ **do begin**
    $u \leftarrow$ sample randomly from $N(0, \sigma)$;
    $x_i \leftarrow x_i + u$;
    $x_i \leftarrow \max\{x_i, l_i\}$;   (∗ lower limit l_i of range of x_i ∗)
    $x_i \leftarrow \min\{x_i, u_i\}$;   (∗ upper limit u_i of range of x_i ∗)
  end
end

The parameter σ controls the spread of the random numbers and corresponds to the standard deviation of the normal or Gaussian distribution. It may be used to control to some degree whether exploration of the search space should be favored (large σ) or whether a local optimization should be performed (exploitation, small σ).
  An extension of Gaussian mutation is so-called **self-adaptive Gaussian mutation**. Instead of a single standard deviation σ, which is the same for all chromosomes, self-adaptive Gaussian mutation endows each chromosome x with its own standard deviation σ_x, which is used to mutate its genes. In addition, not only the genes of the chromosome x, but also its standard deviation σ_x is adapted in a mutation operation. The following algorithm formalizes the operation:

Algorithm 12.3 (Self-adaptive Gaussian Mutation)

procedure mutate_gsa (**var** x: array of real, **var** σ_x: real);
begin   (∗ chromosome-specific standard deviation σ_x ∗)
  $u \leftarrow$ sample randomly from $N(0, 1)$;
  $\sigma_x \leftarrow \sigma_x \cdot \exp(u/\sqrt{\text{length}(x)})$;
  for $i \in \{1, \ldots, \text{length}(x)\}$ **do begin**
    $u \leftarrow$ randomly choose according to $N(0, \sigma_x)$;
    $x_i \leftarrow x_i + u$;
    $x_i \leftarrow \max\{x_i, l_i\}$;   (∗ lower limit $l[i]$ of range of x_i ∗)
    $x_i \leftarrow \min\{x_i, u_i\}$;   (∗ upper limit $u[i]$ of range of x_i ∗)
  end
end

The idea of self-adaptive Gaussian mutation is to exploit the process of evolutionary adaptations not only to find good solution candidates, but at the same time to optimize the mutation step widths (evolution-strategic principle, see Sect. 11.2). Intuitively, we may say that chromosomes with a "suitable" standard deviation—that is, a standard deviation that causes steps of "suitable width" in the region of the search space in which the chromosome is located—are more likely to produce good offspring. As a consequence, this standard deviation will spread in the population, at least among individuals located in the same region of the search space.

A generalization of (one-)bit mutation to chromosomes consisting of (more or less) arbitrary computational objects is so-called **standard mutation**. It simply replaces the current allele of a randomly chosen gene with another randomly chosen allele. An example for a chromosome consisting of integer numbers (as we used it, for example, in Sect. 11.4) is shown in Fig. 12.10: in the third gene the allele 4 is replaced with the allele 6. If desired, multiple genes may be mutated (cf. the n-queens problem in Sect. 11.4, especially Fig. 11.5 on page 247).

Like bit mutation, standard mutation receives as a parameter either a mutation rate p_m, which indicates the probability that a gene gets mutated, or a number n of genes that are to be mutated. The new allele is simply chosen from the other possible alleles using a uniform distribution (equal probabilities).

The mutation operator **transposition** or **pair swap** exchanges the alleles of two randomly chosen genes in a chromosome with each other (see Fig. 12.11). Of course, this operator can only be applied is the two affected genes have the same set of pos-

Fig. 12.10 Example of standard mutation: an allele of a gene is replaced by another allele

Fig. 12.11 Example of transposition: two genes exchange their alleles in a chromosome

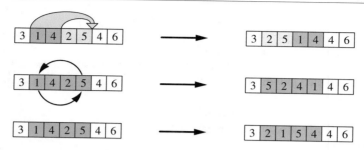

Fig. 12.12 Mutation operators on subsequences: shift, inversion, and arbitrary permutation

sible alleles. Otherwise a chromosome describing an individual outside of the search space may result, which needs special treatment (cf. Sect. 12.1.3). Transposition is an excellent mutation operator in case the chromosomes are permutations of a set of integer numbers (like for traveling salesman problem, see Sect. 11.6), because the set of permutations is obviously closed under it.

Generalizations of transposition are different forms of (constrained) permutations of the alleles of a group of genes, usually forming a subsequence of 3, 4, . . . , k genes in the chromosome. Among these are the **shift** of a subsequence to a new location (which may also be seen as a **cyclic permutation**), the **inversion** of a subsequence (that is, reversing the order of the alleles) and finally applying an arbitrary **permutation** to the alleles in a subsequence. Figure 12.12 illustrates these operators for chromosomes consisting of integer arrays. Clearly, all of these operators require that the sets of alleles of the genes in the affected subsequence are equal. They may be parametrized with the length of the subsequence or with a probability distribution over such lengths. Since all of these operators merely permute the alleles in a subsequence, they clearly preserve the property that a chromosome is a permutation of a set of numbers and thus are nicely applicable to cases where this is required to make the search space closed under the mutation operators (like the traveling salesman problem, see Sects. 11.6 and 12.1.3).

12.3.2 Crossover Operators

Genetic operators that involve two-parent individuals are generally referred to as *crossover* operators. The best-known crossover operator is so-called **one-point crossover** (see Fig. 12.13): a cut point is chosen at random and the gene sequences on one side of the cut point are exchanged between the two (parent) chromosomes.

A straightforward extension of one-point crossover is **two-point crossover** (see Fig. 12.14). In this case two cut points are chosen at random and the section between the two cut points is exchanged between the (parent) chromosomes.

A natural generalization of these forms is n**-point crossover**, for which n cut points are chosen. Offspring is created by alternately exchanging and maintaining the parental gene sequences between two consecutive cut points.

Fig. 12.13   Example of one-point crossover

Fig. 12.14   Example of two-point crossover

As an example and because we draw on it in later algorithms (cf. Sect. 13.1), we formulate the one-point crossover as an algorithm (note that the value of "allele" depends on the encoding; it may be a bit, an integer, a real-valued number, etc.):

Algorithm 12.4 (One-point Crossover)

procedure crossover_1point (**var** r, s: array of allele);
begin   (* exchange part of two chromosomes *)
  $c \leftarrow$ random element of $\{1, \dots, \text{length}(s) - 1\}$;   (* choose cut point *)
  for $i \in \{0, \dots, c - 1\}$ **do begin**   (* traverse the section to exchange *)
    $t \leftarrow r_i; r_i \leftarrow s_i; s_i \leftarrow t$; **end**   (* swap genes up to the cut point *)
end

Instead of choosing a certain number of cut points, **uniform crossover** determines for each gene independently whether it is exchanged or not based on an exchange probability p_x (see Fig. 12.15; a "+" means that the genes are exchanged, a "−" that they are kept). Note that uniform crossover is *not* equivalent to $(L-1)$-point crossover (where $L = \text{length}(s)$), as $(L - 1)$-point crossover alternately exchanges and keeps genes. Uniform crossover rather chooses the number of cut points randomly. (Note, however, that the different numbers are *not* equally likely: numbers close to $L/2$ are more likely than small or large numbers of cut points.)

  Shuffle crossover shuffles the genes randomly before applying an arbitrary two-parent operator and then de-shuffles the genes again. Most commonly, shuffle crossover is implemented with one-point crossover, as shown in Fig. 12.16. It should be noted that shuffle crossover is *not* equivalent to uniform crossover. With uniform crossover the number of exchanged genes is binomially distributed with the parameter p_x. With shuffle crossover in combination with one-point crossover, however, every number of gene exchanges between the chromosomes is equally likely. This makes shuffle crossover one of the most recommendable crossover operators.

  If the chromosomes are permutations, none of the crossover operators we studied up to now is suitable, because none guarantees that the children are permutations. For

Fig. 12.15 Example of uniform crossover. For every gene it is determined independently whether it is exchanged (+) or not (−)

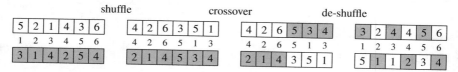

Fig. 12.16 Example of shuffle crossover (which is a modified one-point crossover)

Fig. 12.17 Example of uniform order-based crossover

permutations, special so-called permutation-preserving crossover operators should be used, of which we discuss two variants here.

Uniform order-based crossover resembles uniform crossover, since it also decides for each gene independently whether it should be kept or not (based on an keep probability p_k). It differs from uniform crossover in how the genes are handled that are not to be kept, that is, in how the gaps between the genes to keep are filled. These gaps cannot be filled by simply taking the corresponding genes from the other parent (as in uniform crossover), as this could create a chromosome that is not a permutation. Rather the missing numbers are found and inserted into the gaps in the order (hence "order-based") in which they occur in the other parent.

An example is shown in Fig. 12.17: the plus signs mark the genes that are to be kept. This leaves three gaps to be filled. In the top chromosomes the numbers 3, 6, and 7 are missing, which occur in the bottom parent in the order 3, 7, 6. In this order they are entered into the gaps of the top chromosome. In the bottom chromosome the numbers 2, 5, and 7 are missing, which occur in the top parent in the order 5, 7, 2. In this order they are entered into the gaps of the bottom chromosome.

Note that not only the space of permutations is closed under this operator, but that this operator also preserves order information. Clearly, the kept genes are in the same order, while the genes with which the gaps are filled are in the order in which they occur in the other parent. This property can be useful for certain problems and corresponding encodings of the solution candidates.

An alternative permutation-preserving crossover operator is so-called **edge recombination**. It has been developed specifically for the traveling salesman problem (see Sect. 11.6), in which the round trips are encoded as permutations of the cities. In this method the chromosome is seen as a graph or, more precisely, as a chain or ring

Table 12.4   Example of an edge table for edge recombination (left, cf. Fig. 12.18) and constructing offspring from such an edge table by edge recombination (right, cf. Fig. 12.19)

	Neighbor											
Allele	In A	in B	Aggregated	Allele	Neighbors	Choice: 6	5	2	7	4	3	1
1	3,5	6,4	3,4,5,6	1	3,4,5,6	3,4,5	3,4	3,4	3,4	3		
2	5,7	7,5	5*,7*	2	5*,7*	5*,7*	7*	**7***	–	–	–	–
3	6,1	4,7	1,4,6,7	3	1,4,6,7	1,4,7	1,4,7	1,4,7	1,4	1	**1**	–
4	7,6	1,3	1,3,6,7	4	1,3,6,7	1,3,7	1,3,7	1,3,7	1,3	**1,3**	–	–
5	1,2	2,6	1,2*,6	5	1,2*,6	1,2*	**1,2***	–	–	–	–	–
6	4,3	5,1	1,3,4,5	6	1,3,4,5	**1,3,4,5**	–	–	–	–	–	–
7	2,4	3,2	2*,3,4	7	2*,3,4	2*,3,4	2*,3,4	3,4	**3,4**	–	–	–

$$A: \boxed{6}\,\boxed{3}\,\boxed{1}\,\boxed{5}\,\boxed{2}\,\boxed{7}\,\boxed{4} \qquad\qquad B: \boxed{3}\,\boxed{7}\,\boxed{2}\,\boxed{5}\,\boxed{6}\,\boxed{1}\,\boxed{4}$$

Fig. 12.18   Example parent chromosomes for edge recombination

of edges: each allele is connected by edges to its neighbors in the chromosome. In addition, the first and the last allele are connected by an edge. The crossover operation consists in recombining the edges of two-parent rings (hence the name of the method). It preserves neighborhood information rather than order information.

Edge recombination is a rather complex method and proceeds in two steps. In the first step, an **edge table** is constructed as follows: for every allele its neighbors (in both parents) are listed (including the last allele as a neighbor of the first and *vice versa*). If an allele has the same neighbor in both parents (where the side is irrelevant), this neighbor is listed only once, but it is marked to indicate that it has to be treated specially. As an example, Table 12.4 shows on the left an edge table for the two-parent chromosomes depicted in Fig. 12.18. In the column "aggregated" duplicate neighbors are listed only once and marked with a star.

In the second step, a child is constructed, as demonstrated in Table 12.4 on the right. For the first allele of the child the first allele of a randomly chosen parent is taken. That is, with the example parents shown in Fig. 12.18, we may start either with 6 (first allele of A) or with 3 (first allele of B). For this example we choose 6. The chosen allele is deleted from all neighbor lists in the edge table and its own list of neighbors is retrieved (see the third column of the table on the right of Table 12.4; the list of neighbors is shown in bold print). From this neighbor list an allele is chosen respecting the following precedences:

1. marked neighbors (i.e., neighbors that occur in both parents),
2. neighbors with the shortest neighborhood list (marked neighbors count once),
3. any neighbor,
4. any allele not yet in the child.

If there are multiple choices in the highest applicable precedence class, a random allele from this class is chosen (i.e., ties are broken arbitrarily). In Table 12.4 allele 6

6	5	2	7	4	3	1

Fig. 12.19 Example child constructed from the parents in Fig. 12.18 by edge recombination

has the neighbors 1, 3, 4, and 5. None of these is marked, so we move on to the second precedence class. Here the neighbor lists of 1, 3, and 4 all contain four elements, while the neighbor list of 5 contains only three (due to the fact that allele 2 is marked in this list). Therefore we have to choose allele 5, which thus becomes the second allele of the child. The process is then repeated with this allele: it is deleted from all entries in the edge table and a neighbor from its list of neighbors is chosen according to the above precedences (see the fourth column of the table on the right of Table 12.4; the list of neighbors is again shown in bold print). Since allele 2 is marked in the neighbor list, it becomes the third allele of the child. The neighbor selection process is then iterated until the child is complete, as shown in the remaining columns of the table on the right of Table 12.4. It finally yields the child chromosome depicted in Fig. 12.19.

In analogy to this process, a second child may be constructed from the first allele of the other parent (here: 3, since we started with 6). However, this is rarely done; rather only a single child is constructed from every pair of parent chromosomes.

Note that the precedence rules prefer marked alleles to non-marked in order to increase the chances that edges that are present in both parents are also present in the child. Alleles with short neighbor lists are preferred in order to reduce the chances that in a later step the fallback class (any allele not yet in the child) has to be invoked, because this introduces a new edge that is not present in either parent. The rationale is very simple: short neighbor lists run a higher risk of becoming empty due to allele selections, so one should choose from them earlier than from longer lists.

Note also that edge recombination may be employed just as well if the first and the last allele of a chromosome are *not* to be regarded as neighbors. In this case the corresponding edges (that close the ring or cycle) are simply excluded from the edge table. On the other hand, if we regard the first and last allele as neighbors, then we may, in principle, select an arbitrary starting allele, and not just one of the first alleles of the parents. This constraint (or its equivalent, namely, having to choose the *last* allele of one of the parents as a starting point) is needed only if the first and the last allele are *not* regarded as neighbors, so that the child is constructed from one of the end points of the chains that are represented by the parents.

12.3.3 Multi-parent Operators

Diagonal crossover is a recombination operator that can be applied to three or more parents and that can be seen as a generalization of one-point crossover. For k parent chromosomes, one randomly chooses $k-1$ distinct cut points in $\{1, \ldots, L-1\}$, where L is the length of the chromosomes. For $i = 2, \ldots, k$ the ith section (between the $(i-1)$th and the ith cut point, where the kth cut point is the end of the chromosomes)

Fig. 12.20 Example of diagonal crossover

is then shifted cyclically $(i - 1)$ steps across the k chromosomes. As an example, Fig. 12.20 shows diagonal crossover for three parent chromosomes and hence two cut points. Diagonal crossover is said to lead to a very good exploration of the search space, especially for a large number of parents (around 10–15).

12.3.4 Characteristics of Recombination Operators

Recombination operators are often categorized based on whether they have certain properties, which can aid in selecting the best operators for a given problem. Common properties that are used to characterize recombination operators with this purpose include whether they exhibit positional bias and/or distributional bias.

A recombination operator exhibits **positional bias** if the probability that two genes are jointly inherited from the same parent depends on the (relative) position of these genes in the chromosome. Positional bias is generally an undesirable property, because it can make the exact arrangement of the different genes in a chromosome crucial for the success or failure of an evolutionary algorithm. If genes for certain traits are located in an unfortunate way w.r.t. the positional bias of a crossover operator, it may be very difficult to optimize the allele combination of these genes.

A simple example of a recombination operator that exhibits positional bias is one-point crossover: the probability that two genes are jointly inherited is the higher, the closer the genes are together in the chromosome. The reason is that genes that are close together have only few possible cut points between them. Only if one of these cut points is chosen, they are separated from each other. Since all cut points are equally likely, the probability that two genes are jointly inherited is inversely proportional to the number of cut points between the genes and thus to their distance in the chromosome. As an extreme case consider the first and the last gene of a chromosome. They can never be jointly inherited under one-point crossover, because *any* choice of a cut point separates them from each other. On the other hand, two genes that are neighbors in a chromosome are separated by one-point crossover only with a probability of $1/(L - 1)$, where L is the length of the chromosome.

A recombination operator exhibits **distributional bias** if the probability that a certain number of genes is exchanged between the parent chromosomes is not the same for all possible numbers of genes. Distributional bias is often undesirable, because it causes partial solutions of different lengths to have different chances of progressing to the next generation. However, distributional bias is usually less critical (that is, more easily tolerable) than positional bias.

A simple example of a recombination operator that exhibits distributional bias is uniform crossover. Since for every gene it is decided with probability p_x and independently of all other genes whether it is exchanged or not, the number K of exchanged genes is binomially distributed with the parameter p_x. That is, we have

$$P(K = k) = \binom{L}{k} p_x^k \, (1 - p_x)^{L-k},$$

where L is the total number of genes. Consequently, very small and very large numbers are less likely than numbers close to Lp_x. In contrast to this, one-point crossover exhibits no distributional bias: all cut points are equally likely and since the genes on one side of the cut point are exchanged, all numbers of exchanged genes are obviously equally likely. An example of a crossover operator that exhibits neither positional nor distributional bias is *shuffle crossover* based on one-point crossover.

12.3.5 Interpolating and Extrapolating Recombination

All recombination operators that we discussed up to now merely recombine alleles that already exist in the parent chromosomes, but do not create any new alleles. As a consequence, their effectiveness depends crucially on the diversity of the population. If there is only little variation in the population, recombination operators cannot create sufficiently different offspring and thus the search may be confined to certain limited regions of the search space that can be reached with the individuals of the (initial) population. On the other hand, if a population is very diverse, such recombination operators can explore the search space well.

Especially in the realm of numeric parameter optimization, however, a different kind of recombination operator becomes possible, which can blend the traits of the parents in such a way that offspring with new traits is created, encoded by alleles that are not present (at least in exactly this form) in the parents. An example of such an operator is interpolating recombination, which blends alleles of the parents with a randomly chosen mixing parameter. A more concrete example for chromosomes that are real-valued arrays is **arithmetic crossover**, which can be seen as interpolating between the points that are represented by the parent chromosomes:

Algorithm 12.5 (Arithmetic Crossover)

function crossover_arith $(r, s$: array of real$)$: array of real;
begin
  s' ← new array of real with length(r) elements;
  u ← choose randomly from $U([0, 1])$;
  for $i \in \{1, \ldots, \text{length}(r)\}$ **do**
    $s'_i \leftarrow u \cdot r_i + (1 - u) \cdot s_i$;
  return s';
end

It should be noted, though, that an exclusive application of such a blending operator can cause the so-called **Jenkins nightmare**, that is, the complete disappearance of all variation in a population, because the blending operation introduces a strong tendency to average all parameters (that is, all genes) that are present in the population. Therefore arithmetic crossover should be combined—at least in the early generations of an evolutionary algorithm, in which exploration is vital—with a strongly random-based, diversity-preserving mutation operator.

An alternative is extrapolating recombination operators, which try to infer information from several individuals. Intuitively, they create a prognosis in what direction from the examined parent individuals one can expect fitness improvements. Hence extrapolating recombination may leave the region of the search space in which the individuals are located, from which the prognosis is derived. Extrapolating recombination is one of only few recombination methods that may take fitness values into account. A simple example of an extrapolating operator is arithmetic crossover with $u \in U([-1, 2])$. It should be noted, though, that an extrapolating recombination operator usually cannot compensate a lack of diversity in a population.

References

1. Davidor Y (1990) Lamarckian sub-goal reward in genetic algorithm. In: Proceedings of European conference on artificial intelligence (ECAI, Stockholm, Sweden). Pitman, London/Boston, United Kingdom/USA, pp 189–194
2. Michalewicz Z (1996) Genetic algorithms + Data structures = Evolution programs, 3rd (extended) edn. Springer, New York, NY, USA

Fundamental Evolutionary Algorithms

13

The preceding chapter presented all relevant elements of evolutionary algorithms, namely, guidelines of how to choose an encoding for the solution candidates, procedures how to select individuals based on their fitness, and genetic operators with which modified solution candidates can be obtained. Equipped with these ingredients we proceed in this chapter to introducing basic forms of evolutionary algorithms, including classical genetic algorithms (in which solution candidates are encoded as bit strings, see Sect. 13.1), evolution strategies (which focus on numerical optimization, see Sect. 13.2) and genetic programming (which tries to derive function expressions or even (simple) program structures with evolutionary principles, see Sect. 13.3). Finally, we take a look at related population-based approaches (like ant colony and particle swarm optimization, see Chap. 14).

13.1 Genetic Algorithms

In nature, all genetic information is described in an essentially quaternary code, based on the four *nucleotides* adenine, cytosine, guanine and thymine, which are stringed together in a DNA sequence on a backbone of phosphate-deoxyribose (so-called *primary structure* of a nucleic acid). Although there are also higher level structures (like the fact that the nucleotides are organized in so-called *codons*, which are triplets of nucleotides), this is the basis of the genetic code. Transferring this structure to computer science, it seems natural to base all encoding on the ultimately binary structure of information in a computer. That is, we use chromosomes that are bit strings. This is the distinctive feature of so-called **genetic algorithms** (GA).

In principle, of course, all evolutionary algorithms running on a computer can be seen as genetic algorithms in this sense, simply because a computer ultimately encodes all information in bits, that is, in zeros and ones. For example, the chromosomes we considered for the *n*-queens problem, which are arrays of integers,

© Springer Nature Switzerland AG 2022
R. Kruse et al., *Computational Intelligence*, Texts in Computer Science,
https://doi.org/10.1007/978-3-030-42227-1_13

are stored using a binary representation of these numbers. To be concrete, the chromosome $(4, 2, 0, 6, 1, 7, 5, 3)$, which describes a solution of the 8-queens problem, could be stored (using 3 bits for each number) as 100 010 000 110 001 111 101 011. The difference between a genetic algorithm and the evolutionary algorithm we considered for the n-queens problem consists in the fact that in a genetic algorithm we consider merely the bit string that a chromosome is and ignore any higher level structure. (In the above concrete example: the fact that bit triplets are interpreted together, each indicating a file (column) in which a queen is placed.) That is, we completely separate the encoding from the genetic mechanisms, while in an evolutionary algorithm certain aspects of the encoding are considered, for instance, to restrict the genetic operators. (In the above example: in the evolutionary algorithm for the 8-queens problem, we allow as cut points for a crossover operator only the points between bit triplets, because these are the interpretable information units, while in a genetic algorithm we allow cuts between arbitrary bits.)

A typical genetic algorithm works like this:

Algorithm 13.1 (Genetic Algorithm)

function genalg (f: function, μ: int, p_x: real, p_m: real) : array of bit;
begin
  $t \leftarrow 0$;   (∗ initialize the generation counter ∗)
  pop(t) ← create a population of μ random bit strings; (∗ μ must be even ∗)
  evaluate pop(t) with the fitness function f;   (∗ compute initial fitness ∗)
  while termination criterion is not fulfilled **do begin**
    $t \leftarrow t + 1$;   (∗ count the created generation ∗)
    pop(t) ← ∅;   (∗ build the next generation ∗)
    pop' ← select μ individuals $s_1, \ldots, s_\mu$ from pop(t)
          with *roulette wheel selection*;
    for $i \leftarrow 1, \ldots, \mu/2$ **do begin**   (∗ process individuals in pairs ∗)
      u ← choose random number from $U([0, 1))$;
      if $u \le p_x$ **then** crossover_1point(s_{2i-1}, s_{2i}); **end**
      mutate_bits(s_{2i-1}, p_m);   (∗ crossover rate p_x and ∗)
      mutate_bits($s_{2i},\ \ p_m$);   (∗ mutation rate p_m ∗)
      pop(t) ← pop(t) ∪ $\{s_{2i-1}, s_{2i}\}$;   (∗ add (modified) individuals ∗)
    end   (∗ to the next generation ∗)
    evaluate pop(t) with the fitness function f;
    environmental selection (pop(t), pop($t - 1$));
    (∗ select individuals for the next population ∗)
  end   return best individual from pop(t); (∗ return the solution ∗)
end

That is, a genetic algorithm follows essentially the scheme of a general evolutionary algorithm (see Algorithm 11.1 on page 242), only that it uses bit strings as its chromosomes and applies genetic operators to these chromosomes that ignore any higher level structure of the encoding. A genetic algorithm requires mainly three parameters: the population size μ, for which an even number is chosen in order to simplify the implementation of a random application of the crossover operator, the crossover probability p_x, with which it is decided for a pair of chromosomes whether they undergo crossover or not, and the mutation probability p_m, with which it is decided for each bit of a chromosome whether it is flipped or not.

13.1.1   The Schema Theorem

So far we answered the question why evolutionary algorithms work only by providing plausibility and intuition-based arguments. Due to their restriction to bit strings, genetic algorithms, however, allow for a more formal investigation, which was first proposed in [23]. This leads to the famous **schema theorem**.

Since genetic algorithms work on bit strings only, they allow us to confine our considerations to binary chromosomes. More precisely, we consider **schemata**, that is, partly specified binary chromosomes. We then investigate how the number of chromosomes matching a schema evolves over several generations of a genetic algorithm. The objective of this investigation is to derive a rough stochastic statement that describes how a genetic algorithm explores the search space.

In order to keep things simple, we confine ourselves (following [23]) to bit strings of a fixed length L. In addition, we generally assume the specific form of a genetic algorithm as it was presented in pseudocode in the preceding section. That is, we assume that chromosomes enter the intermediate population pop′ by fitness proportionate selection (to be precise, by roulette wheel selection as it was introduced in Sect. 12.2.1) and that one-point crossover (see Sect. 12.3.2), applied with probability p_x to chromosome pairs, and bit mutation (see Sect. 12.3.1), using the mutation probability p_m, are employed as the genetic operators.

We start with the necessary technical definitions of *schema* and *matching*.

Definition 13.1 (*Schema*) A *schema* h is a character string of length L over the alphabet $\{0, 1, *\}$, that is, $h \in \{0, 1, *\}^L$. The character $*$ is called *wildcard character* or *don't-care symbol*.

Definition 13.2 (*Matching*) A (binary) chromosome $c \in \{0, 1\}^L$ *matches* a schema $h \in \{0, 1, *\}^L$, written as $c \triangleleft h$, if and only if it coincides with h at all positions where h is 0 or 1. Positions at which h is $*$ are not taken into account (which explains the names "don't care symbol" or "wildcard character" for the character $*$).

As an illustration consider the following simple example: let h be a schema of length $L = 10$ and c_1, c_2 two different chromosomes of this length, which look like this:

$$h = **0*11*10*$$
$$c_1 = 1100111100$$
$$c_2 = 1111111111$$

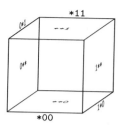

Fig. 13.1 Geometric representation of schemata as hyperplanes in a unit hypercube

Clearly, chromosome c_1 matches schema h, that is, $c_1 \lhd h$, because c_1 differs from h only at positions where h is $*$. On the other hand, chromosome c_2 does not match h, that is, $c_2 \not\lhd h$, because c_2 contains 1s at positions where h is 0 (positions 3 and 9).

Generally, there are 2^L possible chromosomes and 3^L different schemata. Every chromosome matches $\sum_{i=0}^{L} \binom{L}{i} = 2^L$ schemata (because any choice of i positions, at which the bit in the chromosome is replaced with a $*$, yields a schema that the chromosome matches). Based on this fact, [23] started from the idea that the observation of one chromosome corresponds to the observation of many schemata at the same time. This is what Holland called **implicit parallelism**.

Note that a population of size μ can, in principle, match close to $\mu 2^L$ different schemata. However, the number of actually matched schemata is usually a lot smaller, especially in later generations of a genetic algorithm, because the selective pressure produces similar chromosomes. (Assuming that similar chromosomes have similar fitness, selecting individuals with a high fitness has a tendency to select similar chromosomes, cf. Fig. 12.7 on page 267 as an illustration.)

In order to understand schemata better, let us take a look at two interpretations. Geometrically, a schema can be seen as describing a hyperplane in a unit hypercube. Not general hyperplanes, though, but only hyperplanes that are parallel or orthogonal to the sides of the hypercube. This is illustrated in Fig. 13.1 for three dimensions: the schema $*00$ describes the edge connecting the corners 000 and 100 (bottom front). The left face of the cube is captured by the schema $0**$. The schema $***$, which consists of wildcard characters only, describes the whole cube.

An alternative interpretation considers the domain of the fitness function. Suppose we are given a unary fitness function $f : [0, 1] \to \mathbb{R}$ and that the argument of this function is encoded as a binary number (in the usual way—for reasons of simplicity, we disregard here that such a code introduces **Hamming cliffs** (see Sect. 12.1.1), which can be avoided by using a Gray code). In this case a schema corresponds to a "strip pattern" in the domain of the function f. This is illustrated in Fig. 13.2 for the two schemata $0** \ldots *$ (left diagram) and $**1* \ldots *$ (right diagram).

In order to carry out our plan of tracking the evolution of chromosomes that match a schema, we have to examine how selection and applying genetic operators (that is, one-point crossover and bit mutation) influence these chromosomes. We do so in three steps. In the first step we consider the effect of selection, in the second step the effect of one-point crossover, and in the third step the effect of bit mutation.

Fig. 13.2 Representation of schemata as the domain of a function. On the left the schema $0**\dots*$ is shown, on the right the schema $**1*\dots*$

We distinguish the populations in these steps (and quantities referring to them) by splitting the transition from time t to time $t + 1$ into the steps t (original population), $t + \Delta t_s$ (population after selection, that is, intermediate population pop'), $t + \Delta t_s + \Delta t_x$ (population after selection and crossover), $t + \Delta t_s + \Delta t_x + \Delta t_m = t + 1$ (population after selection, crossover, and mutation, which is identical to the new population at time $t + 1$). In these steps we are mainly interested in the (expected) number of chromosomes that match a schema h. We denote these numbers by $N(h, t')$, where $t' \in \{t, t + \Delta t_s, t + \Delta t_s + \Delta t_x, t + 1\}$. Our objective is to derive a (stochastic) relationship between $N(h, t)$ and $N(h, t + 1)$.

In order to capture the effect of selection, we have to consider what fitness the chromosomes have that match a schema h. With fitness proportionate selection, the expected number of offspring of a chromosome s is $\mu \cdot f_{\text{rel}}(s)$. Hence the expected number of chromosomes that match schema h after selection is

$$N(h, t + \Delta t_s) = \sum_{s \in \text{pop}(t), s \vartriangleleft h} \mu \cdot f_{\text{rel}}^{(t)}(s).$$

In order to express this number without referring to individual chromosomes s, it is convenient to define the *mean relative fitness* of chromosomes that match a schema.

Definition 13.3 (*Mean Relative Fitness*) The *mean relative fitness* of chromosomes that match schema h in the population $\text{pop}(t)$ is

$$f_{\text{rel}}^{(t)}(h) = \frac{\displaystyle\sum_{s \in \text{pop}(t), s \vartriangleleft h} f_{\text{rel}}^{(t)}(s)}{N(h, t)}.$$

With this definition we can write $N(h, t + \Delta t_s)$ as

$$N(h, t + \Delta t_s) = N(h, t) \cdot \mu \cdot f_{\text{rel}}^{(t)}(h).$$

By inserting the definition of relative fitness (see page 264), we can transform the expression $\mu \cdot f_{\text{rel}}(h)$ on the right-hand side according to

$$\mu \cdot f_{\text{rel}}(h) = \frac{\displaystyle\sum_{s \in \text{pop}(t), s \vartriangleleft h} f_{\text{rel}}(s)}{N(h, t)} \cdot \mu = \frac{\frac{\sum_{s \in \text{pop}(t), s \vartriangleleft h} f(s)}{\sum_{s \in \text{pop}(t)} f(s)}}{N(h, t)} \cdot \mu = \frac{\frac{\sum_{s \in \text{pop}(t), s \vartriangleleft h} f(s)}{N(h, t)}}{\frac{\sum_{s \in \text{pop}(t)} f(s)}{\mu}} = \frac{\overline{f_t(h)}}{\overline{f_t}},$$

where $\overline{f_t(h)}$ is the mean fitness of the chromosomes that match schema h in generation t and $\overline{f_t}$ is the mean fitness of all chromosomes in generation t. Therefore we can write the expected number of chromosomes that match schema h after selection as

$$N(h, t + \Delta t_s) = N(h, t) \cdot \frac{\overline{f_t(h)}}{\overline{f_t}}.$$

In order to incorporate the effect of the genetic operators, we need measures with which we can compute probabilities that the match to a schema is preserved (or destroyed). For one-point crossover we have to consider how likely it is that the involved parent chromosomes are cut in such a way that all fixed elements of a schema under consideration are inherited from the same parent. In this case the created offspring matches the schema if the corresponding parent does. Otherwise the match to the schema can get lost as the following example demonstrates:

$$
\begin{aligned}
h &= ***0*|1*1** & ***0*1*1** &= h \\
h \triangleright c_1 &= 00000|11111 & \rightarrow \quad 1111111111 &= c_1' \not\triangleleft h \\
h \not\triangleright c_2 &= 11111|00000 & \rightarrow \quad 0000000000 &= c_2' \not\triangleleft h
\end{aligned}
$$

Chromosome c_1 matches the schema h, but c_2 does not. A one-point crossover at the point marked by | creates two offspring chromosomes, both of which do not match the schema. With a different cut point, however, offspring c_1' matches the schema:

$$
\begin{aligned}
h &= ***|0*1*1** & ***0*1*1** &= h \\
h \triangleright c_1 &= 000|0011111 & \rightarrow \quad 1110011111 &= c_1' \triangleleft h \\
h \not\triangleright c_2 &= 111|1100000 & \rightarrow \quad 0001100000 &= c_2' \not\triangleleft h.
\end{aligned}
$$

Obviously, whether an offspring chromosome matches the schema can depend crucially on the location of the cut point relative to the fixed characters of the schema. This gives rise to the notion of the *defining length* of a schema:

Definition 13.4 (*Defining Length of a Schema*) The *defining length* deflen(h) of a schema h is the difference between the position of the last and the first non-* in h.

For instance, deflen($**0*11*10*$) $= 9 - 3 = 6$, because the first character that is not a wildcard, here a 0, is at position 3 and the last character that is not a wildcard, here also a 0, is at position 9. The defining length is the difference of these numbers.

In one-point crossover all possible cut points are equally likely. Therefore the probability that the cut point splits a chromosome in such a way that some of the fixed characters of a schema lie on one side of the cut and some on the other (and therefore are *not* all inherited from the same parent chromosome) is $\frac{\text{deflen}(h)}{L-1}$, because there are deflen(h) cut points between the first and the last fixed character and $L - 1$ possible cut points in total (as L is the length of the chromosomes). In contrast, the probability that all fixed characters of the schema lie on the same side of the cut (and therefore *are* all inherited from the same parent, thus ensuring that the corresponding child matches the schema if the parent does) is $1 - \frac{\text{deflen}(h)}{L-1}$.

In order to derive an expression for $N(h, t + \Delta t_s + \Delta t_x)$, we have to consider whether a chromosome is subject to crossover (otherwise its matching status w.r.t. the schema obviously remains unchanged) and if it is, whether the cut point lies in such a way that the fixed characters of the schema are inherited from the same parent (then the matching status of this parent is transferred to the child). W.r.t. the latter alternative we also have to take into account that a cut point that lets a child inherit the fixed characters of a schema partially from a parent matching the schema and partially from the other may still match the schema, since it may happen that the characters inherited from the other parent match the schema. In particular, this situation occurs if both parents happen to match the schema. Finally, offspring chromosomes that match a schema can be created from parents, both of which do not match the schema, as can be seen from the following example:

$$
\begin{aligned}
h &= {*}{*}{*}0{*}|1{*}1{*}{*} & {*}{*}{*}0{*}1{*}1{*}{*} &= h \\
h \not\triangleright c_1 &= 00010|11111 & \rightarrow \quad 1110111111 &= c_1' \triangleleft h \\
h \not\triangleright c_2 &= 11101|00100 & \rightarrow \quad 0001000100 &= c_2' \not\triangleleft h.
\end{aligned}
$$

Clearly the reason why a match to the schema is created is that both of the parent chromosomes match the schema *partially* and that these partial matches are combined. Note, however, that at most one child can match the schema in such a case.

As a result of the above considerations we write

$$
N(h, t + \Delta t_s + \Delta t_x) = \underbrace{(1 - p_x) \cdot N(h, t + \Delta t_s)}_{A} + \underbrace{p_x \cdot N(h, t + \Delta t_s) \cdot (1 - p_{\text{loss}})}_{B} + C,
$$

where p_x is the probability that a chromosome is subject to crossover and p_{loss} is the probability that after applying one-point crossover the offspring does not match schema h anymore. A is the expected number of chromosomes that match schema h and *are not* subject to one-point crossover (and therefore still match the schema h). B is the expected number of chromosomes that *are* subject to one-point crossover and still match the schema h afterward. Finally, C is the expected number of chromosomes matching schema h that are gained by favorable recombinations of chromosomes that do not match schema h themselves.

Since the term C is almost impossible to estimate properly, we simply neglect it. As a consequence, we obtain only a lower bound for $N(h, t + \Delta t_s + \Delta t_x)$, which, however, is sufficient for our purposes. In order to derive an expression for p_{loss}, we draw on the fact that a loss of match is possible only if the randomly chosen cut point falls in such a way that the fixed characters of the schema are *not* all inherited from the same parent. As argued above, this probability is $\frac{\text{deflen}(h)}{L-1}$. Even in these cases, however, the result may still match the schema (see above). As it is difficult to obtain an expression for *all* possible cases in which a match is preserved, we confine ourselves to those cases in which both parents match the schema and therefore the location of the cut point is irrelevant. This provides us with the expression

$$
p_{\text{loss}} \leq \frac{\text{deflen}(h)}{L - 1} \cdot \left(1 - \frac{N(h, t + \Delta t_s)}{\mu} \right).
$$

The first factor captures the probability that the choice of the cut point is potentially harmful and the second factor captures the probability that the other parent does *not* match the schema and hence there is a certain chance that the result does not match the schema. Clearly, however, this product yields only an upper bound for p_{loss}, because it assumes that any one-point crossover with a potentially harmful cut point and another parent that does not match the schema destroys the match to the schema. This is certainly not the case, as the following example demonstrates:

$$
\begin{aligned}
h &= {*}{*}{*}0{*}|1{*}1{*}{*} & {*}{*}{*}0{*}1{*}1{*}{*} &= h \\
h \rhd c_1 &= 00000|11111 & \rightarrow \quad 1110111111 &= c_1' \lhd h \\
h \not\rhd c_2 &= 11101|00100 & \rightarrow \quad 0000000100 &= c_2' \not\lhd h.
\end{aligned}
$$

Even though chromosome c_2 does not match the schema h and the cut point of one-point crossover falls in such a way that some of the fixed characters of the schema h are inherited from c_1 and some from c_2, the offspring chromosome c_1' matches the schema h. The reason is, of course, that c_2 matches the schema h *partially*. However, capturing partial matches in this analysis is extremely difficult and therefore we confine ourselves to the upper bound for p_{loss} stated above.

Plugging the upper bound for p_{loss} into the formula for $N(h, t + \Delta t_s + \Delta t_x)$ yields

$$
\begin{aligned}
N&(h, t + \Delta t_s + \Delta t_x) \\
&\geq (1 - p_x) \cdot N(h, t + \Delta t_s) \\
&\quad + p_x \ \cdot N(h, t + \Delta t_s) \cdot \left(1 - \frac{\mathrm{deflen}(h)}{L - 1} \cdot \left(1 - \frac{N(h, t + \Delta t_s)}{\mu}\right)\right) \\
&= N(h, t + \Delta t_s)\left(1 - p_x + p_x \cdot \left(1 - \frac{\mathrm{deflen}(h)}{L - 1} \cdot \left(1 - \frac{N(h, t + \Delta t_s)}{\mu}\right)\right)\right) \\
&= N(h, t + \Delta t_s) \cdot \left(1 - p_x \frac{\mathrm{deflen}(h)}{L - 1} \cdot \left(1 - \frac{N(h, t + \Delta t_s)}{\mu}\right)\right) \\
&= N(h, t) \cdot \frac{\overline{f_t(h)}}{\overline{f_t}} \cdot \left(1 - p_x \frac{\mathrm{deflen}(h)}{L - 1} \cdot (1 - N(h, t) \cdot f_{rel}(h))\right),
\end{aligned}
$$

where we exploited in the last step the relationship $N(h, t + \Delta t_s) = N(h, t) \cdot \mu \cdot f_{rel}(h)$. Note that we obtain only an inequality, because we used an upper bound for p_{loss} and because we neglected potential gains from recombinations of chromosomes that do not match the schema h (captured above by the term C, which is missing here).

Having incorporated the effect of crossover, we now turn to mutation. The effect of bit mutation can easily be captured by the *order* of a schema:

Definition 13.5 (*Order of a Schema*) The *order* $\mathrm{ord}(h)$ of a schema h is the number of zeroes and ones in h, that is, $\mathrm{ord}(h) = \#(h, 0) + \#(h, 1) = L - \#(h, *)$, where the operator $\#$ counts the number of occurrences of its second argument in its first.

For instance, $\mathrm{ord}({*}{*}0{*}11{*}10{*}) = 5$, because the chromosome contains 2 zeros and 3 ones and thus a total of 5 fixed (that is, not *don't care*) characters.

With the notion of the order of a schema, we can express the probability that a match to a schema does *not* get lost due to a bit mutation of a chromosome as $(1 - p_m)^{\mathrm{ord}(h)}$. The reason is that a single bit gets flipped with the probability p_m and thus remains unchanged with probability $1 - p_m$ (see Algorithm 12.1 on page 280). If any of the fixed characters in the schema h, of which there are $\mathrm{ord}(h)$, is flipped in the chromosome, the chromosome does not match the schema anymore. We do not care about the $L - \text{ord}(h)$ remaining bits, because the chromosome matches the schema regardless of their value. Since the bit flips are decided independently, the probability that none of the fixed bits in the schema is flipped is $(1 - p_m)^{\text{ord}(h)}$.

As a consequence, we can express the effect of bit mutation as

$$N(h, t + 1) = N(h, t + \Delta t_s + \Delta t_x + \Delta t_m)$$
$$= N(h, t + \Delta t_s + \Delta t_x) \cdot (1 - p_m)^{\text{ord}(h)}.$$

Note that alternative mutation models are easy to handle as well. For example, if at most one-bit is flipped in a chromosome (so-called one-bit mutation, see Sect. 12.3.1), then the effect can be described by

$$N(h, t + 1) = N(h, t + \Delta t_s + \Delta t_x + \Delta t_m)$$
$$= N(h, t + \Delta t_s + \Delta t_x) \cdot \left(1 - \frac{\text{ord}(h)}{L}\right),$$

where $\text{ord}(h)/L$ is the probability that a fixed character in the schema h is flipped in the chromosome (assuming that all bits are equally likely).

Plugging in the result for $N(h, t + \Delta t_s + \Delta t_x)$ that we derived above, we finally obtain **schema theorem** (for bit mutation):

$$N(h, t + 1) \geq N(h, t) \cdot \frac{\overline{f_t(h)}}{\overline{f_t}} \left(1 - p_x \frac{\text{deflen}(h)}{L - 1} \left(1 - \frac{N(h, t)}{\mu} \cdot \frac{\overline{f_t(h)}}{\overline{f_t}}\right)\right) (1 - p_m)^{\text{ord}(h)}.$$

The general form of this relationship between $N(h, t + 1)$ and $N(h, t)$ is clearly

$$N(h, t + 1) \geq N(h, t) \cdot g(h, t).$$

Simplifying, we may therefore say that the number of chromosomes that match a schema h is multiplied in each generation by some factor and thus develops exponentially in the course of several generations. If $g(h, t) > 1$, the number of matching chromosomes grows exponentially, if $g(h, t) < 1$, it decreases exponentially. Since the number of matching chromosomes cannot decrease for *all* schemata (simply because the population size is constant and the contained chromosomes must match some schemata), there must be schemata for which the number of matching chromosomes grows (unless the number of matching chromosomes stays the same for *all* schemata, which, however, implies that the population is essentially constant).

By considering the factors of $g(h, t)$, we can therefore try to derive properties of schemata for which the number of matching chromosomes grows particularly quickly (that is, for which $g(h, t)$ is large). Since $g(h, t)$ is a product, every factor should be as large as possible. Therefore such schemata should have

- high mean fitness   (due to the factor $\overline{f_t(h)}/\overline{f_t}$),
- small defining length (due to the factor $1 - p_x \text{deflen}(h)/(L - 1) \dots$), and
- low order   (due to the factor $(1 - p_m)^{\text{ord}(h)}$).

Such schemata are also called **building blocks**, due to which the schema theorem is sometimes also referred to as the **building block hypothesis**: the evolutionary search focuses on promising building blocks of solution candidates.

It should be kept in mind, though, that the schema theorem or the building block hypothesis applies in the derived form only to bit strings, fitness proportionate selection, one-point crossover and bit mutation. If we use different genetic operators, building blocks may be characterized by other properties than order or defining length. However, a high mean fitness is always among the characteristic features, since all selection methods favor such chromosomes, although differently strongly and not always in direct proportion to the fitness values.

It should also be noted that the schema theorem is open to many different lines of criticism. It widely neglects the interplay of different schemata, as well as the possibility of *epistasis* (the whole derivation implicitly assumes (very) low epistasis). It works with expected values that are strictly valid only for an infinite population size (which obviously cannot be achieved in practice, where effects of stochastic drift need to be taken into account). The factor $g(h, t)$ is clearly not constant, but changes over time due to its dependence on the population at time t, so that the claim of an exponential behavior over several generations is slightly dubious (especially, since saturation effects can be expected) etc.

13.1.2 The Two-Armed Bandit Argument

The schema theorem implies that a genetic algorithm achieves a near-optimal balance between exploration of the search space and exploitation of good solution candidates. As an argument for this claim, Holland used the **two-armed bandit** model as an analogy, that is, a slot machine with two independent arms [23, 30]. The two arms have different expected (per trial) payoffs μ_1 and μ_2 with variances σ_1^2 and σ_2^2, respectively, all of which are unknown. Without loss of generality we may assume $\mu_1 > \mu_2$, though. It is also unknown, however, which of the two arms of the slot machine has the higher payoff (that is, it is unknown whether μ_1 is assigned to the left or to the right arm). Suppose we may play N games with such a slot machine. What is the best strategy to maximize our winnings?

If we knew which arm has the greater payoff, we would simply use that arm for all N games, thus clearly maximizing our expected winnings. However, since we do not know which arm is better, we must invest some trials into gathering information about which arm might be the one with the higher payoff. For example, we may choose to use $2n$ trials, $2n < N$, for this task, in which we play both arms equally often (that is, we use each arm n times). Afterward we evaluate which arm has given us the higher average payoff per trial (exploration). In the remaining $N - 2n$ trials we then exclusively play the arm that has the higher observed payoff (exploitation). Our original question can thus be reformulated as: how should we choose n relative to N in order to maximize our (expected) winnings or, equivalently, to minimize our (expected) loss relative to always having chosen the better arm? In other words, how should we balance exploration (initial $2n$ trials) and exploitation (final $N - 2n$ trials)?

Clearly, there are two types of losses involved here: (1) inevitable loss in the information gathering phase, in which we play the worse arm n times (regardless of which arm is the worse one, since we play both arms equally often), and (2) loss due to the fact that we determine the better arm only based on an empirical payoff estimate, which may point us to the wrong arm as the better one. The former refers to the fact that in the $2n$ trials we devote to exploration we necessarily lose

$$L_1(N, n) = n(\mu_1 - \mu_2),$$

because we do n trials with the arm with lower payoff μ_2 instead of using the arm with the higher payoff μ_1. The loss from the remaining $N - 2n$ trials can only be given in stochastic terms. Let p_n be the probability that the average payoffs per trial, as determined empirically in the first $2n$ trials, actually identify the correct arm. (The index n of this probability is due to the fact that it obviously depends on the choice of $2n$: the larger $2n$, the higher the probability that the empirical payoff estimate from the $2n$ exploration trials identifies the correct arm.) That is, with probability p_n we use the arm actually having the higher payoff μ_1 for the remaining $N - 2n$ trials, while with probability $1 - p_n$ we use the arm actually having the lower payoff μ_2 in these trials. In the former case there is no additional loss (beyond what is incurred in the exploration phase), while in the latter case we lose

$$L_2(N, n) = (N - 2n)(\mu_1 - \mu_2)$$

in the exploitation phase, because we choose the wrong arm (that is, the one actually having the lower payoff μ_2). Therefore the expected total loss is

$$L(N, n) = \underbrace{L_1(N, n)}_{\text{exploration loss}} + (1 - p_n) \underbrace{L_2(N, n)}_{\text{incorrect exploitation loss}}$$

$$= n(\mu_1 - \mu_2) + (1 - p_n)(N - 2n)(\mu_1 - \mu_2)$$

$$= (\mu_1 - \mu_2)(np_n + (1 - p_n)(N - n)).$$

The final form nicely captures that in case the better arm is correctly identified (probability p_n), we lose winnings from n times using the worse arm in the exploration phase, while in case the better arm is incorrectly identified (probability $1 - p_n$), we lose winnings from $N - n$ trials with the worse arm (n of which happen in the exploration phase and $N - 2n$ happen in the exploitation phase).

We now have to minimize the loss function $L(N, n)$ w.r.t. n. The main problem here is to express the probability p_n in terms of n (since p_n clearly depends on n: the longer the exploration phase, the higher the chance that it yields a correct decision). Going into details is beyond the scope of this book, so we only present the final result [23, 30]: n should be chosen according to

$$n \approx c_1 \ln \left(\frac{c_2 N^2}{\ln(c_3 N^2)} \right),$$

where c_1, c_2 and c_3 are certain positive constants. By rewriting this expression, we can turn it into [23, 30]

$$N - n \approx e^{n/2c_1} \sqrt{\frac{\ln(c_3 N^2)}{c_2}} - n.$$

Since with growing n the term $e^{n/2c_1}$ dominates the expression on the right-hand side, this equation can be simplified (accepting further approximation) by

$$N - n \approx e^{cn}.$$

In other words, the total number of trials $N - n$ that are executed with the arm that is observed to be better should increase exponentially compared to the number of trials n that are executed with the arm that is observed to be worse.

This result, though obtained for a two-armed bandit, can be transferred to multi-armed bandits. In this more general form it is then applied to the schemata that are considered in the schema theorem: the arms of the bandit correspond to different schemata, their payoff to the (average) fitness of chromosomes matching them. A chromosome in a population that matches a schema is seen as a trial of the corresponding bandit arm. Recall, however, that a chromosome matches many schemata, thus exhibiting an inherently parallel exploration of the space of schemata.

As we saw in the preceding section, the schema theorem states that the number of chromosomes matching schemata with better than average fitness grows essentially exponentially over several generations. The two- or multi-armed bandit argument now says that this is an optimal strategy to balance exploration of schemata (playing all arms of the bandit) and their exploitation (playing the arm or arms that have been observed to be better than the others).

13.1.3 The Principle of Minimal Alphabets

The **principle of minimal alphabets** is sometimes invoked to claim that binary encodings, as used by genetic algorithms, are "optimal" in a certain sense. The core idea is that the number of possible schemata should be maximized relative to the size of the search space (or the population size), so that the parallelism inherent in the search for schemata is maximally effective. That is, with the chromosomes of the population a number of schemata should be covered that is as large as possible.

If chromosomes are defined as strings of length L over an alphabet $\mathcal{A}$, then the ratio of the number of schemata to the size of the search space is $(|\mathcal{A}| + 1)^L / |\mathcal{A}|^L$. Clearly, this ratio is maximized if the size $|\mathcal{A}|$ of the alphabet is minimized. Since the smallest usable alphabet has $|\mathcal{A}| = 2$, binary codings optimize this ratio.

A more intuitive form of this argument was put forward by [17]: the larger the size of the alphabet, the more difficult it is to find meaningful schemata, because a schema is matched by a larger number of chromosomes. Since a schema averages over the fitness of the matching chromosomes, the quality of a schema may be tainted by some bad chromosomes (low fitness) that happen to match the same schema. Therefore one should strive to minimize the number of chromosomes that are matched by a schema. Since a schema h matches $(L - \mathrm{ord}(h))^{|\mathcal{A}|}$ chromosomes, we should use an alphabet of minimal size. Again, since the smallest usable alphabet has $|\mathcal{A}| = 2$, binary codings can be expected to be optimal.

Whether these arguments are convincing is debatable. At least one has to admit that there is a trade-off between maximizing the number of schemata relative to the search space and expressing the problem in a more natural manner with the help of larger alphabets [18]. Furthermore, a strong argument *in favor* of larger alphabets has been put forward by [1].

13.2 Evolution Strategies

Evolution strategies (ES) [37] are the oldest form of an evolutionary algorithm. They focus on numerical optimization problems and therefore work exclusively with chromosomes that are arrays of real-valued numbers. Their name points to the evolution-strategic principles we mentioned in Sect. 11.2: in (natural) evolution not only the organisms are optimized, but also the mechanisms of evolution. These include parameters like reproduction and mortality rates, life spans, vulnerability to mutations, mutation step sizes, etc. Apart from the focus on numerical optimization problems, it is a distinctive feature of evolution strategies that in many forms they adapt mutation step sizes as well as their direction.

To be more precise, we are given a function $f : \mathbb{R}^n \to \mathbb{R}$, for which we want to find an optimal argument vector $x = (x_1, \ldots, x_n)$, that is, an argument vector that yields a maximum or minimum of the function f. Chromosomes are therefore such vectors of real-valued numbers. Evolution strategies often (though not always) abandon **crossover**, that is, there may be no recombination of chromosomes. Rather they focus on **mutation** as the core variation operator. Mutation in evolution strategies consists generally in adding a random vector r, each element r_i, $i = 1, \ldots, n$, of which is the realization of a normally distributed random variable with mean zero (independent of the element index i) and variances σ_i^2 or standard deviations σ_i. The variances σ_i^2 may or may not depend on the element index i (one variance for the whole vector or a specific variance for each element) and may or may not depend on the generation counter t (time-dependent or independent variance).

As we study in more detail below, the variances may also be coupled to the chromosome and may be subject to mutation themselves. In this way an adaptation of the mutation step sizes and step directions can be realized. Intuitively, we may say that chromosomes with a "suitable" mutation variance—that is, a variance that causes steps of "suitable width" in the region of the search space in which the chromosome is located—are more likely to produce good offspring. As a consequence, this variance can be expected to spread in the population, at least among individuals located in the same region of the search space. It should be noted, though, that the adaptation of the mutation parameters is thus indirect and therefore cannot be expected to be as effective and efficient as the optimization of the function arguments itself. Nevertheless it can help the search process considerably.

13.2.1 Selection

Selection in evolution strategies follows a strict **elite principle**: only the best individuals enter the next generation. There is no random element involved, like, for instance, in the various forms of fitness proportionate selection (see Sect. 12.2.1), which gives better individuals only better chances to enter the next generation, but does not rule out entirely that individuals with a low fitness have offspring.

Even though the elite principle is fixed, there are two different forms of selection, which are distinguished by whether only offspring or parents and offspring together are considered in the selection process. Let μ be the number of individuals in the parent generation and λ the number of offspring individuals that were created by mutation. In the so-called **plus strategy** the parents and children are pooled for the selection process, that is, the selection works on $\mu + \lambda$ individuals (hence an evolution strategy working with this scheme is also called $(\mu + \lambda)$-ES). In contrast to this, in the so-called **comma strategy** (also called (μ, λ)-ES) selection considers only the offspring individuals. In both cases, the μ best individuals (either from the $\mu + \lambda$ pooled individuals for the plus strategy or only from the $\lambda > \mu$ offspring individuals in the comma strategy) are selected for the next generation.

In both the plus and the comma strategy the number λ of offspring individuals (usually) exceeds the number μ of parent individuals (considerably). This approach implements the principle of birth surplus or overproduction, see Sect. 11.2. It is motivated by the fact that usually the majority of mutations are harmful.

Note that the comma strategy actually requires $\lambda \gg \mu$ or at least $\lambda > \mu$, so that there are enough individuals to choose from. (Clearly, $\lambda < \mu$ leads to shrinking populations and $\lambda = \mu$ ignores the fitness of the individuals, since all offspring individuals have to be chosen regardless of their fitness.) However, for the plus strategy it is usually also advisable to create many more individuals than there are parents, in order to reduce the risk of getting stuck in a local optimum, which is caused by the strict elite principle. A typical choice (for both strategies) is $\mu : \lambda = 1 : 7$.

In order to counteract that the plus strategy gets stuck in a local optimum, it is sometimes replaced for some generations with the comma strategy, which increases the diversity of the population again. On the other hand, since in the comma strategy all parent individuals are definitely lost, it is advisable to keep track of the best individual encountered so far in this strategy, while in the plus strategy the best individual is automatically preserved by the strict elite principle.

As a simple example we consider the special case of the $(1 + 1)$-ES. The initial "population" $x_0 \in \mathbb{R}^n$ consists of a single random vector of real numbers. The created offspring individual is $x_t^* = x_t + r_t$, where $r_t \in \mathbb{R}^n$ is random vector of real numbers sampled from a normal distribution. Selection consists in setting

$$x_{t+1} = \begin{cases} x_t^* & \text{if } f(x_t^*) \geq f(x), \\ x_t & \text{otherwise.} \end{cases}$$

Further generations are created until a termination criterion is met. Clearly, this procedure is identical to hill climbing, as we discussed it in Sect. 11.5.2.

As a consequence, we may interpret a more general plus strategy (with $\mu > 1$, but still with $\lambda = \mu$) as a kind of parallel hill climbing. Instead of considering only one current point, the search is performed simultaneously at several places in the search space, always pursuing the μ most promising paths. Note, however, the difference to executing μ hill climbing processes in parallel: in an evolution strategy, both parent and child of a hill climbing pair may enter the next generation (thus creating a fork in the search), for which another pair is extinguished completely (neither parent nor child enter the next generation). As a consequence, there is an exchange of information (about fitness values) between the hill climbing processes, which focuses the search more strongly on promising regions. With $\lambda > \mu$ the search is even more efficient, because several paths are explored from the same parent. However, this may also have the effect of increasing the risk to get stuck in a local optimum.

13.2.2 Global Variance Adaptation

As we mentioned at the beginning of this section, a distinctive feature of evolution strategies is that they try to adapt mutation step sizes. In the simplest form, there is only one global variance σ^2 (or standard deviation σ) that controls the mutations of all chromosomes. This variance is adapted in the course of the generations, so that the mean convergence rate is (approximately) optimized.

In order to obtain a rule how to adapt a global variance, [37] determined the optimal variance σ^2 for the two functions $f_1(x_1, \ldots, x_n) = a + bx_1$ and $f_2(x_1, \ldots, x_n) = \sum_{i=1}^{n} x_i^2$ by determining the probabilities for a successful (that is, improving) mutation. These probabilities are $p_1 \approx 0.184$ for f_1 and $p_2 \approx 0.270$ for f_2. From this result [37] heuristically inferred the so-called $\frac{1}{5}$ success rule: under the plus strategy the mutation step size (as expressed in σ or σ^2) is appropriate if approximately $\frac{1}{5}$ of the offspring are better than the parents.

With the $\frac{1}{5}$ success rule, the standard deviation σ is adapted as follows: if more than $\frac{1}{5}$ of the children are better than the parents, the variance should be increased. On the other hand, if less than $\frac{1}{5}$ of the children are better than the parents, the variance should be reduced. The rationale is that under the assumption that similar individuals have similar fitness, smaller modifications are more likely to create better individuals than larger modifications. In order to obtain a simple rule, the standard deviation σ is increased by multiplying it with a user-specified factor $\alpha > 1$, and reduced by dividing it by the same factor. We thus obtain the following procedure:

Algorithm 13.2 (**Global Variance Adaption**)

function varadapt_global $(\sigma, p_s, \theta, \alpha: \text{real}) : \text{real};$
begin   (∗ standard deviation σ, success rate p_s ∗)
   if $p_s > \theta$ **then return** $\sigma \cdot \alpha;$ (∗ threshold $\theta = \frac{1}{5}$, modification factor $\alpha > 1$ ∗)
   if $p_s < \theta$ **then return** $\sigma/\alpha;$
   return $\sigma;$
end

Here p_s stands for the fraction of the children that are better than the parents, that is, the success rate of the mutations. Note that this rate may also be seen as a measure for the balance of exploration and exploitation. If the success rate is too large, exploitation of good individuals dominates, which can lead to effects of premature convergence (cf. Sect. 12.2.2). On the other hand, if the success rate is too low, exploration dominates, which can lead to slow convergence (though never to vanishing selection pressure, due to the strict elite principle employed in evolution strategies).

Note that for larger populations, the $\frac{1}{5}$ success rule is sometimes too optimistic. In addition, in analogy to simulated annealing (see Sect. 11.5.3), one may define a function that increases the threshold over time (and thus introduces a tendency to reduce the variance). This is the reason for making θ a parameter of the above function instead of fixing it at the value $\frac{1}{5}$ (even though this is frequently chosen).

The complete algorithm for a comma strategy that works with global variance adaption every k generations is shown below. For a plus strategy one merely has to replace the statement "pop' $\leftarrow \emptyset$" with "pop' $\leftarrow$ pop($t - 1$)."

Algorithm 13.3 (Adaptive-ES)

function evostrat_global (f: function, μ, λ, k: int, θ, α: real) : object;
begin   (∗ objective function f, population size μ ∗)
  $t \leftarrow 0$;   (∗ number of offspring λ, modification frequency k ∗)
  $s \leftarrow 0$;   (∗ threshold $\theta = \frac{1}{5}$, modification factor $\alpha > 1$ ∗)
  $\sigma \leftarrow$ value for the initial step size;
  pop(t) $\leftarrow$ create a population with μ individuals;
  evaluate pop(t) with the function f;
  while termination criterion is not fulfilled **do begin**
    $t \leftarrow t + 1$;   (∗ count the created generation ∗)
    pop' $\leftarrow \emptyset$;   (∗ for plus strategy pop' $\leftarrow$ pop($t - 1$) ∗)
    for $i = 1, \ldots, \lambda$ **do begin**
      $x \leftarrow$ select random parent uniformly from pop(t);
      $y \leftarrow$ copy of x;   (∗ create a mutated child ∗)
      mutate_gauss(y, σ);
      if $f(y) > f(x)$ **then** $s \leftarrow s + 1$; **end**
      pop' $\leftarrow$ pop' $\cup \{y\}$; (∗ count the successful mutations ∗)
    end
    pop(t) $\leftarrow$ select best μ individuals from pop';
    if t mod $k = 0$ **then**   (∗ every k generations ∗)
      $\sigma \leftarrow$ varadapt_global($\sigma, s/k\lambda, \theta, \alpha$);
      $s \leftarrow 0$;   (∗ adapt the variance and ∗)
    end   (∗ reinitialize the success counter ∗)
  end
  return best individual in pop(t);
end

Note that for a comma strategy it may be advisable to keep track of the best solution found in the process (since parents are discarded), while a plus strategy automatically does so due to the strict elite principle employed for selection.

13.2.3 Local Variance Adaptation

In contrast to its global counterpart (see above), **local variance adaptation** employs chromosome-specific variances. That is, a chromosome not only consists of an array of real-valued numbers that are the arguments of the function f to optimize, but includes a standard deviation or even an array of standard deviations (one for each function argument), which prescribe chromosome-specific (and gene-specific) mutation step sizes. In the evolution process, not only the function arguments are modified, but the standard deviations are adapted as well. It is plausible that chromosomes with "bad" standard deviation(s) create a lot of "bad" offspring, which is filtered out in the selection process. Chromosomes with "good" standard deviations(s), on the other hand, can be expected to create a larger number of "good" offspring. As a consequence, "good" standard deviation(s) should spread in the population, even though they do not influence the fitness of an individual directly.

A complete self-adaptive algorithm with chromosome-specific mutation step widths, which employs the self-adaptive Gaussian mutation (as it was introduced in Algorithm 12.3 on page 281), is shown below. For a plus strategy one merely has to replace the statement "pop$'$ $\leftarrow$ $\emptyset$" with "pop$'$ $\leftarrow$ pop$(t-1)$."

Algorithm 13.4 (Self-Adaptive-ES)

function adaptive_evostr (f: function, μ, λ: int) : object;
begin   (* objective function f, population size μ *)
  $t \leftarrow 0$;   (* number of offspring λ *)
  pop(t) $\leftarrow$ create population with μ individuals;
  evaluate pop(t) with the function f;
  while termination criterion not fulfilled **do begin**
    $t \leftarrow t + 1$;   (* count the created generation *)
    pop$'$ $\leftarrow$ $\emptyset$;   (* for plus selection: pop$'$ $\leftarrow$ pop(t) *)
    for $i = 1, \ldots, \lambda$ **do begin**
      (x, σ_x) $\leftarrow$ randomly select parent uniformly from pop(t);
      (y, σ_y) $\leftarrow$ copy of (x, σ_x);
      mutate_gsa(y, σ_y);   (* mutate individual and variance *)
      pop$'$ $\leftarrow$ pop$'$ $\cup \{(y, \sigma_y)\}$;
    end
    evaluate pop$'$ with the function f;
    pop(t) $\leftarrow$ select best μ individuals from pop$'$;
  end
  return best individual in pop(t);
end

Note that for element-specific mutation step sizes (that is, a vector of standard deviations per chromosome instead of merely a single standard deviation per chromosome), a somewhat more complex mutation operator is needed.

A commonly used rule for adapting element-specific mutation step sizes (that is, there is one standard deviation σ_i for each function argument) is

$$\sigma_i' = \sigma_i \cdot \exp(r_1 \cdot u_0 + r_2 \cdot u_i).$$

Here $u = (u_0, u_1, \ldots, u_n)$ is a vector, each element of which is sampled from a standard normal distribution with mean 0 and variance 1 (that is, from $N(0, 1)$) and n is the number of arguments of the function f to optimize (and thus the length of the chromosome x). Bäck and Schwefel [6] recommend to choose $r_1 = 1/\sqrt{2n}$ and $r_2 = 1/\sqrt{2\sqrt{n}}$, while [34] proposes to use $r_1 = 0.1$ and $r_2 = 0.2$. In addition, a lower bound (greater than the mandatory lower bound 0 as variances cannot be negative) is specified for the mutation step widths.

13.2.4 Covariances

In the standard form of variance adaption the variances of different vector elements are independent of each other. That is, the **covariance matrix** of the mutation operator is a diagonal matrix. As a consequence, the mutation operator is only able to prefer directions in the search space that are parallel to the coordinate axes of the search space. An oblique direction cannot be represented, even though it may be better than an axes-parallel direction. As a simple example consider a two-dimensional search space. If the best mutations change the two arguments about equally strongly and in the *same* direction, this cannot be represented by merely using independent variances for the arguments. The best approximation is equal variances for both arguments, which, however, allow with the same probability that both arguments are changed by about the same amount, but in *opposite* directions.

This problem can be solved by introducing not only element-specific variances, but covariances for them as well. For instance, in the simple two-dimensional search space example, we may use a covariance matrix like

$$\Sigma = \begin{pmatrix} 1 & 0.9 \\ 0.9 & 1 \end{pmatrix}.$$

With this covariance matrix changes of the two arguments in the same direction (either both are increased or both diminished) are much more likely than changes in opposite directions (increasing one and reducing the other).

As an illustration of the effect of a covariance matrix, which introduces **correlations** between the changes of the arguments, Fig. 13.3 shows uncorrelated, weakly positively, strongly positively, and strongly negatively correlated mutations in a two-dimensional space. Note that in these diagrams the variances in the coordinate directions are the same (which may be a result of normalizing the dimensions by their variance, which is exactly what distinguishes correlation from covariance). If one

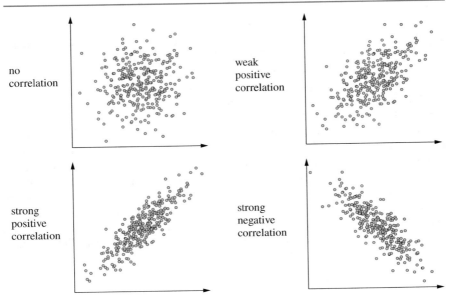

Fig. 13.3 Illustration of covariance and correlation

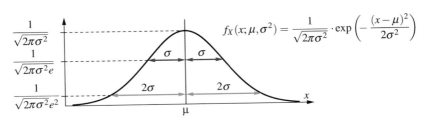

Fig. 13.4 Interpretation of the standard deviation σ for a one-dimensional normal distribution

allows for different variances in the directions of the axes, it becomes clear that correlated mutations allow us to prefer mutation directions in the search space that is (arbitrarily) oblique to the coordinate axes.

In order to understand the meaning of a covariance matrix better, recall the one-dimensional case. A variance is often a bit difficult to interpret: due to its quadratic nature, it is not so easily related to the underlying quantity. This problem is usually handled by computing the standard deviation, which is the square root of the variance and which has the same unit as the underlying quantity. Assuming that the described random process is governed by a normal (or Gaussian) distribution, the standard deviation is easily interpreted, as demonstrated in Fig. 13.4.

This convenient interpretation in the one-dimensional case naturally raises the question whether it can be transferred to two or more dimensions in order to obtain an intuitive interpretation of a covariance matrix. The problem is essentially the same: due to its quadratic nature, a covariance matrix is difficult to interpret. If we could form a "square root" of it, we may obtain an analog of standard deviation, which may be easier to interpret. Thus we face the problem to compute the "square root" of a matrix. Not a general one, though, but a covariance matrix, which has the convenient properties of being symmetric and positive definite.

More technically, let S be an $n \times n$ matrix, that is, $S = (s_{ij})_{1 \leq i \leq m, 1 \leq j \leq n}$. S is called **symmetric** if and only if $\forall 1 \leq i, j \leq n : s_{ij} = s_{ji}$ (or, equivalently, if $S^\top = S$, where $S^\top$ is the transpose of the matrix S). S is called **positive-definite** if and only if for all n-dimensional vectors $v \neq 0$ it is $v^\top S v > 0$. Intuitively, mapping a vector with a positive-definite matrix may stretch or shrink it and may rotate it (by less than $\pi/2$), but does not reflect it about the origin. For symmetric positive-definite matrices an analog of a square root may be computed, for instance, with so-called **Cholesky decomposition** [19]. With this method we find a lower (or left) triangular matrix L (that is, L has non-zero elements only on and below (or left) of the diagonal, while all other elements are zero) such that $LL^\top = S$. By simply spelling out this equation, we easily obtain the elements of L as

$$l_{ii} = \left(s_{ii} - \sum_{k=1}^{i-1} l_{ik}^2 \right)^{\frac{1}{2}},$$

$$l_{ji} = \frac{1}{l_{ii}} \left(s_{ij} - \sum_{k=1}^{i-1} l_{ik} l_{jk} \right), \qquad j = i+1, i+2, \ldots, n.$$

As an example, we consider the special case with only two dimensions. In this case, an analog of a square root can be computed for a covariance matrix

$$\Sigma = \begin{pmatrix} \sigma_x^2 & \sigma_{xy} \\ \sigma_{xy} & \sigma_y^2 \end{pmatrix} \qquad \text{as} \qquad L = \begin{pmatrix} \sigma_x & 0 \\ \frac{\sigma_{xy}}{\sigma_x} & \frac{1}{\sigma_x} \sqrt{\sigma_x^2 \sigma_y^2 - \sigma_{xy}^2} \end{pmatrix}.$$

The matrix L is much easier to interpret than the original covariance matrix Σ: it is a (linear) mapping that describes the deviation from an isotropic behavior (that is, direction independent, from the Greek ἴσος: equal and τρόπος: direction, rotation) that is described by the covariance matrix (as compared to a unit matrix). As an illustration, Fig. 13.5 shows how a unit circle is mapped with the lower triangular matrix resulting for the covariance matrix

$$\Sigma = \begin{pmatrix} 1.5 & 0.8 \\ 0.8 & 0.9 \end{pmatrix} \qquad \text{and thus} \qquad L \approx \begin{pmatrix} 1.2248 & 0 \\ 0.6532 & 0.6880 \end{pmatrix}.$$

The mapping with L has a meaning for an n-dimensional normal (or Gaussian) distribution that is analogous to the meaning of standard deviation for a one-dimensional normal (or Gaussian) distribution. To understand this, let us consider the probability density of an n-dimensional normal distribution, that is,

$$f_X(x; \mu, \Sigma) = \frac{1}{\sqrt{(2\pi)^m |\Sigma|}} \cdot \exp\left(-\frac{1}{2} (x - \mu)^\top \Sigma^{-1} (x - \mu) \right),$$

where μ is an n-dimensional expected value vector and Σ is an $n \times n$ covariance matrix. This probability density is illustrated for $n = 2$ in Fig. 13.6: the left diagram

Fig. 13.5 Mapping of the unit circle with the help of Cholesky decomposition

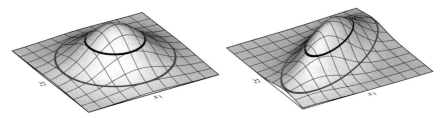

Fig. 13.6 Lines of equal probability density for a standard normal (or Gaussian) distribution (left, variance 1 for both axes, no covariance) and a general normal distribution with covariances (right)

shows a standard normal distribution (unit covariance matrix), while the right diagram shows a general normal distribution (with the above covariance matrix).

For a one-dimensional normal distribution the standard deviation measures the spread (or dispersion) of the probability density by stating the distance between the mean and the points having specific probability densities relative to the height of the mode of the density (see Fig. 13.4). If, in an analogous manner, we mark points with the same probability density for a two-dimensional distribution, we obtain closed curves, which are indicated by gray lines in Fig. 13.6: the darker gray corresponds to σ, the lighter gray to 2σ (in analogy to Fig. 13.4).

As shown in Fig. 13.6, these curves are circles for a standard normal distribution (left), while they are ellipses for a normal distribution with covariances (right). These ellipses are obtained by mapping the circles of the standard normal distribution with the matrix **L** that results from a Cholesky decomposition of the covariance matrix of the general normal distribution. Since the diagrams in Figs. 13.5 and 13.6 are computed with the same covariance matrix, this is particularly obvious.

If we do not require the "square root" of a covariance matrix to be a lower (or left) triangular matrix (as in Cholesky decomposition), the "square root" is, in general, not unique. That is, there may be multiple matrices which, if multiplied with their own transpose, produce the given covariance matrix. Thus it is not surprising that Cholesky decomposition is not the only method with which an analog of standard deviation can be computed. A particularly nice alternative is **eigen decomposition** [19], because it yields a decomposition that consists of elements that make it even better interpretable. Its disadvantage is, however, that it is more costly to compute than Cholesky decomposition [36]. Fortunately, though, this is irrelevant for our

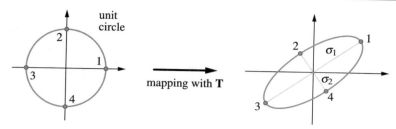

Fig. 13.7 Mapping of the unit circle with the help of eigen decomposition

purposes, because we are mainly interested in interpreting a covariance matrix with the help of finding an analog of a square root of it, and not necessarily in actually computing it in an evolution strategy.

Eigen decomposition is based on the fact that any symmetric and positive-definite matrix S can be written as [19]

$$S = R \, \mathrm{diag}(\lambda_1, \ldots, \lambda_n) \, R^{-1},$$

where the $\lambda_j \geq 0$, $j = 1, \ldots, n$, are the so-called **eigenvalues** of S and the columns of R are the (normalized, that is, length 1) **eigenvectors** of S. The eigenvectors are pairwise orthogonal to each other. As a consequence, $R^{-1} = R^\top$, or, in other words, R is a rotation matrix. With such a decomposition, we can write $S = TT^\top$ with

$$T = R \, \mathrm{diag}\left(\sqrt{\lambda_1}, \ldots, \sqrt{\lambda_m}\right).$$

Eigen decomposition has the clear advantage that the resulting mapping can be interpreted as a scaling of the axes (namely, by the roots of the eigenvalues, denoted by $\sqrt{\lambda_i}$, $i = 1, \ldots, n$) and a rotation (encoded in the rotation matrix R). For the special case with only two dimensions ($n = 2$), eigen decomposition leads to

$$T = \begin{pmatrix} c & -s \\ s & c \end{pmatrix} \begin{pmatrix} \sigma_1 & 0 \\ 0 & \sigma_2 \end{pmatrix}, \qquad \begin{aligned} \sigma_1 &= \sqrt{c^2\sigma_x^2 + s^2\sigma_y^2 + 2sc\sigma_{xy}}, \\ \sigma_2 &= \sqrt{s^2\sigma_x^2 + c^2\sigma_y^2 - 2sc\sigma_{xy}}, \end{aligned}$$

where $s = \sin\phi$, $c = \cos\phi$ and $\phi = \frac{1}{2}\arctan\frac{2\sigma_{xy}}{\sigma_x^2 - \sigma_y^2}$. How a unit circle is mapped with T is shown (in complete analogy to Fig. 13.5) in Fig. 13.7 for

$$\Sigma = \begin{pmatrix} 1.5 & 0.8 \\ 0.8 & 0.9 \end{pmatrix} \quad \text{and thus} \quad T \approx \begin{pmatrix} 1.1781 & -0.3348 \\ 0.8164 & 0.4832 \end{pmatrix},$$

computed from $\phi \approx 0.606 \approx 34.7°$, $\sigma_1 = \sqrt{\lambda_1} \approx 1.4333$ and $\sigma_2 = \sqrt{\lambda_2} \approx 0.5879$. Note that the mapping differs from the one shown in Fig. 13.5—even though the ellipse as a whole is the same—as can be seen from the images of the points labeled 1, 2, 3, and 4, which are in different positions.

A mutation operator employing a covariance matrix is used, for example, in the **covariance matrix adaptation evolution strategy** (CMA-ES) [21], which is among the most highly recommended variants of evolution strategies. Without going into mathematical details and neglecting many subtleties, this algorithm works roughly as follows: an initial population is created by sampling from a standard normal distribution with a chosen mean and variance. In the course of the algorithm, the mean vector, a covariance matrix (initially a unit matrix), and two scaling factors for the mutation are updated. Essentially, the latter three quantities describe a probability distribution from which the mutation vectors are sampled, while the mean vector is mainly used in their update. As a consequence, this approach can be seen as closely related to so-called **estimation of distribution algorithms** [28, 29], which are genetic algorithms that build and sample probabilistic models of promising solution candidates.

Subsequent populations are generated by applying a mutation operator that combines an isotropic and an anisotropic component (where the Greek prefix $\acute{\alpha}\nu$- expresses negation, that is, "anisotropic" means *not* direction independent and thus direction dependent), which are governed by the two scaling factors. The mean vector, the covariance matrix, and the two scaling factors are updated in each iteration in such a way that the likelihood that the mutation operator repeats beneficial mutations is increased. Essentially, this is achieved by updating the current covariance matrix with a (properly normalized) covariance matrix that is computed from the current mean vector and the individuals of the new population. Note that the covariance matrix only captures the direction dependence of the mutations, while the actual mutation step width is controlled by the two scaling factors. Technical details as well as a comparison to other methods can be found in [21].

Note that the approach can be seen as related to estimating the inverse Hessian matrix of the function to optimize, as it is used, for instance, in the Quasi-Newton method in classical optimization [36]. As such, the covariance matrix adaptation evolution strategy uses the population mainly to explore the neighborhood of the mean vector it maintains (and which can be seen as a representative of the currently preferred solution candidate) in order to optimize the mutation direction and step width. Therefore it is more closely related to local search methods as we discussed them in Section 11.5 than to approaches that use a population rather to achieve a broader exploration of the search space, looking for optima (and not just improvement directions) in many places at the same time.

Although such an approach may also be considered, covariance matrices do not lend themselves well to a local variance adaptation scheme, because this requires to incorporate $\frac{n(n+1)}{2}$ mutation parameters into the chromosomes, thus increasing the size of the chromosomes inordinately. In addition, it is debatable whether the indirect adaptation of so many parameters works effectively and efficiently.

13.2.5 Recombination Operators

Evolution strategies are often executed without a crossover operator . If a crossover operator is employed, it is commonly defined as a random selection of components from two parents, that is, analogous to uniform crossover (see Sect. 12.3.2):

$$\begin{matrix} (\mathbf{x_1}, x_2, x_3, \ldots, \mathbf{x_{n-1}}, x_n) \\ (y_1, \mathbf{y_2}, \mathbf{y_3}, \ldots, y_{n-1}, \mathbf{y_n}) \end{matrix} \quad \Rightarrow \quad (x_1, y_2, y_3, \ldots, x_{n-1}, y_n).$$

In principle, any other crossover operator discussed in Sect. 12.3.2, like 1-, 2- or n-point crossover etc. is applicable as well. An alternative is **blending**, as implemented, for example, by arithmetic crossover (see Sect. 12.3.5):

$$\begin{matrix} (x_1, \ldots, x_n) \\ (y_1, \ldots, y_n) \end{matrix} \quad \Rightarrow \quad \tfrac{1}{2}(x_1 + y_1, \ldots, x_n + y_n).$$

In case this crossover operator is used, one should bear in mind, though, that it carries the danger of **Jenkins Nightmare**, that is, a total disappearance of any diversity in a population due to the averaging effect of blending.

13.3 Genetic Programming

With **genetic programming** (GP) it is tried to evolve symbolic expression or even computer programs with certain properties. The purpose of these expressions or programs is usually to associate certain inputs with certain outputs in order to solve a given problem. Genetic programming can be seen as a very general way to learn or to create computer programs, even as complex as programs playing checkers [16]. Its application areas are huge, because many problems can be seen as a search for a program, for example, controller development, scheduling, knowledge representation, symbolic regression, decision tree induction, etc. [33].

Up to now we considered only chromosomes that were arrays of a fixed length, for example, arrays of bits for genetic algorithms or arrays of real-valued numbers for evolution strategies. For genetic programming we abandon the restriction to a fixed length and allow chromosomes that differ in their length. To be more precise, the chromosomes of genetic programming are functional terms and programs, which are commonly called **genetic programs** (also often abbreviated by "GP").

The formal basis of genetic programming is a grammar that describes the language of the genetic programs. Following the standard approach in formal languages, we define two sets, namely, the set $\mathcal{F}$ of **function symbols and operators** and the set $\mathcal{T}$ of **terminal symbols** (constants and variables). These sets are problem-specific and thus comparable to the encoding that we considered w.r.t. other types of chromosomes (see Sect. 12.1). $\mathcal{F}$ and $\mathcal{T}$ should be limited in size in order to restrict the search space to a feasible size, but "rich" enough to enable a problem solution.

As an illustration, we take a look at two examples of symbol sets. Suppose we want to learn a Boolean function that maps n binary inputs to associated binary outputs. In this case the following symbol sets are a natural choice:

$$\mathcal{F} = \{\text{and, or, not, if} \ldots \text{then} \ldots \text{else} \ldots, \ldots\},$$
$$\mathcal{T} = \{x_1, \ldots, x_n, 1, 0\} \quad \text{or} \quad \mathcal{T} = \{x_1, \ldots, x_n, \text{true, false}\}.$$

If the task is symbolic regression,[1] the following sets appear to be appropriate:

$$\mathcal{F} = \{+, -, *, /, \sqrt{}, \sin, \cos, \log, \exp, \ldots\},$$

$$\mathcal{T} = \{x_1, \ldots, x_m\} \cup \mathbb{R}.$$

A desirable property of $\mathcal{F}$ is that all functions in it are domain complete, that is, the functions in $\mathcal{F}$ should accept any possible input value. If this is not the case, the genetic program may cause an error and thus may not be able to complete its execution, which causes problems in the evolution program. Simply examples of functions that are not domain complete are division, which causes an error if the divisor is zero, or a logarithm, which usually accepts only positive arguments.

If the function set $\mathcal{F}$ contains functions that are not domain complete, we have to solve an optimization problem with constraints: we must ensure that the expressions and programs are structured in such a way that a function is never applied to an unacceptable input. For this we could employ repair mechanisms or may introduce a penalty term for the fitness function (cf. Sect. 12.1.3).

As an alternative, we may render all functions in $\mathcal{F}$ domain complete by implementing **protected versions** of error-prone functions. For example, we may define a protected division that returns zero or the (appropriately signed) maximally representable value if the divisor is zero, or a protected nth root that operates on the absolute value of its argument, or a protected logarithm that yields $\log(x) = 0$ for all $x \le 0$. Along the same lines, we may also reinterpret data types and thus render functions that are defined for one data type applicable to the other. For example, if the chosen function set $\mathcal{F}$ contains Boolean as well as numeric functions, we may define 0 as "false" and any value not equal to 0 as "true" (this is the convention, for example, in the programming languages C and C++), thus making it possible to apply Boolean functions to numeric arguments. By fixing a value for a "true" result of a Boolean function, for example, 1 or -1, numeric functions become applicable to Boolean arguments. If necessary, we may also define a conditional operator (if …then …else …) that executes the else-part unless the condition is a proper Boolean "true," in which case only the then-part is executed.

Another important property is the completeness of the function sets $\mathcal{F}$ and $\mathcal{T}$ w.r.t. the functions (mapping inputs to outputs) they can represent. Genetic programming can only effectively solve a given problem if $\mathcal{F}$ and $\mathcal{T}$ are sufficient to find an appropriate program. For example, in Boolean propositional logic $\mathcal{F} = \{\land, \neg\}$ and $\mathcal{F} = \{\rightarrow, \neg\}$ are complete sets of operators, because any Boolean function with any number of arguments can be represented by appropriate combinations of the operators in these sets. However, $\mathcal{F} = \{\land\}$ is not a complete set, because even the simple negation of an argument cannot be represented. Finding the smallest complete set of operators for a given set of functions to represent is (usually) NP-hard. As a consequence, $\mathcal{F}$ usually contains more functions than are actually necessary. However, this is not necessarily a disadvantage, since richer function sets may allow for simpler and thus more easily interpretable solutions.

---

[1] Regression finds a function from a given class to given data by minimizing the sum of squared deviations and is also called the *method of least squares* see Sect. 10.2.

Fig. 13.8 Parse tree of the
symbolic expression
$(+ (* 3 x) (/ 8 2))$

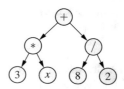

Given the sets $\mathcal{F}$ and $\mathcal{T}$ genetic programs—that is, the chromosomes of genetic programming—can be created. They consist of elements from $C = \mathcal{F} \cup \mathcal{T}$ and possibly opening and closing parenthesis in order to clarify precedence of operators (if necessary) or to group a function symbol and its arguments. However, the genetic programs are not arbitrary sequences of these symbols. Rather we confine ourselves to so-called "well-defined" expressions, that is, expressions that can be created by following certain rules, so that the expressions are interpretable. These rules specify a grammar, which thus defines the language of genetic programs.

Most commonly, this grammar uses a **recursive definition** based on a prefix notation of the functions and operators, similar to the style of functional programming languages like Lisp or Scheme. In these languages, all expressions are either atoms or nested lists. The first list element corresponds to the function symbol or the operator and the subsequent elements are the function arguments or the operands.

More formally, valid symbolic expressions are defined as follows:

- Constants and variables (i.e., the elements of $\mathcal{T}$) are symbolic expressions.
- If $t_1, \ldots, t_n$ are symbolic expressions and $f \in \mathcal{F}$ is an n-ary function symbol, then $(f t_1 \ldots t_n)$ is a symbolic expression.
- Any other character sequence is not a symbolic expression.

For example, the character sequence "$(+ (* 3 x) (/ 8 2))$" is a valid symbolic expression, which, in more conventional terms, means $3 \cdot x + \frac{8}{2}$. In contrast to this, the character sequence "$2\ 7 * (3\ /$" is not a valid or "well-defined" symbolic expression.

For the following discussion, it is convenient to represent a symbolic expression by its so-called **parse tree**. Parse trees are commonly used, for example, in compilers, especially for arithmetic expressions. Such a parse tree for the symbolic regression which we used as an example above is shown in Fig. 13.8.

To find good symbolic expressions for a given problem, genetic programming follows the same general procedure of an evolutionary algorithm as we presented it in Algorithm 11.1 on page 242. That is, first we create an **initial population** of random symbolic expressions, which are *evaluated* by computing their fitness values. Here the fitness is a measure how well a genetic program maps certain vectors of input values to their corresponding output values. For example, suppose we want to find a symbolic expression that computes a Boolean function, which is given as a set of pairs of input and output vectors, all elements of which are either *true* or *false*. In this case the fitness function could simply be the number of correctly computed outputs summed over the input–output pairs. If the task is symbolic regression, the fitness function may simply be the sum of squared deviations from the desired output values. For example, if we are given a data set consisting of pairs (x_i, y_i), $i = 1, \ldots, n$, where x is the input and y the output variable, the fitness may be computed a $f(c) = \sum_{i=1}^{n} (g(x_i) - y_i)^2$, where g stands for the genetic program that maps a input x to an output value. **Selection** is implemented by any of the methods that have been studied in Sect. 12.2. Finally **genetic operators**, usually only crossover, are applied to the selected individuals and the resulting individuals are evaluated again. The procedure of selecting individuals, applying genetic operators, and evaluating the (new) individuals is repeated until a termination criterion is met, for example, that an expression exceeding a user-specified fitness threshold has been found. In the following we discuss the individual steps in more detail.

13.3.1 Initialization

Creating an initial population is somewhat more complex in genetic programming that in the algorithms we studied up to now, because we cannot simple create random sequences of function symbols, constants, variables, and parentheses. We have to respect the recursive definition of a valid or "well-defined" expression. The most natural approach to create a random genetic program for the initial population is a recursive procedure that simply follows the recursive definition. In addition, since it is also convenient for the later evaluation of genetic programs, we actually do not create sequences of symbols, but directly their corresponding parse trees.

In order to limit the size of the parse trees (and to ensure that the recursive procedure terminates), we may specify a maximum tree height or a maximum number of nodes. If this maximum tree height or maximum number of nodes is reached, all unfilled function arguments are chosen only from the terminal symbols to avoid further recursion. A simple form of this procedure looks like this [26]:

Algorithm 13.5 (Initialize-Grow)

function init_grow $(d, d_{\max}: \text{int}) : \text{node};$
begin (* current depth d, maximal depth $d_{\max}$ *)
  if $\quad d = 0$ (* avoid mere constants or variables *)
  then $n \leftarrow$ draw from $\mathcal{F}$ using a uniform distribution;
  elseif $d \geq d_{\max}$ (* stop at the maximal tree height *)
  then $n \leftarrow$ draw from $\mathcal{T}$ using a uniform distribution;
  else $\quad n \leftarrow$ draw from $\mathcal{F} \cup \mathcal{T}$ using a uniform distribution; **end**
  forall $c \in$ arguments of n **do** (* if arguments/operands are needed, *)
    $c \leftarrow$ init_grow$(d + 1, d_{\max})$; (* create sub-expressions recursively *)
  return n; (* if $n \in \mathcal{T}$, n has no arguments *)
end (* finally return the created node *)

Instead of simply pooling $\mathcal{F}$ and $\mathcal{T}$ if $d < d_{\max}$, we may explicitly specify the probabilities with which a function symbol or a terminal symbol is chosen. With such a parameter the size and complexity of the trees can be controlled to some degree: it is the higher, the more likely it is that a function symbol is chosen.

An common alternative to the above procedure is a "full" initialization, which always grows the parse trees to the specified maximum tree height [26]:

Algorithm 13.6 (Initialize-Full)

function init_full $(d, d_{max}: \text{int}) : \text{node};$
begin   (* current depth d, maximal depth d_{max} *)
 if $d \geq d_{max}$ **then**   (* stop at the maximal tree height *)
  $n \leftarrow$ draw from $\mathcal{T}$ using a uniform distribution;
 else   (* below the maximal tree height *)
  $n \leftarrow$ draw from $\mathcal{F}$ using a uniform distribution;
  for $c \in$ arguments of n **do**   (* always choose a function symbol *)
   $c \leftarrow$ init_full$(d + 1, d_{max})$;   (* and create its arguments recursively *)
  end
 return n;   (* finally return the created node *)
end

Each of the two methods *grow* and *full* presented above may be used as the only parse tree generation method, or they may be combined, creating half of the initial population with the method *grow* and the other half with the method *full*. Furthermore, it is advisable to vary the maximum tree depth between 1 and some user-specified maximum, so that trees of different height are created. This approach commonly called *ramped half-and-half* initialization [26], where the term "ramped" refers to the varying maximum tree height, which is implemented by increasing this parameter by one with each created tree (pair).

Algorithm 13.7 (Initialize-Ramped-Half-and-Half)

function init_halfhalf $(\mu, d_{max}: \text{int}) : \text{set of node};$
begin   (* maximal depth d_{max} *)
 $P \leftarrow \emptyset;$ (* population size μ (even multiple of d_{max}) *)
 for $i \leftarrow 1 \dots d_{max}$ **do begin**
  for $j \leftarrow 1 \dots \mu/(2 \cdot d_{max})$ **do begin**
   $P \leftarrow P \cup$ init_grow$(0, i);$
   $P \leftarrow P \cup$ init_full$(0, i);$
  end   (* initialize half the trees with *grow* *)
 end   (* and the other half with *full* *)
 return P;   (* return the created population *)
end

This ramped half-and-half method has the advantage that it ensures a good diversity of the population, with trees of different height and complexity.

13.3.2 Genetic Operators

The initial population of genetic programming (as of any evolutionary algorithm) usually has a very low fitness, because it is highly unlikely that a random generation of parse trees yields even a remotely adequate solution of any non-trivial problem. In order to obtain better solution candidates, genetic operators are applied. For convenience, we describe these genetic operators (crossover and mutation) by showing how they modify the parse trees underlying the genetic programs.

In genetic programming, **crossover** consists in an exchange of two sub-expressions (and thus sub-trees of the parse trees). A simple example for a Boolean function problem is shown in Fig. 13.9, with the parents at the top and the produced offspring at the bottom. The exchanged sub-trees are encircled.

The mutation operator replaces a sub-expression (that is, a sub-tree of the parse tree) by a randomly created sub-expression. To create a new sub-expression, we may simply draw on the initialization method init_grow presented above, for which we may choose the maximum tree height, for example, as a (small) modification of the old sub-tree height. A simple example, again for a Boolean function problem, is shown in Fig. 13.10. It is common to restrict mutation to replacing small sub-trees

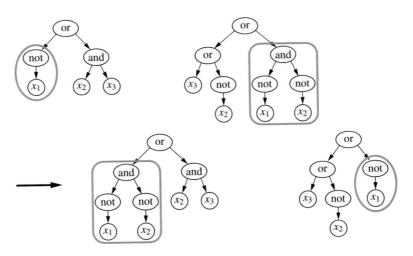

Fig. 13.9 Crossover of two sub-expressions or sub-trees

Fig. 13.10 Mutation of a sub-expression or sub-tree

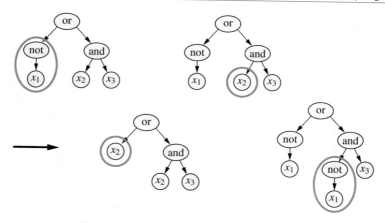

Fig. 13.11 Advantage of genetic programming crossover compared to array-based crossover operators: applied to two identical chromosomes, two different children can be created

(with a height no larger than three or four), so that an actually similar individual is created. An unrestricted mutation could, in principle, replace the whole parse tree, which is equivalent to introducing an entirely new individual.

However, if the population is sufficiently large, so that the "genetic material" present in it guarantees adequate diversity (mainly of function and terminal symbols), mutation is often abandoned and crossover becomes the only genetic operator.[2] The reason is that crossover is—compared to other evolutionary algorithms that work with arrays of fixed length—a much more powerful operator. For example, if we apply it to two identical individuals, the simple fact that different sub-trees may be chosen creates the possibility that two different individuals result. An example demonstrating this is shown in Fig. 13.11: even though the two (identical) parents are fairly simple parse trees, a crossover can create considerable modifications. In contrast to this, any of the operators discussed in Sect. 12.3.2 cannot create any variation if they are applied to two identical chromosomes.

13.3.3 Application Examples

As an illustration of genetic programming, we briefly discuss two applications: learning the Boolean function of a multiplexer and symbolic regression.

---

[2] Note that this is exactly opposite to evolution strategies (see Sect. 13.2), in which crossover is often abandoned and mutation is the only genetic operator.

Fig. 13.12   Configuration of
a Boolean 11-bit multiplexer.
Since the address lines a_0 to
a_2, interpreted as a binary
number, have value 6, the
data line d_6 is passed through
to the output

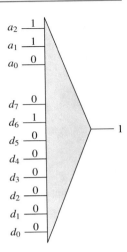

The 11-Multiplexer Problem

A classical example of genetic programming is to learn a Boolean 11-bit multiplexer [26], that is, a multiplexer with 8 data and 3 address lines. A sketch of such a multiplexer, together with one valid input vector and corresponding output, is shown in Fig. 13.12. The purpose of a multiplexer is to pass the value of the data line to the output that is indicated by the address lines. In the example of Fig. 13.12, the values of the address lines represent the number 6 (in a binary encoding) and thus the value of data line d_6 is passed through to the output. In total, there are $2^{11} = 2048$ possible input configurations of an 11-bit multiplexer. Each of these configurations is associated with a single bit output.

In order to learn a symbolic expression that describes the function of an 11-bit multiplexer with genetic programming, we choose the following symbol sets:

$$\mathcal{T} = \{a_0, a_1, a_2, d_0, \ldots, d_7\} \quad \text{and} \quad \mathcal{F} = \{\text{not}/1, \text{and}/2, \text{or}/2, \text{if}/3\}.$$

That is, the set $\mathcal{T}$ of terminal symbols contains the 11 input variables and the function set $\mathcal{F}$ contains four simple Boolean operators, with which these variables can be combined to compute the output. The numbers after the "/" indicate the arity of these operators. That is, the negation takes a single argument, "and" and "or" take two arguments, and "if" takes three arguments, which correspond to the condition, the then-part, and the else-part. Obviously, since we are dealing merely with Boolean values, all functions are domain complete. In addition, the function set is complete in the sense that one can represent any Boolean function with it.

As the fitness function we choose $f(s) = 2048 - \sum_{i=1}^{2048} e_i$, where e_i is the error for the ith input configuration. That is, $e_i = 0$ if the computed output for configuration i coincides with the desired output for this configuration and $e_i = 1$ otherwise. As the termination criterion we may choose that we obtain an individual with fitness 2048. Since such an individual produces the correct output for all 2048 configurations (that is, it makes no mistakes), it is clearly a solution of the problem.

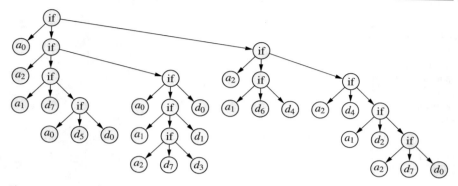

Fig. 13.13 Solution of the 11-multiplexer after 9 generations

In order to solve the 11-bit multiplexer problem, [26] used a population size of $\mu = 4,000$. This number of individuals is needed as with a lower number it becomes unlikely that a good program can be found in a search space as large as the one encountered here. (Note that the expressions may, in principle, become arbitrarily complex unless measures are taken to limit their size.) The initial tree depth (maximum depth in the initial population) was set to 6 and the maximum tree depth to 17 (no parse trees with a depth larger than 17 can be generated).

In the experiment reported by [26], the fitness values of the initial population ranged between 768 and 1280, with an average fitness of 1063. Note that the expected value is 1024 since a random output is correct for about 50% of all configurations. Hence this is a plausible result for an initial population of genetic programs, which produces essentially random output as it has not yet been geared towards the task at hand. 23 expressions of the initial population had a fitness of 1280, with one of them corresponding to a 3-multiplexer, namely (if a_0 d_1 d_2). This is important, because such individuals provide building blocks for the whole 11-bit multiplexer.

In his experiment, [26] used fitness proportionate selection method. 90% of the individuals (that is, 3,600) were modified by crossover and the remaining 10% (that is, 400) were left unchanged. According to [26], after only 9 generations the solution depicted in Fig. 13.13 was found, which actually has the (maximally possible) fitness of 2048. However, due to its complexity, this solution is somewhat difficult to interpret for human beings (although a little effort shows that it actually computes the Boolean function of an 11-bit multiplexer).

In order to obtain a better solution, the expression is *edited*. **Editing** can be seen as an asexual (genetic) operation that is applied to single individuals. It serves the purpose to simplify an expression and is divided into general and special editing.

General editing evaluates a sub-tree and replaces it with its result if the sub-tree yields a constant result, for example, because its leaves contain only constants or because its structure makes it independent of its variable arguments. An example of the latter is the expression "(or x_5 (not x_5))" which is a tautology and thus may be replaced by the constant *true*, despite the fact that x_5 is a variable.

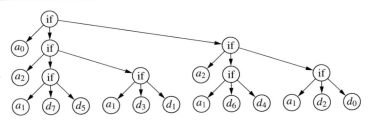

Fig. 13.14 The 11-bit multiplexer solution of Fig. 13.13 after special editing

Special editing exploits certain equivalences that do not change the value of an expression, but may make it possible to express the same function in a simpler way. Simple examples of such identities in Boolean logic are $A \wedge A \equiv A$, $B \vee B \equiv B$, where A and B are arbitrary Boolean expressions, or De Morgan's laws, that is, $\neg(A \vee B) \equiv (\neg A \vee \neg B)$ and its dual. An example for a numerical expression is $\sqrt{x^4} \equiv x^2$.

In principle, editing may already be performed during the evolutionary search, namely by using it as a genetic operator (alongside crossover and mutation). In this case it reduces the number of unnecessarily inflated individuals at the expense of diversity in the population. However, more commonly editing is used to simplify the solutions found by genetic programming in order to obtain a result that is easier to interpret. For the 11-bit multiplexer example, an edited solution (derived from the one shown in Fig. 13.13 after several editing steps) is shown in Fig. 13.14.

Note that this (edited) solution has a hierarchical structure. That is, it divides the 11-multiplexer problem into two smaller problems, namely, two 6-multiplexers with two address bits and 4 data bits, which in turn are divided into two 3-multiplexers with one address bit and 2 data bits. At the top level, the address variable a_0 is used to distinguish between the odd data lines d_7, d_5, d_3, and d_1 (the part of the top-level if) and the even data lines d_6, d_4, d_2, and d_0 (else-part of the top-level if). On the next level the address variable a_2 distinguishes (in both branches) between the higher pair in each quadruple and the lower pair. At the bottom level, the address variable a_1 makes the final decision between the remaining two data lines.

Although the population size was large enough with 4,000 generations, we do not want to conceal our doubts that finding a solution to the 11-bit multiplexer problem in only 9 generations is a typical result. Judging from own experiments with genetic programming (for a related problem), we feel that it usually takes longer to find a solution and that solutions are much more complex than the one shown in Fig. 13.13. Possibly this is due to the fact that we did not use editing, neither in the search nor to simplify our solutions, or that our methods of restricting the complexity of the parse trees were not effective enough. However, we certainly feel that genetic programming is not quite as straightforward as we may have presented it here, but may involve a lot of tweaking to obtain satisfactory results.

Symbolic Regression with Constants

Finding expressions that compute Boolean functions with genetic programming is comparatively simple (though certainly not a trivial task), because there are at most (if one chooses to use them) two constants: *true* and *false*. For symbolic regression—that is, fitting a real-valued function to given data points—the situation is much more complex. Suppose, for example, the geometric problem that we are given the radius r and the area A of several circles and that we desire to find the relationship between the radius and the area. (That is, we want to learn the function $A = \pi r^2$ from the sample data we are given—of course, without presupposing this function or any specific functional form of the relationship.)

Clearly, the choice of the function set $\mathcal{F}$ is not a problem. Choosing merely the basic mathematical operators (that is, $+$, $-$, $*$, $/$) may already be sufficient and we may easily add functions like square root, logarithm, some trigonometric functions without creating any serious problems: $\mathcal{F}$ can always be kept finite. However, the set of terminal symbols is less easy to choose. For this particular problem (finding an expression for the relationship of the area of a circle on its radius), we do not lose the possibility of finding the solution by using a finite set $\mathcal{T}$, because basic mathematical constants like π and e can (and should) be made elements of $\mathcal{T}$. However, more general cases, already starting with finding a mere regression line $y = ax + b$ for given pairs (x_i, y_i) causes problems: clearly we cannot make all numbers in $\mathbb{R}$ elements of $\mathcal{T}$, so that we have any values we may need for a and b available, simply because genetic programming, as we described it up to now, requires $\mathcal{T}$ to be finite. Because $\mathcal{T}$ needs to be finite, even $\mathbb{N}$ or $\mathbb{Q}$ already cannot be added to $\mathcal{T}$. However, how can we then be sure that we have the constants necessary to properly describe the relationship in question? That is, how can we make sure that $\mathcal{F}$ and $\mathcal{T}$ together are rich enough to cover all functions we may want to explore in the search?

The solution to this problem is to introduce so-called **ephemeral random constants** [26]. That is, instead of adding all elements of $\mathbb{R}$ to $\mathcal{T}$, we add only an special terminal symbol, for instance, the *symbol* "$\mathbb{R}$" (not the *set* of real numbers). If in the process of initializing a genetic program, we choose this special symbol as a constant, it is not used literally, but we rather sample a random value from a meaningful interval, thus creating a constant from a potentially infinite set without having these constants all literally contained in $\mathcal{T}$.

If ephemeral random constants are employed, it may be advisable to extend the mutation operator, so that it can also add a random value (sampled, for instance, from a normal distribution with a certain standard deviation) to a constant in a genetic program. The reason is that the standard mutation can only replace a constant with constant that is newly created from scratch (new sub-tree of height 1) and thus may differ completely from the constant that was present before. However, in order to achieve a gradual improvement of solution candidates in symbolic regression, it is more appropriate to modify already existing constants only slightly.

13.3.4 The Problem of Introns

Unless explicit countermeasures are taken, genetic programming exhibits a tendency
to produce larger and more complex individuals with every generation. The main
reason for this phenomenon is so-called **introns**. In biology, introns are parts of the
DNA sequence that do not carry any information in the sense that they do not code for
any phenotypical trait. Introns may be either inactive (they lack a reading trigger and
thus are possibly obsolete) or actually functionless nucleotide sequences beyond
a gene or in-between genes (also known as *junk DNA*). In genetic programming,
introns can occur, for example, if sub-expressions like "(if 2 > 1 then ...else ...)"
are created. Here the else-part of the conditional statement is an intron, because it
can never be executed (unless the tested condition is changed).

Note that a mutation or a crossover operation that only affects an intron is fitness-
neutral, because the intron is never executed and the fitness only depends on the
active program code (that is, the sub-expressions that are actually evaluated for some
input data). As a consequence, introns can grow arbitrarily in size and complexity,
since this complexity does not carry any fitness penalty. In nature, the situation is
somewhat different, because a larger genome usually means a larger metabolic overhead for copying it etc. As a consequence, some organisms, for which the metabolic
overhead is a serious fitness aspect (especially certain forms of bacteria), have very
"streamlined" genomes that contain no or almost no introns.

It is generally advisable to prevent the creation of introns as much as possible,
because they inflate the chromosomes unnecessarily, increase the processing time
and make it more difficult to interpret found solutions. These disadvantages can
be seen as analogies to the metabolic overhead that introns carry in nature. This
analogy also suggests the idea to introduce a fitness penalty for large and complex
chromosomes, which may, for example, be a function of the height or the number of
nodes of the parse tree. An alternative countermeasure is **editing**, which we already
discussed above in the example of the 11-bit multiplexer. If it is used as a genetic
operator, it serves the purpose to keep the chromosomes simple, although at the price
of reducing the variety of "genetic material" in a population.

Other methods that have been suggested to reduce the creation of introns are
modified genetic operators like, for instance, **brood recombination**. This operator
creates many children from the same two parents by applying a crossover operator
with different parameters. Only the best child of the brood enters the next generation.
This method is particularly useful if combined with a fitness penalty, because then
it favors children that achieve the same result with simpler means (that is, with a
less complex chromosome). **Intelligent recombination** chooses crossover points
purposefully, which can help to prevent the creation of introns.

13.3.5 Extensions

In the course of the further development of genetic programming, many extensions
and improvements have been suggested. For example, in order to automatically define

new functions, for instance, one may introduce **encapsulation**: potentially good sub-expressions are protected from getting destroyed by crossover and mutation. One way to achieve this is to define new functions for certain sub-expressions (of a chromosome with high fitness) and to add corresponding new symbols to the function set $\mathcal{F}$. The arity of the new function is the number of (different) variables in the leaves of its sub-tree or the number of all different symbols in the leaves (including constants), so that occurring constants can be replaced by different values in an instantiation. Other extensions include **iterations** and **recursion**, which introduce more powerful programming constructs. However, a detailed discussion of these forms and other extensions is beyond the scope of this book. An interested reader can find more information, for example, in [8].

13.4 Multi-objective Optimization

In everyday life we frequently encounter situations that cannot be described in the simple form of the optimization problems as we defined them in Definition 11.1 on page 235. In particular, we often face the task to select from a set of options that satisfy several criteria (objectives) to different degrees. Often enough, these objectives are even conflicting, that is, trying to improve one causes another objective to be less well satisfied. Consider, for example, the task of finding an apartment. You may desire:

- large floorspace,
- a certain number of bedrooms, bathrooms, etc., a garage, a balcony,
- low rent (excluding service charges),
- low service charges,
- short distance to work,
- short distance to shopping centers and public facilities,
- good environment (low noise, air pollution, etc.),
- good neighborhood, etc.

Clearly, several of these objectives are conflicting. Usually, the larger the floorspace and the quality of the facilities, the higher the rent and the service charges. In addition, the rent usually also reflects the quality of the location. A short distance to work or to shopping centers may conflict with the objective of living in a quiet place etc.

Similar situations occur when you are considering to buy almost any consumer good. Generally, quality and price are conflicting objectives (getting high quality at a low price is rarely possible). Design and usability are also often conflicting (what looks better is often less convenient to use).

13.4.1 Weighted Combination of Objectives

Formally, multi-objective optimization can be described by k objective functions

$$f_i : \Omega \to \mathbb{R}, \qquad i = 1, \ldots, k.$$

Our objective is to find an element of the search space for which all functions yield a value that is as high as possible. The simplest approach to this problem is to combine all k objective functions into a single function and thus to reduce it to a standard optimization problem. For example, we may compute a weighted sum

$$f(s) = \sum_{i=1}^{k} w_i \cdot f_i(s),$$

where the absolute values of the weights specify the relative importance we attach to the different objectives (taking their range of values into account), and their signs may be used to account for both objectives to maximize and to minimize.

Unfortunately, an approach based on combining several objectives in this way has severe drawbacks: apart from the fact that it may not be easy to choose proper weights, we thus lose the possibility to adapt our relative preferences based on the properties of potential solutions we obtain (which is something we definitely do in decision making processes like the search for an apartment as we considered it above). However, the problem is even more fundamental: in general, we face here the problem of having to **aggregate preferences**: each objective defines a preference order of the solution candidates and we have to aggregate these preference orders over the different objectives to obtain an ordering of the solution candidates.

Note that the same problem generally occurs in elections: each voter has a preference order for the candidates, and these preference orders must be aggregated in order to determine the result of an election. Unfortunately, as shown by [2], there is no aggregation function that has all desirable properties. This result is also known as **Arrow's paradox**. Although the consequences of Arrow's impossibility theorem [2] can, in principle, be avoided by using **scaled preference assignments**, the needed scaling functions introduce an additional degree of freedom that makes the aggregation specification even more difficult.

13.4.2 Pareto-Optimal Solutions

An alternative approach to combining multiple objectives in optimization is to try to find all or at least many *Pareto-optimal* solutions.

Definition 13.6 *(Pareto optimality)* An element $s \in \Omega$ is called *Pareto-optimal* w.r.t. the objective functions f_i, $i = 1, \ldots, k$ if there does not exist any element $s' \in \Omega$ for which the following two properties hold:

$$f_i(s') \geq f_i(s)$$

for all $1 \leq i \leq k$ and there is an i with $1 \leq i \leq k$ such that

$$f_i(s') > f_i(s).$$

Intuitively, Pareto optimality (named after the Italian economist V. Pareto) means that the satisfaction of any objective cannot be improved without harming another.

The notion of Pareto optimality may also be defined in two steps as follows:

Definition 13.7 An element $s_1 \in \Omega$ **dominates** an element $s_2 \in \Omega$ if and only if for all $1 \leq i \leq k$

$$f_i(s_1) \geq f_i(s_2).$$

Definition 13.8 An element $s_1 \in \Omega$ **strictly dominates** an element $s_2 \in \Omega$ if s_1 dominates s_2 and there is an i with $1 \leq i \leq k$ such that

$$f_i(s_1) > f_i(s_2).$$

Definition 13.9 An element $s_1 \in S$ is called **Pareto-optimal** if it is *not* strictly dominated by any element $s_2 \in \Omega$.

The notions of "dominates" and "strictly dominates" introduced here will be very useful below to describe the procedure of some algorithms.

Clearly, an advantage of searching for Pareto-optimal solutions is that the objective functions need not be combined (and thus there is no need to specify any weights or any aggregation function). In addition, we preserve the possibility to adapt our view of how important a criterion is relative to the others based on the solutions we obtain. However, the disadvantage is that there is rarely only just one Pareto-optimal solution and thus that there is no unique solution of the optimization problem. This is demonstrated in Figure 13.15, which shows three different forms of the so-called **Pareto frontier**, which is the set of Pareto-optimal solutions, for two criteria. The gray areas contain all solution candidates, which are located according to the values that the functions f_1 and f_2 assign to them (that is, the gray area is the search space). The dark gray lines in the two diagrams on the left and the dark gray dot in the diagram on the right are the Pareto frontiers. Clearly, the points marked in dark gray are exactly the solution candidates that are not strictly dominated by any other

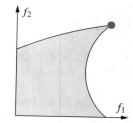

Fig. 13.15 Illustration of Pareto-optimal solutions, i.e., the so-called *Pareto frontier*. All points of the search space are situated in the gray area (with the functions f_1 and f_2 providing the coordinates). Pareto-optimal solutions are located on the part of the border that is drawn in dark gray. With the exception of the right diagram, there are multiple Pareto-optimal solutions

solution candidate. Only in the diagram on the right we have a unique solution. In the other two diagrams several candidate solutions are Pareto-optimal.

13.4.3 Finding Pareto Frontiers with Evolutionary Algorithms

Although we so far assumed that we are given a single function to optimize, evolutionary algorithms may be applied to multi-objective optimization problems. In this case the goal is to find a selection of solution candidates on or at least close to the Pareto frontier, which covers this frontier sufficiently well. That is, we desire that not all solutions candidates are located in the same part of the Pareto frontier, while other parts are not covered by any solutions candidates.

The goal to cover the Pareto frontier sufficiently well rules out an approach that combines the objective functions by weights, so that we may simply apply one of the methods we already discussed. Although this approach will certainly yield a solution that is on or at least close to the Pareto frontier, it covers only a single point. Even if we return not just one, but, say, the best r individuals of the final population, it is unlikely that they are distributed over the Pareto frontier, because the weights prefer a specific point on the Pareto frontier to any other. As an illustration consider the diagrams in Fig. 13.15 again: assigned weights can be visualized as a straight line through the origin with a slope that is given by the ratio of the weights. The solutions found with a weighted combination of the objective functions are located close to the intersection of this straight line with the Pareto frontier. Provided, of course, that there is an intersection, which may not be the case in the two diagrams on the right.

An obvious alternative is the so-called **Vector Evaluated Genetic Algorithm** (VEGA) [38], which works as follows: for every $i \in \{1, \ldots, k\}$, where k is the number of objectives, $\frac{|pop|}{k}$ individuals are chosen based on the fitness function f_i. Intuitively, this can be seen as pooling sub-populations, each of which evolves w.r.t. a different objective function. The clear benefit of this approach is its simplicity and low computational costs. Its major drawback is that solutions that meet all criteria moderately well, but none of them particularly well have a significant selection disadvantage. As a consequence, with a Pareto frontier that is shaped as shown in Fig. 13.16, the search focuses on the "corners." Taking genetic drift into account, it may even happen that eventually the evolutionary algorithm converges to a randomly chosen corner, which is clearly undesirable.

Fig. 13.16 Problem of the VEGA approach: the search focuses on the "corners," while solution candidates that satisfy all criteria moderately well are neglected

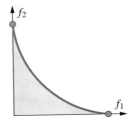

A better approach consists in exploiting the notion that some solution candidates *dominate* others as we introduced it above. The basic idea is to divide the individuals in the population into ranked sets by iteratively removing non-dominated individuals. To be more specific, the division process works as follows:

1. Find all non-dominated candidate solutions of the population.
2. Assign the highest rank to these candidate solutions and remove them from the population.
3. Repeatedly determine and remove non-dominant candidate solutions, assigning them progressively lower ranks, until the population is empty.

Based on the assigned ranks, we may then perform rank-based selection to choose the individuals for the next generation (cf. Sect. 12.2.6). This approach is usually combined with niche techniques (cf. Sect. 12.2.10) to distinguish individuals having the same rank. Their purpose is, of course, to ensure a proper distribution of the individuals over the area close to the Pareto frontier. For example, we may employ **power law sharing**: the fitness assigned to an individual is the lower, the more individuals in the population have similar function values. Note that in contrast to Sect. 12.2.10 the similarity measure for the individuals is based here on the objective function values and not on the structure of the chromosome.

An alternative and the most popular approach is the **Non-dominated Sorting Genetic Algorithm-II** (NSGA-II) [11]. Instead of the rank-based selection, this algorithm relies on a scheme that is closely related to *tournament selection* (cf. Sect. 12.2.7), where the winner of the tournament is determined by the dominance term. In addition, a crowding distance mechanism is drawn on to ensure a good spread of solutions along the obtained approximated front. This approach works as follows: An offspring population $Q(t)$ is generated from the parent population $P(t)$ using genetic operators. Then the non-dominated sorting mechanism is applied to both of these populations: We first identify the non-dominated solutions from $(P(t) \cup Q(t))$, remove them from the populations and store them in a set called $\mathcal{F}_1$. Then the non-dominated solutions from the remaining $(P(t) \cup Q(t) \setminus \mathcal{F}_1)$ are identified and removed to an extra set $\mathcal{F}_2$. This procedure is repeated until there is no solution left, i.e., $P(t) \cup Q(t) = \emptyset$. In this way, the sets $\mathcal{F}_1, \mathcal{F}_2, \cdots$ sort the parent and offspring populations into several fronts. After the non-dominated sorting mechanism, the new population $P(t + 1)$ is filled by the different fronts starting from $\mathcal{F}_1$. As the new population has a limited size, the solutions of the $\mathcal{F}_i$ which cannot totally fit into the new population, go through a selection mechanism called crowding distance mechanism. As the solutions in one front are all indifferent to each other, one can differentiate between them by using the so-called crowding distance values (cd). The solutions in the crowded areas (in the objective space) get a low chance (small crowding distance) to survive the selection. The NSGA-II Algorithm works as follows. In the first iteration ($t = 0$), the initial population ($P(0)$) is sorted into different non-dominated fronts using the function *NonDomSorting*. Afterward each individual is assigned a fitness value which is equal to its front number. Then we apply genetic operators to the population and produce the offspring population $Q(0)$ (ProduceOffspring($P(t), \mathcal{F}$)).

Algorithm 13.8 (NSGA-II)

function NSGA-II ($F_1, \ldots, F_k$: function, μ: int) : set of object;
begin   (∗ objective functions $F_1, \ldots, F_k$ ∗)
  $t \leftarrow 0$;   (∗ population size μ ∗)
  $P(t) \leftarrow$ create population with μ individuals;
  $Q(t) \leftarrow \emptyset$;
  while termination criterion not fulfilled **do begin**
    evaluate $P(t)$ with the functions $F_1, \ldots, F_k$;
    $\mathcal{F} \leftarrow$ NonDomSorting($P(t)$)(∗ $\mathcal{F} = (\mathcal{F}_1, \mathcal{F}_2, \ldots)$ all non-dominated fronts ∗)
    $Q(t) \leftarrow$ ProduceOffspring($P(t), \mathcal{F}$);
    $R(t) \leftarrow P(t) \cup Q(t)$;
    $\mathcal{F} \leftarrow$ NonDomSorting($R(t)$);
    $P(t + 1) \leftarrow \emptyset$ and $i = 1$;
    until $| P(t + 1) | + | \mathcal{F}_i | \leq \mu$
      $P(t + 1) = P(t + 1) + \mathcal{F}_i$;
      $i \leftarrow i + 1$;
    end
    Sort $(\mathcal{F}_i, cd)$;   (∗ Sort $\mathcal{F}_i$ according to the descending cd values ∗)
    $P(t + 1) \leftarrow P(t + 1) \cup \mathcal{F}_i[1 : (\mu - | P(t + 1) |)]$;
    $t \leftarrow t + 1$;
    end ;
  return non-dominated individuals from $P(t)$;
end

Another approach called **strength Pareto evolutionary algorithm 2** (SPEA2) [44], stores the non-dominated individuals separately in an archive. This archive usually has a finite size, so that new individuals can only be added if (dominated) individuals are deleted from it. If it cannot be filled completely, because not enough non-dominated individuals are known, it is filled with dominated individuals:

Algorithm 13.9 (SPEA2)

function spea2 ($F_1, \ldots, F_k$: function, $\mu, \tilde{\mu}$: int) : set of object;
begin   (∗ objective functions $F_1, \ldots, F_k$ ∗)
  $t \leftarrow 0$;   (∗ population size μ, archive size $\tilde{\mu}$ ∗)
  $P(t) \leftarrow$ create population with μ individuals;
  $R(t) \leftarrow \emptyset$;
  while termination criterion not fulfilled **do begin**
    evaluate $P(t)$ with the functions $F_1, \ldots, F_k$;
    for $A \in P(t) \cup R(t)$ **do**
      noDom(A) $\leftarrow |\{B \in P(t) \cup R(t) \mid A >_{\text{dom}} B\}|$;

for $A \in P(t) \cup R(t)$ **do begin**

  $d \leftarrow$ distance of A and its $\sqrt{\mu + \tilde{\mu}}$ nearest individuals in $P(t) \cup R(t)$;

  $A.F \leftarrow \frac{1}{d+2} + \sum_{B \in P(t) \cup R(t), B >_{\text{dom}} A} \text{noDom}(B)$;

end

$R(t + 1) \leftarrow \{A \in P(t) \cup R(t) \mid A \text{ is non-dominated}\}$;

while $|R(t + 1)| > \tilde{\mu}$ **do**

  remove from $R(t + 1)$ the individual A that has the smallest value $A.F$;

if $|R(t + 1)| < \tilde{\mu}$ **then**

  fill $R(t + 1)$ with the best dominated individuals from $P(t) \cup R(t)$; **end**

$t \leftarrow t + 1$;

if termination criterion not fulfilled **then**

  $P(t) \leftarrow$ select from $P(t - 1)$ with tournament selection;

  apply recombination and mutation to $P(t)$;

end

end

return non-dominated individuals from $R(t + 1)$;

end

SPEA2 is a fairly ordinary evolutionary algorithm with a combined evaluation function. The archive, which is always kept at a user-specified size, may contain non-dominated individuals and is also used to compute the fitness. A niche technique is employed that evaluates at a certain number of nearest neighbors.

Although there is a wide variety of evolutionary algorithms for multi-objective optimization, most of them start to have trouble to approximate the Pareto frontier well if more than three objectives are given. One of the reasons is, naturally, that the size of the Pareto frontier grows with the number of objectives, thus making it more difficult for algorithms to cover it sufficiently well or even to find solution candidates sufficiently close to it. This problem may be mitigated by presenting solutions to a user during the search and letting the user choose the direction the search space in which the search should continue (semi-automatic search).

The outcome of the above multi-objective evolutionary algorithms is a set of solutions. To measure the quality of the solutions in terms of convergence to and the spread along the Pareto frontier, we need to use evaluation metrics. Usually, there is no single metric capturing both the diversity (and spread) and the convergence. Therefore, we need to use two or three evaluation metrics. The choice of an evaluation metric depends on whether we already know the Pareto frontier of the problem or not. In certain cases, for example, if we want to examine the functionality of a new algorithm, test problems such as ZDT [43], DTLZ [12], or WFG[24] can be used. For such problems, the Pareto frontiers are known and Generational Distance (GD) and Inverted Generational Distance (IGD) metrics [40] can be used to measure the quality of the produced solutions.

Suppose that we have a set of Pareto-optimal solutions P and a set of obtained solutions S by the algorithm. The GD metric measures the average distance of the solutions in S from P as follows:

$$GD(P, S) = \frac{(\sum_{i=1}^{|S|} d_i^q)^{1/q}}{|S|}. \tag{13.1}$$

The parameter d_i is the Euclidean distance (in the objective space) between the solution $i \in S$ and the nearest member of P (for $q = 2$). Intuitively, the algorithm which has a small value of GD has a good convergence. The IGD metric is very similar to GD, but this time we measure the average distances of the solutions in P to S as follows:

$$GD(P, S) = \frac{(\sum_{i=1}^{|P|} d_i^q)^{1/q}}{|P|}. \tag{13.2}$$

The parameter d_i is the Euclidean distance (in the objective space) between the solution $i \in P$ and the nearest member of S (for $q = 2$). Intuitively, the algorithm which has a small value of IGD has a good Diversity.

In case that the Pareto frontier of a problem is unknown, as it is the case for most of the real-world problems, we can use metrics such as the set coverage metric (C-metric) [42] or the Hypervolume metrics [15,42]. In C-metric, we compare two sets of solutions. These could be the solutions of two consecutive generations of an algorithm. In this way we can measure the improvement over generations. Suppose that we have two sets of non-dominated solutions A and B. The C-metric counts the number of solutions in B that are dominated by A:

$$C(A, B) = \frac{|b \in B|\exists a \in A : a \prec b|}{|B|} \tag{13.3}$$

C-metric must be calculated for both $C(A, B)$ and $C(B, A)$. The Hypervolume metric measures both the convergence and diversity of the solutions as shown in Fig. 13.17. This metric provides a qualitative measure of convergence as well as diversity in a combined sense. However, it is usually reported together with the results from another metric to provide a good evaluation of the solutions. It measures the surface (for 2 objectives), volume (for 3 objectives) or the Hypervolume (for more than 3 objectives), which is dominated by the solutions and is located between them and a reference point. The reference point must be set by the user. It calculates the

Fig. 13.17 Hypervolume metric. The shaded area indicates the Hypervolume

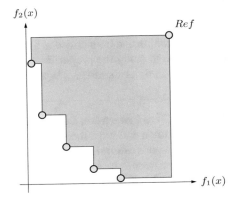

volume (in the objective space) covered by members of S. For each solution $i \in S$, a hypercube v_i is constructed with a reference point ref and the solution i as the diagonal corners of the hypercube. A union of all Hypervolume (HV) is calculated as follows:

$$HV(S) = \cup_{i=1}^{|S|} v_i \tag{13.4}$$

The drawbacks of this approach are that we need to set this reference point and the measurement gets computationally expensive for problems with large number of objectives. The reference point can be found by constructing a vector of the worst objective function values.

13.5 Special Applications and Techniques

With this chapter we close our discussion of evolutionary algorithms by giving an overview of an application of and two special techniques for this kind of metaheuristics. In Sect. 13.5.1 we consider behavioral simulation for the iterated prisoners dilemma with an evolutionary algorithm. In Sect. 13.4 we study evolutionary algorithms for multi-criteria optimization, especially in the presence of conflicting criteria, which instead of returning a single solution try to map out the so-called **Pareto frontier** with several solution candidates. Finally, we take a look at parallelized versions of evolutionary algorithms in Sect. 13.5.2.

13.5.1 Behavioral Simulation

Up to now we applied evolutionary algorithms only to numerical or discrete optimization problems. In this section, however, we employ them for behavioral simulation, in order to study strategies for social interaction as well as so-called population dynamics. The basis for the presented approach is **game theory**, which has proven to be highly useful to analyze social and economic situations and is one of the most important theoretical foundation of economics. The rationale of game theory is to model agents and their actions as game moves in a formally specified framework. Here we consider how the behavioral strategy underlying the moves of an agent in a specific situation can be encoded in a chromosome, so that an evolutionary algorithm can be used to study the properties of successful strategies as well as the influence of the distribution of strategies in a population.

13.5.1.1 The Prisoner's Dilemma

The best-known and most thoroughly studied problem of the game theory is the so-called **prisoner's dilemma**. This delicate problem is commonly described as follows: suppose two people robbed a bank and were arrested. Unfortunately, the available circumstantial evidence is not sufficient for a conviction because of the bank robbery.

Table 13.1 Special payoff matrix of the prisoner's dilemma

A \ B	keeps silent	confesses
keeps silent	−1 \ −1	−1 \ −10 ... 0
confesses	0 \ −10	−10 \ −5 ... −5

There is, however, sufficient evidence for a conviction because of a lesser criminal offense (say, illegal possession of firearms). Suppose this lesser criminal offense carries a sentence of one year in prison. In order to attain a sentence for the bank robbery, the prosecutor offers both prisoners to become a key witness: if one of them confesses to the bank robbery (and thus incriminates the other), he/she is exempted from punishment, while the other prisoner will be punished with the full force of the law, which means 10 years imprisonment. The problem resides in the fact that both prisoners are offered this possibility and thus both may be tempted to confess. However, if both confess, the key witness rule is inapplicable and thus both will be punished. Since they both pleaded guilty, though, they receive a mitigated sentence, meaning that both of them have to spend 5 years in prison.

A popular tool to analyze situations like the prisoner's dilemma formally is a **payoff matrix**. Such a matrix states for all possible pairs of action choices the payoff each of the agents receives. For the constellation of the prisoner's dilemma, we may use the payoff matrix shown in Table 13.1, in which the number of years in prison is used as the payoff—stated as a negative number to indicate that the larger the (absolute) value, the worse the outcome for the two agents.

Seen from a global point of view, it is best if both prisoners keep silent, because this minimizes the total number of years they have to spend in prison. However, there is a temptation to confess, because a confession, provided the other prisoner keeps silent, leads to a better outcome for the confessing prisoner. Unfortunately, this carries the risk to end up in a situation in which both have to spend 5 years in prison. As a consequence, there is a tendency that only a suboptimal payoff result is achieved. More technically, a double confession is the so-called **Nash equilibrium** [32] of this payoff matrix: neither agent can improve its payoff by changing its action, while the other agent maintains the same action. An improvement is only possible if *both* agents change their action choice. Nash [32] showed that every payoff matrix has at least one Nash equilibrium under the condition that mixed strategies are allowed. A mixed strategy is one in which probabilities other than 1 and 0 may be assigned to the different options. That is, a mixed strategy is an unrestricted probability distribution over the different options a player can choose from.

Generalizing from the special situation stated above, the prisoner's dilemma can be defined with a general payoff matrix as it is shown in Table 13.2. The letters mean: **R** reward for mutual cooperation, **P** punishment for mutual defection, **T** temptation

Table 13.2 General payoff matrix of the prisoner's dilemma

A \ B	cooperate	defect
cooperate	R \ R	S \ T
defect	T \ S	P \ P

to defect, **S** sucker's payoff. Note that the exact values of **R**, **P**, **T** and **S** are not important as long as the following two inequalities hold

$$\mathbf{T} > \mathbf{R} > \mathbf{P} > \mathbf{S} \quad \text{and} \quad 2 \cdot \mathbf{R} > \mathbf{T} + \mathbf{S}. \tag{13.5}$$

The first condition states that the payoff for a defect in case the other agent cooperates must be better than the reward for cooperation, so that there is actually a temptation to defect. In addition, cooperation must lead to a better outcome than defection. Finally, mutual defection should still be preferable to being exploited (sucker's payoff), so that there is an incentive to avoid getting exploited. The second condition is needed to make ongoing cooperation preferable to alternating exploitation. With these conditions, mutual defection is a Nash equilibrium of the payoff matrix.

Many situations in everyday life are analogous to the prisoner's dilemma: we often face the choice to exploit an interaction partner and thus to improve our payoff or to cooperate with the chance to achieve an globally better result, but at the risk of getting exploited ourselves. It is often argued that the rational choice in the prisoner's dilemma is to defect in order to ensure that one is not exploited. However, we observe cooperating behavior on a large scale, not only among humans, but among animals as well. Therefore [3] raised the question under what conditions cooperation emerges in a world of egoists without any central authority. Several centuries earlier, in the book *Leviathan* [22], a famous classic of political philosophy, it had already been claimed: under no conditions whatsoever! Before governmental order and thus a directing central authority existed, the state of nature was dominated by egoistic individuals that competed against each other in such a reckless way that life was "solitary, poor, nasty, brutish, and short" [22]. However, on an international level we observe that countries cooperate (economically as well as politically), even though there is de facto no central authority (unless one sees the United Nations as such, despite their lack of directive power). In addition, animals show cooperative behavior without any form of government.

13.5.1.2 The Iterated Prisoner's Dilemma

References [3,4] proposed to find an explanation why we observe cooperation despite the unfavorable Nash equilibrium of the prisoner's dilemma by looking at the **iterated version** of this game. The reason is that in real life, interactions are rarely isolated, but take place in a social context that makes it likely that we will interact with the same agent again in the future. The idea is that the expectation of future iterations of the game may reduce the temptation to defect, since this opens up the possibility of retaliation. This could introduce an incentive to cooperate.

Table 13.3 Axelrod's payoff
matrix for the iterated
prisoner's dilemma

A \ B	cooperate	defect
cooperate	3 \ 3	0 \ 5
defect	5 \ 0	1 \ 1

These considerations lead to the following two questions:

- Is cooperation created in the iterated prisoner's dilemma?
- What is the best strategy in the iterated prisoner's dilemma?

In order to obtain a concrete framework [3,4] specified the **payoff matrix** shown in Table 13.3. The chosen values for the quantities R, P, T and S are the smallest non-negative integer numbers that satisfy the two conditions stated in Eq. 13.5. R. Axelrod then invited scientists from diverse disciplines (psychology, social and political sciences, economics, mathematics) to encode what they believed to be an optimal strategy for the iterated prisoner's dilemma with this payoff matrix. The programs were to have access to all games already played against the same agent, that is, they could try to exploit information gained about the behavioral strategy of their opponents from the result of earlier games. Each program was to be evaluated by the total payoff it achieved in a round-robin tournament (that is, every participant plays against every other participant).

With this framework, R. Axelrod conducted two tournaments. In the *first tournament*, 14 programs and one random player[3] competed against each other in a round-robin tournament with 200 matches per pairing. The winner was a surprising simple program submitted by the mathematician A. Rapoport, which implemented the common sense strategy of *tit for tat*: cooperate in the first move; in subsequent moves simply copy the move of the opponent from the preceding match. The name "tit for tat" for this strategy derives from the simple fact that it retaliates in case its opponent does not cooperate by defecting itself in the next match.

In order to substantiate the result, R. Axelrod published the program code of all participants of this tournament together with the payoff results and invited a second tournament. The idea was that by analyzing the results of the first tournament, insights about what constitutes a good strategy may be gained, so that better programs could be designed. R. Axelrod then conducted a *second tournament* in which 62 programs and one random player participated.[4] The tournament conditions were identical, that is, a round-robin tournament with 200 matches per pairing. Surprisingly, the winner was also the same, namely, *tit for tat*.

The result may be surprising, because *tit for tat* does *not* win generally against any other strategy. For example, against a strategy that defects in all moves (and

---

[3] All programs were written in the programming language *Fortran*.
[4] All programs were written in the programming languages *Fortran* and *Basic*.

thus behaves "rationally" as seen from the point of view of the uniterated prisoner's dilemma) it loses due to the exploitation it suffers in the first match. However, if there are agents in the population with whom it can cooperate, it can gain an overall advantage. Another problem of *tit for tat* is that it may react inadequately to mistakes. Suppose two instances of *tit for tat* play against each other, cooperating nicely at first, but in some move one instance accidentally plays defect. Clearly, the *tit for tat* strategy is unable to recover from the ensuing train of mutual retaliations.

As a consequence, it may be worthwhile to consider *tit for two tat* as an alternative: this strategy starts retaliating only after having been exploited twice. This behavior maintains cooperation even after an accidental defect. On the other hand, if the *tit for two tat* strategy is fixed and known to the opponent, it can be exploited: an opponent that defects in every other move gains a very clear advantage.

13.5.1.3 A Genetic Algorithm Approach

Although the two tournaments saw participants written by capable and renowned researchers, one may still harbor doubts whether the space of possible strategies was sufficiently well explored. In addition, the winner may depend on the selection of participants. In order to obtain better even substantiated results, [5] approached the problem by simulating and evolving populations of strategies for the prisoner's dilemma with a genetic algorithm.

In this approach a strategy is encoded by considering all possible sequences of three consecutive games. Since three games mean six moves (three moves by either opponent), each of which can either cooperate or defect, there are $2^6 = 64$ possible sequences. Thus the chromosome consists of 64 bits, which simply state what move should be played in the next game if the corresponding sequence of three matches was observed (see Table 13.4). In addition, each chromosome is endowed with 6 bits that encode the course of the game "before" the first move. This is necessary, so that a uniform strategy is specified that also covers the first three moves, in which the history is shorter than three matches. In total, every chromosome consists of 70 binary genes that are either C (cooperate) or D (defect).

The genetic algorithm works as follows: the initial population is created by randomly sampling bit sequences of length 70. The individuals of a population are

Table 13.4 Encoding of strategies for the prisoner's dilemma. The first element of every pair corresponds to the own move, the second one to the opposing move.

	1st match	2nd match	3rd match	Next move
1st bit: response to	(C,C),	(C,C),	(C,C):	C
2nd bit: response to	(C,C),	(C,C),	(C,D):	D
3rd bit: response to	(C,C),	(C,C),	(D,C):	C
⋮	⋮	⋮	⋮	⋮
64th bit: response to	(D,D),	(D,D),	(D,D):	D

evaluated by pairing each individual with sufficiently many opponents that are randomly selected from the population. (Depending on the size of the population a full round robin tournament may be prohibitively costly, since its cost is quadratic in the population size. Note also that the experiment was conducted in 1987 and thus at a time where computing power was much more limited than today.) In each pairing, 200 matches were played (like in the tournaments reported about above). The fitness of an individual is the average payoff it gained per pairing. Individuals are selected for the next generation according to the simplified expectation value model mentioned in Sect. 12.2.5: let $\mu_f(t)$ be the average fitness of the individuals in the population at time t and $\sigma_f(t)$ its standard deviation. Individuals s with $f(s, t) < \mu_f(t) - \sigma_f(t)$ do not receive offspring; individuals s with $\mu_f(t) - \sigma_f(t) \leq f(s, t) \leq \mu_f(t) + \sigma_f(t)$ have one child and individuals s $f(s, t) > \mu_f(t) + \sigma_f(t)$ have two children. As genetic operators, standard mutation and one-point crossover were applied. The algorithm was than run for a certain number of generations and the best individuals of the final population were examined.

From the result, [5] identified the following general patterns:

- **Don't rock the boat.** Cooperate after three times cooperate.
  (C,C), (C,C), (C,C) → C
- **Be provokable.** Play defect after a sudden defect of the opponent.
  (C,C), (C,C), (C,D) → D
- **Accept an apology.** Cooperate after mutual exploitation.
  (C,C), (C,D), (D,C) → C
- **Forget.** (Do not be resentful.) Cooperate after cooperation has been restored after one defect (also without retaliation).
  (C,C), (C,D), (C,C) → C
- **Accept a rut.** Play defect after three times defect of the opponent.
  (D,D), (D,D), (D,D) → D

Clearly the *tit for tat* strategy, which won the two tournaments with human-designed programs, has all of these properties. The *tit for two tat* strategy only lacks the "be provokable" property as it reacts only after two exploitation. As already mentioned above, this makes it vulnerable if the strategy is fixed and known to the opponent.

Note that this result should still not be taken as an argument that *tit for tat* is generally the best strategy. A lone individual playing *tit for tat* in a population of individuals that constantly play defect cannot strive. In order to gain the upper hand, *tit for tat* needs a (sufficiently large) sub-population of individuals with which it can cooperate. Its growth may be facilitated, though, if individuals can choose whom to play with based on their experience, so that they can prefer for future pairings individuals with whom they cooperated successfully in the past. Clearly, this is an important aspect that is also decisive in the real world.

13.5.1.4 Extensions

In order to generalize and possibly improve the results obtained by [5] longer match histories may be considered (exploiting the greater computing power we have available nowadays). The strategies may also be described by Moore machines (a specific type of finite state machine) or even general programs that are evolved with the principles of genetic programming (cf. Sect. 13.3).

The prisoner's dilemma may also be extended in many different ways to render its setup more realistic and to capture more and more general situations. We already mentioned above that individuals usually choose with whom to play again and are not forced to repeatedly play against opponents that exploited them. In addition, in the real world, we often find ourselves in situations in which more than two agents are involved, thus leading to a multiplayer version of the prisoner's dilemma. Furthermore, in the real world the consequences of actions are not always perfectly observable. That is, it may not always be perfectly clear whether the last move of our opponent was actually a defect, even though it appeared to be (as judged, for example, by the payoff we received). In order to avoid getting stuck in a train of mutual retaliations, variations of the *tit for tat* strategy may be needed. A very simple method to achieve this is the introduction of a random component, at least if a pairing is stuck in a rut of mutual defect: after a random number of steps try to restore cooperation by cooperating even though the last moves of the opponent were defects. Note that this may require a random component (in order to make the moment unpredictable at which cooperation is played), which cannot be encoded with the scheme we described in the preceding section.

13.5.2 Parallelization

Evolutionary algorithms are fairly expensive optimization methods. In order to obtain sufficiently good solutions, one often has to work with both a large population (a few thousand up to several tens of thousands of individuals) and a large number of generations (a couple of hundreds). Although this drawback is often compensated by a slightly better solution quality compared to other approaches, the execution time of an evolutionary algorithm can be unpleasantly long. One way to improve this situation is **parallelization**, that is, to distribute the necessary operations on several processors (exploiting that essentially all modern processors have multiple cores and multi-processor machines are also becoming more frequent). In this section we discuss which steps can be parallelized (sufficiently easily) and what additional, specialized techniques are inspired by a parallel organization of the algorithm.

13.5.2.1 Parallelizable Operations

Creating an initial population is often easy to parallelize, because usually the chromosomes of the initial population are created randomly and independently of each other. The attempt to prevent duplicate individuals may, however, pose obstacles to a parallel execution. Parallelizing this step is of fairly little importance overall, though, because the initial population is created just once.

The **evaluation** of chromosomes is also easily parallelizable because usually an individual is evaluated independently of any other individual. Since for many important problems the evaluation of the chromosomes is the most costly task, this is a decisive advantage. Even in the evolutionary algorithm used to study the prisoner's dilemma (cf. Sect. 13.5.1.1), we can process pairings in parallel. In order to **compute (relative) fitness values** or a ranking of the individuals, however, we need a central agent that collects and processes evaluations. As a consequence, whether the **selection** of the individuals for the next generation is parallelizable, depends heavily on the chosen selection method: the *expected value model* and *elitism* all require to consider the population as a whole, thus need a central agent, and therefore are difficult to parallelize. *Roulette wheel and rank-based selection* can be parallelized after the initial step of computing the relative fitness values or sorting the individuals according to their fitness has been carried out. The initial step, however, needs a central agent that collects and processes all fitness values. *Tournament selection* is usually best suited for a parallel execution, especially for small tournament sizes, because all tournaments are independent and thus can be held in parallel.

Genetic operators can easily be applied in parallel, since they affect only one (mutation) or two chromosomes (crossover), and are independent of any other chromosomes. Even if multi-parent operators (like diagonal crossover) are used and thus more chromosomes are affected at the same time, different crossover procedures can be executed in parallel. If combined with tournament selection, a steady-state evolutionary algorithm can thus be parallelized very well.

Whether a **termination criterion** can be parallelized depending on the specific form of the termination criterion. The simple test whether a certain number of generations is reached does not cause any problems. However, termination criteria like

- the best individual of the population exceeds a user-specified fitness threshold, or
- the best individual has not changed (a lot) over a certain number of generations

need a central agency that collects this information about the individuals.

13.5.2.2 Island Model and Migration

Even if we require a selection method that causes some troubles for parallelization (like fitness proportionate selection, which needs a central agency at least to compute the relative fitness values), we may achieve a parallel execution by simply processing several independent populations, each on its own processor. Drawing on an obvious analogy from nature, each population can be seen as inhabiting an island, which explains the name **island model** for such an architecture. A pure island model is equivalent to executing the same evolutionary algorithm multiple times, which may just as well be done in a serial fashion. Usually it yields results that are somewhat worse than those of a single run with a larger population.

However, with a parallel execution, one may consider exchanging individuals between the island populations at certain fixed points in time (doing so in every generation creates too much communication overhead). Again drawing on an obvious analogy from nature, such an approach is commonly called **migration**. The idea underlying this method is that transferring genetic material between the islands improves the exploration properties of the island populations, without needing direct recombinations of chromosomes from different islands.

For the mechanisms that **control the migration between islands**, many different proposals exist. In the *random model* pairs of islands are chosen randomly, which then exchange some of their inhabitants. In this model any two island can, in principle, exchange individuals. This freedom is reduced in the *network model*, in which the islands are arranged into a network or graph. Individuals can migrate only between islands that are connected by an edge in the graph. Typical graph structures include rectangular and hexagonal grids in two or three dimensions. Along which of the edges individuals are exchanged is determined randomly.

Instead of merely exchanging individuals and thus genetic material, island populations may also be seen as competing with each other (*contest model*). In this case the evolutionary algorithms that are applied on the islands differ in approaches and/or parameters. The effect of the contest is that the population size of an island is increased or decreased according to the average fitness of its individuals. Usually a lower bound for the population size is set, so that islands cannot become empty.

13.5.2.3 Cellular Evolutionary Algorithms

Cellular evolutionary algorithms are a form of parallelization that is also called *"isolation by distance."* They work with a large number of (virtual) processors, each of which handles a single individual (or only a small number of individuals). The processors are arranged in an rectangular grid, usually in the shape of a torus in order to avoid boundary effects. Selection and crossover are restricted to adjacent processor, that is, to processors that are connected by an edge of the grid. *Selection* means that a processor chooses the best chromosome of the (four) processors adjacent to it (or one of these chromosomes randomly based on their fitness). The processor then performs *crossover* of the selected chromosome with its own. The better child resulting from such a crossover replaces the chromosome of the processor (local elite principle). A processor may also *mutate* its chromosome, the result of which, however, replaces the old chromosome only if it is better (local elite principle again). In such an architecture, groups of adjacent processors are created that maintain similar chromosomes. This mitigates the usually destructive effect of the crossover.

13.5.2.4 Combination with Local Search Methods

The approach of [31] is a combination of an evolutionary algorithm and hill climbing. After an individual has been created by a mutation or a crossover operation, it is optimized with hill climbing: random mutations are applied and kept if they are advantageous. Otherwise they are retracted and a different mutation is tried. Obviously, the local hill climbing optimization can easily be parallelized, because it is executed independently by the individuals.

Furthermore, individuals search for a crossover partner not in the whole population, but only in their (local) neighborhood, thus easing a parallel execution. Note that this requires a distance measure for the individuals and relates the approach to niche techniques (cf. Sect. 12.2.10). The offspring (the crossover products) perform local hill climbing. The individuals of the next generation are selected with a *local elite principle* (cf. Sect. 12.2.8), that is, the best two individuals among the four involved individuals (two parents and two children) replace the parents.

References

1. Antonisse J (1989) A new interpretation of schema notation that overturns the binary encoding constraint. In: Proceedings of 3rd international conference on genetic algorithms. Morgan Kaufmann, San Francisco, CA, USA, pp 86–97
2. Arrow KJ (1951) Social choice and individual values. Wiley, New York, NY, USA
3. Axelrod R (1980) More effective choice in the Prisoner's Dilemma. J. Confl Resolut 24:379–403. SAGE Publications, New York, NY, USA
4. Axelrod R (1984) The evolution of cooperation. Basic Books, New York, NY, USA
5. Axelrod R (1987) The evolution of strategies in the iterated Prisoner's Dilemma. In: Davis L (ed) Genetic algorithms and simulated annealing. Morgan Kaufmann, San Francisco, CA, USA, pp 32–41
6. Bäck T, Schwefel H-P (1993) An overview of evolutionary algorithms for parameter optimization. Evol Comput 1(1):1–23. MIT Press, Cambridge, MA, USA
7. Baluja S (1994) Population-based Incremental Learning: A Method for Integrating Genetic Search Based Function Optimization and Competitive Learning. Technical Report CMU-CS-94-163, School of Computer Science, Carnegie Mellon University, Pittsburgh, PA, USA
8. Banzhaf W, Nordin P, Keller RE, Francone FD (1998) Genetic programming–an introduction: on the automatic evolution of computer programs and its applications. Morgan Kaufmann, San Francisco, CA, USA
9. Brownlee J (2011) Clever algorithms: nature-inspired programming recipes. Lulu Press, Raleigh, NC, USA
10. Dawkins R (1982) The extended phenotype: the long reach of the gene. Oxford University Press, Oxford, United Kingdom 1982; new edition 1999
11. Deb K, Pratap A, Agarwal S, Meyarivan T (2002) A fast and elitist multiobjective genetic algorithm: NSGA-II. IEEE Trans Evol Comput 6(2):182–197
12. Deb K, Thiele L, Laumanns M, Zitzler E (2002) Scalable multi-objective optimization test problems. Proc 2002 Congress Evol Comput 1(1):825–830. IEEE
13. Dorigo M (1992) Optimization, Learning and Natural Algorithms. PhD Thesis, Politecnico di Milano, Milan, Italy
14. Dorigo M, Stützle T (2004) Ant colony optimization. Bradford/MIT Press, Cambridge, MA, USA
15. Fleischer M (2003) The measure of pareto optima—applications to multi-objective metaheuristics. In: Fonseca CM et al (eds) Evolutionary Multi-Criterion Optimization (EMO). Springer, Berlin, pp 519–533
16. Fogel DB (2001) Blondie24: playing at the edge of AI. Morgan Kaufmann, San Francisco, CA, USA
17. Goldberg DE (1989) Genetic algorithms in search optimization and machine learning. Addison Wesley, Reading, MA, USA

18. Goldberg DE (1991) Theory of virtual alphabets. In: Proceedings of 1st workshop on parallel problem solving in nature (PPSN, (1991) Dortmund, Germany), LNCS 496:13–22. Springer, Heidelberg, Germany, p 1991

19. Golub GH, Van Loan CF (1996) Matrix computations, 3rd edn. The Johns Hopkins University Press, Baltimore, MD, USA

20. Goss S, Aron S Deneubourg J-L, Pasteels JM (1989) Self-organized Shortcuts in the Argentine Ant. Naturwissenschaften 76:579–581. Springer, Heidelberg, Germany

21. Hansen N (2006) The CMA evolution strategy: a comparing review. In: [Lozano et al. 2006], pp 1769–1776

22. Hobbes T (1651) Leviathan. Or the matter, forme and power of a commonwealth ecclesiastical and civil. 1651. Reprinted as: Shapiro I (ed) Leviathan. Or The matter, forme, and power of a common-wealth ecclesiastical and civil. Yale University Press, New Haven, CT, USA 2010

23. Holland JH (1975) Adaptation in Natural and artificial systems: an introductory analysis with applications to biology, control, and artificial intelligence. University of Michigan Press, Ann Arbor, MI, USA (1975); 2nd edition: MIT Press. MA, USA, Cambridge, p 1992

24. Huband S, Hingston P, Barone L, While L (2006) A review of multiobjective test problems and a scalable test problem toolkit. IEEE Trans Evol Comput 10(5):477–506. IEEE

25. Kennedy J, Eberhart R (1995) Particle swarm optimization. Proc IEEE Int Conf Neural Netw 4:1942–1948. IEEE Press, Piscataway, NJ, USA

26. Koza JR (1992) Genetic programming: on the programming of computers by means of natural selection. MIT Press, Boston, MA, USA

27. Knowles J, David C (1999) The pareto archived evolution strategy: a new baseline algorithm for pareto multiobjective optimisation. In: Proceedings of IEEE congress on evolutionary computation (CEC 1999, Washington, DC), vol 1. IEEE Press, Piscataway, NJ, USA, pp 98–105

28. Larrañaga P, Lozano JA (eds) (2002) Estimation of distribution algorithms: a new tool for evolutionary computation. Kluwer Academic Publishers, Boston, USA

29. Lozano JA, Larrañaga T, Inza O, Bengoetxea E (eds) (2006) Towards a new evolutionary computation. Advances on estimation of distribution algorithms. Springer, Berlin/Heidelberg, Germany

30. Michell M (1998) An introduction to genetic algorithms. MIT Press, Cambridge, MA, USA

31. Mühlenbein H (1989) Parallel genetic algorithms, population genetics and combinatorial optimization. In: Proceedings of 3rd international conference on genetic algorithms (CEC 1989, Fairfax, VA). Morgan Kaufmann, San Francisco, CA, USA, pp 416–421

32. Nash JF (1951) Non-cooperative games. Ann Math 54(2):286–295. Princeton University, Princeton, NJ, USA

33. Nilsson NJ (1998) Artificial intelligence: a new synthesis. Morgan Kaufmann, San Francisco, CA, USA

34. Nissen V (1997) Einführung in evolutionäre Algorithmen: Optimierung nach dem Vorbild der Evolution. Vieweg, Braunschweig/Wiesbaden, Germany

35. Pelikan M, Goldberg DE, Cantú-Paz EE (2000) Linkage problem, distribution estimation, and bayesian networks. Evol Comput 8:311–340. MIT Press, Cambridge, MA, USA

36. Press WH, Teukolsky SA, Vetterling WT, Flannery BP (1992) Numerical recipes in C: the art of scientific computing, 2nd edn. Cambridge University Press, Cambridge, United Kingdom

37. Rechenberg I (1973) Evolutionstrategie: Optimierung technischer Systeme nach Prinzipien der biologischen Evolution. Fromman-Holzboog, Stuttgart, Germany

38. Schaffer JD (1985) Multiple objective optimization with vector evaluated genetic algorithms. In: Proceedings of 1st international conference on genetic algorithms. L. Erlbaum Associates, Hillsdale, NJ, USA, pp 93–100

39. Srinivas N, Deb K (1994) Multiobjective optimization using nondominated sorting in genetic algorithms. Evol Comput 2(3):221–248. MIT Press, Cambridge, MA, USA

40. Van Veldhuizen DA, Lamont GB (2000) On measuring multiobjective evolutionary algorithm performance. In: Proceedings of the 2000 congress on evolutionary computation. IEEE, pp 204–211

41. Weicker K (2007) Evolutionäre Algorithmen, 2nd edn. Teubner, Stuttgart, Germany
42. Zitzler E (1999) Evolutionary algorithms for multiobjective optimization: methods and applications. Shaker Verlag
43. Zitzler E, Deb K, Thiele L (2000) Comparison of multiobjective evolutionary algorithms: empirical results. Evol Comput 8(2):173–195. MIT Press, Cambridge, MA, USA
44. Zitzler E, Laumanns M, Thiele L (2001) SPEA2: Improving the Strength Pareto Evolutionary Algorithm. Technical Report TIK-Rep. 103. Department Informationstechnologie und Elektrotechnik, Eidgenössische Technische Hochschule Zürich, Switzerland

Computational Swarm Intelligence

14

Swarm Intelligence (SI) is about a collective behavior of a population of individuals. The main properties of such populations are that all of the individuals have the same simple rule from which the global collective behavior cannot be predicted. Moreover, the individuals can only communicate within their local neighborhoods. The outcome of this local interaction defines the collective behavior which is unknown to single individuals. The world of Computational Swarm Intelligence contains several approaches for dealing with optimization problems. Particle Swarm Optimization (PSO) [16] and Ant Colony Optimization (ACO) [9] will be addressed in this chapter. After the introduction, we explain the basic principles of computational swarm intelligence for PSO in Sect. 14.2 upon which the following Sects. 14.3 to 14.5 are built. The second part of the chapter, Sect. 14.6, is about the Ant Colony Optimization method.

14.1 Introduction

Swarm Intelligence is a known phenomenon from nature: a group of animals or swarms of insects like ants and flies, flocks of birds, and schools of fish. These examples have the common features for a collective behavior: simple individuals, simple rules for all, and local interactions. In nature we can even find heterogeneous swarms which contain individuals of various abilities like ant workers and ant queen. Nevertheless, the major common feature is that the global and collective behavior cannot be concluded based on the rules of single individuals. This phenomenon is known as *emergence*. In the literature emergence is explained by "The whole is greater than the sum of its parts." Very often swarm intelligence is an example for self-organization in nature [3]. Self-organized biological systems generate some patterns or structures over time. A prominent example here is an ant road which is built by foraging ants while finding the shortest path to food. Another example is the

© Springer Nature Switzerland AG 2022
R. Kruse et al., *Computational Intelligence*, Texts in Computer Science,
https://doi.org/10.1007/978-3-030-42227-1_14

flight formation of birds or the so-called murmuration of thousands of starlings in a fantastic acrobatic flight. In physics and chemical systems, dissipative structures are built in nonlinear systems. Even the molecular structures can be seen as an emergence phenomenon [18].

The field computational swarm intelligence is inspired by nature and studies the above features in two ways of local interactions using direct and indirect communications. The main approach in swarm intelligence which involves direct communication is called Particle Swarm Optimization (PSO). Inspired by flocks of birds and schools of fish, this approach was introduced by Kennedy and Eberhard [16] in 1995. PSO and its variants are getting very popular in solving optimization problems. Apart from the easy implementation and less amount of parameters than the other nature-inspired approaches, the original PSO showed promising results in terms of fast convergence. Nevertheless, there had to be more attempts to avoid stagnation in the local optima. Indirect communication or "Stigmergy" has been used in Ant Systems. Inspired by colonies of ants, Dorigo and Stützle [9] proposed a model which contained a swarm of individuals and an environment as a carrier for the information. The individuals use the environment for the indirect communication. In the following sections, we address the two major topics of PSO and ACO.

14.2 Basic Principles of Computational Swarm Intelligence

In this section, the basic principles for computational swarm intelligence using direct communication such as Particle Swarm Optimization (PSO) are explained. Let us assume that we have a population of N individuals, which move in a n-dimensional real-valued search space defined by $\Omega \subseteq \mathbb{R}^n$. In this model, each individual i has a certain position $x_i(t) \in \Omega$ at time step t. The movement of the individuals in a swarm can be modeled based on the three behaviors introduced by Reynolds [26]: cohesion, separation, and alignment.

- Cohesion (or aggregation) can be described as a behavior that the individuals tend to gather together. For this they measure their distances to their neighbors and if they are farther than a given predefined distance, they move towards the neighbors.
- Separation is the opposite behavior by which the individuals keep a minimum distant d to their neighbors and get repelled once they get closer than this limit.
- Alignment means that the individuals move in the same direction as the average direction of the individuals in their neighborhood. It is the only behavior among the three that does not require the measurement of the distances to the neighbors.

In the following we concentrate on aggregation and separation. The movement in the swarm can be described by Kinematic equations using discrete time scale:

$$x_i(t+1) = x_i(t) + v_i(t+1)$$
$$v_i(t+1) = wv_i(t) + g_i \tag{14.1}$$

v_i refers to the velocity of the individual i and w indicates the so-called inertia weight. g_i denotes the forces which model the cohesion and separation behaviors for the individual i in the swarm. As the above equations are vector-based, we consider every dimension separately in the following.

Before we define g_i in details, let us consider the following example with $N = 2$ and $w = 0$. The individual i measures its distance to individual j and calculates its new position $x_i(t + 1)$ according to the aggregation and separation behaviors. If it is very close to j, it computes a so-called *repulsion* force and if it is far from j, it needs to compute an *attraction* force. An example is depicted below:

Repulsion ⟵—O—⟶ Attraction O

i j

For simplicity, suppose that we only consider the movement in the horizontal direction. Since $x_i(t) < x_j(t)$ in terms this direction, i must add a positive constant value $f^i(i, j)$ to its current position $x_i(t)$ to get closer to j and a negative constant value $-f_i(i, j)$ must be added if it wants to get repelled:

$$x_i(t + 1) = x_i(t) + f_i(i, j), \text{ if } \mid x_i(t) - x_j(t) \mid > d$$
$$x_i(t + 1) = x_i(t) - f_i(i, j), \text{ if } \mid x_i(t) - x_j(t) \mid < d,$$

where $\mid x_i(t) - x_j(t) \mid$ indicates the Euclidean distance between i and j and d is a predefined constant. Now, we can change these equations to a generic form.

Table 14.1 shows several examples for attraction, repulsion or both attraction and repulsion functions. For instance, a linear attraction function can be defined as a function which depends on the distances between the two individuals. The intensity of attraction is linearly reduced by decreasing values of the distance. Such a function is illustrated in Table 14.1(a) and can be described as

$$f_i(i, j) = -k_a \left(x_i(t) - x_j(t) \right),$$

where k_a is a constant value indicating the slope of the linear function. For $0 \leq k_a \leq 1$, the individual i can only reach j, while for larger values than 1, i can even jump over j. A linear repulsion is the opposite behavior as shown in (b). Constant attraction and repulsion functions are independent from the distances between i and j as shown in (c) and (d). The Figures (e) and (f) illustrate two functions which contain both attraction and repulsion. The individual i gets attracted to j for large distances and it gets repelled for smaller distances. The function in (e) has an almost linear attraction value of a and a nonlinear and unbounded repulsion, where the function in (f) has a nonlinear but bounded repulsion. These two functions are symmetrical with respect to the origin. Taking a symmetrical function as in (f), one can show [12] that the center of the swarm indicated by $\overline{x_c}(t)$ at time t stays stationary:

$$\overline{x_c}(t) = \frac{1}{N} \sum_{i=1}^{N} x_i(t)$$
$$\overline{x_c}(t + 1) - \overline{x_c}(t) = 0.$$

Table 14.1 Examples of attraction and repulsion functions

$$f_i(i, j) = -k_a \left(x_i(t) - x_j(t) \right)$$

(a) Linear attraction

$$f_i(i, j) = k_r \left(x_i(t) - x_j(t) \right)$$

(b) Linear repulsion

$$f_i(i, j) = -k_a \frac{(x_i(t) - x_j(t))}{|x_i(t) - x_j(t)|}$$

(c) Constant attraction

$$f_i(i, j) = k_r \frac{(x_i(t) - x_j(t))}{|x_i(t) - x_j(t)|}$$

(d) Constant repulsion

$$f_i(i, j) = - \left(x_i(t) - x_j(t) \right) \left(a - \frac{b}{|x_i - x_j|^2} \right)$$

(e) Nonlinear attraction and unbounded repulsion

$$f_i(i, j) = - \left(x_i(t) - x_j(t) \right) \left(a - b \exp \frac{-|x_i(t) - x_j(t)|^2}{c} \right)$$

(f) Nonlinear attraction and bounded repulsion

This means that the swarm moves in the search space and individuals change their positions as shown in Fig. 14.1, but the center does not vary over time.

Fig. 14.1 Taking the attraction and repulsion function from Table 14.1(f), the center of swarm stays stationary

The selection of an appropriate attraction and repulsion function can help to obtain a certain stable aggregation behavior in a swarm [12]. In order to consider the interaction of individual i in the swarm of size $N > 2$, we need to calculate the mutual interactions between i and every single other individual in the population. In this case the g_i in Eq. 14.1 (on page 348) is calculated as follows:

$$g_i = \sum_{j=1, j \neq i}^{N} f_i(i, j).$$

The above concept gives us the ability to move a swarm towards any desired point in the search space. In other words, swarms can track a certain point, individual or object in the search space, by using an attraction function. This feature is an important aspect when using swarms for optimization purposes. For instance, we can let the swarm track an object with the position $x^{tr}(t)$, while enforcing a cohesion behavior, as follows:

$$v_i(t + 1) = w v_i(t) + \sum_{j=1, j \neq i}^{N} f_i(i, j) - k_a \left(x_i(t) - x^{tr}(t) \right).$$

This concept considers local or global **interactions** among the individuals and there is no information exchange between them.

14.2.1 Swarms in Known Environments

Based on the Eqs. 14.1, we can let a swarm to move in a certain environment (search space). Suppose that we define a profile for the environment as $\sigma(x)$ in which the swarm is moving. Obviously, the environment has the same dimensionality as the individuals. The swarm movement in such an environment can be described as

$$v_i(t + 1) = w v_i(t) + g_i - \nabla_x \sigma.$$

As an example we can take a plane environment such as $\sigma(x) = ax + b$. According to the above equation, the swarm moves towards the negative of the gradient of this function $(-a)$ while considering the cohesion part related to g_i. The center of the swarm in an environment is influenced by the profile of the environment and is not stationary anymore. For instance, the center of the swarm when using a symmetrical attraction and repulsion function as (f) from Table 14.1 is $-\nabla_x \sigma$. Usually the profile of the environment is not known or cannot easily be modeled. Therefore the above equation can only be used as a theoretical background for the next sections.

14.2.2 Swarms in Unknown Environments

Moving in unknown environments, i.e., environments without a known profile, requires **direct communication** between individuals, which is more than only the local interactions in the form of attraction and repulsion from the above sections. Suppose that the goal of the swarm is to find an optimal (minimum or maximum) point in a given environment with unknown profile σ. For moving towards the optimal solution, the individuals must exchange their local information about the value of the profile at their own positions. For instance, individual i knows $\sigma(x_i(t))$ for its own position x_i at time t. This can be realized by using sensors on individuals which can sense the profile of the environment at the spot they move on. Given a swarm with N individuals and positions $x_i(t)$, $1 \leq i \leq N$ at time t, the swarm can move within the environment by using an attraction function:

$$v_i(t + 1) = wv_i(t) - \phi k_a \left(x_i(t) - x^{(best)}(t) \right),$$

where $x^{(best)}(t)$ refers to the best individual in terms of the profile of the environment in the neighborhood of individual i. ϕ is a random value $\in [0, 1]$ and $k_a \leq 1$ is a constant positive value for a linear attraction towards the best. This is shown in Algorithm 14.1.

In this algorithm, finding the best individual at the position $x^{(best)}$ requires the information exchange among the individuals in the neighborhood. The movement in the swarm considers tracking the best in the neighborhood. Here we do not consider the cohesion in the swarm. The random value for ϕ in combination with a linear attraction with slope of k_a helps to get close to a certain area around the $x^{(best)}$ in the population. In the next iterations, another individual might be the best in the neighborhood. The search for the minimum stops if one of these termination conditions has reached: if a maximum amount of iteration has reached or if there is no improvement of the global best individual over several iterations.

Algorithm 14.1 (**Swarm movement in unknown environments**)

function SwarmMove (N: int, k_a: real constant);
begin   (∗ N: number of individuals ∗)
  $t \leftarrow 0$;
  for $i \in \{1, \dots, N\}$ **do begin**   (∗ initialize the individuals ∗)
    $v_i \leftarrow 0$;   (∗ initialize velocity and position ∗)
    $x_i \leftarrow$ choose a random point of $\Omega = \mathbb{R}^n$;
  end
  repeat   (∗ update the swarm ∗)
    $t \leftarrow t + 1$;
    for $i \in \{1, \dots, N\}$ **do begin**
      $x^{(\text{best})}(t) \leftarrow$ Find the best individual in the neighborhood of i;
      if $\sigma(x_i) \geq \sigma\left(x^{(\text{best})}\right)$ **then** $x^{(\text{best})} \leftarrow x_i$; **end**
      $\phi \leftarrow$ sample random number uniformly from $[0, 1]$;
      $v_i(t + 1) \leftarrow w \cdot v_i(t) - \phi k_a \left(x_i(t) - x^{(\text{best})}(t)\right)$
      $x_i(t + 1) \leftarrow x_i(t) + v_i(t)$;
    end
  until termination criterion is fulfilled;
  return $x^{(\text{best})}$;   (∗ return the best point found ∗)
end

14.3   Particle Swarm Optimization

Similar to the above concept, a swarm can be used to move in a search space and optimize a given function $f(x)$ (the function can be seen as the profile for the environment):

$$\text{optimize}\ \ f(x)$$
$$s.t.\ \ x \in \Omega = \mathbb{R}^n.$$

This approach is called particle swarm optimization (PSO) [16] which is inspired by the behavior of animal species that search for food in swarms, schools, or flocks, for example, bird or fish. This search is characterized by individuals exploring the environment in the vicinity of the swarm up to some distance, but always returning to the swarm as well. In this way information about discovered food sources is conveyed to the other members of the swarm, namely, when the discovering member returns to the food source and other members follow it. Eventually the whole swarm may move toward the food source (provided it is attractive enough).

  Particle swarm optimization can be seen as a method that combines gradient-based search (for example, gradient descent, cf. Sect. 11.5.1, and hill climbing, cf. Sect. 11.5.2) with population-based search (like evolutionary algorithms). Like gradient descent, it requires that the search space Ω is real-valued, that is, $\Omega \subseteq \mathbb{R}^n$. Hence the function to be optimized must be of the form $f : \mathbb{R}^n \to \mathbb{R}$.

Algorithm 14.2 (Particle Swarm Optimization (Maximization))

function PSO (N: int, w, C_1, C_2: real);
begin   (∗ N: number of particles ∗)
  $t \leftarrow 0$;
  for $i \in \{1, \ldots, N\}$ **do begin**   (∗ initialize the particles ∗)
    $v_i \leftarrow 0$;   (∗ initialize velocity ∗)
    $x_i \leftarrow$ choose a random point of $\Omega = \mathbb{R}^n$; (∗ initialize position ∗)
    $x_i^{(\text{personal})} \leftarrow x_i$;   (∗ initialize the personal memory ∗)
  end
  $x^{(\text{global})} \leftarrow x_1$ (∗ Find the global best solution ∗)
  for $i \in \{2, \ldots, N\}$ **do begin**
    if $f(x_i) \geq f\left(x^{(\text{global})}\right)$ **then** $x^{(\text{global})} \leftarrow x_i$; **end**
  end
  repeat   (∗ update the swarm ∗)
    for $i \in \{1, \ldots, N\}$ **do begin**   (∗ update the personal and global memory ∗)
      if $f(x_i) \geq f\left(x^{(\text{personal})}\right)$ **then** $x^{(\text{personal})} \leftarrow x_i$; **end**
      if $f(x_i) \geq f\left(x^{(\text{global})}\right)$ **then** $x^{(\text{global})} \leftarrow x_i$; **end**
    end
    for $i \in \{1, \ldots, N\}$ **do begin**   (∗ update the particles ∗)
      $\phi_1 \leftarrow$ sample random number uniformly from $[0, 1]$;
      $\phi_2 \leftarrow$ sample random number uniformly from $[0, 1]$;
      $v_i(t+1) \leftarrow w \cdot v_i(t) + \phi_1 C_1 \left(x_i^{(\text{personal})}(t) - x_i(t)\right) + \phi_2 C_2 \left(x^{(\text{global})}(t) - x_i(t)\right)$;
      $x_i(t+1) \leftarrow x_i(t) + v_i(t)$; (∗ update velocity and position ∗)
    end
    $t \leftarrow t + 1$;   (∗ count the update step ∗)
  until termination criterion is fulfilled;
  return $x^{(\text{global})}$;   (∗ return the best solution found ∗)
end

The rationale of the search is to use a "swarm" of N candidate solutions to aggregate information from the individual solution to guide the search. Every candidate solution corresponds to a "particle" having a position x_i in the search space and a velocity v_i, $i = 1, \ldots, N$. In each step the position and the velocity of the ith particle are updated according to the following equations:

$$v_i(t+1) = wv_i(t) + \phi_1 C_1 \left(x_i^{(\text{personal})}(t) - x_i(t)\right) + \phi_2 C_2 \left(x^{(\text{global})}(t) - x_i(t)\right)$$
$$x_i(t+1) = x_i(t) + v_i(t),$$

where w indicates the inertia weight, C_1 and C_2 are two constant values and ϕ_1 and ϕ_2 are random values in the range $[0, 1]$. $x_i^{(\text{personal})}$ is the **personal memory** of each individual (i.e., particle). It is the best solution in the search space that this particle has visited up to time step t. That is for maximization problems,

$$x_i^{(\text{personal})} = x_i \left(\arg\max_{u=1}^t f\left(x_i(u)\right)\right).$$

Similarly, $x^{(\text{global})}$ is the **global memory** of the swarm as a whole. This term represents the social aspect in the swarm and requires communication between the particles. The global memory is the position of best particle in the swarm (population) at step t. That is,

$$x^{(\text{global})}(t) = x_j^{(\text{local})}(t) \quad \text{with} \quad j = \arg\max_{i=1}^N f\left(x_i^{(\text{local})}\right).$$

Obviously this depends on the way we define the global knowledge in the swarm. In fact, there are several schemes for finding the globally best solution all of which depend on the communication topology. The well-known topologies are *fully connected network* (as presented above) and *ring topology* (every individual is only communicating with two others in the swarm). The general procedure of particle swarm optimization for a maximization problem is shown in Algorithm 14.2.

A demonstration program illustrating particle swarm optimization with the above algorithm for a two-dimensional function with many local maxima can be found at:

http://www.borgelt.net/psopt.html

Since this program visualizes each particle with a "tail" of previous positions, the movement of the particles can be observed well.

14.3.1 Influence of the Parameters

The parameters ϕ_1 and ϕ_2, with which the strength of "attraction" towards the personal and the global memory are controlled, are randomly chosen (in the range $[0, 1]$) in every step. The so-called learning factors denoted by C_1 and C_2 are two constants defining the amount of linear attraction [4]. The inertia weight w can be set to a constant value (< 1, e.g., 0.4). The small values of w enforces the swarm to a very quick convergences towards the local optima, while the large values has the opposite behavior. Figure 14.2 illustrates this behavior. Large values of w make the movements steps larger and larger over iterations which leads to an explosive behavior. In this case the solution i cannot get close enough to the global best solution. The small values of w result in a very quick convergence towards the global best solution in the population. Based on this behavior, we can change w over time so that it starts with large values (adding some randomness in the beginning of the optimization) and decreases over time. That is, as time advances, the influence of the (current) velocity of the particle decreases and thus the (relative) influence of the attraction of the personal and the global memory increases. Generally, it can be observed that after starting with fairly high speed (caused by the attraction towards the global memory) the particles become slower and slower and finally come (almost) to a halt at the best point discovered by the swarm.

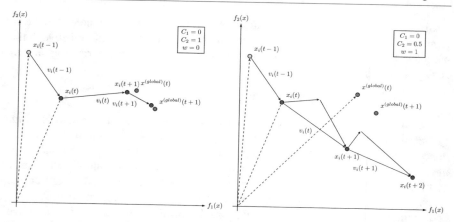

Fig. 14.2 Influence of the inertia weight: $w = 0$ leads to a quick convergence towards the global memory (left), while with $w = 1$ it gets very difficult to get close to the global memory over iterations

14.3.2 Turbulence Factor

Turbulence factor is a **diversity control mechanism**, which is meant to prevent a premature convergence to suboptimal solutions. A naive turbulence factor is to change the position of a randomly selected particle to a random position. Another approach is to select the particles with very small velocities and replace the to random positions in the search space. In this way, these particles which could not move due to a very small (or even equal to zero velocity) help explore the search space and eventually avoid the premature stagnation in the local optima.

14.3.3 Boundary Handling

When designing PSO to solve an optimization problem with a given search space bounded to a range $[x_{min}, x_{max}]$, boundary handling methods need to be applied to the particles which have left the boundaries [14]. This is particularly important if, as is often the case, the optimal solutions lie close to the boundary of the search space. It has been shown that as the dimensionality of the search space increases, the probability of particles moving out of the search space can dramatically grow [15]. The remedy to this behavior is to define a maximum velocity value of v_{max}. Additionally one can use a boundary handling method in the case that a particle has left the border, replace it (Table 14.2).

- On the border:
  if $x_i > x_{max}$ **then** $x_i = x_{max}$; or **if** $x_i < x_{min}$ **then** $x_i = x_{min}$
- or at a random position in the search space:

Table 14.2 Boundary handling methods (a) to (c). The red line in (d) shows the amount of inversed velocity

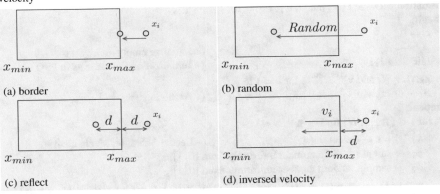

(a) border

(b) random

(c) reflect

(d) inversed velocity

if $x_i > x_{max}$ **or if** $x_i < x_{min}$ **then** $x_i =$ Rand(x_{min}, x_{max})
- or at its reflected position. In this case we measure the distance of the particle to the border and replace it in the search space with the same distance to the border: **if** $x_i > x_{max}$ **then** $d = x_i - x_{max}$; $x_i = x_{max} - d$ or

if $x_i < x_{min}$ **then** $d = x_{min} - x_i$; $x_i = x_{min} + d$

Figure 14.2a–c shows these three boundary handling methods for only one horizontal dimension. One important aspect in boundary handling is to change the velocity of the replaced particle. Either the new velocity of the particle will be set to zero or the velocity value can be inversed proportional to the value d from above: $v_i = -(v_i - d)$ as shown in Fig. 14.2d.

14.4 Multi-objective Particle Swarm Optimization

As already mentioned in Sect. 13.4, in multi-objective optimization there are several functions to be optimized simultaneously. The solution of such problems is a set of Pareto-optimal solutions. The main goal of Multi-Objective Particle Swarm Optimization (MO-PSO) methods is to find a diverse set of approximated Pareto-optimal solutions [25]. In order to achieve this, we need to add two major extensions to PSO such as leader selection mechanism and archiving. Algorithm 14.3 illustrates a generic framework for MO-PSO. In addition to the population P_t, there is another population denoted as *Archive* (A_t) which is meant to keep the non-dominated solutions. The archive gets updated using an archiving mechanism after each iteration.

Algorithm 14.3 (**Generic Framework for MO-PSO**)

function MO-PSO (N: int, w, C_1, C_2: real);
begin
   $t \leftarrow 0$;
   $A_t \leftarrow empty$ (∗ initialize empty archive ∗)
   Initialize P_t: (∗ Initialization the same as in PSO ∗)
      $x_i(t)$, $v_i(t)$ and $x_i^{(\text{personal})}$, for $i = 1$ to N;
      $A_t \leftarrow \text{Archiving}(A_t, P_t)$;   (∗ Archiving ∗)
   repeat
      for $i \in \{1, \ldots, N\}$ **do begin**
         $x_i^{(\text{best})}(t) \leftarrow \text{Selection}(A_t)$;   (∗ Leader Selection Mechanism ∗)
         $\phi_1 \leftarrow$ sample random number uniformly from $[0, 1]$;
         $\phi_2 \leftarrow$ sample random number uniformly from $[0, 1]$;
         $v_i(t + 1) \leftarrow w \cdot v_i(t) + \phi_1 C_1 \left(x_i^{(\text{personal})}(t) - x_i(t) \right) + \phi_2 C_2 \left(x^{(\text{global})}(t) - x_i(t) \right)$;
         $x_i(t + 1) \leftarrow x_i(t) + v_i(t)$;   (∗ update velocity and position ∗)
         $x_i^{(\text{personal})}(t + 1) \leftarrow \text{Update} \left(x_i^{(\text{personal})}(t), x_i(t + 1) \right)$;
                                        (∗ update personal best memory ∗)
      end
      $A_t \leftarrow \text{Archiving}(A_t, P_t)$;   (∗ Archiving ∗)
      $t \leftarrow t + 1$;
   until termination criterion is fulfilled;
   return A_t;   (∗ return the archive ∗)
end

In the above algorithm, P_t refers to the swarm (population of individuals) and A_t indicates the archive which also delivers the output for the algorithm. The functions "Selection" and "Archiving" refer to the leader selection mechanism and archiving methods explained in Sects. 14.4.1 and 14.4.2.

The function "Update" is used to *update* the personal best memory. The simplest approach for updating the personal best is that it only gets updated if it is dominated by x_i. Therefore, if it is indifferent to x_i (i.e., x_i and $x_i^{(\text{personal})}$ do not dominate each other) nothing will happen and the oldest non-dominated position will be kept in the memory. Another approach can be to always replace the personal memory with x_i except if it dominates x_i. This means if x_i is indifferent to $x_i^{(\text{personal})}$, $x_i^{(\text{personal})}$ is updated to the value of x_i and therefore we keep the newest non-dominated position in the memory. In this case, the particle will not be dragged back to previously explored regions but is not much influenced by $x_i^{(\text{best})}$ as long as it keeps finding non-dominated solutions [2].

14.4.1   Leader Selection Mechanism

One of the main challenges in using PSO for multi-objective problems is that there is no single optimal solution and therefore there is no single $x_i^{(\text{best})}(t)$. In this case, there is a set of non-dominated solutions from which each individual has to select one as the *leader* towards which it gets attracted. The way the particles select the leader is called *Leader Selection Mechanism*. Usually the non-dominated solutions (often stored in an archive A_t) in each iteration are the good candidates for being selected as a leader. The simplest leader selection mechanism is a random selection [5]. Each particle randomly selects one of the archive members. As the goal of multi-objective optimization is to find a diverse set of solutions, the following approaches are shown to be more effective than the random leader selection mechanism.

14.4.1.1   Ranking method

In this approach, the solutions in the archive are ranked based on the number of solutions in the population which are being dominated by them. The archive member who dominates a large number of solutions will be selected with a lower probability than a solution which dominates less solutions. This is due to the fact that around the solutions which dominate a low number of other solutions must be explored more than around those who already dominate a large number.

14.4.1.2   Sigma Method

The main idea behind this approach is that all the solutions must be directed towards the Pareto front [20]. Taking a minimization problem with two objective functions $f_1(x)$ and $f_2(x)$, first a sigma value for the non-dominated solutions in the archive are computed as follows:

$$\sigma_i = \frac{f_1^2(x_i) - f_2^2(x_i)}{f_1^2(x_i) + f_2^2(x_i)}, \quad \text{for all } i \in A_t$$

Figure 14.3 (left) shows an example. A unique property of the sigma values is that all the solutions lying on one line have the same sigma value. For example all the solutions with $f_1 = 0$ have $\sigma = -1$ independent from their f_2 values. Similarly all the solutions with $f_1 = f_2$ have $\sigma = 0$ which means that a solution with $f_1 = f_2 = 6$ has the same sigma value as a solution with $f_1 = f_2 = 10$. After computing the sigma values for the non-dominated solutions in the archive, the sigma values for the individuals in the population must be computed. Each individual selects the solution from the archive with the closest sigma value to its own, as its global solution. Figure 14.3 (right) illustrates an example. This approach forces the individuals to move towards the origin of the objective space. The population is supposed to stop at the Pareto front.

14.4.1.3   Indicator-Based Method

In this method, "Hypervolume" measurement is employed as an indicator for selecting the leaders [23]. Hypervolume is defined as the area or volume dominated by the non-dominated solutions and is located between the non-dominated solutions and

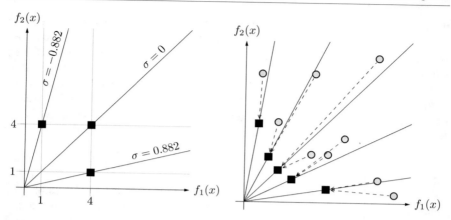

Fig. 14.3 Sigma leader selection method: Computing sigma values for archive members (left). The individuals from the population select the archive members with the closest sigma value (right).

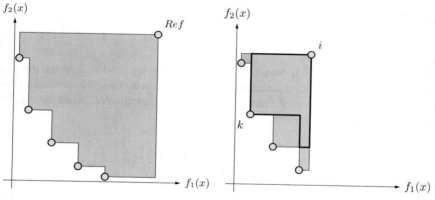

Fig. 14.4 Indicator-based leader selection method: Hypervolume is shown in gray (left). The individual i selects k as the leader, since k has the largest contribution (shown in black) to the hypervolume (right).

a given reference point [11,27]. Figure 14.4 (left) shows an example. In this leader selection mechanism the individual (who is looking for a leader) is chosen as a reference point and the hypervolume is computed for the entire archive with respect to the individual. The archive member whose contribution to the hypervolume is maximum is chosen as the leader for the individual. In case the individual is not dominated by the archive members, there is no hypervolume and the random leader selection is being used. Figure 14.4 (right) shows an example. The individual i is considered to be the reference point. The contribution of the archive member k to the hypervolume shown in black is the highest. Individual i selects k as its leader.

14.4.2 Archiving

The archive A_t is meant to keep all the non-dominated solutions over the iterations. Therefore at each iteration the newly found solutions in the population P_t and the archive members A_t must be compared in terms of the domination criterion with each other in order to update the solutions in the archive. Considering a population size of N and the archive size of N_1, the computational effort for updating the archive for a multi-objective problem with m objectives can easily grow. The following approach is known to be very efficient [7]. We combine the solutions from A_t into P_t (with the size $N + N_1 = M$) and make A_t empty. The non-dominated solutions from P_t must be stored in A_t. The first element of P_t is inserted in A_t. The second individual in P_t is compared with the only member of A_t. If one of them dominates the other one only the non-dominated one is kept. If both are indifferent to each other, both are kept. In the next step the third element from P_t will be compared with A_t and so on. This is shown in Algorithm 14.4.

Usually the archive A_t has a limited size. Limiting the size of the archive is very useful due to the following reasons. First, from the decision maker's point of view, presenting the entire non-dominated set is useless when the size increases a certain limit. Second, it is computationally very expensive to find the leaders from an unlimited number of solutions in the archive. Therefore, if the size of the non-dominated solutions exceeds a predefined limit, the so called *archiving* mechanism is used. Archiving is also known as a Diversity-Preserving Method due to the fact that most of the approaches try to keep a diverse set of solutions in the archive. There are two goals for the archiving mechanisms: (1) The archive must be kept domination-free, i.e., only non-dominated solutions are allowed to be kept. Therefore every time the population gets updated, the archive must also be updated and the dominated solutions must be deleted. (2) If the number of non-dominated solutions increases the size limit of the archive, it must be pruned. In the following three existing approaches are usually used.

Algorithm 14.4 (Updating the archive)

function UpdateArchive (P: set , A: set); (∗ population P and the archive A ∗)
begin
  $M \leftarrow |\,P\,| + |\,A\,|$;
  $P \leftarrow P \cup A$; (∗ combine the two sets ∗)
  $A \leftarrow empty$; (∗ empty archive ∗)
  $i = 1$; and $A \leftarrow \{p_i\} \cup A$ (∗ insert the first element of P to A ∗)
  for $j \in \{2, \ldots, M\}$ **do begin** (∗ j iterates over P ∗)
    for $k \in \{1, \ldots, |\,A\,|\}$ **do begin** (∗ k iterates over A ∗)
      if p_j dominates a_k (∗ remove a_k and insert p_j ∗)
      then $A \leftarrow A \setminus \{a_k\} \wedge A \leftarrow A \cup \{p_j\}$;

else if a_k dominates p_j **then** break;
   else $A \leftarrow A \cup \{p_j\}$; **end**   (∗ keep a_k and insert p_j ∗)
  end
end
  return (A) (∗ return the archive ∗)
end

14.4.2.1 Clustering

The main aim of the clustering method is to prune a non-dominated set and generate a representative subset, which maintains the characteristics of the original set. The clustering technique [27] used in SPEA is a hierarchical clustering method, which works iteratively by joining adjacent clusters until the required size of the set is obtained. This method is first introduced by Morse [19] and is named *average linkage method*. At the beginning, each solution in the set is a cluster. Then the two clusters, which are closest to each other, are joined to make a bigger cluster. This is done iteratively, until the required number of clusters is achieved. The two clusters are selected according to the nearest neighbor criterion to make a new cluster. At the end, the solution with minimal average distance to other solutions inside each cluster is kept and the others are removed. This is shown in Fig. 14.5 (left).

14.4.2.2 Crowding Distance

This approach was originally proposed in NSGA-II [7]. The main idea is to find the crowded areas in the objective space by assigning a crowding values to the archive members. The crowding value represents the sum of the distances between each archive member i and its two neighbors $i - 1$ and $i + 1$ in the objective space. Therefore, small values are assigned to solutions in the crowded areas while those in isolated areas get larger values. Additionally, in order to keep the extreme solutions,

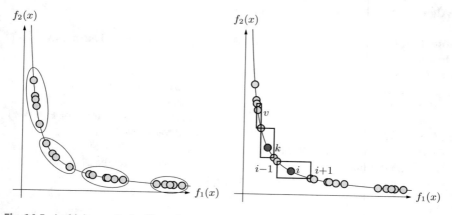

Fig. 14.5 Archiving methods: Clustering (left) and Crowing Distance (right)

the highest crowding values (e.g., ∞) are assigned to these solutions independent from their crowding values. Figure 14.5 (right) shows an example. The solution i has a larger crowding values than the solution denoted by k. Algorithm 14.5 shows the procedure for computing the crowding distances. f_j^{max} and f_j^{min} refer to the maximum and minimum values of the objective function j and are used to normalize the values.

Algorithm 14.5 (Crowding Distance)

function CD (A: set;)   (* the archive A *)
begin
  $N \leftarrow | A |$;
  for $i \in \{1, \ldots, N\}$ **do begin**
    $d_i \leftarrow 0$
  end
  for $j \in \{1, \ldots, m\}$ **do begin**
    $A_s \leftarrow \text{sort}(A, j)$;   (* Sort A in terms of j *)
    $d_1 = d_N \leftarrow \inf$;   (* extreme solutions get the highest value *)
    for $i \in \{2, \ldots, N - 1\}$ **do begin**
      $d_i \leftarrow d_i + \frac{f_j(i+1) - f_j(i-1)}{f_j^{max} - f_j^{min}}$ (* Compute distance to both of the neighbors *)
    end
  end
  return d;   (* return the crowding distance values *)
end

14.4.2.3 ϵ-domination

This approach uses the idea from the ϵ-domination. Suppose we have a multi-objective *minimization* problem. A solution $x_1 \in \Omega$ is said to ϵ-*dominate* a solution $x_2 \in \Omega$ for some $\epsilon > 0$ (denoted $x_1 \prec_\epsilon x_2$) if:

1. $f_i(x_1)/(1 + \epsilon) \leq f_i(x_2) \ \forall i = 1, \ldots, m$.
2. $f_i(x_1)/(1 + \epsilon) < f_i(x_2)$ for at least one $i = 1, \ldots, m$.

Figure 14.6 shows the concept of ϵ-domination. Accordingly, the ϵ-approximate Pareto front can be defined as follows. Let $F \subseteq \Re^m$ be a set of solutions and $\epsilon > 0$. The ϵ-approximate Pareto front $F_\epsilon \subseteq F$ contains all solutions $x_1 \in F$ which are not ϵ-dominated by any other solution $x_2 \in F$:

$$\forall x_2 \in F : \exists x_1 \text{ such that } x_1 \prec_\epsilon x_2.$$

We have to note that the set F_ϵ is not unique and can contain a certain amount of solutions depending on the ϵ-value [17,24]. For any finite ϵ and any set F with objective values $1 \leq f_i \leq K, \ \forall i \in \{1, \ldots, m\}$, there exists a set F_ϵ containing *at most* $|F_\epsilon|$ solutions (ϵ is considered to be the same for all objectives):

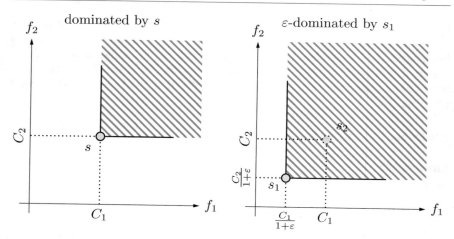

Fig. 14.6 Domination and ϵ-domination in the objective space

$$|F_\epsilon| = O\left[\left(\frac{\log K}{\log(1 + \epsilon)}\right)^{m-1}\right]$$

This implies that there is an upper bound for the solutions in the set F_ϵ, hence the set has a limited size. By using ϵ-dominance instead of domination the size of the approximated Pareto front can be limited to an upper bound [21]. It is obvious that the size of the archive depends on the ϵ-value. Hence by using this ϵ-dominance, we can limit the size of the archive without investing computational effort as in clustering or crowding methods. Applying the ϵ-dominance in MO-PSO techniques also has influence on the convergence and diversity of the solutions. While the other archiving approaches can deliver a good spread of the solutions along the Pareto front, the spread of the solutions by using this approach can be influenced by the shape (the degree of convexity) of the front.

14.5   Many-objective Particle Swarm Optimization

Multi-objective optimization problems with more than 3 objectives are called many-objective problems and are very difficult to be solved using the definition of domination, as the majority of the population members (even in the first iteration) are non-dominated. This is particularly a challenge for MO-PSO as the leader selection mechanisms cannot guide the population to get closer to the Pareto front. The existing Many-Objective Particle Swarm Optimization (Ma-PSO) methods either use several swarms or use ranking methods instead of the domination in the leader selection mechanism. In the latter case, the leader selection mechanisms use a ranking method and select the best ranked solutions as the leaders. More precisely, each solution in the population selects one of the archive members using a selection mechanism (such as tournament or fitness proportional) based on the ranks of the archive members. In the following we study two ranking methods.

14.5.1 Ranking Non-dominated Solutions

Suppose that we have a set of N non-dominated solutions. The so-called weighted average ranking [1,6] assigns a rank to a non-dominated solution by counting the number of objectives in which it is better than the other non-dominated solutions. Consider the two non-dominated solutions x_i and x_j. We compute a vector a_{ij} with m elements as follows:

$$a_{ijk} = \begin{cases} 1, & \text{if } f_k(x_i) < f_k(x_j) \\ 0, & \text{if } f_k(x_i) = f_k(x_j) \\ -1, & \text{if } f_k(x_i) > f_k(x_j), \end{cases}$$

where $k = 1, \cdots, m$ and the corresponding rank $R_1(x_i)$ for the non-dominated solution x_i is

$$R_1(x_i) = \sum_{k=1}^{m} \sum_{\substack{j=1, \\ j \neq i}}^{N} \left((N+1) - a_{ijk} \right).$$

This ranking mechanism considers the differences between the objective values in terms of being better or not. For each solution, we measure the number of objectives that are better than the other solutions and the rank is the sum of them. If a solution is better in most of the objectives, it will obtain a lower (better) rank.

14.5.2 Distance Based Ranking

This ranking method ranks a non-dominated solution in terms of both its number of objectives which are better than the others and additionally the distances between its objective values to the others [22]. In other words, this ranking mechanism favors diversity among the archive members. First we compute a vector $d_{ij} = (|f_1(x_i) - f_1(x_j)|, \ldots, |f_k(x_i) - f_k(x_j)|)$, which measures the absolute values of the differences between the objectives of the two solutions x_i and x_j. Then, the rank of each solution can be computed as

$$R_2(x_i) = \sum_{k=1}^{m} \sum_{\substack{j=1, \\ j \neq i}}^{N} d_{ijk}.$$

Here, in the contrary to R_1 ranks from above, the solutions with high (better) ranks indicate the preferred solutions which play a key role in keeping the information about the uncrowded areas in the objective space.

14.6   Ant Colony Optimization

Ant colony optimization (ACO) [8,9] is inspired by the path-finding behavior of certain species of ants. Since food from a discovered source has to be transported to the nest, many species of ants form "transport trails," which are marked by "odor signatures" of secreted chemical substances called **pheromones** (from the Greek $\phi \acute{\epsilon} \rho \epsilon \iota \nu$: to bear, to carry and $\acute{o} \rho \mu \acute{\eta}$: impetus). Since most ants are practically blind, pheromones are (besides sound and touch) their main means of communication. By following pheromone traces left by their fellows, ants find their way to a discovered food source. The amount of secreted pheromone signals both the quality and the amount of the discovered food.

The process of leaving traces in the environment, which trigger actions of other individuals, is commonly called **stigmergy** (from the Greek $\sigma \tau \acute{\iota} \gamma \mu \alpha$: mark, sign and $\acute{\epsilon} \rho \gamma o \nu$: work, action). Stigmergy enables ants to discover and follow **shortest paths** without having any global overview of the situation. All they need to adapt their behavior to the global requirements are local information. Intuitively, starting from a random exploration, the shortest path receives more pheromone than other paths, because it is traversed by more individuals in the same amount of time.

A striking illustration of this phenomenon is the so-called **double-bridge experiment** [13]. In this experiment the nest of an ant colony of the species *Iridomyrmex humilis* was connected to a food source by a double bridge, the two branches of which had different length (see Fig. 14.7 for a sketch). Since the ants are practically blind,

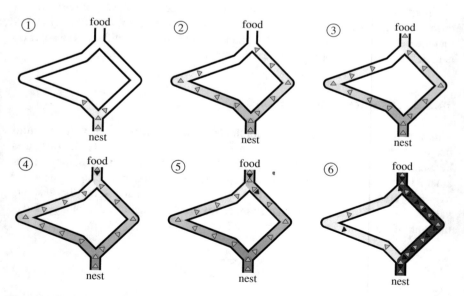

Fig. 14.7 The double-bridge experiment. The shading indicates the amount of pheromone.

they are unable to see which side of the bridge is shorter. Due to the construction of the bridge they also cannot derive any information from the angle at which the two branches fork off the initial path (see Fig. 14.7: both branches start with 45°; the longer branch changes its direction only later).

In most of the trials [13] conducted, almost all ants took the shorter branch of the bridge after only a few minutes. This phenomenon can be explained as follows (see Fig. 14.7, in which the amount of deposited pheromone is indicated by shades of gray): at the beginning, both branches are chosen by the same number of ants (that is, the branches are chosen with the same probability), because there is no pheromone on either of them (Steps 1 and 2 in Fig. 14.7). However, the ants following the shorter branch reach the food source earlier (simply because a shorter path means less travel time, Step 3). Ants returning from the food source observe more pheromone on the shorter branch, because more ants have already reached the food source on this branch (and more ants mean more secreted pheromone, Steps 4 and 5). This leads to an increasing preference of the shorter branch, so that after some time the shorter branch is chosen almost exclusively (step 6). The core principle here is that the shorter path is systematically reinforced, which is also called **auto-catalysis**: The more pheromone is on a path, the more ants choose this path; more ants traveling a path deposit more pheromone on it and so on.

Note that the shortest path can be found only if the ants deposit pheromone in *both directions*, that is, on the way from the nest to the food *and* on the way from the food to the nest. Suppose, for instance, that they deposited pheromone only on the way to the food source. Although the first ants returning from the food source choose the shorter path (because there is more pheromone on this path, see the discussion above), the amount of pheromone on this path is not systematically increased, because according to our assumption the ants do not deposit pheromone on their way back from the food source. The initial pheromone difference is rather eventually equalized by the ants that arrive (though somewhat later) on the longer path. The same argument applies for the opposite direction, at least if we assume that the ants cannot remember which path they came on and thus choose the return path randomly based on the amount of pheromone: in this case the pheromone difference would have to be brought about by ants returning on the shorter path and thus arriving earlier at the nest. Although initially ants starting out from the nest after the first ants have returned from the food source observe a pheromone difference, this difference is eventually equalized by the ants returning on the longer branch. As a consequence, no preference for the shorter path can develop.

Of course, due to random fluctuations in selecting a path (the ants choose essentially randomly with probabilities corresponding to the amount of pheromone), the search may still converge to one of the branches under such conditions. That is, eventually almost all ants may choose the same branch of the bridge. However, whether this is the shorter or the longer branch is entirely random.

Furthermore, note that (under the original conditions, that is, pheromone is deposited in both directions) the shortest path is found only if both branches exist from the beginning and neither contains any pheromone: a preference for an established path (marked with pheromone) is maintained. This plausible claim is supported by a second bridge experiment [13]: in an initial setup the nest and the food source were connected only by the longer branch of the bridge. The second, shorter branch was added only after some time. In this setup the majority of the ants kept using the longer branch, which they had established in the early stage of the experiment. Only in very rare cases the ants switched to the shorter path (likely caused by a strong random fluctuation in the path selections).

The described natural principle can be transferred to computer-aided optimization by considering the problem of finding shortest paths in weighted graphs, for example, the shortest path between two given vertices. Each ant constructs a candidate solution. It starts at one of the given vertices and then moves from vertex to vertex, choosing the edge to follow according to a probability distribution that is proportional to the amount of pheromone it observes on the edges.

Unfortunately, a fundamental problem of such a straightforward approach are cycles traversed by the ants, because they introduce a tendency to reinforce themselves: if an ant traverses a cycle, the pheromone deposited by it makes it likely that the ant traverses the same cycle again. This drawback is counteracted by depositing pheromone only after the complete path has been constructed. In addition, before pheromone is deposited, any cycles that a path may contain are removed.

Another potential problem is that the search may focus on solution candidates that are constructed early in the process. Since these solution candidates receive pheromone, there is a tendency to stick to them (or minor variations of them). This can lead to **premature convergence**, similar to what we studied in Sect. 12.2.2 for evolutionary algorithms. In order to handle this problem, it is common to introduce **pheromone evaporation** (which plays only a minor role in nature). Other beneficial extensions and improvements include making the amount of pheromone that is deposited dependent on the quality of the constructed candidate solution and the introduction of heuristics to improve the edge selection, for example, by considering not only the pheromone, but also the weight of the edge.

With these considerations it should be clear that ant colony optimization is well suited to tackle the traveling salesman problem (cf. Sect. 11.6). We represent this problem by an $n \times n$ matrix $\mathbf{D} = (d_{ij})_{1 \leq i, j \leq n}$, where n is the number of cities and the d_{ij} are the distances between the cities i and j. Obviously, $\forall i \in \{1, \ldots, n\} : d_{ii} = 0$. However, $\mathbf{D}$ need not be symmetric, that is, it may be $d_{ij} \neq d_{ji}$ for some i and j. We desire to find a round trip through all cities that has minimum length. Formally, we try to find a permutation π of the numbers $\{1, \ldots, n\}$ that minimizes the sum of the edge weights of a tour that visits the cities in the order given by π.

Corresponding to the distance matrix $\mathbf{D}$, pheromone deposits are represented by an $n \times n$ matrix $\Phi = (\phi_{ij})_{1 \leq i, j \leq n}$. Intuitively, a matrix element $\phi_{ij}, i \neq j$, states how desirable it is to visit city j directly after city i, while the ϕ_{ii} are not used (formally, we may set them to 0 for all i). Like $\mathbf{D}$, the matrix Φ need not be symmetric. The matrix elements $\phi_{ij}, 1 \leq i, j \leq n, i \neq j$, are initialized with the same arbitrary small value ϵ. That is, at the beginning every edge carries the same amount of pheromone and thus there is no preference between the edges. As an alternative, the pheromone deposits may be initialized with values that are inversely proportional to the edge weights. Following the general description given above, the pheromone values are used to determine the probability with which the next city to visit is chosen in the tour construction process. They are updated based on the quality of constructed round trips by depositing pheromone on the edges of good trips.

More formally, we proceed as follows: each ant constructs a solution by traversing a (random) Hamiltonian cycle. In order to avoid that an already visited city is revisited, we endow each ant with a "memory," consisting of the set C of indices of the cities that have not been visited yet. (Note that this deviates from the biological prototype!) An ant constructs a round trip (Hamiltonian cycle) as follows:

1. The ant starts in an arbitrary city (randomly chosen).
2. The ant observes the pheromones on the connections from its current city i to any city j it has not yet visited. Then it chooses to move to city j with the probability

$$p_{ij} = \frac{\phi_{ij}}{\sum_{k \in C} \phi_{ik}},$$

   where C is the set of indices of cities that have not been visited and ϕ_{ij} is the amount of pheromone on the connection from city i to city j.
3. The ant repeats step 2 until it has visited all cities.

After a round trip has been constructed, represented as a permutation π of the city indices, the pheromone matrix Φ is updated according to

$$\forall i \in \{1, \ldots, n\} : \quad \phi_{\pi(i)\pi((i \bmod n)+1)} \leftarrow \phi_{\pi(i)\pi((i \bmod n)+1)} + Q(\pi),$$

where Q is a function that measures the quality of the solution. A natural choice for Q for the traveling salesman problem is the (scaled) inverse trip length

$$Q(\pi) = c \cdot \left(\sum_{i=i}^{n} d_{\pi(i)\pi((i \bmod n)+1)} \right)^{-1},$$

where c is a user-specified constant that controls the strength of the pheromone changes. Intuitively, this means that the shorter the trip (and thus the better the solution), the more pheromone is deposited on its edges.

In addition, after μ ants the pheromone is updated by **pheromone evaporation**:

$$\forall i, j \in \{1, \ldots, n\} : \quad \phi_{ij} \leftarrow (1 - \eta) \cdot \phi_{ij},$$

where η is an evaporation factor (fraction of pheromone that evaporates).

The complete algorithm for the traveling salesman problem looks like this:

Algorithm 14.6 (**Ant-Colony-Optimization-TSP**)

function aco_tsp (**D**: matrix of real, μ: int, η: real) : list of int;
begin   (* **D** $= (d_{ij})_{1 \le i, j \le n}$: distance matrix *)
  $\Phi \leftarrow (\phi_{ij})_{1 \le i, j \le n}$ (* μ: number of ants, η: evaporation *)
    with $\phi_{ij} = \epsilon$ for all i, j;   (* initialize the pheromone matrix *)
  $\pi_{\text{best}} \leftarrow$ random permutation of $\{1, \dots, n\}$;
  repeat   (* search loop *)
    for $a \in \{1, \dots, \mu\}$ **do begin**   (* construct candidate solution *)
      $C \leftarrow \{1, \dots, n\}$;   (* set of cities to visit *)
      $i \leftarrow$ randomly choose start city from C;
      $\pi \leftarrow (i)$;   (* start tour at the chosen city and *)
      $C \leftarrow C \setminus \{i\}$;   (* remove it from the unvisited cities *)
      while $C \neq \emptyset$ **do begin**   (* not all cities have been visited *)
        $j \leftarrow$ choose next city of the trip from C
          with probability $p_{ij} = \phi_{ij} / \sum_{k \in C} \phi_{ik}$;
        π.append(j);   (* add the chosen city to the tour and *)
        $C \leftarrow C \setminus \{j\}$;   (* remove it from the unvisited cities *)
        $i \leftarrow j$;   (* finally go to the selected city *)
      end
      update pheromone matrix Φ with π and $Q(\pi)$;
      if $Q(\pi) > Q(\pi_{\text{best}})$ **then** $\pi_{\text{best}} \leftarrow \pi$; **end**
    end   (* update the best tour *)
    update pheromone matrix Φ with evaporation η;
  until termination criterion is fulfilled;
  return π_{best};   (* return the best tour found *)
end

This basic algorithm may be extended in several ways. For instance, we may introduce a preference of cities that are close, in analogy to a nearest neighbor heuristics: choose to move from city i to city j with probability

$$p_{ij} = \frac{\phi_{ij}^{\alpha} \tau_{ij}^{\beta}}{\sum_{k \in C} \phi_{ik}^{\alpha} \tau_{ik}^{\beta}},$$

where C contains the indices of cities that have not been visited yet and $\tau_{ij} = d_{ij}^{-1}$.

A "greedy" approach introduces a stronger tendency to choose the best edge (than is already present due to the fact that is has the highest probability) with a user-specified probability p_{exploit}. That is, the ant moves from city i to city j_{best} with

$$j_{\text{best}} = \arg\max_{j \in C} \phi_{ij} \quad \text{or} \quad j_{\text{best}} = \arg\max_{j \in C} \phi_{ij}^{\alpha} \tau_{ij}^{\beta},$$

with probability p_{exploit}, while it chooses the next city according to the rule given above (that is, based on probabilities p_{ij}) with probability $1 - p_{\text{exploit}}$.

Furthermore, we may draw on an **elite principle** by reinforcing the best-known round trip: after every iteration of the search (that is, after every μ ants), additional pheromone is deposited on the edges of the best round trip that has been found so far. The amount of pheromone can conveniently be specified as the number of ants that traverse this tour (in addition to the tours constructed in the normal process).

Further variants include **rank-based updating**, in which pheromone is deposited only on the edges of the best m solution candidates of the last iteration (consisting of the runs of μ ants), and maybe also on the best solution candidate found so far. This approach can be seen as analogous to rank-based selection (cf. Sect. 12.2.6), whereas the standard approach is analogous to fitness proportionate selection (cf. Sect. 12.2.1). **Strict elite principles** are extreme forms of rank-based updating: pheromone is deposited only on the best solution candidate of the last iteration or even only on the best solution found so far. However, this approach carries the risk of premature convergence and thus of getting stuck in a local optimum.

In order to avoid extreme values of the pheromone deposits, it can be advisable to introduce **lower and upper bounds** for the amount of pheromone on an edge. They correspond to lower and upper bounds for the probability of selecting an edge and thus help to enforce a better exploration of the search space (though at the price of slower convergence). A similar effect can be achieved by **restricted evaporation**: pheromone evaporates only from edges that have been traversed in the last iteration. This reduces the risk of pheromone deposits becoming very small.

Improvements of the standard approach, which are meant to lead to better solution candidates, are local improvements of the round trip (like removing edge crossings, which obviously cannot be optimal). More generally, we may consider simple operations as they could be used in a hill climbing approach (cf. Sect. 11.5.2) and thus try (in a limited number of steps) to optimize solution candidates locally. Among such operations are exchange of cities that are visited in consecutive steps, permutation of adjacent triplets, "inverting" a part of a round trip (cf. Sect. 11.6), etc. More costly local optimization should only be considered to improve the best solution candidate before it is returned from the search procedure.

A program illustrating ant colony optimization for the traveling salesman problem (containing several of the mentioned improvements) can be found at

http://www.borgelt.net/acopt.html

In order to apply ant colony optimization to other optimization problems, the problem has to be formulated as a search in a graph. In particular, it must be possible to describe a solution candidate as a set of edges. However, these edges need not form a path. As long as there is an iterative procedure with which the edges of the set can be chosen (based on pheromone-described probabilities), ant colony optimization is applicable. Even more generally, ant colony optimization is applicable if solution candidates are constructed with the help of a series of (random) decisions, where every decision extends a (partial) solution. The reason is that the sequence of decisions can be interpreted as a path in a **decision graph** (also called **construction graph**). The ants explore paths in this decision graph and try to find the best (shortest, cheapest) path, which yields a best set or sequence of decisions.

As a final remark we consider the convergence properties of ant colony optimization. Referring to a "standard procedure," in which (1) pheromone evaporates with a constant factor from all edges, (2) new pheromone is deposited only on the edges of the best found candidate solution (strict elite principle), and (3) the pheromone values are bounded from below by ϕ_{min}, the search converges in probability to the optimal solution [9]. That is, if we let the number of computation steps go to infinity, then the probability that the optimal solution is found approaches 1. If the lower bound ϕ_{min} goes to 0 "sufficiently slowly" (for instance, $\phi_{min} = \frac{c}{\ln(t+1)}$ where t is the iteration step and c is some constant), one can even show that with the number of iterations going to infinity, every ant in the colony will construct the optimal solution with a probability approaching 1.

References

1. Bentley P, Wakefield J (1997) Finding acceptable Pareto-optimal solutions using multiobjective genetic algorithms. Soft Comput Eng Des Manuf. Springer
2. Branke J, Mostaghim S (2006) about selecting the personal best in multi-objective particle swarm optimization. In: 9th international conference on parallel problem solving from nature—PPSN IX, Proceedings, pp 523–532
3. Camazine S, Franks NR, Sneyd J, Bonabeau E, Deneubourg J-L, Theraula G (2001) Self-organization in biological systems. Princeton University Press, Princeton, NJ, USA
4. Clerc M, Kennedy J (2002) The particle swarm: Explosion, stability, and convergence in a multi-dimensional complex space. IEEE Trans Evolut Comput 6(1):58–73
5. Coello Coello CA, Lechuga MS (2002) MOPSO: a proposal for multiple objective particle swarm optimization. In: IEEE proceedings, world congress on computational intelligence, pp 1051–1056
6. Corne D, Knowles J (2007) Techniques for highly multiobjective optimisation: Some non-dominated points are better than others. In: Genetic and evolutionary computation conference (GECCO)
7. Deb K, Pratap A, Agarwal S, Meyarivan T (2002) A fast and elitist multiobjective genetic algorithm: NSGA-II. IEEE TransEvol Comput 6(2):182–197
8. Dorigo M (1992) Optimization, Learning and Natural Algorithms. PhD Thesis, Politecnico di Milano, Milan, Italy
9. Dorigo M, Stützle T (2004) Ant colony optimization. Bradford/MIT Press, Cambridge, MA, USA
10. Eberhart RC, Kennedy J (2001) Swarm intelligence. Morgan Kaufmann
11. Fleischer M (2003) The measure of pareto optima—applications to multi-objective metaheuristics. In: Fonseca CM et al (eds) Evolutionary Multi-Criterion Optimization (EMO). Springer, Berlin, pp 519–533
12. Gazi V, Passino K (2011) Swarm stability and optimization. Springer, Berlin
13. Goss S, Aron S, Deneubourg J-L, Pasteels JM (1989) Self-organized shortcuts in the argentine ant. Naturwissenschaften 76:579–581. Springer, Heidelberg, Germany
14. Helwig Sabine, Branke Jürgen, Mostaghim Sanaz (2013) Experimental analysis of bound handling techniques in particle swarm optimization. IEEE Trans Evol Comput 17(2):259–271
15. Helwig S, Wanka R (2007) Particle swarm optimization in high-dimensional bounded search spaces. In: Swarm intelligence symposium. IEEE, pp 198–205

16. Kennedy J, Eberhart R (1995) Particle swarm optimization. In: Proceedings IEEE international conference on neural networks, vol 4. IEEE Press, Piscataway, NJ, USA, pp 942–1948

17. Laumanns M, Thiele L, Deb K, Zitzler E (2002) Archiving with guaranteed convergence and diversity in multi-objective optimization. In: Genetic and Evolutionary Computation Conference (GECCO), pp 439-447

18. Mátyus E, Hutter J, Müller-Herold U, Reiher M (2011) On the emergence of molecular structure. J Phys Rev A 53(5):052512-1-5

19. Morse JN (1980) Reducing the size of the nondominated set: pruning by clustering. Comput Oper Res 7(1–2):55–66

20. Mostaghim S, Teich J (2003) Strategies for finding good local guides in multi-objective particle swarm optimization. In: Swarm intelligence symposium. IEEE, pp 26–33

21. Mostaghim S, Teich J (2003) The role of ϵ-dominance in multi objective particle swarm optimization methods. In: IEEE congress on evolutionary computation, Procedings, pp 1764–1771

22. Mostaghim S, Schmeck H (2008) Distance based ranking in many-objective particle swarm optimization. In: 10th international conference on parallel problem solving from nature—PPSN X, Procedings, pp 753–762

23. Padhye Nikhil, Branke Jürgen, Mostaghim Sanaz (2009) Empirical comparison of MOPSO methods—guide selection and diversity preservation. In: IEEE congress on evolutionary computation, Procedings, pp 2516–2523

24. Papadimitriou CH, Yannakakis M (2000) On the approximability of trade-offs and optimal access of web sources. In: IEEE symposium on foundations of computer science

25. Reyes-Sierra M, Coello CC (2006) Multi-objective particle swarm optimizers: a survey of the state-of-the-art. Int J Comput Intell Res

26. Reynolds CW (1987) Flocks, herds, and schools: a distributed behavioral model. Comput Graph 21(4); ACM SIGGRAPH '87 Conference Proceedings. Anaheim, California, pp 25–34

27. Zitzler E (1999) Evolutionary algorithms for multiobjective optimization: methods and applications. Shaker Verlag

Part III
Fuzzy Systems

Introduction to Fuzzy Sets and Fuzzy Logics

15

Many propositions about the real world are not either true or false, rendering classical logic inadequate for reasoning with such propositions. Furthermore, most concepts used in human communication do not have crisp boundaries, rendering classical sets inadequate to represent such concept. The main aim of fuzzy logic and fuzzy sets is to overcome the disadvantages of classical logic and classical sets.

15.1 Natural Languages and Formal Models

Classical logic and mathematics assume that we can assign one of the two values, *true* or *false*, to each logical proposition or statement. If a suitable formal model for a certain problem or task can be specified, conventional mathematics provides powerful tools which help us to solve the problem. When we describe such a formal model, we use a terminology which has much more stringent rules than natural language. This specification often requires more work and effort, but by using it we can avoid misinterpretations. Furthermore, based on such models we can prove or reject hypotheses or derive unknown correlations.

However, in our everyday life, formal models do not concern the interhuman communication. Human beings are able to assimilate easily linguistic information without thinking in any type of formalization of the specific situation. For example, a person will have no problems to accelerate slowly while starting a car, if he is asked to do so. If we want to automate this action, it will not be clear at all, how to translate this advice into a well-defined control action. It is necessary to determine a concrete statement based on an unambiguous value, i.e., step on the gas at the velocity of half a centimeter per second. On the other hand, this kind of information will not be adequate or very helpful for a person.

Therefore, automated control is usually not based on a linguistic description of heuristic knowledge or knowledge from one's own experience, but it is based on a

© Springer Nature Switzerland AG 2022
R. Kruse et al., *Computational Intelligence*, Texts in Computer Science,
https://doi.org/10.1007/978-3-030-42227-1_15

formal model of the technical or physical system. This method is definitely a suitable approach, especially if there is a good model to be determined.

However, a completely different technique is to use knowledge formulated in natural language directly for the design of the control strategy. In this case, a main problem will be the translation of the verbal description into concrete values, i.e., assigning "step on the gas slowly" into "step on the gas at the velocity of a centimeter per second" as in the above-mentioned example.

When describing an object or an action, we usually use uncertain or imprecise concepts. In natural language, we hardly ever find exactly defined concepts like supersonic speed for the velocity of a passing airplane. Supersonic speed characterizes an unambiguous set of velocities, because the speed of sound is a fixed entity and therefore it is unambiguously clear whether an airplane flies faster than sound or not. Frequently used imprecise concepts, like *fast*, *very big*, *small*, and so on, make it impossible to decide unambiguously whether a given value satisfies such a imprecise concept or not. One of the reasons for this is that imprecise concepts are usually context dependent. Talking about airplanes *fast* has a different meaning than using this characteristic while referring to cars. But also if we agree that we are talking about cars it is not easy to distinguish clearly between fast and non-fast cars. The difficulty here is not to find a value telling us whether a car (or its top speed) is fast or not, but we have to presuppose that such a value does exist. It is more likely that we will be reluctant to fix such a value because there are velocities, we can classify as fast for a car and there are some we can classify as not fast, and in between there is a wide range of velocities which are considered as more or less fast.

15.2 Fuzzy Sets

The idea of fuzzy sets is to solve this problem by avoiding the sharp separation of conventional sets into two values—complete membership or complete non-membership. Instead, fuzzy sets can handle partial membership. So in fuzzy sets we have to determine to what degree or extent an element is a member of this fuzzy set. Therefore, we define

Definition 15.1 A *fuzzy subset* or simply a *fuzzy set* μ of a set X (the universe of discourse) is a mapping $\mu : X \to [0, 1]$, which assigns to each element $x \in X$ a *degree of membership* $\mu(x)$ to the fuzzy (sub)set μ. The set of all fuzzy (sub)sets of X is denoted $\mathcal{F}(X)$.

A conventional set $M \subseteq X$ can be viewed as a special fuzzy set by identifying it with its **characteristic function** or **indicator function**.

$$I_M : X \to \{0, 1\}, \qquad x \mapsto \begin{cases} 1 \text{ if } x \in M \\ 0 \text{ otherwise} \end{cases}$$

Seen in this way, fuzzy sets can be considered as generalized characteristic functions.

Example 15.1 Figure 15.1 shows the characteristic function of the set of velocities which are higher than 170 km/h. This set does not represent an adequate model of all high velocities. The jump at the value of 170 causes that 169.9 km/h would not be a high velocity but 170.1 km/h would be. Therefore, a fuzzy set (Fig. 15.2) seems to be more adequate to model the concept *high velocity*.   □

Some authors use the term fuzzy set only for an imprecise concept $\mathcal{A}$ like *high velocity* and call the membership function $\mu_{\mathcal{A}}$ that models the imprecise concept, a characterizing or membership function of the fuzzy set or the imprecise concept $\mathcal{A}$. When operating with fuzzy sets, there is no advantage to make this distinction. From a philosophical point of view, one might be interested in distinguishing between an abstract imprecise concept and its concrete model in the form of a fuzzy sets. Since we do not want to initiate a philosophical discussion in this book, we stick to our restricted definition of a fuzzy set as a (membership) function, yielding values in the unit interval.

Besides the formal definition of a fuzzy set as a mapping to the unit interval, there are also other notations which are preferred by some authors, but we will not use them in this book. In some publications, a fuzzy set is written as a set of pairs of the elements of the underlying set and the corresponding degrees of membership in the form of $\{(x, \mu(x)) \mid x \in X\}$ following the fact that in mathematics a function is usually formally defined as a set of pairs, each consisting of one argument of the function and the image of this argument. A little bit more misleading is the notation of a fuzzy set as a formal sum $\sum_{x \in X} x/\mu(x)$ with an at most countable reference set X or as an "integral" $\int_{x \in X} x/\mu(x)$ for an uncountable reference set X.

We want to emphasize here that fuzzy sets are formalized in the framework of "conventional" mathematics, just as probability theory is formulated in the framework of "conventional" mathematics. In this sense, fuzzy sets do not open the door to a "new" kind of mathematics, but define merely a new branch of mathematics.

Fig. 15.1 The characteristic function of the set of velocities that are higher than 170 km/h

Fig. 15.2 The fuzzy set μ_{hv} of high velocities

Knowing that a strictly two-valued view is not suitable to model imprecise concepts adequately, which can be handled by human beings easily, we have introduced the concept of fuzzy sets on a purely intuitive basis. In applications, we have to specify precisely, how to interpret degrees of membership. The meanings of 1 as complete membership and 0 as complete non-membership are obvious, but we also have to answer the question, how to interpret a degree of membership of 0.7 and why the degree of 0.7 is better suited than the degree of 0.8 for membership for a certain element.

Gradual membership is a different idea than the concept of probability. Formally, a fuzzy set μ must not be regarded as a probability distribution or density, because, in general, μ does not satisfy the condition that is required in probability theory for density functions. Also the degree of membership $\mu(x)$ of an element x to the fuzzy set μ should not be interpreted as the probability that x belongs to μ (namely, that the integral over the whole domain is unity). Probability theory is designed for modeling uncertainty, whereas fuzzy sets are used to model imprecise concepts.

The standard method for quantification of uncertainty is probability theory. There are two broad categories of probability interpretations which can be called physical (objective, frequentistic) and evidential (subjective, epistemic, logical) probabilities. Note that there are additionally lots of other useful methods for quantification of epistemic information (e.g., belief) that turn out to be helpful in real applications. In our chapter on Bayesian networks, we use an evidential interpretation for handling uncertainty. It is important to note that for successful real applications it is necessary to specify the exact meaning of an uncertainty degree.

It is not surprising that in fuzzy set theory we also have many different interpretations of a fuzzy membership degree, a large part of which are highly application dependent. The straightforward way is to interpret a membership degree as a "truth-value within a many-valued logic" in other to model partial truth. But there are many other interpretations for membership degrees such as the "degree of possibility," the "degree of preference," or the "degree of similarity to reference values." Refer to Sect. 15.5 for more details.

In Chaps. 15–17, we interpret fuzzy membership degrees in the straightforward view as partial truth values of a many-valued logic. Other interpretations, e.g., similarity and possibility are studied later on.

15.3   Representation of Fuzzy Sets

After having introduced fuzzy sets formally as functions from a universe of discourse to the unit interval, we now discuss different methods for specifying concrete fuzzy sets and adequate techniques for representing fuzzy sets as well as for storing them in a computer.

15.3.1 Definition Based on Functions

If the universe of discourse $X = \{x_1, \ldots, x_n\}$ is a finite, discrete set of objects x_i, a fuzzy set μ can, in general, only be specified by the degrees of membership $\mu(x)$ for each element $x \in X$, i.e., in the form of $\mu \cong \{(x_1, \mu(x_1)), \ldots, (x_n, \mu(x_n))\}$.

In most of the cases, we will consider fuzzy sets here, the universe of discourse X will be the domain of a real-valued variable, i.e., a subset of the real line, usually an interval. Then a fuzzy set μ is a real function taking values in the unit interval and can be illustrated by drawing its graph. With a purely graphical definition of fuzzy sets membership degrees of the single elements can only be specified up to a certain, quite rough precision leading to difficulties and errors in further calculations. Thus, the graphical representation is only suitable for illustration purposes.

Usually fuzzy sets are used for modeling expressions—sometimes also called *linguistic* expressions in order to emphasize the relation to natural language, e.g., "about 3," "of middle height," or "very tall" which describe an imprecise value or an imprecise interval. Fuzzy sets associated with such expressions should monotonically increase up to a certain value and monotonically decrease from this value. Equivalently, for all $u, v \in \mathbb{R}$ and $\lambda \in [0, 1]$ it holds

$$A(\lambda u + (1 - \lambda)v) \geq \min\{\mu(u), \mu(v)\}.$$

Such fuzzy sets μ are called **convex**.

Figure 15.3 shows three convex fuzzy sets which could model the expressions "about 3," "of middle height," and "very tall." In Fig. 15.4, we see a non-convex fuzzy set. Note that the convexity of a fuzzy set μ does not imply that μ is also convex as a real function.

For applications, it is very often sufficient to consider only a few basic forms of convex fuzzy sets, so that a fuzzy set can be specified uniquely by a few

Fig. 15.3 Three convex fuzzy sets

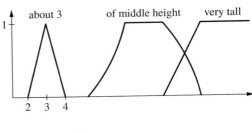

Fig. 15.4 A non-convex fuzzy set

parameters. Typical examples of such parametric fuzzy sets are **triangular functions** (cf. Fig. 15.5)

$$\Lambda_{a,b,c} : \mathbb{R} \to [0, 1], \qquad x \mapsto \begin{cases} \frac{x-a}{b-a} & \text{if } a \leq x \leq b \\ \frac{c-x}{c-b} & \text{if } b \leq x \leq c \\ 0 & \text{otherwise,} \end{cases}$$

where $a < b < c$ holds.

Triangular functions are special cases of **trapezoidal functions** (cf. Fig. 15.5)

$$\Pi_{a',b',c',d'} : \mathbb{R} \to [0, 1], \qquad x \mapsto \begin{cases} \frac{x-a'}{b'-a'} & \text{if } a' \leq x \leq b' \\ 1 & \text{if } b' \leq x \leq c' \\ \frac{d'-x}{d'-c'} & \text{if } c' \leq x \leq d' \\ 0 & \text{otherwise,} \end{cases}$$

where $a' < b' \leq c' < d'$ holds. We also permit the following parameter combinations: $a' = b' = -\infty$ or $c' = d' = \infty$. The resulting trapezoidal functions are shown in Fig. 15.6. For $b' = c'$, we have $\Pi_{a',b',c',d'} = \Lambda_{a',b',d'}$.

If we want to use smooth functions instead of piecewise linear functions like triangular or trapezoidal ones, *bell curves* in the form of

$$\Omega_{m,s} : \mathbb{R} \to [0, 1], \qquad x \mapsto \exp\left(\frac{-(x-m)^2}{s^2}\right)$$

might be a possible choice. We have $\Omega_{m,s}(m) = 1$. The parameter s determines the width of the bell curve.

A fuzzy set is called *normal*, if there is at least one element of X with membership degree 1. The *support* of a fuzzy set consists of all elements of X with membership degree bigger than 0.

Fig. 15.5 The triangular function $\Lambda_{a,b,c}$, the trapezoidal function $\Pi_{a',b',c',d'}$, and the bell curve $\Omega_{m,s}$

Fig. 15.6 The trapezoidal functions $\Pi_{-\infty,-\infty,a,b}$, $\Pi_{a,b,c,d}$ and $\Pi_{c,d,\infty,\infty}$

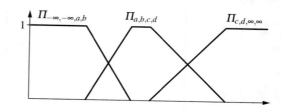

15.3.2 α-Cuts

The representation of a fuzzy set as a function from the universe of discourse to the unit interval, assigning a membership degree to each element, is called *vertical view*. Another possibility to describe a fuzzy set is the *horizontal view*. For each value α of the unit interval, we consider the set of elements having a membership degree of at least α to the fuzzy set.

Definition 15.2 Let $\mu \in \mathcal{F}(X)$ be a fuzzy set over the universe of discourse X and let $0 \le \alpha \le 1$. The (usual) set

$$[\mu]_\alpha = \{x \in X \mid \mu(x) \ge \alpha\}$$

is called α-*cut* of the fuzzy set μ.

Figure 15.7 shows the α-cut $[\mu]_\alpha$ of the fuzzy set μ for the case that μ is a trapezoidal function. In this case, the α-cut is a closed interval. For an arbitrary fuzzy set μ over the real numbers, we have that μ is convex as a fuzzy set if all its α-cuts are intervals. Figure 15.8 shows an α-cut of a non-convex fuzzy set consisting of two disjoint intervals.

The α-cuts of a fuzzy set have the important property of characterizing the fuzzy set uniquely. When we know the α-cuts $[\mu]_\alpha$ of a fuzzy set μ for all $\alpha \in [0, 1]$, we can determine the degree of membership $\mu(x)$ of any element x to μ by the equation

$$\mu(x) = \sup \left\{ \alpha \in [0, 1] \mid x \in [\mu]_\alpha \right\}. \tag{15.1}$$

Geometrically speaking, a fuzzy set is the upper envelope of its α-cuts.

Fig. 15.7 The α-cut $[\mu]_\alpha$ of the fuzzy set μ

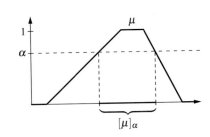

Fig. 15.8 α-cut $[\mu]_\alpha$ of the fuzzy set μ consisting of two disjoint intervals

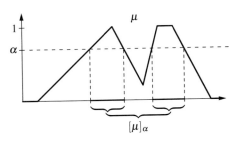

Characterizing a fuzzy set through its α-cuts will allow us in Sect. 15.6 and Fig. 15.16 to work level-wise with operations on fuzzy sets on the basis of usual sets.

The connection between a fuzzy set and its α-cuts is frequently used for the internal representation of fuzzy sets in computers. But only the α-cuts for a finite amount of selected values α, e.g., 0.25, 0.5, 0.75, 1, are saved. In order to determine the degree of membership of an element x to the fuzzy set μ, Eq. 15.1 can be used, where the supremum is only taken over a finite number of values for α. Thus, we discretize the degrees of membership and obtain an approximation of the original fuzzy set. Figure 15.10 shows the α-cuts $[\mu]_{0.25}$, $[\mu]_{0.5}$, $[\mu]_{0.75}$, and $[\mu]_1$ of the fuzzy set μ defined in Fig. 15.9. If we only use these four α-cuts in order to represent μ, we obtain the fuzzy set

$$\tilde{\mu}(x) = \max\left\{\alpha \in \{0.25, 0.5, 0.75, 1\} \mid x \in [\mu]_\alpha\right\}$$

in Fig. 15.11 as an approximation for μ.

Confining us to a finite number of α-cuts in order to represent or save a fuzzy set corresponds to a discretization of the membership degrees. Besides this vertical discretization we can also discretize the domain (horizontal discretization). Depending on the considered problem, we have to choose how fine the discretization should be chosen in both directions. Therefore, no general rules for discretization can be specified. In general, a refined discretization of the membership degrees seldom leads to significant improvements of a fuzzy system. One reason for this is that the fuzzy

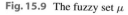

Fig. 15.9 The fuzzy set μ

Fig. 15.10 The α-cuts of the fuzzy set μ for $\alpha = 0.25, 0.5, 0.75, 1$

Fig. 15.11 The approximation of the fuzzy set μ resulting from the α-cuts

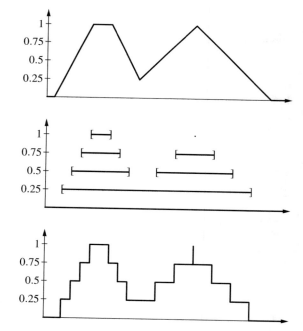

sets are usually determined heuristically or can only be specified roughly. Another reason is that human experts tend to use a limited amount of differentiation levels or degrees of acceptance or membership in order to judge a situation.

15.4 Fuzzy Logic

The notion *fuzzy logic* has two different meanings. In some cases, the term fuzzy logic refers to fuzzy logic in the broader sense, including all applications and theories where fuzzy sets or concepts are involved.

Fuzzy logic in the narrow sense, which is the topic of this section, considers fuzzy systems from the point of view of multi-valued logic and is devoted to issues connected to logical calculi and the associated deduction mechanisms.

We cannot provide a complete introduction to fuzzy logic as a multi-valued logic [11]. A detailed study of this aspect is found in [3]. In this section, we will introduce those notions of fuzzy logic which are necessary or useful to understand the main concepts. We mainly need the concepts of (fuzzy) logic to introduce the set-theoretical operations for fuzzy sets. The basis for operations like union, intersection, and complement are the logical connectives disjunction, conjunction, and negation, respectively. Therefore, we briefly repeat some fundamental concepts from classical logic in order to generalize them to the field of fuzzy logic.

15.4.1 Propositions and Truth Values

Classical propositional logic deals with the formal handling of statements (propositions) to which one of the two truth values 1 (for *true*) or 0 (for *false*) can be assigned. We represent these propositions by Greek letters φ, ψ, etc. Typical propositions, for which the formal symbols φ_1 and φ_2 may stand, are

$$\varphi_1 : \text{Four is an even number.}$$
$$\varphi_2 : 2 + 5 = 9.$$

The truth value which is assigned to a proposition φ is denoted by $[\![\varphi]\!]$. For the above propositions, we obtain $[\![\varphi_1]\!] = 1$ and $[\![\varphi_2]\!] = 0$. If the truth values of single propositions are known, we can determine the truth values of combined propositions using truth tables that define the interpretation of the corresponding logical connectives. The most important logical connectives are the logical AND $\wedge$ (conjunction), the logic OR $\vee$ (disjunction), the negation NOT $\neg$, and the IMPLICATION $\rightarrow$.

The conjunction $\varphi \wedge \psi$ of two propositions φ and ψ is true, if and only if both φ and ψ are true. The disjunction $\varphi \vee \psi$ of φ and ψ obtains the truth value 1 (true), if and only if at least one of the two propositions is true. The implication $\varphi \rightarrow \psi$ is only false, if the antecedent φ is true and the consequent ψ is false. The negation $\neg \varphi$ of the proposition φ is false, if and only if φ is true. These definitions are shown

Table 15.1 The truth value table for conjunction, disjunction, implication, and negation

$[\varphi]$	$[\psi]$	$[\varphi \wedge \psi]$
1	1	1
1	0	0
0	1	0
0	0	0

$[\varphi]$	$[\psi]$	$[\varphi \wedge \psi]$
1	1	1
1	0	1
0	1	1
0	0	0

$[\varphi]$	$[\psi]$	$[\varphi \wedge \psi]$
1	1	1
1	0	0
0	1	1
0	0	1

$[\varphi]$	$[\neg\varphi]$
1	0
0	1

in the truth value tables for conjunction, disjunction, implication, and negation in Table 15.1.

This definition implies that the propositions

$$\text{Four is an even number AND } 2 + 5 = 9.$$

and

$$\text{Four is an even number IMPLICATION } 2 + 5 = 9.$$

are false, whereas the propositions

$$\text{Four is an even number OR } 2 + 5 = 9.$$

and

$$\text{NOT } 2 + 5 = 9.$$

are true. Formally expressed, this means that we have $[\![\varphi_1 \wedge \varphi_2]\!] = 0$, $[\![\varphi_1 \rightarrow \varphi_2]\!] = 0$, $[\![\varphi_1 \vee \varphi_2]\!] = 1$, and $[\![\neg\varphi_2]\!] = 1$.

The assumption that a statement is either true or false is suitable for mathematical issues. But for many expressions formulated in natural language such a strict separation between true and false statements would be unrealistic and would lead to counterintuitive consequences. If somebody promises to come to an appointment at 5 o'clock, his statement would have been false, if he came 1 minute later. Nobody would call him a liar, although, strictly speaking, his statement was not true. Even more complicated is the statement of being at a party at about 5. The greater the difference between the arrival and 5 o'clock the "less true" the statement is. A sharp definition of an interval of time corresponding to "about 5" is impossible.

Humans are able to formulate such "fuzzy" statements, understand them, draw conclusions from them, and work with them. If someone starts an approximately 4-hour drive at around 11 o'clock and is going to have lunch for about half an hour, we can use these imprecise pieces of information and conclude at what time more or less the person will arrive. A formalization of this simple issue in a logical calculus, where statements can be either true or false only, is not adequate.

Table 15.2 Truth tables of conjunction, disjunction, implication, and equivalence for several different three-valued logics

		Łukasiewicz				Bochvar				Kleene				Heyting				Reichenbach			
a	b	$\wedge$	$\vee$	$\rightarrow$	$\leftrightarrow$	$\wedge$	$\vee$	$\rightarrow$	$\leftrightarrow$	$\wedge$	$\vee$	$\rightarrow$	$\leftrightarrow$	$\wedge$	$\vee$	$\rightarrow$	$\leftrightarrow$	$\wedge$	$\vee$	$\rightarrow$	$\leftrightarrow$
0	0	0	0	1	1	0	0	1	1	0	0	1	1	0	0	1	1	0	0	1	1
0	$\frac{1}{2}$	0	$\frac{1}{2}$	1	$\frac{1}{2}$	$\frac{1}{2}$	$\frac{1}{2}$	$\frac{1}{2}$	$\frac{1}{2}$	0	$\frac{1}{2}$	1	$\frac{1}{2}$	0	$\frac{1}{2}$	1	0	0	$\frac{1}{2}$	1	$\frac{1}{2}$
0	1	0	1	1	0	0	1	1	0	0	1	1	0	0	1	1	0	0	1	1	0
$\frac{1}{2}$	0	0	$\frac{1}{2}$	$\frac{1}{2}$	$\frac{1}{2}$	$\frac{1}{2}$	$\frac{1}{2}$	$\frac{1}{2}$	$\frac{1}{2}$	0	$\frac{1}{2}$	$\frac{1}{2}$	$\frac{1}{2}$	0	$\frac{1}{2}$	0	0	0	$\frac{1}{2}$	$\frac{1}{2}$	$\frac{1}{2}$
$\frac{1}{2}$	$\frac{1}{2}$	$\frac{1}{2}$	$\frac{1}{2}$	1	1	$\frac{1}{2}$	$\frac{1}{2}$	$\frac{1}{2}$	$\frac{1}{2}$	$\frac{1}{2}$	$\frac{1}{2}$	$\frac{1}{2}$	$\frac{1}{2}$	$\frac{1}{2}$	$\frac{1}{2}$	1	1	$\frac{1}{2}$	$\frac{1}{2}$	1	1
$\frac{1}{2}$	1	$\frac{1}{2}$	1	1	$\frac{1}{2}$	$\frac{1}{2}$	$\frac{1}{2}$	$\frac{1}{2}$	$\frac{1}{2}$	$\frac{1}{2}$	1	1	$\frac{1}{2}$	$\frac{1}{2}$	1	1	$\frac{1}{2}$	$\frac{1}{2}$	1	1	$\frac{1}{2}$
1	0	0	1	0	0	0	1	0	0	0	1	0	0	0	1	0	0	0	1	0	0
1	$\frac{1}{2}$	$\frac{1}{2}$	1	$\frac{1}{2}$	$\frac{1}{2}$	$\frac{1}{2}$	$\frac{1}{2}$	$\frac{1}{2}$	$\frac{1}{2}$	$\frac{1}{2}$	1	$\frac{1}{2}$	$\frac{1}{2}$	$\frac{1}{2}$	1	$\frac{1}{2}$	$\frac{1}{2}$	$\frac{1}{2}$	1	$\frac{1}{2}$	$\frac{1}{2}$
1	1	1	1	1	1	1	1	1	1	1	1	1	1	1	1	1	1	1	1	1	1

The classical propositional logics have been extended by several researchers to three-valued logics, in which the truth values 1 (for *true*), 0 (for *false*), and $\frac{1}{2}$ (for *indeterminate*) are allowed [5,11,13,18,23,31]. The conceptual form and basic ideas of three-valued logic was initially published by Jan Łukasiewicz. In these logics, the truth tables for conjunction, implication, and equivalence are kept for the values 0 and 1, the truth tables for the negation in these logics is always defined by $\neg 0 = 1$, $\neg 1 = 0$, and $\neg \frac{1}{2} = \frac{1}{2}$. For the other combinations, there are several options that differ from logic to logic, see Table 15.2 for a few examples. There are also lots of many-valued logics with a finite number of truth values.

In the context of fuzzy systems, we are mainly interested in logics with truth values within the unit interval: There does not exist a standard interpretation of the truth degrees. How they are to be understood depends on the actual field of application. It is general usage to assume that there are two particular truth degrees, usually denoted by 0 and 1. These particular truth degrees act, respectively, like the traditional truth values *falsum* and *verum*—but sometimes also like *absolutely false* and *absolutely true*. Two well-known logics with truth values in the unit interval were developed by [9,24].

In the Łukasiewicz logic L_∞ for truth degrees, the following operators are used:

$$\neg x := 1 - x \qquad \qquad \text{(negation)}$$
$$x \wedge y := \min\{x, y\} \qquad \text{(weak conjunction)}$$
$$x \vee y := \max\{x, y\} \qquad \text{(weak disjunction)}$$
$$x \otimes y := \max\{0, x + y - 1\} \qquad \text{(strong conjunction)}$$
$$x \oplus y := \min\{1, x + y\} \qquad \text{(strong disjunction)}$$
$$x \rightarrow y := \min\{1, 1 - x + y\} \qquad \text{(implication)}$$

Using these definitions, the truth values of formulas can be evaluated recursively. Note that all truth functions of the basic connectives are continuous. A formula is a *tautology* if it evaluates to 1 under each valuation of propositional variables by real numbers in [0, 1].

In the Gödel logic G_∞ the operators are

$$\neg x = \begin{cases} 1 & \text{if } x = 0 \\ 0 & \text{if } x > 0 \end{cases}$$

$$x \wedge y = \min\{x, y\}$$

$$x \vee y = \max\{x, y\}$$

$$x \to y = \begin{cases} 1 & \text{if } x \leq y \\ y & \text{if } x > y. \end{cases}$$

Again, all formulas can be evaluated recursively, but the negation is not continuous.

In natural language using fuzzy statements or information is not an exception but normal. In a recipe nobody would replace the statement "Take a pinch of salt" by "Take 80 grain of salt." A driver will not calculate the distance he will need for stopping his car abruptly on a wet road by using another friction constant in some mathematical formula to calculate this distance. He will consider the rule: the wetter the road, the longer the distance needed for breaking.

In order to model this human information processing in a more appropriate way, we use gradual truth values for statements. This means a statement can not only be true (truth value 1) or false (truth value 0) but also more or less true expressed by a value between 0 and 1.

The connection between fuzzy sets and imprecise or fuzzy statements can be described in the following way. A fuzzy set models a property that elements of the universe of discourse can have more or less. For example, let us consider the fuzzy set μ_{hv} of high velocities from Fig. 15.2 on page 375. The fuzzy set represents the property or the predicate *high velocity*. That means the degree of membership of a specific velocity v to the fuzzy set of high velocities corresponds to the "truth value" which is assigned to the statement "v is a high velocity." In this sense, a fuzzy set determines the corresponding truth values for a set of statements. In order to understand how to operate with fuzzy sets, it is first of all useful to consider classical crisp propositions. Note that this is only one way to interpret a membership degree. There are several other options. In applications, it is necessary to agree on one interpretation, because each interpretation uses other mathematical methods.

Dealing with combined propositions like "160 km/h is a high velocity AND the stopping distance is about 110 m" requires the extension of the truth tables for logical connectives like conjunction, disjunction, implication, or negation. The truth tables shown in Table 15.1 determine a truth function for each logic connective. For conjunction, disjunction, and implication, this truth function assigns to each combination of two truth values (the truth value assigned to φ and ψ) one truth value (the truth value of the conjunction, disjunction of φ and ψ or the implication $\varphi \to \psi$). The truth function assigned to the negation has only one truth value as argument. If we denote the truth function by w_* associated with the logical connective $* \in \{\wedge, \vee, \to, \neg\}$, w_* is a binary or unary function. This means

$$w_\wedge, w_\vee, w_\to : \{0, 1\}^2 \to \{0, 1\}, \quad w_\neg : \{0, 1\} \to \{0, 1\}.$$

For fuzzy propositions, where the unit interval $[0, 1]$ replaces the binary set $\{0, 1\}$ as the set of possible truth values, we have to assign truth functions to the logic connectives accordingly. These truth functions have to be defined on the unit square or the unit interval.

$$w_\wedge, w_\vee, w_\rightarrow : [0, 1]^2 \rightarrow [0, 1], \quad w_\rightarrow : [0, 1] \rightarrow [0, 1].$$

A minimum requirement we demand of these functions is that, limited to the values 0 and 1, they should provide the same values as the corresponding truth function associated with the connectives of classical logic. This requirement says that a combination of fuzzy propositions which are actually crisp (non-fuzzy), i.e., their truth values are 0 or 1, coincide with the usual combination of classical crisp propositions.

The most frequently used truth functions for conjunction and disjunction in fuzzy logic are the minimum or maximum. That means $w_\wedge(\alpha, \beta) = \min\{\alpha, \beta\}$, $w_\vee(\alpha, \beta) = \max\{\alpha, \beta\}$. Normally the negation is defined by $w_\neg(\alpha) = 1 - \alpha$. In his seminal work [38], Lotfi Zadeh introduced the concept of fuzzy sets and used these functions for operating with fuzzy sets.

The implication is often understood in the sense of the **Łukasiewicz implication**

$$w_\rightarrow(\alpha, \beta) = \min\{1 - \alpha + \beta, 1\}$$

or the **Gödel implication**

$$w_\rightarrow(\alpha, \beta) = \begin{cases} 1 & \text{if } \alpha \leq \beta \\ \beta & \text{otherwise.} \end{cases}$$

15.4.2 t-Norms and t-Conorms

Until now we have interpreted the truth values from the unit interval in a purely intuitive way as gradual truths. So, choosing the truth functions for the logical connectives in the above-mentioned way seems to be plausible but it is not unique at all. There are several many-valued logics with different semantics, which makes a choice of a suitable logic and its operators difficult. Instead of trying to find more or less arbitrary functions, we might better use an axiomatic approach where we define some reasonable properties a truth function should satisfy and thus confining the possible truth functions. We discuss this axiomatic approach in detail for the conjunction.

Let us consider the function $t : [0, 1]^2 \rightarrow [0, 1]$ as a potential candidate for the truth function of a conjunction for fuzzy propositions. The truth value of a conjunction of several propositions should not depend on the order in which the propositions are considered. In order to guarantee this property t has to be commutative and associative, which means

$$t(\alpha, \beta) = t(\beta, \alpha) \tag{T1}$$

$$t(t(\alpha, \beta), \gamma) = t(\alpha, t(\beta, \gamma)) \tag{T2}$$

should hold for all α, β, and γ.

The truth value of the conjunction $\varphi \wedge \psi$ should not be less than the truth value of the conjunction $\varphi \wedge \chi$, if χ has a lower truth value than ψ. Therefore, we require some monotonicity condition of t:

$$\text{From } \beta \leq \gamma \text{ follows } t(\alpha, \beta) \leq t(\alpha, \gamma) \tag{T3}$$

for all α, β, and γ. Because of the commutativity (T1), (T3) implies that t is non-decreasing in both arguments.

Furthermore, we require that the truth value of a proposition φ will be the same as the truth value of the conjunction of φ with any true proposition ψ. For the truth function t, this leads to

$$t(\alpha, 1) = \alpha, \text{ for all } \alpha. \tag{T4}$$

Definition 15.3 A function $t : [0, 1]^2 \rightarrow [0, 1]$ is called a *t-norm (triangular norm)*, if the axioms (T1)–(T4) are satisfied.

In the framework of fuzzy logic, we should always choose a t-norm as the truth function for conjunction [19,20]. From the property (T4) follows that for every t-norm t, we have $t(1, 1) = 1$ and $t(0, 1) = 0$. From $t(0, 1) = 0$, we obtain $t(1, 0) = 0$ using the commutativity property (T1). Furthermore, because of the monotonic property (T3) and $t(0, 1) = 0$ we must have $t(0, 0) = 0$. In this way, every t-norm restricted to the values 0 and 1 coincides with the truth function given by the truth table of the usual conjunction.

We can verify easily that the discussed truth function $t(\alpha, \beta) = \min\{\alpha, \beta\}$ for the conjunction is a t-norm. Other examples of t-norms are

$$\textit{Łukasiewicz} - norm : t(\alpha, \beta) = \max\{\alpha + \beta - 1, 0\}$$

$$\textit{algebraic product} : \quad t(\alpha, \beta) = \alpha \cdot \beta$$

$$\textit{drastic product} : \quad t(\alpha, \beta) = \begin{cases} 0 & \text{if } 1 \notin \{\alpha, \beta\} \\ \min\{\alpha, \beta\} & \text{otherwise.} \end{cases}$$

The minimum, the algebraic product, and the Łukasiewicz t-norm are illustrated as three-dimensional function graphs in Fig. 15.12.

These few examples show that the spectrum of the t-norms is very broad. The limits are given by the drastic product, which is the smallest t-norm and which is

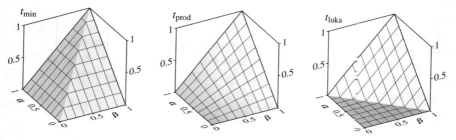

Fig. 15.12 The t-norms $t_{\min}$ (minimum), t_{prod} (algebraic product), and t_{luka} (Łukasiewicz)

discontinuous, and the minimum, which is the greatest t-norm. Besides this, the minimum can be considered as a special t-norms, since it is the only idempotent t-norm which means that only the minimum satisfies the property $t(\alpha, \alpha) = \alpha$ for all $\alpha \in [0, 1]$.

Only the idempotence of a t-norm can guarantee that the truth values of the proposition φ and $\varphi \wedge \varphi$ coincide, which at first sight seems to be a canonical requirement, letting the minimum seem to be the only reasonable choice for the truth functions for the conjunction in the context of fuzzy logic. However, the following example shows that the idempotency property is not always desirable.

Example 15.2 A buyer has to decide between the houses A and B. The houses are very similar in most aspects. So, he makes his decision considering the criteria good price and good location. After careful consideration he assigns the following "truth values" to the decisive aspects:

	Statement	Truth value $[\![\varphi_i]\!]$
φ_1	The price of house A is good.	0.9
φ_2	The location of house A is good.	0.6
φ_3	The price of house B is good.	0.6
φ_4	The location of house B is good.	0.6

He chooses house $x \in \{A, B\}$ for which the proposition "The price of house x is good AND The location of house x is good" yields the greater truth value. This means that the buyer will choose house A if $[\![\varphi_1 \wedge \varphi_2]\!] > [\![\varphi_3 \wedge \varphi_4]\!]$ holds, and house B otherwise. When we determine the truth value of the conjunction by the minimum, we would obtain the value 0.6 for both of the houses and thus the houses would be regarded as equally good. But this is counterintuitive because house A has definitely a better price than house B and the locations are equally good. However, when we choose a non-idempotent t-norm, e.g., the algebraic product or the Łukasiewicz t-norm, as truth function for the conjunction, we will always favor house A. $\square$

Besides the discussed examples for the t-norms there are many others. In particular, there are whole families of t-norms which can be defined in a parametric way. For example, the Weber family

$$t_\lambda(\alpha, \beta) = \max \left\{ \frac{\alpha + \beta - 1 + \lambda \alpha \beta}{1 + \lambda}, 0 \right\}$$

which determines a t-norm for each $\lambda \in (-1, \infty)$. For $\lambda = 0$, it results in the Łukasiewicz t-norm.

In most practical applications only the minimum, the algebraic product and the Łukasiewicz t-norm are chosen. Therefore, we will not consider the enormous variety of other t-norms here. For further readings on t-norms, we refer the reader to [6, 21]. Note that triangular norms have been originally used in probabilistic metric spaces [33].

In the same way as we have defined t-norms as possible truth functions for the conjunction, we can define candidates for truth functions for the disjunction. Just like the t-norms they should satisfy the properties (T1)–(T3). Instead of (T4) we ask for

$$s(\alpha, 0) = \alpha, \text{ for all } \alpha, \tag{T4'}$$

which means that the truth value of a proposition φ will be the same as the truth value of the disjunction of φ with any false proposition ψ.

Definition 15.4 A function $s : [0, 1]^2 \rightarrow [0, 1]$ is called *t-conorm* (*triangular conorm*) if the axioms (T1)–(T3) and (T4') are satisfied.

t-norms and t-conorms are dual concepts in the following sense. Each t-norm induces a t-conorm s by

$$s(\alpha, \beta) = 1 - t(1 - \alpha, 1 - \beta), \tag{15.2}$$

and vice versa, from a t-conorm s we obtain the corresponding t-norm by

$$t(\alpha, \beta) = 1 - s(1 - \alpha, 1 - \beta). \tag{15.3}$$

Equations 15.2 and 15.3 correspond to De Morgan's Laws

$$[\![\varphi \vee \psi]\!] = [\![\neg(\neg\varphi \wedge \neg\psi)]\!] \quad \text{and} \quad [\![\varphi \wedge \psi]\!] = [\![\neg(\neg\varphi \vee \neg\psi)]\!]$$

if we compute the negation using the truth function $[\![\neg\varphi]\!] = 1 - [\![\varphi]\!]$.

The t-conorms we obtain from the t-norms minimum, Łukasiewicz t-norm, algebraic, and drastic product by applying Eq. 15.2 are

$$
\begin{aligned}
&maximum: &&s(\alpha, \beta) = \max\{\alpha, \beta\} \\
&\text{Łukasiewicz } t - conorm: &&s(\alpha, \beta) = \min\{\alpha + \beta, 1\} \\
&algebraic\,sum: &&s(\alpha, \beta) = \alpha + \beta - \alpha\beta \\
&drastic\,sum: &&s(\alpha, \beta) = \begin{cases} 1 & \text{if } 0 \notin \{\alpha, \beta\} \\ \max\{\alpha, \beta\} & \text{otherwise.} \end{cases}
\end{aligned}
$$

The minimum, the algebraic sum, and the Łukasiewicz t-conorm are illustrated as three-dimensional function graphs in Fig. 15.13.

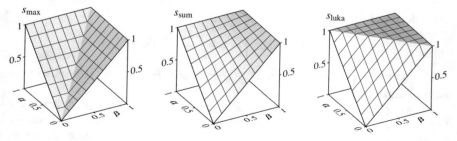

Fig. 15.13 The t-conorms s_{max} (maximum), s_{sum} (algebraic sum), and s_{luka} (Łukasiewicz)

The duality between t-norms and t-conorms implies immediately that the drastic sum is the greatest, the maximum is the smallest t-conorm, and the maximum is the only idempotent t-conorm. Analogously to t-norms we can define parametric families of t-conorms, such as

$$s_\lambda(\alpha, \beta) = \min\left\{\alpha + \beta - \frac{\lambda\alpha\beta}{1+\lambda}, 1\right\}$$

which is the Weber family of the t-conorms.

Operating with t-norms and t-conorms we should be aware that not all laws we know for classical conjunction and disjunction hold also for t-norms and t-conorms. For instance, minimum and maximum are not merely the only idempotent t-norms or t-conorms, but also the only pair defined by duality (cf. Eq. 15.2) which satisfies the distribution laws.

In the example of the man who wanted to buy a house, we could see that the idempotency of a t-norm is not always desirable. The same holds for t-conorms. Let us consider the propositions $\varphi_1, \ldots, \varphi_n$ which shall be connected in a conjunctive or disjunctive manner. The significant disadvantage of the idempotency is the following. Applying the conjunction in terms of the minimum, the resulting truth value of the connection of propositions depends only on the truth value of the proposition to which the least truth value is assigned. Applying the disjunction in the sense of the maximum only the proposition with the greatest truth value determines the truth value of the disjunction of the propositions. We can avoid this disadvantage, if we give up idempotency. Another approach is to use an averaging function, see Sect. 15.4.3.

In addition to the connection between t-norms and t-conorms, we can also find connections between t-norms and implications. A continuous t-norm t induces the *residuated implication* $\vec{t}$ by the formula

$$\vec{t}(\alpha, \beta) = \sup\{\gamma \in [0, 1] \mid t(\alpha, \gamma) \leq \beta\}.$$

Thus, we obtain by residuation the Łukasiewicz implication from the Łukasiewicz t-norm and the Gödel implication from the minimum.

Later we will need the corresponding **biimplication** $\overleftrightarrow{t}$ which is defined by the formula

$$\overleftrightarrow{t}(\alpha, \beta) = \vec{t}\big(\max\{\alpha, \beta\}, \min\{\alpha, \beta\}\big) \qquad (15.4)$$
$$= t\big(\vec{t}(\alpha, \beta), \vec{t}(\beta, \alpha)\big)$$
$$= \min\{\vec{t}(\alpha, \beta), \vec{t}(\beta, \alpha)\}.$$

This formula is motivated by the definition of the biimplication or equivalence in classical logic in terms of

$$\llbracket \varphi \leftrightarrow \psi \rrbracket = \llbracket (\varphi \rightarrow \psi) \wedge (\psi \rightarrow \varphi) \rrbracket.$$

Besides the logical operators like conjunction, disjunction, implication, or negation in (fuzzy) logic, there also exist the quantifiers $\forall$ (all) and $\exists$ (exists).

The universal quantifier $\forall$ and the existential quantifier $\exists$ are closely related to the conjunction and the disjunction, respectively. Let us consider the universe of discourse X and the predicate $P(x)$. For instance, X could be the set $\{2, 4, 6, 8, 10\}$

and $P(x)$ the predicate "x is an even number." If the set X is finite, e.g., $X = \{x_1, \ldots, x_n\}$, the statement $(\forall x \in X)(P(x))$ is equivalent to the statement $P(x_1) \wedge \ldots \wedge P(x_n)$. Therefore, in this case, it is possible to define the truth value of the statement $(\forall x \in X)(P(x))$ on the basis of the conjunction which means

$$[\![\forall x \in X : P(x)]\!] = [\![P(x_1) \wedge \ldots \wedge P(x_n)]\!].$$

If we assign the minimum to the conjunction as truth value function, we obtain

$$[\![\forall x \in X : P(x)]\!] = \min\{[\![P(x)]\!] \mid x \in X\}$$

which can be extended to an infinite universe of discourse X by

$$[\![\forall x \in X : P(x)]\!] = \inf\{[\![P(x)]\!] \mid x \in X\}.$$

Other t-norms than the minimum are normally not used for the universal quantifier, since the non-idempotent property leads easily to the truth value zero in the case of an infinite universe of discourse.

The same consideration about the existential quantifier leads to its definition

$$[\![\exists x \in X : P(x)]\!] = \sup\{[\![P(x)]\!] \mid x \in X\}.$$

If the universe of discourse for the existential quantifier is finite, the propositions $\exists x \in X : P(x)$ and $P(x_1) \vee \ldots \vee P(x_n)$ are equivalent.

Example 15.3 As an example we consider the predicate $P(x)$ with the interpretation "x is a high velocity." Let the truth value $[\![P(x)]\!]$ be given by the fuzzy set of the high velocities from Fig. 15.2 on page 375 which means $[\![P(x)]\!] = \mu_{hv}(x)$. So, we have, for instance, $[\![P(150)]\!] = 0$, $[\![P(170)]\!] = 0.5$ and $[\![P(190)]\!] = 1$. Thus, the statement $\forall x \in [170, 200] : P(x)$ ("All velocities between 170km/h and 200km/h are high velocities") has the truth value

$$
\begin{aligned}
[\![\forall x \in [170, 200] : P(x)]\!] &= \inf\{[\![P(x)]\!] \mid x \in [170, 200]\} \\
&= \inf\{\mu_{hv}(x) \mid x \in [170, 200]\} \\
&= 0.5.
\end{aligned}
$$

Analogously we obtain $[\![\exists x \in [100, 180] : P(x)]\!] = 0.75$. $\square$

15.4.3 Aggregation Functions

The aggregation (or fusion) of several input values of the unit interval into one most informative value is a basic task in handling quantitative information [26]. Properties of the aggregation functions are mostly related to the field of their application, such as multicriteria decision-making, multi-valued logics, or probability theory.

We used t-norms for modeling a kind of **conjunctive aggregation**: The result of the fusion function T for the n values of the unit interval is always smaller than or equal to the minimum, i.e.,

$$T(\alpha_1, \ldots, \alpha_n) \leq \min\{\alpha_1, \ldots, \alpha_n\}.$$

t-norms were originally used in the framework of probabilistic metric spaces [33]. The name triangular norm refers to the fact that in the framework of probabilistic metric spaces t-norms are used to generalize triangle inequalities of ordinary metric spaces. Another class of conjunctive aggregation functions is studied in the context of statistics: the dependency structure of a bivariate random vector (X, Y) can be captured by so-called copulas. For each (X, Y), there is a copula $C : [0, 1]^2 \to [0, 1]$ such that $P(X \leq x \wedge Y \leq y) = C(P(X \leq x), P(Y \leq y))$. A copula is a binary operation C on the unit interval that satisfies

$$C(0, x) = C(x, 0) = 0,$$
$$C(1, x) = C(x, 1) = x, \text{ and}$$
$$\text{if } x < y \text{ and } u < v \text{ then } C(y, v) - C(y, u) - C(x, v) + C(x, u) \geq 0$$

for each $x, y, u, v \in [0, 1]$.

Copulas have been used as truth functions of fuzzy conjunctions which are not necessary associative [28].

In Example 15.2 on page 387 of the man who wanted to buy a house, we can also think of using a "compensatory" operator which is a compromise between conjunction and disjunction. An example for such a compensatory operator is the Gamma operator [40]

$$\Gamma_\gamma(\alpha_1, \ldots, \alpha_n) = \left(\prod_{i=1}^{n} \alpha_i \right) \cdot \left(1 - \prod_{i=1}^{n} (1 - \alpha_i) \right)^\gamma$$

with the parameter $\gamma \in [0, 1]$. For $\gamma = 0$, this results in the algebraic product, for $\gamma = 1$ we obtain the algebraic sum. Another compensatory operator is of course the arithmetical mean.

The arithmetic mean belongs to the class of **averaging aggregation** functions F, which share the property

$$\min \{\alpha_1, \ldots, \alpha_n\} \leq F(\alpha_1, \ldots, \alpha_n) \leq \max \{\alpha_1, \ldots, \alpha_n\}.$$

The ordered weighted averaging operator introduced in [35] provides a parametrized family of aggregation operators which have been used in many applications [36].

The OWA (ordered weighted averaging) operator is defined by

$$F(\alpha_1, \ldots, \alpha_n) = \sum_{i=1}^{n} w_i b_i,$$

where b_i is the ith largest of the α_j and w_i are a collection of weights such that $w_i \in [0, 1]$ and $\sum_{i=1}^{n} w_i = 1$.

A convenient vector expression of this aggregation operator can be obtained if we let $\mathbf{W}$ be an n-dimensional vector whose components are the w_i and let $\mathbf{B}$ be an n-dimensional vector whose components are the b_i. We call $\mathbf{W}$ the weighting vector and $\mathbf{B}$ the ordered argument vector. Using these vectors we can express $F(\alpha_1, \ldots, \alpha_n) = \mathbf{W}^T \mathbf{B}$.

By selecting different manifestations of $\mathbf{W}$ we can implement different aggregations. Particularly notable among the operators that can be obtained are the maximum, minimum, and the arithmetic mean. These are, respectively, obtained by the vectors $W^\star$ where $w_1 = 1$ and $w_i = 0$ for $i \neq 1$, $\mathbf{W}_\star$, where $w_n = 1$ and $w_i = 0$ for $i \neq n$, and $\mathbf{W}_A$ where $w_i = \frac{1}{n}$. [35,36] discusses various different examples of weighting vectors.

15.5 Semantics of Membership Degrees

Probability theory and fuzzy set theory serve for modeling two different phenomena: probability is used for the quantification of the (subjective or objective) uncertainty whether an event may happen. Fuzziness models the knowledge about situations in which a property or statement is only satisfied to some degree. Often both phenomena can be observed simultaneously.

15.5.1 Membership Degrees as Truth Degrees

The straightforward interpretation of a fuzzy membership degree is the idea of a truth degree within a multi-valued logic. An essential assumption we have used in the section is that of **truth functionality**. This means that the truth value of the combination of several propositions depends only on the truth values of the propositions, but not on the individual propositions. This assumption holds in classical logic but not, for example, in the context of probability theory or probabilistic logic. In probability theory, it is not enough to know the probability of two events in order to determine the probability that both events will occur simultaneously or at least one of them will occur. For this we also need information about the dependency of these events. In the case of independence, the probability that both events occur is the product of the single probabilities, and the probability that at least one of the events will occur is the sum of the probabilities if the two events exclude each other. We cannot determine these probabilities without knowing the independence of the events.

We should be aware of the assumption of truth functionality in the framework of fuzzy logic, because in real applications it is often not satisfied. Coming back to the example of the man buying a house, we gave reasons for using non-idempotent t-norms. If we use these t-norms, like, for instance, the algebraic product, for propositions like "The price of the house A is good AND …AND the price of house A is good," this combined proposition can obtain a very small truth value. Depending on how we interpret the conjunction, this effect might be desirable or might lead to inconsistency. If we understand the conjunction in its classical sense, a conjunctive combination of a proposition with itself should be equivalent to itself which is not satisfied for non-idempotent t-norms. Another possibility is to understand the conjunction as a list of pro and con arguments for a thesis or as a proof. The repeated use of the same (fuzzy) argument within a proof might result in a loss of credibility and thus idempotency is not desirable, even for a conjunction of a proposition with itself.

In such application, it is appropriate to use a different interpretation of a fuzzy membership degree, and actually there are several application-dependent semantics of a membership degree. To mention only the most important interpretation, membership degrees are used to measure partial truth, the similarity to references values, information about preferences, and a degrees of possibility.

15.5.2   Membership Degrees as Similarity to a Reference Value

An often used interpretation of a fuzzy membership degree is the similarity to a reference value: to obtain a numeric value for a gradual membership in this interpretation it is (most) natural to compare the object under consideration with one that definitely belongs to the concept under consideration (and possibly also with one that definitely does *not* belong to this concept). The membership value can then be derived from the similarity of the two objects: the more similar an object is to an object that unambiguously belongs to the considered concept one, the higher the membership degree. And analogously, the more similar an object is to one that can definitely be excluded from a concept, the lower the membership degree [34]. As in most cases, in which similarity has to be formalized, we revert to distance measures: the similarity between two objects is higher, the smaller their distance.

For example, let us consider the fuzzy set μ_{hv} of high velocities from Fig. 15.1 on page 375. The term *high velocity* (e.g., in the context of cars on German highways) cannot be described adequately with a sharp boundary at 170km/h. However, a velocity of 200km/h may be described as "high" without any doubt. But in between there exist many possible velocities for which it is not so clear whether they should be labeled as "high velocity" or not. To model this situation we could use fuzzy sets similar to those in Fig. 15.2: velocities which are clearly "high" receive the membership degree 1.0, velocities which are clearly not "high" receive the membership degree 0.0. To velocities in between, degrees of membership are assigned depending on their distance to clearly high (and clearly not high) velocities. In the simplest case (and in this example), we use linear interpolation, i.e., the distance to the most extreme unambiguous value is directly converted into a membership degree. Of course, we have to consider the question how to obtain these unambiguous values (that is, the reference values for similarity or distance calculations) if we use such an approach. In most cases, however, this is not a critical issue since close to these boundaries membership degrees are still (very) high. As such they can hardly be distinguished from unambiguous values (which receive a membership degree of 1.0).

In its purest form we find the interpretation of membership degrees as similarity to reference values in fuzzy clustering (see Sect. 22.2). Besides fuzzy clustering, fuzzy control (see Chap. 20) is the area in which an interpretation of membership degrees as similarity is most common.

15.5.3 Membership Degrees as Preferences

In this interpretation, membership degrees convey which values (or objects) should be preferred to others. In this case, membership degrees are to be interpreted ordinally, i.e., they are ordered and we can decide which is higher (and as consequence we prefer the value or object whose membership degree is higher). Depending on the source of the preference ordering it may also be possible, for example, to use metric membership degrees, if preferences are associated with some kind of costs or utilities. In such a case, we prefer the value or object that yields lower costs or higher utility or which is considered to be of superior quality based on some other consideration [8]. The interpretation of membership degrees as preference is often closely related to an interpretation as similarity, since a preference ordering can be derived from the similarity of a given object to an ideal object or an optimal value. The interpretation of membership degrees as preferences [30] is often found in preference modeling and fuzzy decision-making theory [2, 12, 14, 15] as well as fuzzy optimization [22] (especially in the fuzzy variants of linear programming [32]).

15.5.4 Membership Degrees as Possibility

The fourth interpretation of a fuzzy set is used to express possibility. It is used to quantify the state of knowledge of an agent. The objective is to distinguish between what is surprising from what is expected, and what is plausible from what is less plausible. Possibility degrees represent a flexible restriction on what is the actual state with the following convention: $\mu(u) = 0$ means that u is rejected as impossible, $\mu(u) = 1$ means that u is totally possible, and the larger $\mu(u)$ is, the more plausible u is. This interpretation is often used in intelligent data analysis, as we will discuss in more detail in Chap. 22.

15.5.5 Consistent Interpretations of Fuzzy Sets in Applications

Unfortunately, a stringent interpretation of fuzzy sets is rarely maintained in applications and this is one of the reasons why inconsistencies may occur if fuzzy sets are applied only on an intuitive basis. If we interpret fuzzy sets as generalized characteristic functions, there is no ultimate reason for choosing the unit interval as the canonical extension of the set $\{0, 1\}$. In principle, it is possible that any linearly ordered set or—more generally—a lattice L might be better suited than the unit interval. In this case, one speaks of L fuzzy sets. But even if we agree on the unit interval as the set of possible degrees of membership, we should explain in which sense or as what kind of structure we understand it. Although a metric interpretation of the unit interval is generally very tempting (due to its mathematical properties), it can be viewed as a purely ordinal scale, ignoring its metric structure. That is, only the linear ordering of the numbers is considered for expressing preferences, their numerical values are disregarded. In this case, the interpretation of a number between 0 and 1 as a degree of membership makes sense only if it is compared (in the sense of greater or smaller) to another degree of membership, but we cannot consider the numerical difference between any two degrees of membership. Thus can we express that one element belongs more to a fuzzy set than another element, but we cannot say by how much more (even though the fact that the membership degrees are numbers, and thus we can formally subtract them from each other, seems to allow for such a statement).

A problem resulting from this purely ordinal view of the unit interval is the incomparability of degrees of membership stated by different persons. The same difficulty occurs if we compare grades. Two examinees receiving the same grade from different examiners may have performed very differently. However, under normal circumstances, the scale of grades used for examinations is not seen as a purely ordinal scale. Pointing out which performance or which amount of mistakes leads to which grade is an attempt to make it possible to compare grades given by different examiners. With the canonical metric quantifying the distance between two numbers and operations like addition and multiplication the unit interval has a considerably richer structure than the linear ordering of the numbers. As a consequence, a metric interpretation of numeric membership degrees is always highly tempting and therefore it is usually better (at least from a psychological point of view) to understand the unit interval as a metric scale in order to obtain a more concrete interpretation of the degrees of membership and to avoid the confusion that metric operations like subtraction, even though they look natural enough on a number scale, are not allowed if the scale is actually only ordinal. We will discuss the issue of semantics of degrees of membership and fuzzy sets in more detail in the application-oriented chapters.

15.6 Operations on Fuzzy Sets

The preceding sections describe how imprecise concepts can be modeled using fuzzy sets. In order to operate with imprecise concepts or apply some kinds of deduction mechanisms to them, we need suitable operations for fuzzy sets. Therefore, in this section operations like union, intersection, or complement well known from classical set theory will be extended. The idea is to choose a suitable fuzzy logic, to interpret the membership degrees of the fuzzy sets as truth values of the chosen fuzzy logic, and to use the operator of the logic pointwise.

15.6.1 Intersection

The underlying concept of generalizing fundamental set-theoretic operations to fuzzy sets is explained in detail for the intersection of (fuzzy) sets. For the other operations, a generalization can be carried out in a straightforward manner analogously to intersection. For two ordinary sets M_1 and M_2, we have that an element x belongs to the intersection of the two sets, if and only if it belongs to both M_1 and M_2. Whether

x belongs to the intersection depends only on the membership of x to M_1 and M_2 but not on the membership of any other element $y \neq x$ to M_1 and M_2. Formally speaking, this means

$$x \in M_1 \cap M_2 \quad \Longleftrightarrow \quad x \in M_1 \wedge x \in M_2. \tag{15.5}$$

For two fuzzy sets μ_1 and μ_2, we also assume that the degree of membership of an element x to the intersection of the two fuzzy sets depends only on the membership degrees of x to μ_1 and μ_2. We interpret the degree of membership $\mu(x)$ of an element x to the fuzzy set μ as truth value $[\![x \in \mu]\!]$ of the fuzzy proposition "$x \in \mu$" that x is an element of μ. In order to determine the membership degree of an element x to the intersection of the fuzzy sets μ_1 and μ_2, we have to calculate the truth value of the conjunction "x is an element of μ_1 AND x is an element of μ_2" following the equivalence in Eq. 15.5. The previously discussed concepts of fuzzy logic have told us, how we can define the truth value of the conjunction of two fuzzy propositions. Therefore, it is necessary to choose a suitable t-norm t as the truth function for the conjunction. Thus, we define the intersection of two fuzzy sets μ_1 and μ_2 (w.r.t. the t-norm t) as the fuzzy set $\mu_1 \cap_t \mu_2$ with

$$(\mu_1 \cap_t \mu_2)(x) = t \left(\mu_1(x), \mu_2(x) \right).$$

If we interpret the degree of membership $\mu(x)$ of an element x to the fuzzy set μ as truth value $[\![x \in \mu]\!]$ of the fuzzy proposition "$x \in \mu$" that x is an element of μ, the definition of the intersection of two fuzzy sets can be written in the following way:

$$[\![x \in (\mu_1 \cap_t \mu_2)]\!] = [\![x \in \mu_1 \wedge x \in \mu_2]\!],$$

where we assign the t-norm t as truth function for the conjunction.

By defining the intersection of fuzzy sets on the basis of a t-norm, the properties of the t-norm are inherited to the intersection operator: the axioms (T1) and (T2) make the intersecting of fuzzy sets commutative and associative, respectively. The monotonicity property (T3) guarantees that replacing a fuzzy set μ_1 by a larger fuzzy set μ_2, which means $\mu_1(x) \leq \mu_2(x)$ for all x, can only lead to a larger intersection:

$$\mu_1 \leq \mu_2 \text{ implies } \mu \cap_t \mu_1 \leq \mu \cap_t \mu_2.$$

Axiom (T4) guarantees that the intersection of a fuzzy set with an ordinary set, respectively, its characteristic function, results in the original fuzzy set limited to the ordinary set with which we intersected it. If $M \subseteq X$ is an ordinary subset of X and $\mu \in \mathcal{F}(X)$ a fuzzy set of X, we have

$$(\mu \cap_t I_M)(x) = \begin{cases} \mu(x) & \text{if } x \in M \\ 0 & \text{otherwise.} \end{cases}$$

If not otherwise stated, the intersection of two fuzzy sets will be computed w.r.t. the minimum t-norm. In this case, or when it is clear to which t-norm we refer, we will write $\mu_1 \cap \mu_2$ instead of $\mu_1 \cap_t \mu_2$ for the case of $t = \min$.

Fig. 15.14 The fuzzy set $\mu_{170-190}$ of the velocities not much less than 170 km/h and not much greater than 190 km/h

Fig. 15.15 Intersection $\mu_{hv} \cap_t \mu_{170-190}$ of the fuzzy sets μ_{hv} and $\mu_{170-190}$ calculated with the minimum (solid black line) and the Łukasiewicz t-norm (dashed gray line)

Example 15.4 We consider the intersection of the fuzzy set μ_{hv} of high velocities from Fig. 15.2 on page 375 and the fuzzy set $\mu_{170-190}$ of the velocities not much less than 170 km/h and not much greater than 190km/h from Fig. 15.14. Both of them are trapezoidal functions:

$$\mu_{hv} = \Pi_{150,180,\infty,\infty}, \quad \mu_{170-190} = \Pi_{160,170,190,200}.$$

Figure 15.15 shows the intersection of the two fuzzy sets on the basis of the minimum (solid line) and the Łukasiewicz t-norm (dashed line).   □

15.6.2 Union

From the representation in Eq. 15.5, we have derived the definition for the intersection of two fuzzy sets. Analogously we can define the union of two fuzzy sets on the basis of

$$x \in M_1 \cup M_2 \quad \Longleftrightarrow \quad x \in M_1 \vee x \in M_2.$$

This leads to

$$(\mu_1 \cup_s \mu_2)(x) = s(\mu_1(x), \mu_2(x)),$$

as the union of the two fuzzy sets μ_1 and μ_2 w.r.t. the t-conorm s. In the interpretation of the membership degree $\mu(x)$ of an element x to the fuzzy set μ as truth value $[\![x \in \mu]\!]$ of the fuzzy proposition "$x \in \mu$" we can define the union in the form of

$$[\![x \in (\mu_1 \cup_t \mu_2)]\!] = [\![x \in \mu_1 \vee x \in \mu_2]\!],$$

where we assign the t-conorm s as the truth function for the disjunction. As in the case of the intersection, we will write $\mu_1 \cup \mu_2$ instead of $\mu_1 \cup_s \mu_2$ when we use the maximum t-conorm or when it is clear which t-conorm we refer to.

15.6.3 Complement

The complement of a fuzzy set is derived from the formula

$$x \in \overline{M} \quad \Longleftrightarrow \quad \neg(x \in M)$$

for ordinary sets where $\overline{M}$ stands for the complement of the (ordinary) set M. If we assign the truth function $w_\neg(\alpha) = 1 - \alpha$ to the negation, we obtain the fuzzy set

$$\overline{\mu_1}(x) = 1 - \mu(x),$$

as the complement $\overline{\mu}$ of the fuzzy set μ. This is also in accordance with

$$[\![x \in \overline{\mu}]\!] = [\![\neg(x \in \mu)]\!].$$

Figure 15.16 illustrates the intersection (top right), union (bottom right), and complement (bottom left) of two fuzzy sets (top left).

Like the complement for ordinary sets, the complementing of fuzzy sets is an involution, which means that $\overline{\overline{\mu}} = \mu$ holds. In classical set theory, we have that the intersection of a set with its complement yields the whole universe of discourse. In the context of fuzzy sets, these two laws are weakened to $(\mu \cap \overline{\mu})(x) \le 0.5$ and $(\mu \cup \overline{\mu})(x) \ge 0.5$ for all x. Figure 15.17 illustrates this phenomenon.

If the intersection and the union are defined on the basis of minimum or maximum, we can use the representation of fuzzy sets by α-cuts as introduced in Sect. 15.3 in order to compute the resulting fuzzy set. We have

$$[\mu_1 \cap \mu_2]_\alpha = [\mu_1]_\alpha \cap [\mu_2]_\alpha \quad \text{and} \quad [\mu_1 \cup \mu_2]_\alpha = [\mu_1]_\alpha \cup [\mu_2]_\alpha$$

for all $\alpha \in [0, 1]$. According to these equations, the α-cuts of the intersection and the union of two fuzzy sets are the intersection or the union of the α-cuts of the single fuzzy sets.

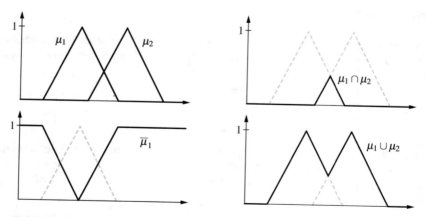

Fig. 15.16 Intersection, union, and complement for fuzzy sets

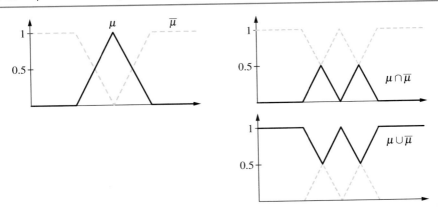

Fig. 15.17 Union and intersection of a fuzzy set with its complement

15.6.4 Covering and Partition

Let X be a set and let $\mu_1, \ldots, \mu_n$ be fuzzy sets of X. We say that these fuzzy sets form a *fuzzy covering* of X if any element of X belongs to at least one of the fuzzy sets $\mu_1, \ldots, \mu_n$ with a non-zero membership degree, i.e.,

$$\mu_1(x) + \ldots + \mu_n(x) > 0 \quad \forall x \in X.$$

We furthermore define the concept of a *fuzzy partition* that is often used for approximate reasoning: a fuzzy covering of X is called a fuzzy partition, if it consists of normal, convex fuzzy sets $\mu_1, \ldots, \mu_n$ satisfying the condition

$$\mu_1(x) + \ldots + \mu_n(x) \leq 1 \quad \forall x \in X.$$

A typical fuzzy partition of $\mathbb{R}$ is depicted in Fig. 15.18.

15.6.5 Linguistic Modifiers

Besides the complement as a unary operation on fuzzy sets, derived from the corresponding operation for ordinary sets, there are more fuzzy set specific unary opera-

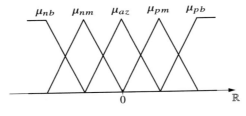

Fig. 15.18 Fuzzy partition of $\mathbb{R}$ with fuzzy sets μ_{nb} for "negative big," μ_{nm} for "negative medium," μ_{az} for "approximately zero," μ_{pm} for "positive medium," and μ_{pb} for "positive big" numbers

Fig. 15.19 The fuzzy sets
μ_{hv}, μ_{vhv}, and μ_{mhv} of high,
very high, and more or less
high velocities

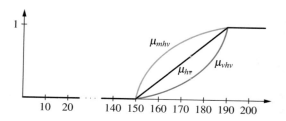

tions that have no counterpart in classical set theory. Normally, a fuzzy set represents an imprecise concept like "high velocity," "young," or "tall." From such concepts, we can derive other imprecise concepts applying **linguistic modifiers** ("linguistic hedges") like "very" or "more or less."

As an example we consider the fuzzy set μ_{hv} of high velocities from Fig. 15.2 on page 375. How should the fuzzy set μ_{vhv} representing the concept of the "*very* high velocities" look like? A very high velocity is also a high velocity but not vice versa. Thus, the membership degree of a specific velocity v to the fuzzy set μ_{vhv} should not exceed its membership degree to the fuzzy set μ_{hv}. We can achieve this by understanding the linguistic modifier "very," similar to the negation, as a unary operator and assigning a suitable truth function to it, for instance, $w_{\text{very}}(\alpha) = \alpha^2$. In this way, we obtain $\mu_{vhv}(x) = (\mu_{hv}(x))^2$. Now, a velocity which is to a degree of 1 a high velocity is also a very high velocity. A velocity which is not a high velocity (membership degree of 0) is not a very high velocity either. If the membership degree of a velocity to μ_{hv} is between 0 and 1, it is also a very high velocity but with a lower membership degree.

In the same way, we can assign a truth function to the modifier "more or less." This truth function should increase the degree of membership, for instance, $w_{\text{more or less}}(\alpha) = \sqrt{\alpha}$.

Figure 15.19 shows the fuzzy set μ_{hv} of high velocities and the resulting fuzzy sets μ_{vhv} of very high velocities and μ_{mhv} of more or less high velocities.

15.7   Fuzzy Sets of Type 2

In classical fuzzy set theory, a membership degree is expressed by a real number in the unit interval. However, in some applications, experts prefer to use linguistic expressions instead of numbers as membership degrees. Furthermore, it is sometimes necessary to integrate imprecision of the membership degree into the modeling. In this situation, one can use membership degrees that are elements of a general lattice L, due to which we then speak of L-fuzzy sets. This extension, which was already proposed in [10], provides us with more flexibility for the specification of membership degrees.

Definition 15.5 Let $(L, \leq)$ be a lattice, and let sup and inf be its join and meet operators, respectively. A lattice is called *complete* if any non-empty subset of L has a join (sup) and a meet (inf). A *negation* on a complete lattice is an operator not $: L \rightarrow L$ such that $\text{not}(\text{not}(l)) = l$ for all $l \in L$, and if $l \leq l'$ then $\text{not}(l) \geq \text{not}(l')$, for all $l, l' \in L$. A lattice is called *distributive* if for all x, y, and z in L the following holds:

$$\inf \{x, \sup \{y, z\}\} = \sup \{\inf \{x, y\}, \inf \{x, z\}\}$$

.

The extension of fuzzy sets from the unit interval [0, 1] to a general lattice L and the operations can be easily defined as follows:

Definition 15.6 An L-fuzzy set on the universe X is a mapping $\mu : X \rightarrow L$. The operations on the class $\mathcal{F}_L(X)$ of L-fuzzy sets are defined pointwise as follows:

$$(\mu \wedge \mu')(x) := \inf \{\mu(x), \mu'(x)\} \qquad (intersection)$$
$$(\mu \vee \mu')(x) := \sup \{\mu(x), \mu'(x)\} \qquad (union)$$
$$(\neg\mu)(x) := \text{not}\mu(x) \qquad (complement).$$

Example 15.5 Let $([0, 1], \leq)$ be a complete distributive lattice. Classical fuzzy sets can be seen as L-fuzzy sets with $L = [0, 1]$. The family of fuzzy sets over X with the inclusion $(\mathcal{F}(X), \leq)$ also forms a complete distributive lattice. This allows us to define "second order fuzzy sets."

L-fuzzy sets can be also used in situations, in which the user prefers to express his knowledge about a membership degree of fuzzy sets by imprecise statements like "the membership degree is between 0.7 and 0.9." A simple way is to express such information is to use an interval-valued fuzzy sets $\mu : X \rightarrow L([0, 1])$, where $L([0, 1]) := \{[\underline{x}, \overline{x}] : 0 \leq \underline{x} \leq \overline{x} \leq 1\}$ denotes the set of all closed sub-intervals of [0, 1].

Definition 15.7 An interval-valued fuzzy set μ on a universe $X \neq \emptyset$ is a mapping $\mu : X \rightarrow L([0, 1])$ such that the membership degree of $x \in X$ is given by

$$\mu(x) = \left[\underline{\mu}(x), \overline{\mu}(x)\right] \in L([0, 1]),$$

where $\underline{\mu}(x) : X \rightarrow [0, 1]$ and $\overline{\mu}(x) : X \rightarrow [0, 1]$ are mappings defining the lower and upper bounds of the membership interval $\mu(x)$.

Example 15.6 Let μ be the interval-valued fuzzy set on $\mathbb{R}$ depicted in Fig. 15.20. The aim is to describe an expert's perception of "approximately 100" by assigning to each x an interval $\left[\underline{\mu}(x), \overline{\mu}(x)\right]$. The membership at 95 is a value that belongs to the interval [0.5, 0.75]. The interval representation is used since the expert cannot precisely say what that number is. Thus she/he provides merely bounds for that number.

Fig. 15.20 Interval-valued fuzzy set

Definition 15.8 The standard operations on interval-valued fuzzy sets are

$$(\mu \wedge \mu')(x) = \left[\min \left\{ \underline{\mu}(x), \underline{\mu}'(x) \right\}, \min \left\{ \overline{\mu}(x), \overline{\mu}'(x) \right\} \right] \qquad (\textit{intersection})$$

$$(\mu \vee \mu')(x) = \left[\max \left\{ \underline{\mu}(x), \underline{\mu}'(x) \right\}, \max \left\{ \overline{\mu}(x), \overline{\mu}'(x) \right\} \right] \qquad (\textit{union})$$

$$(\neg\mu)(x) = \left[1 - \overline{\mu}(x), 1 - \underline{\mu}(x) \right] \qquad (\textit{complement})$$

for all $x \in X$.

There are applications in fuzzy decision analysis in which the user is more comfortable with imprecise statements like "The membership degree μ is 0.7, while the degree of non-membership v is 0.2." Note that in this statement the degrees of membership and non-membership do not sum to 1, as one would expect in standard fuzzy set theory, but only to 0.9. We may say that, in this case, we are hesitant with respect to what the exact membership value is depending on whether we look from the membership or the non-membership side. The value $1 - \mu - v = 0.1$ is called the *hesitance* [12,37]. A simple method for modeling mathematically such a type of knowledge is to use a pair of mappings (μ, v) that fulfill the conditions $\mu : X \to [0, 1]$, $v : X \to [0, 1]$ and $\mu(x) + v(x) \leq 1$ for all $x \in X$. Such a pair is called Intuitionistic Fuzzy Set [1] or Bipolar Fuzzy Set [7].

A standard fuzzy set μ can be interpreted as a pair $(\mu, 1 - \mu)$. Mathematically Intuitionistic fuzzy sets can be seen as L-fuzzy sets that are based on the lattice $L = \{ (x, y) \in [0, 1]^2 : x + y \leq 1 \}$, where the inequality relation $\leq_L$ generating the lattice structure is defined by $(x, y) \leq_L (x', y')$ if and only if $x \leq x'$ and $y' \leq y$ for all $x, x', y, y' \in [0, 1]$. The basic operations are defined as follows:

$$(\mu, v) \cap (\mu', v') := (\min \{\mu, \mu'\}, \max \{v, v'\}) \qquad (\textit{intersection})$$

$$(\mu, v) \cup (\mu', v') := (\max \{\mu, \mu'\}, \min \{v, v'\}) \qquad (\textit{union})$$

$$\neg(\mu, v) := (v, \mu) \qquad (\textit{complement}).$$

In order to motivate the introduction of further extensions of fuzzy set we use a different visualization of the interval-valued fuzzy set $\mu : X \to L([0, 1])$ of Example 15.6 in Fig. 15.21. For each $x \in X$ we depict the indicator function $\mu(x) : [0, 1] \to [0, 1]$ of the interval $\left[\underline{\mu}(x), \overline{\mu}(x) \right]$. The shaded area (as a relation) gives a good visualization of all possible fuzzy sets that fit to the given information. This information is also visualized in Fig. 15.21.

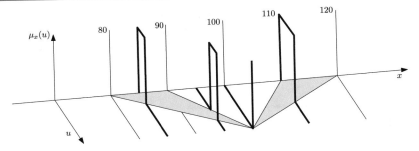

Fig. 15.21 Isometric representation of an interval-valued fuzzy set

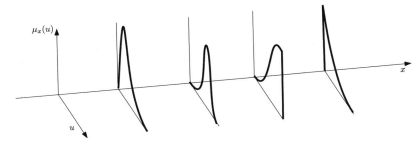

Fig. 15.22 Different slices of a type-2 fuzzy set

A further extension of expressivity (but also of complexity) is to allow the description of a membership to be itself a fuzzy set, that is, to consider L-fuzzy sets with $L = \mathcal{F}(X)$. The idea of taking the expert's uncertainty into account when they determine the membership degrees of elements of a fuzzy sets was originally developed in [39]:

Definition 15.9 A type-2 fuzzy set on X is a mapping $\mu : X \to \mathcal{F}([0, 1])$.

Example 15.7 Let $\mu : X \to \mathcal{F}([0, 1])$ be a type-2 fuzzy set depicted in Fig. 15.22. We use the notion $\mu : x \to \mu_x$, where $\mu_x : [0, 1] \to [0, 1]$. The objective here is to describe his perception of "approximately 100" by assigning to each $x \in X$ a fuzzy set of $[0, 1]$.

Type-2 fuzzy sets [25, 39] are particular L-fuzzy sets, so we can define the standard lattice-based algebraic operations as in Definition 15.6. Another option consists in employing the extension principle to define operations [27]. In Sect. 15.5, we stressed that it is important to be aware of the fact that there are several semantics for fuzzy sets such as similarity, preference, possibility, and so on. In the case of "second-order concepts" such as type-2 fuzzy sets or fuzzy random variables (see Sect. 22.5), these semantical issues are even more complicated. It is clear that handling the complexity of these extensions is a considerable challenge. On the other hand, the extended approaches give us more flexibility due to more parameters, see, e.g., [17]. Hence, it is not surprising that there are some applications in which these extended models produce results which are better than those obtained with standard fuzzy sets.

References

1. Atanassov K (1986) Intuitionistic fuzzy sets. Fuzzy Sets Syst 20:87–96
2. Bellman RE, Zadeh LA (1970) Decision-making in a fuzzy environment. Manag Sci 17:141–154
3. Belohlavek R (2017) Fuzzy logic and mathematics. Oxford Press, Oxford
4. Bezdek JC (1993) Fuzzy models—what are they, and why? IEEE Trans Fuzzy Syst 1:1–5. IEEE Press, Piscataway, NJ, USA
5. Bochvar DA (1981) On a three valued logical calculus and its application to the analysis of the classical extended functional calculus (translated from Russian by M. Bergmann). History Philoso Logic 2: 87–112
6. Butnariu D, Klement EP (1993) Triangular norm-based measures and games with fuzzy coalitions. Kluwer, Dordrecht, Netherlands
7. Dubois D, Prade H (2008) An introduction to bipolar representation of information and preference. Int J Intell Syst 23:866–877
8. Fodor JC, Roubens MR (1994) Fuzzy preference modelling and multicriteria decision support. Springer, Heidelberg, Germany
9. Gödel K (1932) Zum intuitionistischen Aussagenkalkül. Anzeiger der Akademie der Wissenschaften in Wien (69): 65f. Vienna, Austria
10. Goguen JA (1967) L-fuzzy sets. J Math Anal Appl 18:145–174
11. Hajek P (1998) Metamathematics of fuzzy logic. Kluwer, Dordrecht
12. Herrera F, Herrera-Viedma E (2000) Linguistic decision analysis: Steps for solving decision problems under linguistic information. Fuzzy Sets Syst 115:67–82
13. Heyting A (1930) Die formalenRegeln der intuitionistischen Logik. Sitzungsberichte der Preussischen Akademie der Wissenschaften, Physikalisch-mathematische Klasse 42–56
14. Hüllermeier E, Brinker K (2008) Learning valued preference structures for solving classification problems. Fuzzy Sets Syst 159:2337–2353
15. Kacprzyk J (1986) Group decision making with a fuzzy majority. Fuzzy Sets Syst 18:105–118
16. Kacprzyk J, Pedrycz W (eds) (2015) Springer handbook of computational intelligence. Springer, Berlin/Heidelberg, Germany
17. Karnik NN, Mendel JM (2001) Centroid of a type-2 fuzzy set. Inf Sci 132:195–220
18. Kleene SC (1952) Introduction to metamathematics. North Holland, Amsterdam
19. Klement EP, Mesiar R, Pap E (2000) Triangular norms. Springer, Berlin
20. Klir GJ, Yuan B (1995) Fuzzy sets and fuzzy logic: theory and applications, vol 4. Prentice Hall, New Jersey
21. Kruse R, Gebhardt J, Klawonn F (1994) Foundations of fuzzy systems. Wiley, Chichester, United Kingdom
22. Luhandjula MK (1989) Fuzzy optimization: an appraisal. Fuzzy Sets Syst 30:257–282
23. Łukasiewicz J (1970) O logice trójwartosciowej. Ruch filozoficzny 5:170–171, (1920) English translation: On three-valued logic. In: Borkowski L (ed) Selected works by Jan Łukasiewicz, pp 87–88. North-Holland, Amsterdam
24. Łukasiewicz J (1930) Philosophische Bemerkungen zu mehrwertigen Systemen des Aussagenkalküls. Comptes rendus des séances de la Société des Sciences et des Lettres de Varsovie, Cl. III, 23, 51–77.Warsaw, Poland

25. Mendel JM, John RI (2002) Type-2 fuzzy sets made simple. IEEE Trans Fuzzy Syst 10:117–127
26. Mesiar R, Kolesarova A, Komornikova M (2015) Aggregation functions on [0, 1]. In: [Kacprzyk and Pedrycz 2015], pp 61–74
27. Mizumoto M, Tanaka K (1976) Some properties of Fuzzy Sets of Type 2. Inf Control 31:312–340
28. Neldsen RB (2006) An introduction to copulas, vol 139. Lecture Notes in Statistics. Springer, NY
29. Novák V, Perfilieva I, Dvorák A (2016) Insight into fuzzy modeling. Wiley, New York
30. Perny P, Roubens M (1998) Fuzzy Preference Modeling. In: Słowiński R (eds) Fuzzy sets in decision analysis, operations research and statistics. The handbooks of fuzzy sets series, vol 1. Springer, Boston, MA
31. Reichenbach H (1935) Wahrscheinlichkeitslehre. Eine Untersuchung über die logischen und mathematischen Grundlagen der Wahrscheinlichkeitsrechnung. Translated: E.H. Hutten and M. Reichenbach. The Theory of Probability. An Iquity into Logical and Mathematical Foundations of the Calculus of Probability. Berkeley, CA, University of California Press
32. Rommelfanger H (1996) Fuzzy linear programming and applications. Euro J Oper Res 92:512–527
33. Schweizer B, Sklar A (1983) Probabilistic metric spaces. North Holland, New York
34. Wang X, De Baets B, Kerre E (1995) A comparative study of similarity measures. Fuzzy Sets Syst 73(2):259–268
35. Yager RR (1988) On ordered weighted averaging aggregation operators in multi-criteria decision making. IEEE Trans Syst Man Cybern 18:183–190
36. Yager RR, Kacprzyk J (1997) The ordered weighted averaging operators: theory and applications. Kluwer, Norwell, MA
37. Ye J (2010) Fuzzy decision making method based on the weighted correlation coefficient under intutuionistic fuzzy environment. Euro J Oper Res 205(1):202–204
38. Zadeh LA (1965) Fuzzy sets. Inf Control 8(3):338–353
39. Zadeh LA (1971) Quantitative fuzzy semantics. Information. Science 3:159–176
40. Zimmermann HJ, Zysno P (1980) Latent connectives in human decision making and expert systems. Fuzzy Sets Syst 4:37–51. Elsevier, Amsterdam, Netherlands

The Extension Principle

16

In Sect. 15.6, we have discussed how set-theoretic operations like intersection, union, and complement can be generalized to fuzzy sets. This section is devoted to the issue of extending the concept of mappings or functions to fuzzy sets. These ideas allow us to define operations like addition, subtraction, multiplication, division, or taking squares as well as set-theoretic concepts like the composition of relations for fuzzy sets. Here we again use the interpretation of partial truth of a membership degree.

16.1 Mappings of Fuzzy Sets

As an example, let us consider the mapping $f : \mathbb{R} \to \mathbb{R}$ with $x \mapsto |x|$. The fuzzy set $\mu = \Lambda_{-1.5,-0.5,2.5}$ shown in Fig. 16.1 models the imprecise concept "about -0.5."
Which fuzzy set should represent "the absolute value of about -0.5" or, in other words, what is the image $f[\mu]$ of the fuzzy set μ? For a usual subset M of the universe of discourse X the image $f[M]$ under the mapping $f : X \to Y$ is defined as the subset of Y that contains all images of elements of M. Technically speaking, this means

$$f[M] = \{y \in Y \mid \exists x \in X : x \in M \wedge f(x) = y\},$$

or, in other words,

$$y \in f[M] \quad \Longleftrightarrow \quad (\exists x \in X)(x \in M \wedge f(x) = y). \tag{16.1}$$

For instance, for $M = [-1, 0.5] \subseteq \mathbb{R}$ and the mapping $f(x) = |x|$ we obtain the set $f[M] = [0, 1]$ as image of M under f.
Equation 16.1 allows us to define the image of a fuzzy set μ under a mapping f. As in Sect. 15.6 on the extension of the set-theoretic operations to fuzzy sets, we use the concepts of fuzzy logic again, which we introduced in Sect. 15.4. For fuzzy sets, Eq. 16.1 means

$$[\![y \in f[\mu]]\!] = [\![\exists x \in X : x \in \mu \wedge f(x) = y]\!].$$

© Springer Nature Switzerland AG 2022
R. Kruse et al., *Computational Intelligence*, Texts in Computer Science,
https://doi.org/10.1007/978-3-030-42227-1_16

Fig. 16.1 The fuzzy set $\mu = \Lambda_{-1.5,-0.5,1.5}$ standing for "about -0.5"

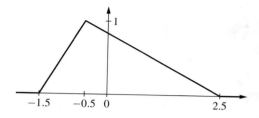

As explained in Sect. 15.4, we evaluate the existential quantifier by the supremum and associate a t-norm t with the conjunction. Therefore, the fuzzy set

$$f[\mu](y) \;=\; \sup\{t\,(\mu(x),\, \llbracket f(x) = y\rrbracket)\,|\, x \in X\} \tag{16.2}$$

represents the image of μ under f. The choice of the t-norm t does not play any role in this case because the statement $f(x) = y$ is either true or false which means $\llbracket f(x) = y\rrbracket \in \{0, 1\}$, and therefore

$$t\,(\mu(x),\, \llbracket f(x) = y)\rrbracket) = \begin{cases} \mu(x) & \text{if } f(x) = y \\ 0 & \text{otherwise} \end{cases}$$

Thus Eq. 16.2 is reduced to

$$f[\mu](y) = \sup\{\mu(x)\,|\, f(x) = y\}. \tag{16.3}$$

This definition says that the membership degree of an element $y \in Y$ to the image of the fuzzy set $\mu \in \mathcal{F}(X)$ under the mapping $f : X \to Y$ is the greatest possible membership degree of all x to μ that are mapped to y under f. This extension of a mapping to fuzzy sets is called **extension principle** (for a function with one argument) [4].

From the example of the fuzzy set $\mu = \Lambda_{-1.5,-0.5,2.5}$ representing the imprecise concept "about -0.5," we obtain the fuzzy set shown in Fig. 16.2 as the image under the mapping $f(x) = |x|$. To illustrate the underlying principle, we determine the membership degree $f[\mu](y)$ for $y \in \{-0.5, 0, 0.5, 1\}$. Because of $f(x) = |x| \geq 0$ no value is mapped to $y = -0.5$ under f. So, we obtain $f[\mu](-0.5) = 0$. There is only one value that is mapped to $y = 0$, i.e., $x = 0$, and hence we have $f[\mu](0) = \mu(0) = 5/6$. For $y = 0.5$, there exist two values ($x = -0.5$ and 0.5) mapped to y, which lead to

$$f[\mu](0.5) = \max\{\mu(-0.5), \mu(0.5)\} = \max\{1, 2/3\} = 1.$$

The values mapped to $y = 1$ are $x = -1$ and $x = 1$, respectively. Therefore, we obtain

$$f[\mu](1) = \max\{\mu(-1), \mu(1)\}$$
$$= \max\{0.5, 0.5\} = 0.5.$$

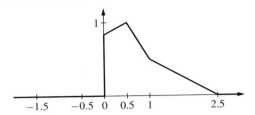

Fig. 16.2 The fuzzy set standing for the imprecise concept "the value of about -0.5"

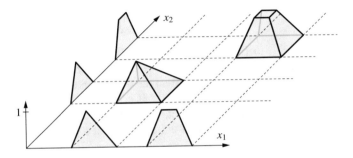

Fig. 16.3 Projection of a fuzzy set into the space X_2

Example 16.1 Let $X = X_1 \times \ldots \times X_n$, $i \in \{1, \ldots, n\}$.

$$\pi_i : X_1 \times \ldots \times X_n \to X_i, \qquad (x_1, \ldots, x_n) \mapsto x_i$$

denotes the projection of the Cartesian product $X_1 \times \ldots \times X_n$ to the ith coordinate space X_i. According to the extension principle defined in Eq. 16.3, the projection of a fuzzy set $\mu \in \mathcal{F}(X)$ to the space X_i is given by

$$\pi_i[\mu](x) = \sup\{ \mu(x_1, \ldots, x_{i-1}, x, x_{i+1}, \ldots, x_n) \mid$$
$$x_1 \in X_1, \ldots, x_{i-1} \in X_{i-1}, x_{i+1} \in X_{i+1}, \ldots, x_n \in X_n\}.$$

Figure 16.3 shows the projection of a fuzzy set which has non-zero membership degrees in two different regions. $\qquad\square$

16.2 Mappings of α-Cuts

The membership degree of an element to the image of a fuzzy set can be computed on the basis of the membership degrees of all elements to the original one that are mapped to the considered element. Another way to characterize the image of a fuzzy set is to determine its α-cuts. Unfortunately, the α-cut of the image of a fuzzy set cannot be derived directly form the corresponding α-cut of the original fuzzy set.

The inclusion $[f[\mu]]_\alpha \supseteq f[[\mu]_\alpha]$ is always valid, but equality of these two sets is only satisfied in special cases. For instance, for the fuzzy set

$$\mu(x) = \begin{cases} x \text{ if } 0 \le x \le 1 \\ 0 \text{ otherwise} \end{cases}$$

we obtain as an image under the mapping

$$f(x) = I_{\{1\}}(x) = \begin{cases} 1 \text{ if } x = 1 \\ 0 \text{ otherwise} \end{cases}$$

the fuzzy set

$$f[\mu](y) = \begin{cases} 1 \text{ if } y \in \{0, 1\} \\ 0 \text{ otherwise.} \end{cases}$$

Hence, we have $\left[f[\mu]\right]_1 = \{0, 1\}$ and $f[[\mu]_1] = \{1\}$ because of $[\mu]_1 = \{1\}$.

Provided that the universe of discourse $X = \mathbb{R}$ is the set of real numbers, the effect that the image of a α-cut is smaller than the corresponding α-cut of the image fuzzy set cannot happen, when the mapping f is continuous and for all $\alpha > 0$ the α-cuts of the original fuzzy set are compact. Therefore, in this case, it is possible to characterize the image fuzzy set by the α-cuts. See [1] for a proof.

Example 16.2 Let us consider the mapping $f : \mathbb{R} \to \mathbb{R}, x \mapsto x^2$. Obviously, the image of a fuzzy set $\mu \in \mathcal{F}(\mathbb{R})$ is given by

$$f[\mu](y) = \begin{cases} \max\{\mu(\sqrt{y}), \mu(-\sqrt{y})\} \text{ if } y \ge 0 \\ 0 \hspace{3.5cm} \text{otherwise.} \end{cases}$$

Let the fuzzy set $\mu = \Lambda_{0,1,2}$ represent the imprecise concept "about 1." The question, what "the square of about 1" is, can be answered by determining the α-cuts of the image fuzzy set $f[\mu]$ from the α-cuts of μ. Here, this is possible because the function f has compact α-cuts and the fuzzy set μ is continuous. Thus, we have $[\mu]_\alpha = [\alpha, 2 - \alpha]$ for all $0 < \alpha \le 1$ and we obtain

$$\left[f[\mu]\right]_\alpha = f\left[[\mu]_\alpha\right] = [\alpha^2, (2 - \alpha)^2].$$

The fuzzy sets μ and $f[\mu]$ are shown in Fig. 16.4. Here, we can observe that the imprecise concept "the square of about 1" does not exactly match to the imprecise concept "about 1." The concept "the square of about 1" is "fuzzier" than "about 1." This effect is very similar to the increase in rounding errors when more and more computation steps are performed.   □

16.3 Cartesian Product and Cylindrical Extension

So far we have only extended mappings with one argument to fuzzy sets. To define operations like addition for fuzzy sets over real numbers, we need a concept how to apply a mapping $f : X_1 \times \ldots \times X_n \to Y$ to a tuple $(\mu_1, \ldots, \mu_n) \in \mathcal{F}(X_1) \times$

Fig. 16.4 The fuzzy sets μ and $f[\mu]$ that represent the vague concepts "about 1" and "square of about 1," respectively

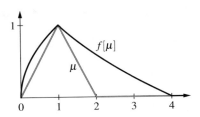

... $\times \mathcal{F}(X_n)$ of fuzzy sets. Interpreting addition as a function in two arguments $f : \mathbb{R} \times \mathbb{R} \to \mathbb{R}$, $(x_1, x_2) \mapsto x_1 + x_2$, this concept will enable us to extend addition and other algebraic operations to fuzzy sets over real numbers.

In order to generalize the extension principle described in Eq. 16.3 to mappings with several arguments, we introduce the concept of the Cartesian product of fuzzy sets. Consider the fuzzy sets $\mu_i \in \mathcal{F}(X_i)$, $i = 1, \ldots, n$. The **Cartesian product** of the fuzzy sets $\mu_1, \ldots, \mu_n$ is the fuzzy set

$$\mu_1 \times \ldots \times \mu_n \in \mathcal{F}(X_1 \times \ldots \times X_n)$$

with

$$(\mu_1 \times \ldots \times \mu_n)(x_1, \ldots, x_n) = \min\{\mu_1(x_1), \ldots, \mu_n(x_n)\}.$$

This definition is motivated by the property

$$(x_1, \ldots, x_n) \in M_1 \times \ldots \times M_n \quad \Longleftrightarrow \quad x_1 \in M_1 \wedge \ldots \wedge x_n \in M_n$$

of the Cartesian product of usual sets and corresponds to the formula

$$[\![(x_1, \ldots, x_n) \in \mu_1 \times \ldots \times \mu_n]\!] = [\![x_1 \in \mu_1 \wedge \ldots \wedge x_n \in \mu_n]\!],$$

where the minimum is chosen as truth function for the conjunction.

A special case of a Cartesian product is the **cylindrical extension** of a fuzzy set $\mu \in \mathcal{F}(X_i)$ to a product space $X_1 \times \ldots \times X_n$. The cylindrical extension is the Cartesian product of μ with the remaining universe of discourses X_j, $j \neq i$ or their characteristic functions:

$$\hat{\pi}_i(\mu) = I_{X_1} \times \ldots \times I_{X_{i-1}} \times \mu \times I_{X_{i+1}} \times \ldots \times I_{X_n},$$

$$\hat{\pi}_i(\mu)(x_1, \ldots, x_n) = \mu(x_i).$$

Obviously, projecting a cylindrical extension results in the original fuzzy set which means $\pi_i\left[\hat{\pi}_i(\mu)\right] = \mu$ provided that the sets $X_1, \ldots, X_n$ are non-empty. In general, $\pi_i\left[\mu_1 \times \ldots \times \mu_n\right] = \mu_i$ holds if the fuzzy sets μ_j, $j \neq i$ are **normal** which means $(\exists x_j \in X_j)(\mu_j(x_j)) = 1$.

16.4 Extension Principle for Multivariate Mappings

Using the Cartesian product, the extension principle for mappings with several arguments can be simplified to the extension principle for functions with one argument. Consider the mapping

$$f : X_1 \times \ldots \times X_n \to Y.$$

Then, the image of the tuple

$$(\mu_1, \ldots, \mu_n) \in \mathcal{F}(X_1) \times \ldots \times \mathcal{F}(X_n)$$

of fuzzy sets under f is the fuzzy set

$$f[\mu_1, \ldots, \mu_n] = f[\mu_1 \times \ldots \times \mu_n]$$

over the universe of discourse Y. That means

$$f[\mu_1, \ldots, \mu_n](y)$$

$$= \sup_{(x_1,\ldots,x_n)\in X_1\times\ldots\times X_n} \{(\mu_1 \times \ldots \times \mu_n)(x_1, \ldots, x_n) f(x_1, \ldots, x_n) = y\}$$

$$= \sup_{(x_1,\ldots,x_n)\in X_1\times\ldots\times X_n} \{\min\{\mu_1(x_1), \ldots, \mu_n(x_n)\} f(x_1, \ldots, x_n) = y\}. \quad (16.4)$$

This formula represents the **extension principle** [5–7].

Example 16.3 Consider the mapping $f : \mathbb{R}\times\mathbb{R} \to \mathbb{R}, (x_1, x_2) \mapsto x_1+x_2$ representing the addition. The fuzzy sets $\mu_1 = \Lambda_{0,1,2}$ and $\mu_2 = \Lambda_{1,2,3}$ model the imprecise concepts "about 1" and "about 2." Applying the extension principle, we obtain the fuzzy set $f[\mu_1, \mu_2] = \Lambda_{1,3,5}$ for the imprecise concept "about 1 + about 2" (cf. Fig. 16.5). We can observe the effect we already know from computing the square of "about 1" (see Example 16.2 and Fig. 16.4). The "fuzziness" of the resulting fuzzy set is greater than these of the original fuzzy sets to be added. □

Analogously to the addition of fuzzy sets we can define subtraction, multiplication, and division using the extension principle. These operations are continuous, therefore we can, like in Example 16.2, calculate the α-cuts of the resulting fuzzy sets directly from the α-cuts of the given fuzzy sets, provided that these are compact. When we have convex fuzzy sets, we carry out interval arithmetic on the corresponding levels. Interval arithmetic [2,3] allows us to operate with intervals instead of real numbers.

Fig. 16.5 The result of the extension principle for "about 1 + about 2"

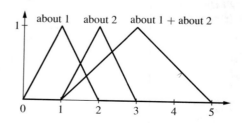

Applying the extension principle we should be aware that we carry out two generalization steps at the same time: the extension of single elements to sets and the change from crisp to fuzzy sets. The extension principle cannot preserve all properties of the original mapping. This effect is not necessarily caused by the extension from crisp to fuzzy sets. Most of the problems are caused by the step from extending a pointwise mapping to (crisp) sets. For example, in contrast to the standard addition, there is no inverse for the addition of fuzzy sets. There is no fuzzy set which added to the fuzzy set for "about 1 + about 2" from Fig. 16.5 which will yield back the fuzzy set "about 1." This phenomenon already occurs in interval arithmetic, so not the "fuzzification" of the addition is the problem, but the extension of the addition from numbers (points) to sets.

References

1. Kruse R, Gebhardt J, Klawonn F (1994) Foundations of fuzzy systems. Wiley, Chichester
2. Moore RE (1966) Interval analysis. Prentice Hall, Englewood Cliffs
3. Moore RE (1979) Methods and applications of interval analysis. Society for Industrial and Applied Mathematics, Philadelphia
4. Nguyen HT (1978) A note on the extension principle for fuzzy sets. J Math Anal Appl 64:369–380
5. Zadeh LA (1975) The concept of a linguistic variable and its application to approximate reasoning, part I. Inf Sci 8:199–249
6. Zadeh LA (1975) The concept of a linguistic variable and its application to approximate reasoning, part II. Inf Sci 8:301–357
7. Zadeh LA (1975) The concept of a linguistic variable and its application to approximate reasoning, part III. Inf Sci 9:43–80

Fuzzy Relations

17

Relations can be used to model dependencies, correlations, or connections between variables, quantities, or attributes. Technically speaking, a (binary) relation over the universes of discourse X and Y is a subset R of the Cartesian product $X \times Y$ of X and Y. The pairs $(x, y) \in X \times Y$ belonging to the relation R are linked by a connection described by the relation R. Therefore, a common notation for $(x, y) \in R$ is also $x R y$.

We generalize the concept of relations to fuzzy relations. Fuzzy relations are useful for representing and understanding fuzzy controllers that describe a vague connection between input and output values. Furthermore, we can establish an interpretation of fuzzy sets and membership degrees on the basis of special fuzzy relations called similarity relations. This interpretation plays a crucial role in the context of fuzzy controllers. Similarity relations will be discussed in Chap. 18.

17.1 Crisp Relations

Before we introduce the definition of a fuzzy relation, we briefly review the fundamental concepts and mechanisms of crisp relations that are needed for understanding fuzzy relations.

Example 17.1 A house has six doors and each of them has a lock which can be unlocked by certain keys. Let the set of doors be $T = \{t_1, \ldots, t_6\}$, the set of keys $S = \{s_1, \ldots, s_5\}$. Key s_5 is the main key and fits to all doors. Key s_1 fits only to door t_1, s_2 to t_1 and t_2, s_3 to t_3 and t_4, s_4 to t_5. This situation can be formally described by the relation $R \subseteq S \times T$ ("fits to"). The pair $(s, t) \in S \times T$ is an element of R if

© Springer Nature Switzerland AG 2022
R. Kruse et al., *Computational Intelligence*, Texts in Computer Science,
https://doi.org/10.1007/978-3-030-42227-1_17

Table 17.1 The relation R: "key fits to door"

R	t_1	t_2	t_3	t_4	t_5	t_6
s_1	1	0	0	0	0	0
s_2	1	1	0	0	0	0
s_3	0	0	1	1	0	0
s_4	0	0	0	0	1	0
s_5	1	1	1	1	1	1

Fig. 17.1 The relation R:
$y \mathbin{\hat{=}} x \pm 0.1$

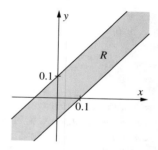

and only if key s fits to door t that means

$$R = \{(s_1, t_1), (s_2, t_1), (s_2, t_2), (s_3, t_3), (s_3, t_4), (s_4, t_5),$$
$$(s_5, t_1), (s_5, t_2), (s_5, t_3), (s_5, t_4), (s_5, t_5), (s_5, t_6)\}.$$

Another way of describing the relation R is shown in Table 17.1. The entry 1 at position (s_i, t_j) indicates that $(s_i, t_j) \in R$ holds, 0 stands for $(s_i, t_j) \notin R$. $\qquad\square$

Example 17.2 Let us consider a measuring instrument which can measure a quantity $y \in \mathbb{R}$ with a precision of ± 0.1. If x_0 is the measured value, we know the true value y_0 lies within the interval $[x_0 - 0.1, x_0 + 0.1]$. This can be described by the relation

$$R = \{(x, y) \in \mathbb{R} \times \mathbb{R} \mid |x - y| \le 0.1\}.$$

A graphical representation of this relation is given in Fig. 17.1.

Mappings or their graphs can be considered as special cases of relations. If the function $f : X \to Y$ is a mapping of X to Y, the graph of f is the relation

$$\mathrm{graph}(f) = \{(x, f(x)) \mid x \in X\}.$$

In order to be able to interpret a relation $R \subseteq X \times Y$ as a graph of a function we need that for each $x \in X$ there exists exactly one $y \in Y$ such that the pair (x, y) is contained in R.

17.2 Application of Relations

So far we have used relations in a merely descriptive way. But similar to functions relations can also be applied to elements or sets. If $R \subseteq X \times Y$ is a relation between the sets X and Y and $M \subseteq X$ is a subset of X, the image of M under R is the set

$$R[M] = \{y \in Y \mid \exists x \in X : (x, y) \in R \text{ AND } x \in M\}. \tag{17.1}$$

$R[M]$ contains the elements from Y which are related to at least one element of the set M.

If $f : X \to Y$ is a mapping, then applying the relation $\text{graph}(f)$ to a one-element set $\{x\} \subseteq X$ we obtain the one-element set which contains the image of x under the function f:

$$\text{graph}(f)[\{x\}] = \{f(x)\}.$$

More generally, we have

$$\text{graph}(f)[M] = f[M] = \{y \in Y \mid \exists x \in X : x \in M \wedge f(x) = y\}$$

for arbitrary subsets $M \subseteq X$.

Example 17.3 Now we use the relation R from Example 17.1 in order to determine which doors can be unlocked if we have keys $s_1, \ldots, s_4$. All we have to do is to calculate all elements (doors) which are related (relation "fits to") to at least one of the keys $s_1, \ldots, s_4$. That means

$$R[\{s_1, \ldots, s_4\}] = \{t_1, \ldots, t_5\}$$

is the set of doors we want to know.

The set $R[\{s_1, \ldots, s_4\}]$ can be determined easily using the matrix in Table 17.1 in the following way. We encode the set $M = \{s_1, \ldots, s_4\}$ as a row vector with five components which contains the entry 1 at the ith place if $s_i \in M$ holds, and 0 in the case of $s_i \notin M$. Thus, we obtain the vector $(1, 1, 1, 1, 0)$. Analogously to the scheme for matrix multiplication of a vector by a matrix, we write the vector to the lower left of the matrix. Then, we transpose the vector and compare it with every column of the matrix. If we find at least one position during this comparison of the vector and a matrix column where the vector and the matrix have the entry 1, we write a one under this column, otherwise a zero. The resulting vector $(1, 1, 1, 1, 1, 0)$ below the matrix specifies the set $R[M]$ we are looking for in an encoded form: It contains a 1 at place i if and only if $t_i \in R[M]$ holds. Table 17.2 illustrates this scheme for relations. □

Example 17.4 We follow up Example 17.2 and assume that we have the information that the measuring instrument indicated a value between 0.2 and 0.4. From this we can conclude that the true value is contained in the set $R\Big[[0.2, 0.4]\Big] = [0.1, 0.5]$ which is illustrated in Fig. 17.2.

Table 17.2 Scheme for the calculation of $R[M]$

					1	0	0	0	0	0
					1	1	0	0	0	0
					0	0	1	1	0	0
					0	0	0	0	1	0
					1	1	1	1	1	1
1	1	1	1	0	1	1	1	1	1	0

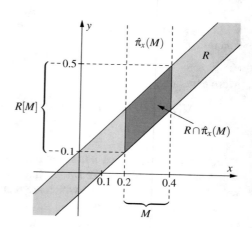

Fig. 17.2 How to determine the set $R[M]$ graphically

In this figure, we can see that we obtain the set $R[M]$ as the projection of the intersection of the relation with the cylindrical extension of the set M which means

$$R[M] = \pi_y \left[R \cap \hat{\pi}_x(M) \right]. \tag{17.2}$$

□

Example 17.5 Logical deduction based on an implication of the form $x \in A \rightarrow y \in B$ can be modeled and computed by relations, too. All we have to do is to encode the rule $x \in A \rightarrow y \in B$ by the relation

$$R = \{(x, y) \in X \times Y \mid x \in A \rightarrow y \in B\} = (A \times B) \cup \bar{A} \times Y. \tag{17.3}$$

X and Y are the sets of possible values that x and y can attain. For the rule "If the velocity is between 90 km/h and 110 km/h , then the fuel consumption is between 6 and 8 liters" (as a logical formula: $v \in [90, 110] \rightarrow b \in [6, 8]$) we obtain the relation shown in Fig. 17.3.

If we know that the velocity has the value v, in the case $90 \leq v \leq 110$ we can conclude that for the consumption b we must have $6 \leq b \leq 8$. Otherwise and without further knowledge and any pieces of information than just the rule and the value for v, we cannot say anything about the value for the consumption which means

Fig. 17.3 Relation for the rule $v \in [90, 110] \to b \in [6, 8]$.

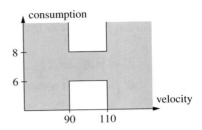

that we obtain $b \in [0, \infty)$. We get the same result by applying the relation R to the one-element set $\{v\}$:

$$R[\{v\}] = \begin{cases} [6, 8] & \text{if } v \in [90, 110] \\ [0, \infty) & \text{otherwise.} \end{cases}$$

More generally, we have: when we know that the velocity attains a value in the set M, we can conclude in the case $M \subseteq [90, 110]$ that the consumption is between 6 and 8 liters, otherwise we only know $b \in [0, \infty)$. This coincides with the result we obtained by applying the relation R to the set M:

$$R[M] = \begin{cases} [6, 8] & \text{if } M \subseteq [90, 110] \\ \emptyset & \text{if } M = \emptyset \\ [0, \infty) & \text{otherwise.} \end{cases}$$

$\square$

17.3 Logical Deduction with Relations

The example above shows how logical deduction can be represented in terms of a relation. Inferring new facts from rules and known facts usually means that we deal with chained deduction steps in the form of $\varphi_1 \to \varphi_2$, $\varphi_2 \to \varphi_3$ from which we can derive $\varphi_1 \to \varphi_3$. A similar principle can be formulated in the context of relations. Consider the relations $R_1 \subseteq X \times Y$ and $R_2 \subseteq Y \times Z$. An element x is indirectly related to an element $z \in Z$ if there exists an element $y \in Y$ such that x and y are in the relation R_1 and y and z are in the relation R_2. We can say that we go from x to z via y. In this way, the composition of the relations R_1 and R_2 can be defined as the relation

$$R_2 \circ R_1 = \{(x, z) \in X \times Z \mid \exists y \in Y : (x, y) \in R_1 \wedge (y, z) \in R_2\} \qquad (17.4)$$

between X and Z. Then, we have for all $M \subseteq X$

$$R_2\big[R_1[M]\big] = (R_2 \circ R_1)[M].$$

For the relations $\text{graph}(f)$ and $\text{graph}(g)$ induced by the mappings $f : X \to Y$ or $g : Y \to Z$, respectively, the composition of these relations is equal to the relation induced by the composition of the two mappings:

$$\text{graph}(g \circ f) = \text{graph}(g) \circ \text{graph}(f).$$

Example 17.6 We extend Example 17.1 of the keys and doors by considering a set $P = \{p_1, p_2, p_3\}$ of three people owning various keys. This is expressed by the relation

$$R' = \{(p_1, s_1), (p_1, s_2), (p_2, s_3), (p_2, s_4), (p_3, s_5)\} \subseteq P \times T.$$

$(p_i, s_j) \in R'$ means that person p_i owns the key s_j. The composition

$$R \circ R' = \{(p_1, t_1), (p_1, t_2), (p_2, t_3), (p_2, t_4), (p_2, t_5),$$
$$(p_3, t_1), (p_3, t_2), (p_3, t_3), (p_3, t_4), (p_3, t_5), (p_3, t_6)\}$$

of the relations R' and R contains the pair $(p, t) \in P \times T$ if and only if person p can unlock door t. For example, using the relation $R \circ R'$ we can determine which doors can be unlocked if the people p_1 and p_2 are present. The corresponding set of doors is

$$(R \circ R')[\{p_1, p_2\}] = \{t_1, \ldots, t_5\} = R\Big[R'[\{p_1, p_2\}]\Big].$$

$\square$

Example 17.7 In Example 17.2, we used the relation

$$R = \{(x, y) \in \mathbb{R} \times \mathbb{R} : |x - y| \le 0.1\}$$

to model the fact that the measured value x represents the true value y with a precision of 0.1. When we can determine the quantity z from the quantity y with a precision of 0.2, we obtain the relation

$$R' = \{(y, z) \in \mathbb{R} \times \mathbb{R} : |x - y| \le 0.2\}.$$

Composition of R' and R results in the relation

$$R' \circ R = \{(x, z) \in \mathbb{R} \times \mathbb{R} : |x - z| \le 0.3\}.$$

If the measuring instrument indicates the value x_0, we can conclude that the value of the quantity z is in the set

$$(R' \circ R)[\{x_0\}] = [x_0 - 0.3, x_0 + 0.3].$$

$\square$

Example 17.8 Example 17.5 demonstrated how an implication of the form $x \in A \to y \in B$ can be represented by a relation. When another rule $y \in C \to z \in D$ is known, in the case of $B \subseteq C$ we can derive the rule $x \in A \to z \in D$. Otherwise, knowing x does not provide any information about z in the context of these two rules. That means we obtain the rule $x \in X \to z \in Z$ in this case. Correspondingly, the composition of the relations R' and R representing the implications $x \in A \to y \in B$ and $y \in C \to z \in D$, respectively, result in the relation associated with the implication $x \in A \to z \in D$ and $x \in A \to z \in Z$, respectively:

$$R' \circ R = \begin{cases} (A \times D) \cup (\bar{A} \times Z) & \text{if } B \subseteq C \\ (A \times Z) \cup (\bar{A} \times Z) = X \times Z & \text{otherwise.} \end{cases}$$

$\square$

17.4 Simple Fuzzy Relations

After giving a general idea about the fundamental terminology and concepts of crisp relations, we can now introduce fuzzy relations.

Definition 17.1 A fuzzy set $\varrho \in \mathcal{F}(X \times Y)$ is called (binary) *fuzzy relation* between the reference sets X and Y.

In this sense, a fuzzy relation is a generalized crisp relation where two elements can be gradually related to each other. The greater the membership degree $\varrho(x, y)$ the stronger is the relation between x and y.

Example 17.9 Let $X = \{s, f, e\}$ denote a set of financial fonds, devoted to shares (s), fixed-interest stocks (f) and real estates (e). The set $Y = \{l, m, h\}$ contains the elements low (l), medium (m), and high (h) risk. The fuzzy relation $\varrho \in \mathcal{F}(X \times Y)$ in Table 17.3 shows for every pair $(x, y) \in X \times Y$ how much the fond x is considered having the risk factor y.

For example, the entry in column m and row e means that the fond dedicated to real estates is considered to have a medium risk with a degree of 0.5. Therefore, $\varrho(e, m) = 0.5$.   □

Example 17.10 The measuring instrument from Example 17.2 had a precision of ± 0.1. However, it is not very realistic to assume that, given that the instrument shows the value x_0, all values from the interval $[x_0 - 0.1, x_0 + 0.1]$ are equally likely to represent the true value of the measured quantity. Instead of the crisp relation R from Example 17.2 for representing this fact, we can use a fuzzy relation, for instance,

$$\varrho : \mathbb{R} \times \mathbb{R} \to [0, 1], \quad (x, y) \mapsto 1 - \min\{10|x - y|, 1\},$$

yielding the membership degree of 1 for $x = y$. The membership degree to the relation decreases linearly with increasing distance $|x - y|$ until the difference between x and y exceeds the value 0.1.   □

In order to operate with fuzzy relations in a similar way as with usual relations, we have to define what the image of a fuzzy set under a fuzzy relation is. This means we have to extend Eq. 17.1 to the framework of fuzzy sets and fuzzy relations.

Table 17.3 The fuzzy relation ϱ: "x is a financial fond with risk factor y"

ϱ	l	m	h
s	0.0	0.3	1.0
f	0.6	0.9	0.1
e	0.8	0.5	0.2

Definition 17.2 For a fuzzy relation $\varrho \in \mathcal{F}(X \times Y)$ and a fuzzy set $\mu \in \mathcal{F}(X)$ the image of μ under ϱ is the $\varrho[\mu]$ fuzzy set

$$\varrho[\mu](y) = \sup\big\{\min\{\varrho(x, y), \mu(x)\} \mid x \in X\big\} \tag{17.5}$$

over the universe of discourse Y.

This definition can be justified in several ways. If ϱ and μ are the characteristic functions of a usual relation R and the crisp set M, respectively, then $\varrho[\mu]$ is the characteristic function of the image $R[M]$ of M under R. In this sense, the definition is a generalization of Eq. 17.1 for sets to fuzzy sets.

Equation 17.1 is equivalent to

$$y \in R[M] \quad \Longleftrightarrow \quad \exists x \in X : (x, y) \in R \land x \in M.$$

We obtain Eq. 17.5 for fuzzy relations from this equivalence by assigning the minimum as truth function to the conjunction and evaluate the existential quantifier by the supremum. This means

$$\varrho[\mu](y) = [\![y \in \varrho[\mu]]\!]$$

$$= [\![\exists x \in X : (x, y) \in \varrho \land x \in \mu]\!]$$

$$= \sup\big\{\min\{\varrho(x, y), \mu(x)\} \mid x \in X\big\}.$$

Definition 17.2 can also be derived from the extension principle. In order to do this, we consider the partial mapping

$$f : X \times (X \times Y) \to Y, \qquad \big(x, (x', y)\big) \mapsto \begin{cases} y & \text{if } x = x' \\ \text{undefined} & \text{otherwise.} \end{cases} \tag{17.6}$$

It is obvious that for a set $M \subseteq X$ and a relation $R \subseteq X \times Y$, we have

$$f[M, R] = f[M \times R] = R[M].$$

When we introduced the extension principle, we did not require that the mapping f, which has to be extended to fuzzy sets, must be defined everywhere. Therefore, the extension principle can also be applied to partial mappings. The extension principle for the mapping in Eq. 17.6, which can be used to compute the image of a set under a relation, leads to the formula specified in Definition 17.2, the formula for the image of a fuzzy set under a fuzzy relation.

Another justification of Definition 17.2 is based on the idea that was exploited in Example 17.4 and Fig. 17.2. Therefore, we computed the image of a set under a relation as projection of the intersection of the cylindrical extension of the set with the relation (cf. Eq. 17.2). Having this equation in mind, we replace the set M by a fuzzy set μ and the relation R by a fuzzy relation ϱ and again obtain Eq. 17.5 if the intersection of fuzzy sets is computed using the minimum and the projection and the cylindrical extension for fuzzy sets are calculated as in Chap. 16.

Example 17.11 On the basis of the fuzzy relations from Example 17.9, we want to estimate the risk of a mixed fond which concentrates on shares but also invests a smaller part of its many into real estates. We represent this mixed fond over the universe of discourse $\{s, r, f\}$ as a fuzzy set μ with

$$\mu(s) = 0.8, \qquad \mu(f) = 0, \qquad \mu(r) = 0.2.$$

In order to determine the risk of this mixed fond we compute the image of the fuzzy set μ under the fuzzy relation ϱ from Table 17.3. We obtain

$$\varrho[\mu](l) = 0.2, \qquad \varrho[\mu](m) = 0.3, \qquad \varrho[\mu](h) = 0.8.$$

Analogously to Example 17.3, the fuzzy set $\varrho[\mu]$ can be determined using a modified scheme. The zeros and ones in Table 17.2 have to be replaced by the corresponding membership degrees. Below each column of the fuzzy relation, we obtain the membership degree of the corresponding value to the image fuzzy set $\varrho[\mu]$ in the following way. We first take the componentwise minimum of the vector representing the fuzzy set μ and the corresponding column of the matrix representing the fuzzy relation ϱ and then we compute the maximum of these minima.

In this sense, the calculation of the image of a fuzzy set μ under a fuzzy relation ϱ is similar to matrix multiplication of a matrix with a vector where the multiplication of the components is replaced by the minimum and the addition by the maximum. □

Example 17.12 We have the information that the measuring instrument from Example 17.10 indicated a value of "about 0.3" which we represent by the fuzzy set $\mu = \Lambda_{0.2, 0.3, 0.4}$. For the true value y we obtain the fuzzy set

$$\varrho[\mu](y) = 1 - \min\{5|y - 0.3|, 1\}$$

as image of the fuzzy set μ under the relation ϱ from Example 17.10. □

Logical deduction is based on the logical implication operator. In the Łukasiewicz logic, the truth value of the implication is $a \rightarrow b = \neg a \vee b = \min\{1, 1 - a + b\}$. This is also the value of the residuated implication $\sup\{x \in [0, 1] \mid (a \wedge x) \leq b\}$ of its t-norm $a \wedge b = \max\{a + b - 1, 0\}$. In the Gödel logic, the truth value of the implication is defined by the residuum of its t-norm:

$$a \rightarrow b = \sup\{x \in [0, 1] \mid (a \wedge x) \leq b\}$$
$$= \sup\{x \in [0, 1] \mid \min\{a, x\} \leq b\}$$
$$= \begin{cases} 1 & \text{if } a \leq b \\ b & \text{if } a > b. \end{cases}$$

Another method for defining a fuzzy implication was used by Lotfi Zadeh. He used the property $a \rightarrow b = \neg a \vee (a \wedge b)$ in classical logic to complete the definition of his standard fuzzy logic $\neg a = 1 - a$, $a \wedge b = \min\{a, b\}$, $a \vee b = \max\{a, b\}$, and $a \rightarrow b = \max\{1 - a, \min\{a, b\}\}$.

Fig. 17.4 The fuzzy set $\varrho[\mu']$

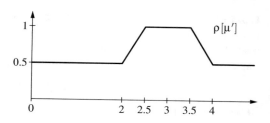

Example 17.13 Example 17.5 illustrated how an implication of the form $x \in A \rightarrow y \in B$ can be represented using a relation. We generalize this method for the case that the sets A and B are replaced by the fuzzy sets μ or ν. Following Eq. 17.3 and using the formula $[\![(x, y) \in \varrho]\!] = [\![x \in \mu \rightarrow y \in \nu]\!]$ where we choose the Gödel implication as truth function for the implication, we define the fuzzy relation

$$\varrho(x, y) = \begin{cases} 1 & \text{if} \mu(x) \leq \nu(y) \\ \nu(y) & \text{otherwise.} \end{cases}$$

The rule "If x is about 2, then y is about 3" leads to the fuzzy relation

$$\varrho(x, y) = \begin{cases} 1 & \text{if} \min\{|3 - y|, 1\} \leq |2 - x| \\ 1 - \min\{|3 - y|, 1\} & \text{otherwise,} \end{cases}$$

if we model "about 2" by the fuzzy set $\mu = \Lambda_{1,2,3}$ and "about 3" by the fuzzy set $\nu = \Lambda_{2,3,4}$. Knowing "x is about 2.5" represented by the fuzzy set $\mu' = \Lambda_{1.5,2.5,3.5}$ we obtain for y the fuzzy set

$$\varrho[\mu'](y) = \begin{cases} y - 1.5 & \text{if } 2.0 \leq y \leq 2.5 \\ 1 & \text{if } 2.5 \leq y \leq 3.5 \\ 4.5 - y & \text{if } 3.5 \leq y \leq 4.0 \\ 0.5 & \text{otherwise,} \end{cases}$$

shown in Fig. 17.4.

The membership degree of an element y_0 to this fuzzy set should be interpreted as how much one can believe that it is possible that the variable y attains the value y_0. This interpretation is a generalization of what we have obtained for the implication based on crisp sets in Example 17.5. In that case, only two sets were possible results of applying the deduction scheme: either the entire universe of discourse, if we could not guarantee that the antecedent of the implication is satisfied, or the set determined specified in the consequent of the implication for the case that the antecedent is satisfied. The first case tells us that—only knowing the single rule—all values for y are still possible, since the rule is not applicable in this case. In the second case, we know that the rule is applicable and only those values specified in the consequent of the rule are considered possible for Y. Extending this framework from crisp to fuzzy sets, the antecedent and the consequent of the implication can be partially satisfied. The consequence is that not only the whole universe of discourse and the fuzzy set in the consequent are possible results, but also fuzzy sets in between. In our example, all values y have a membership degree of at least 0.5 to the fuzzy set $\varrho[\mu']$. The reason for this is that there exists a value, namely, $x_0 = 2.0$ which has a membership degree of 0.5 to the fuzzy set μ' and a membership degree of 0 to μ. This means that the variable x can attain a value with a degree of 0.5, for which we cannot apply the rule, i.e., y can attain any value. The membership degree 1 of the value $x_0 = 2.5$ to the fuzzy set μ' leads to the fact that all values of the interval $[2.5, 3.5]$ have a membership degree of 1 to $\varrho[\mu']$. For $x_0 = 2.5$ we obtain $\mu(2.5) = 0.75$ which means that the antecedent of the implication is satisfied with the degree of 0.75. This implies that in order to satisfy the implication, a membership degree of at least 0.75 is required for the consequent. And the values in the interval $[2.5, 3.5]$ are those with a membership degree of at least 0.75 to the fuzzy set ν.

In a similar way, we can treat membership degrees between 0 and 1 to justify or compute the fuzzy set $\varrho[\mu']$. □

17.5 Composition of Fuzzy Relations

Now we are able to discuss the composition of fuzzy relations. The definition of an image of a fuzzy set under a fuzzy relation was motivated by Eq. 17.1 for crisp sets. Analogously, we define the composition of fuzzy relations based on Eq. 17.4 that describes composition in the case of crisp relations.

Definition 17.3 Let $\varrho_1 \in \mathcal{F}(X \times Y)$ and $\varrho_2 \in \mathcal{F}(Y \times Z)$ be fuzzy relations. The *composition* of the two fuzzy relations is the fuzzy relation

$$(\varrho_2 \circ \varrho_1)(x, z) = \sup \left\{ \min\{\varrho_1(x, y), \varrho_2(y, z)\} \mid y \in Y \right\} \qquad (17.7)$$

between the universes of discourse X and Z.

This definition can be obtained from the equivalence

$$(x, z) \in R_2 \circ R_1 \iff \exists y \in Y : (x, y) \in R_1 \wedge (y, z) \in R_2,$$

assigning the minimum as truth function to the conjunction and evaluating the existential quantifier by the supremum such that we obtain

$$(\varrho_2 \circ \varrho_1)(x, z) = [\![(x, y) \in (\varrho_2 \circ \varrho_1)]\!]$$

$$= [\![\exists y \in Y : (x, y) \in R_1 \wedge (y, z) \in R_2]\!]$$

$$= \sup \left\{ \min\{\varrho_1(x, y), \varrho_2(y, z)\} \mid y \in Y \right\}.$$

Equation 17.7 can also be derived by applying the extension principle to the partial mapping

$$f : (X \times Y) \times (Y \times Z) \to (X \times Y),$$

$$((x, y), (y', z)) \mapsto \begin{cases} (x, z) & \text{if } y = y' \\ \text{undefined} & \text{otherwise,} \end{cases}$$

on which the composition of crisp relations is based because we have

$$f[R_1, R_2] = f[R_1 \times R_2] = R_2 \circ R_1.$$

If ϱ_1 and ϱ_2 are the characteristic functions of the crisp relations R_1 or R_2, then $\varrho_2 \circ \varrho_1$ is the characteristic function of the relation $R_2 \circ R_1$. In this sense, Definition 17.3 generalizes the compositions of crisp relations to fuzzy relations.

For every fuzzy set $\mu \in \mathcal{F}(X)$, we have

$$(\varrho_2 \circ \varrho_1)[\mu] = \varrho_2[\varrho_1[\mu]].$$

Example 17.14 Let us come back to Example 17.11 analyzing the risk of financial fonds. Now, we extend the risk estimation of fonds by the set $Z = \{hl, ll, lp, hp\}$. The elements stand for "high loss," "low loss," "low profit," and "high profit." The fuzzy relation $\varrho' \in \mathcal{F}(Y \times Z)$ in Table 17.4 determines for each tuple $(y, z) \in Y \times Z$ the possibility to have a profit or loss of z under the risk y. The fuzzy relation resulting from the composition of the fuzzy relations ϱ and ϱ' is shown in Table 17.5.

In this case, where the universes of discourse are finite and the fuzzy relations can be represented as tables or matrices, the computation scheme for the composition of fuzzy relations is similar to matrix multiplication where we have to replace the componentwise multiplication by the minimum and the addition by the maximum. For the mixed fond from Example 17.11 which was represented by the fuzzy set μ

$$\mu(s) = 0.8, \qquad \mu(f) = 0, \qquad \mu(r) = 0.2,$$

we obtain

$$(\varrho' \circ \varrho)[\mu](hl) = (\varrho' \circ \varrho)[\mu](ll) = (\varrho' \circ \varrho)[\mu](lp) = (\varrho' \circ \varrho)[\mu](hp) = 0.8$$

as fuzzy set describing the possible profit or loss.   □

Table 17.4 The fuzzy relation ϱ': "Given the risk y the profit/loss z is possible"

ϱ'	hl	ll	lp	hp
l	0.0	0.4	1.0	0.0
m	0.3	1.0	1.0	0.4
h	1.0	1.0	1.0	1.0

Table 17.5 The fuzzy relation $\varrho' \circ \varrho$: "With the yield object x the profit/loss z is possible"

ϱ'	hl	ll	lp	hp
s	1.0	1.0	1.0	1.0
f	0.3	0.9	0.9	0.4
r	0.3	0.5	0.8	0.4

Example 17.15 The precision of the measuring instrument from Example 17.10 was described by the fuzzy relation $\varrho(x, y) = 1 - \min\{10|x - y|, 1\}$ which determines in how far the value y is the true value if x is the value indicated by the measuring instrument. We assume that we cannot exactly read the value from the (analog) instrument and therefore use the fuzzy relation $\varrho'(a, x) = 1 - \min\{5|a - x|, 1\}$. This relation tells us in how far the value x is the value indicated by the measuring instrument when we read the value a. In order to estimate which value could be the true value y of the measured quantity, given we have read the value a, we have to compute the composition of the fuzzy relations ϱ' and ϱ

$$(\varrho \circ \varrho')(a, y) = 1 - \min\left\{\frac{10}{3}|a - y|, 1\right\}.$$

Assuming we have read $a = 0$, we obtain the fuzzy set

$$(\varrho \circ \varrho')[I_{\{0\}}] = \Lambda_{-0.3, 0, 0.3}$$

for the true value y. □

17.6 Fuzzy Relational Equations

Let the fuzzy sets μ on X and ν on Y be given. Suppose we are searching for a fuzzy relation ϱ on $X \times Y$ that satisfies the equality $\nu = \mu \circ \varrho$. Such an equation is called a fuzzy relational equation. Let us first assume that the composition of the fuzzy set and the fuzzy relation is defined by the minimum operator for the conjunction:

$$\mu \circ \varrho : Y \to [0, 1], \quad y \mapsto \sup_{x \in X} \{\min \{\mu(x), \varrho(x, y)\}\}$$

A fuzzy relational equation does not always have a solution. However, if it is solvable, then the Gödel implication

$$\varrho_{\mu, \nu}^{\text{Gödel}} : X \times Y \to [0, 1], \quad (x, y) \mapsto \begin{cases} 1, & \text{if } \mu(x) \le \nu(y), \\ \nu(y), & \text{otherwise} \end{cases}$$

is the greatest solution. In general, there is no single smallest solution with respect to the inclusion of fuzzy relations—the space of solutions forms an upper semi-lattice. The Cartesian product $\mu \times \nu$ of μ and ν defined on $X \times Y$ by

$$(\mu \times \nu)(x, y) := \min \{\mu(x), \nu(y)\}$$

is a "small" solution, if there is a solution at all.

Example 17.16 Let the fuzzy sets μ on $X = \{x_1, x_2, x_3\}$ and ν on $Y = \{y_1, y_2, y_3, y_4\}$ be given:

x	x_1	x_2	x_3
$\mu(x)$	0.9	1.0	0.7

y	y_1	y_2	y_3	y_4
$\nu(y)$	0.1	0.4	0.8	0.7

The Gödel relation $\varrho_{\mu,\nu}^{\text{Gödel}}$ is the greatest solution of $\mu \circ \varrho = \nu$, whereas ϱ_1 and ϱ_2 are two different minimal solutions. The Cartesian product $\mu \times \nu$ is a "small" solution.

$\varrho_{\mu,\nu}^{\text{Gödel}}$	y_1	y_2	y_3	y_4
x_1	1.0	0.4	0.8	0.7
x_2	1.0	0.4	0.8	0.7
x_3	1.0	0.4	1.0	1.0

$\mu \times \nu$	y_1	y_2	y_3	y_4
x_1	0.9	0.4	0.8	0.7
x_2	1.0	0.4	0.8	0.7
x_3	0.7	0.4	0.7	0.7

ϱ_1	y_1	y_2	y_3	y_4
x_1	0.0	0.0	0.0	0.7
x_2	1.0	0.4	0.8	0.0
x_3	0.0	0.0	0.0	0.0

ϱ_2	y_1	y_2	y_3	y_4
x_1	0.0	0.4	0.8	0.0
x_2	1.0	0.0	0.0	0.0
x_3	0.0	0.0	0.0	0.7

These results can be generalized to systems of fuzzy relational equations. Let $\mu_1, \ldots, \mu_r$ be fuzzy sets on X and $\nu_1, \ldots, \nu_r$ be fuzzy sets on Y. We are searching for a (single) fuzzy relation ϱ on $X \times Y$ that satisfies the r fuzzy relational equations

$$\nu_i = \mu_i \circ \varrho, \quad i = 1, \ldots, r$$

If this system has a solution at all, then the fuzzy relation

$$\min_{i \in \{1,\ldots,r\}} \left\{ \varrho_{\mu_i,\nu_i}^{\text{Gödel}} \right\}$$

is the greatest solution of this system of equations.

The central role that is played by the Gödel relation in this context is due to the fact that the minimum operator for the conjunction was chosen. If the minimum was replaced by some other t-norm, a different relation would take the role of the Gödel relation.

References

1. De Baets B (2000) Analytical solution methods for fuzzy relation equations. In: Dubois D, Prade H (eds) The handbook of fuzzy set series, vol 1. Academic Kluwer Publ., Boston, pp 291–340
2. DiNola A, Sessa S, Pedrycz W, Sanchez E (1989) Fuzzy relation equations and their applications to knowledge engineering. Theory and Decision Libr., ser. D, Kluwer Acad. Publ., Dordrecht
3. Gottwald S (2000) Generalised solvability behaviour for systems of fuzzy equations. In: [Novak and Perfilieva 2000], pp 401–430
4. Novak V, Perfilieva I (eds) (2000) Discovering the world with fuzzy logic. Advances in soft computing. Physica, Heidelberg
5. Gottwald S (1993) Fuzzy sets and fuzzy logic. foundations of application - from a mathematical point of view. Vieweg, Braunschweig, Wiesbaden
6. Gottwald S, Bandemer H (1995) Fuzzy sets, fuzzy logic. Fuzzy methods with applications. Wiley, Chichester
7. Li P, Fang SC (2009) A survey on fuzzy relational equations, Part I: classification and solvability. Fuzzy Optim Decis Making 8:179–229

In this chapter, we discuss a special type of fuzzy relations, called similarity relations. They play an important role in interpreting fuzzy controllers and, more generally, can be used to characterize the inherent indistinguishability or vagueness of a fuzzy systems. In this context, fuzzy sets are induced concepts based on a similarity relation. The membership degrees are interpreted as grades of similarity.

18.1 Similarity

Similarity relations are fuzzy relations that specify for pairs of elements or objects how indistinguishable or similar they are. From a similarity relation, we should expect that it is reflexive and symmetric which means that every element is similar to itself (with degree 1) and that x is as similar to y as y to x. In addition to these two minimum requirements, we also ask for the following weakened transitivity condition: if x is similar to y to a certain degree and y is similar to z to a certain degree, then x should also be similar to z to a certain (maybe lower) degree. Technically speaking, we define a similarity relation as follows.

Definition 18.1 A *similarity relation* $E : X \times X \to [0, 1]$ with respect to the t-norm t on the set X is a fuzzy relation over $X \times X$ which satisfies the conditions

$$E(x, x) = 1 \qquad \text{(\textit{reflexivity})} \qquad \text{(E1)}$$
$$E(x, y) = E(y, x) \qquad \text{(\textit{symmetry})} \qquad \text{(E2)}$$
$$t\big(E(x, y), E(y, z)\big) \leq E(x, z) \qquad \text{(\textit{transitivity})} \qquad \text{(E3)}$$

for all $x, y, z \in X$.

© Springer Nature Switzerland AG 2022
R. Kruse et al., *Computational Intelligence*, Texts in Computer Science,
https://doi.org/10.1007/978-3-030-42227-1_18

The transitivity conditions for similarity relations can be understood in terms of fuzzy logic, as it was presented in Sect. 15.4, as follows: The truth value of the proposition

x and y are similar AND y and z are similar

should be at most as great as the truth value of the proposition

x and z are similar

where the t-norm t is used as the truth function for the conjunction AND.

In Example 17.10, we have already seen an example for a similarity relation, i.e., the fuzzy relation

$$\varrho : \mathbb{R} \times \mathbb{R} \to [0, 1], \qquad (x, y) \mapsto 1 - \min\{10|x - y|, 1\}$$

that indicates how indistinguishable two values are using a measuring instrument. We can prove easily that this fuzzy relation is a similarity relation with respect to the Łukasiewicz t-norm $t(\alpha, \beta) = \max\{\alpha + \beta - 1, 0\}$. More generally, an arbitrary pseudometric, i.e., a distance measure $\delta : X \times X \to [0, \infty)$ satisfying the symmetry condition $\delta(x, y) = \delta(y, x)$ and the triangle inequality $\delta(x, y) + \delta(y, z) \geq \delta(x, z)$ induces a similarity relation with respect to the Łukasiewicz t-norm by

$$E^{(\delta)}(x, y) = 1 - \min\{\delta(x, y), 1\}$$

and vice versa, every similarity relation E with respect to the Łukasiewicz t-norm defines a pseudometric by

$$\delta^{(E)}(x, y) = 1 - E(x, y).$$

Furthermore, we have $E = E^{(\delta^{(E)})}$ and $\delta(x, y) = \delta^{(E^{(\delta)})}(x, y)$ if δ is bounded by one, i.e., $\delta(x, y) \leq 1$ holds for all $x, y \in X$, such that similarity relations and pseudometrics (bounded by one) represent dual concepts.

Later on, other examples will provide a motivation to consider similarity relations with respect to other t-norms than the Łukasiewicz t-norm in order to characterize the vagueness or indistinguishability that is inherent in fuzzy systems.

18.2 Fuzzy Sets and Extensional Hulls

If we assume that a similarity relation characterizes a certain indistinguishability, we should expect that elements that are (almost) indistinguishable should behave similar or have similar properties. For fuzzy systems, the (fuzzy) property of being an element of a (fuzzy) set is essential. Therefore, those fuzzy sets play an important role that are coherent with respect to a given similarity relation in the sense that similar elements have similar membership degrees. This property of a fuzzy set is called extensionality and is formally defined as follows.

Definition 18.2 Let $E : X \times X \to [0, 1]$ be a similarity relation with respect to the t-norm t on the set X. A fuzzy set $\mu \in \mathcal{F}(X)$ is called *extensional* with respect to E if

$$t\big(\mu(x), E(x, y)\big) \leq \mu(y)$$

holds for all $x, y \in X$.

In the view of fuzzy logic, the extensionality condition can be interpreted in the sense that the truth value of the proposition

x is an element of the fuzzy set μ AND
x and y are similar (indistinguishable)

should not be smaller than the truth value of the proposition

y is an element of the fuzzy set μ

where the t-norm t as truth function is assigned to the conjunction AND.

A fuzzy set can always be made extensional by adding all elements which are similar to at least one of its elements. If we formalize this idea, we obtain the following definition.

Definition 18.3 Let $E : X \times X \to [0, 1]$ be a similarity relation with respect to the t-norm t on the set X. The extensional hull $\hat{\mu}$ of the fuzzy set $\mu \in \mathcal{F}(X)$ (with respect to the similarity relation E) is given $\hat{\mu}$ by

$$\hat{\mu}(y) = \sup \big\{ t\big(E(x, y), \mu(x)\big) \mid x \in X \big\}.$$

If t is a continuous t-norm, then the extensional hull $\hat{\mu}$ of μ is the smallest extensional fuzzy set containing μ in the sense of $\mu \leq \hat{\mu}$.

In principle, we obtain the extensional hull of a fuzzy set μ under the similarity relation E as the image of μ under the fuzzy relation E as in Definition 17.2. For the extensional hull, the minimum in Eq. 17.5 in Definition 17.2 is replaced by the t-norm t.

Example 18.1 Let us consider the similarity relation $E : \mathbb{R} \times \mathbb{R} \to [0, 1]$ defined by $E(x, y) = 1 - \min\{|x - y|, 1\}$ with respect to the Łukasiewicz t-norm which is induced by the usual metric $\delta(x, y) = |x - y|$ on the real numbers. A (crisp) set $M \subseteq \mathbb{R}$ can be viewed as a special type of fuzzy set when we consider its characteristic function I_M. In this way, we can also compute the extensional hulls of crisp sets.

The extensional hull of a point x_0, which means the one-element set $\{x_0\}$, with respect to the similarity relation E mentioned above is a fuzzy set with a triangular membership function $\Lambda_{x_0-1,x_0,x_0+1}$. The extensional hull of the interval $[a, b]$ is the trapezoidal function $\Pi_{a-1,a,b,b+1}$ (cf. Fig. 18.1). □

Fig. 18.1 Extensional hull
of the point x_0 and the
interval $[a, b]$

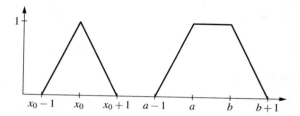

This example establishes an interesting connection between fuzzy sets and similarity relations: triangular and trapezoidal membership functions, which are very popular in most applications of fuzzy systems, can be interpreted as extensional hulls of points or intervals, which means as fuzzy points or fuzzy intervals in a vague environment which is characterized by a similarity relation induced by the standard metric on the real numbers.

18.3 Scaling Concepts

The standard metric on the real numbers allows only very limited forms of triangular and trapezoidal functions as extensional hulls of points and intervals: the slopes where the membership degrees increase and decrease are always 1 and −1, respectively. However, it is reasonable to allow a scaling of the standard metric so that other forms of fuzzy sets as extensional hulls can result. This scaling can have two different meanings.

The degree of similarity of two measured values depends on the measuring unit. Two values measured in kilo-units can have a small distance and might be considered as almost indistinguishable or very similar, while the very same quantities measured in milli-units have a much greater distance and are distinguishable. Of course, the degree of similarity should not depend on the measurement unit. The height of two persons should have the same degree of similarity, no matter whether we measure the height in centimeters, meters, or inches. In order to adapt a similarity relation to a measuring unit, the distance or the real axis has to be scaled by a constant factor $c > 0$, as in Example 17.10. In this way, we obtain scaled metric $|c \cdot x - c \cdot y|$ which induces the similarity relation $E(x, y) = 1 - \min\{|c \cdot x - c \cdot y|, 1\}$.

An extension of the scaling concept is the use of varying scaling factors allowing a local scaling fitting to the problem.

Example 18.2 We want to describe the behavior of an air conditioner using fuzzy rules. It is neither necessary nor desirable to measure the room temperature, to which the air conditioner will react, as exactly as possible. However, the temperature values are of different importance for the air conditioner. Temperatures of 10° C or 15° C, for example, are much too cold, so the air conditioner should heat at full power, and the values 27° C or 32° C are much too hot and the air conditioner should cool at full power. Therefore, there is no need to distinguish between the values 10° C and 15°

C or 27° C and 32° C for the control of the room temperature. Since we do not have to distinguish between 10° C and 15° C, we can choose a very small positive scaling factor—in the extreme case even the factor 0, then these temperatures are not distinguished at all. However, it would not be correct, if we chose a small scaling factor for the entire range of the temperature, because the air conditioner has to distinguish very well between the cold temperature 18.5° C and the warm one 23.5° C.

Instead of a global scaling factor we should choose individual scaling factors for different ranges of the temperature, so that we can make a fine distinction between temperatures which are near the optimum room temperature and a more rough one between temperatures which are much too cold or much too warm. Table 18.1 shows an example of a partition into five temperature ranges, each one having its individual scaling factor.

These scaling factors define a transformation of the range and the distances for the temperature that can be used to define a similarity relation. Table 18.2 shows the transformed distances and the resulting similarity degrees for some pairs of values of the temperature. For each pair, the two values lie in a range where the scaling factor

Table 18.1 Different sensitivities and scaling factors for controlling the room temperature

Temperature (in °C)	Scaling factor	Interpretation
<15	0.00	Exact value insignificant (much too cold temperature)
15–19	0.25	Too cold but near approximately OK, not very sensitive
19–23	1.50	Near the optimum, very sensitive
23–27	0.25	Too hot but near approximately OK, not very sensitive
>27	0.00	Exact value insignificant (much too hot temperature)

Table 18.2 Transformed distances based on the scaling factors and the induced similarity degrees

Pair of variates (x, y)	Scaling factor c	Transf. distance $\delta(x, y) = \|c \cdot x - c \cdot y\|$	Similarity degree $E(x, y) = 1 - \min\{\delta(x, y), 1\}$
(13, 14)	0.00	0.000	1.000
(14, 14.5)	0.00	0.000	1.000
(17, 17.5)	0.25	0.125	0.875
(20, 20.5)	1.50	0.750	0.250
(21, 22)	1.50	1.500	0.000
(24, 24.5)	0.25	0.125	0.875
(28, 28.5)	0.00	0.000	1.000

Fig. 18.2 Transformation of a co-domain with the help of scaling factors

does not change. In order to understand how to determine the transformed distance and the resulting similarity degree for two temperatures that are not in a range with a constant scaling factor, we first analyze the effect of a single scaling factor.

Let us consider an interval $[a, b]$ where we measure the distance between two points based on the scaling factor c. Computing the scaled distance means that we apply stretching ($c > 1$) or shrinking ($0 \leq c < 1$) to the interval according to the factor c and calculate the distances between the points in the transformed (stretched or shrunk) interval. In order to take individual scaling factors for different ranges into account, we have to stretch or shrink each range, where the scaling factor is constant, correspondingly. Gluing these transformed ranges together, we can now measure the distance between points that do not lie in a region with a constant scaling factor. The induced transformation is piecewise linear as shown in Fig. 18.2.

On the basis of the following three examples, we explain and illustrate the calculation of the transformed distance and the resulting similarity degrees. We determine the similarity degree between the values 18 and 19.2. The value 18 is in the interval [15, 19] with constant scaling factor 0.25. This interval of length 4 is now shrunk to an interval of length 1. Therefore, the distance of the value 18 to the right boundary of the interval, i.e., 19, is also shrunk by the factor 0.25, such that the transformed distance between 18 and 19 is exactly 0.25. In order to calculate the transformed distance between 18 and 19.2, we have to add the transformed distance between 19 and 19.2. In this range, the scaling factor is constantly 1.5, so the distance between 19 and 19.2 is stretched by the factor 1.5 and the resulting transformed distance is 0.3. Thus, the transformed distance between the values 18 and 19.2 is $0.25 + 0.3 = 0.55$ which leads to a similarity degree of $1 - \min\{0.55, 1\} = 0.45$.

In the second example, we consider the values 13 and 18. Because of the scaling factor 0, the transformed distance between 13 and 15 is also 0. In the range between 15 and 18, the scaling factor is 0.25 and therefore the transformed distance between 13 and 18 is 0.75. Since the transformed distance between 13 and 15 is zero, the overall transformed distance between 13 and 18 is 0.75 and the similarity degree between 13 and 18 is 0.25.

Finally, we determine the transformed distance and the similarity degree between the values 22.8 and 27.5. Here, we have to take three ranges with different scaling factors into account: the scaling factor between 22.8 and 23 is 1.5, between 23 and 27 it is 0.25, and between 27 and 27.5 it is constantly 0. Therefore, the transformed distances for the pairs (22.8, 23), (23, 27), and (27, 27.5) are 0.3, 1, and 0, respectively. The sum of these distances and therefore the transformed distance between 22.8 and 27.5 is 1.3. The degree of similarity is $1 - \min\{1.3, 1\} = 0$. □

The idea of using different scaling factors for different ranges can be extended by assigning a scaling factor to each single value which determines how to distinguish in the direct environment of this value. Instead of a piecewise constant scaling function as in Example 18.2 any (integrable) scaling functions $c : \mathbb{R} \to [0, \infty)$ can be used. The transformed distance between the values x and y under such a scaling function can be calculated using the following equation [4]:

$$\left| \int_x^y c(s)\,ds \right| . \tag{18.1}$$

18.4 Fuzzy Sets and Similarity Relations

We have seen that fuzzy sets can be interpreted as induced concepts based on similarity relation. Assuming that a similarity relation models a problem-specific indistinguishability, fuzzy sets are induced by crisp values or sets in a canonical way in the form of extensional hulls. In the following, we analyze the opposite view. Given a set of fuzzy sets, can we find a suitable similarity relation such that the fuzzy sets are nothing but extensional hulls of crisp values or sets? The results that we present here will be helpful in analyzing and understanding fuzzy controllers later on. In the context of fuzzy controllers, a number of vague expression modeled by fuzzy sets are used for each considered variable. So for each domain X a set $\mathcal{A} \subseteq \mathcal{F}(X)$ of fuzzy sets is given. As we will see later the inherent indistinguishability of these fuzzy sets can be characterized by similarity relations. The coarsest (greatest) similarity relation for which all fuzzy sets in $\mathcal{A}$ are extensional plays an important role. The following theorem [3] describes how to compute this similarity relation.

Theorem 18.1 *Let t be a continuous t-norm and $\mathcal{A} \subseteq \mathcal{F}(X)$ a set of fuzzy sets. Then*

$$E_{\mathcal{A}}(x, y) = \inf \left\{ \overleftrightarrow{t} \big(\mu(x), \mu(y) \big) \mid \mu \in \mathcal{A} \right\} \tag{18.2}$$

is the coarsest (greatest) similarity relation with respect to the t-norm t for which all fuzzy sets in $\mathcal{A}$ are extensional. $\overleftrightarrow{t}$ is the biimplication from Equation 15.4 associated with the t-norm t.

Coarsest similarity relation means that for every similarity relation E for which all fuzzy sets in $\mathcal{A}$ are extensional, we have that $E_{\mathcal{A}}(x, y) \geq E(x, y)$ holds for all $x, y \in X$.

Equation 18.2 for the similarity relation $E_{\mathcal{A}}$ can be understood and interpreted within the framework of fuzzy logic. We interpret the fuzzy sets in $\mathcal{A}$ as representations of vague properties. Two elements x and y are similar with respect to a single property, if x has the property if and only if y has the property. Modeling this idea within fuzzy logic, this means that we have to interpret "x has the property associated with the fuzzy set μ" as the membership degree of x to μ. Then the similarity degree of x and y, taking only the property associated with the fuzzy set μ into account, is defined by $\overleftrightarrow{t}\big(\mu(x), \mu(y)\big)$. When we can use all properties associated with the fuzzy sets in $\mathcal{A}$ to distinguish x and y, this means: x and y are similar if they are similar with respect to all properties in $\mathcal{A}$. Evaluating "for all" by the infimum, we obtain Equation 18.2 for the similarity degree of two elements.

We have seen in Example 18.1 that typical fuzzy sets with triangular membership functions can be interpreted as extensional hulls of single points or values. This interpretation of fuzzy sets as vague values will be very useful in the context of fuzzy control. This is why we also study the issue, when a set $\mathcal{A} \subseteq \mathcal{F}(X)$ of fuzzy sets can be interpreted as extensional hulls of points.

Theorem 18.2 *Let t be a continuous t-norm and $\mathcal{A} \subseteq \mathcal{F}(X)$ a set of fuzzy sets. Let each $\mu \in \mathcal{A}$ have the property such that there exists an $x_\mu \in X$ with $\mu(x_\mu) = 1$. There exists a similarity relation E, such that for all $\mu \in \mathcal{A}$ the extensional hull of the point x_μ is equal to the fuzzy set μ if and only if*

$$\sup_{x \in X}\{t\big(\mu(x), \nu(x)\big)\} \;\le\; \inf_{y \in X}\{\overleftrightarrow{t}\big(\mu(y), \nu(y)\big)\} \tag{18.3}$$

holds for all $\mu, \nu \in \mathcal{A}$. In this case, $E = E_{\mathcal{A}}$ is the coarsest similarity relation for which all fuzzy sets in $\mathcal{A}$ are extensional hulls of points.

Condition 18.3 says that the degree of non-disjunction of any two fuzzy sets $\mu, \nu \in \mathcal{A}$ must not exceed their degree of equality. The corresponding formulas are obtained by interpreting the following conditions in terms of fuzzy logic:

- Two sets μ and ν are non-disjoint if and only if

$$\exists x : x \in \mu \wedge x \in \nu$$

holds.
- Two sets μ and ν are equal if and only if

$$\forall y : y \in \mu \leftrightarrow y \in \nu$$

holds.

Condition 18.3 in Theorem 18.2 is definitely satisfied when the fuzzy sets μ and ν are disjoint with respect to the t-norm t, that is, we have $t\big(\mu(x), \nu(x)\big) = 0$ for all $x \in X$. A proof for this theorem can be found in [5].

Most of the variables in the context of fuzzy control are numerical or continuous. Similarity relations on the real line can be defined in a simple and understandable way using scaling functions based on the concept of distance, as described in Equation 18.1. When we require that the similarity relation in Theorem 18.2 should be definable in terms of a scaling function, the following result proved in [4] is important.

Theorem 18.3 *Let $\mathcal{A} \subseteq \mathcal{F}(\mathbb{R})$ be a non-empty, at most countable set of fuzzy sets such that for each $\mu \in \mathcal{A}$ the following conditions are satisfied:*

- *There exists an $x_\mu \in \mathbb{R}$ such that $\mu(x_\mu) = 1$.*
- *μ (as real-valued function) is non-decreasing in the interval $(-\infty, x_\mu]$.*
- *μ is non-increasing in the interval $[x_\mu, \infty)$.*
- *μ is continuous.*
- *μ is differentiable almost everywhere.*

There exists a scaling function $c : \mathbb{R} \to [0, \infty)$ such that for all $\mu \in \mathcal{A}$ the extensional hull of the point x_μ with respect to the similarity relation

$$E(x, y) = 1 - \min \left\{ \left| \int_x^y c(s)\, ds \right|, 1 \right\}$$

is equal to the fuzzy set μ if and only if the condition

$$\min\{\mu(x), \nu(x)\} > 0 \implies \left| \frac{d\mu(x)}{dx} \right| = \left| \frac{d\nu(x)}{dx} \right| \tag{18.4}$$

is satisfied for all $\mu, \nu \in \mathcal{A}$ almost everywhere. In this case

$$c : \mathbb{R} \to [0, \infty), \quad x \mapsto \begin{cases} \left| \frac{d\mu(x)}{dx} \right| & \text{if } \mu \in \mathcal{A} \text{ and } \mu(x) > 0 \\ 0 & \text{otherwise} \end{cases}$$

can be chosen as (almost everywhere well defined) scaling function.

Example 18.3 In order to illustrate how extensional hulls of points with respect to a similarity relation induced by a piecewise constant scaling function might look like, we recall the scaling function

$$c : [0, 35) \to [0, \infty), \quad s \mapsto \begin{cases} 0 & \text{if} & 0 \leq s < 15 \\ 0.25 & \text{if} & 15 \leq s < 19 \\ 1.5 & \text{if} & 19 \leq s < 23 \\ 0.25 & \text{if} & 23 \leq s < 27 \\ 0 & \text{if} & 27 \leq s < 35 \end{cases}$$

from Example 18.2. Figure 18.3 shows the extensional hulls of the points 15, 19, 21, 23, and 27 with respect to the similarity relation induced by the scaling function c.

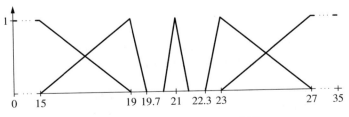

Fig. 18.3 The extensional hulls of the points 15, 19, 21, 23, and 27

Fig. 18.4 The extensional hulls of the points 18.5 and 22.5

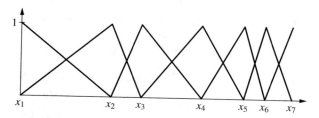

Fig. 18.5 Fuzzy sets for which we can define a scaling function

These extensional hulls have triangular or trapezoidal membership functions. The reason for this is that the points are chosen in such a way that—moving away from the corresponding point—the membership degree drops to zero before the scaling function changes its value. When we choose a point close to a point where the scaling function jumps from one value to another we will not obtain a triangular or trapezoidal membership function for the extensional hull. In the case of a piecewise linear scaling function, we can only guarantee that we obtain piecewise linear membership functions as extensional hulls as they are shown in Fig. 18.4.

It is quite common to use sets of fuzzy sets for fuzzy controllers as they are illustrated in Fig. 18.5. Therefore, we choose values $x_1 < x_2 < \ldots < x_n$ and use triangular functions of the form $\Lambda_{x_{i-1},x_i,x_{i+1}}$ or the trapezoidal functions $\Pi_{-\infty,-\infty,x_1,x_2}$ and $\Pi_{x_{n-1},x_n,\infty,\infty}$ at the limits x_1 and x_n of the corresponding range, i.e.,

$$\mathcal{A} = \{\Lambda_{x_{i-1},x_i,x_{i+1}} \mid 1 < i < n\} \cup \{\Pi_{-\infty,-\infty,x_1,x_2}, \Pi_{x_{n-1},x_n,\infty,\infty}\}.$$

In this case, we can always define a scaling function c such that the fuzzy sets can be interpreted as the extensional hulls of the points $x_1, \ldots, x_n$, namely,

$$c(x) = \frac{1}{x_{i+1} - x_i} \qquad \text{if } x_i < x < x_{i+1}.$$

Now that we have introduced the basic concepts of similarity relations and their relation to fuzzy sets, we are able to discuss in more detail the conceptual background of fuzzy sets.

Using gradual membership degrees is the fundamental principle of fuzzy sets. Similarity relations are based on the fundamental concept of indistinguishability or similarity. The unit interval serves as the domain for gradual memberships as well as for similarity degrees. The values between 0 and 1 are interpreted in a mere intuitive way. There is no clear definition what exactly a membership or similarity degree of 0.8 or 0.9 means or what the quantitative difference is, apart from the fact that 0.9 is greater than 0.8.

Similarity relations with respect to the Łukasiewicz t-norm can be induced by pseudometrics. The concept of metric or distance is, at least in the case of the real line, elementary and does not need any further explanation or motivation.

In this sense, similarity relations, induced by the canonical metric—perhaps with some additional suitable scaling—can be understood as an elementary concept where the similarity degrees are interpreted dual to the concept of distances or metrics.

Fuzzy sets can be understood as concepts derived from similarity relations in the sense of extensional hulls of points or sets.

References

1. Dubois D, Prade H (1988) Possibility theory. Plenum Press, New York, NY, USA
2. Dubois D, Prade H (2007) Possibility theory. Scholarpedia 2(10):2074
3. Klawonn F, Castro JL (1995) Similarity in fuzzy reasoning. Mathware Soft Comput 2:197–228. University of Granada, Granada, Spain
4. Klawonn F (1994) Fuzzy sets and vague environment. Fuzzy Sets Syst 66:207–221. Elsevier, Amsterdam, Netherlands
5. Kruse R, Gebhardt J, Klawonn F (1994) Foundations of fuzzy systems. Wiley, Chichester, United Kingdom

Approximate Reasoning

19

Approximate reasoning is based on the idea of sacrificing soundness or completeness for a better understandability of the results and for a significant speed-up of the reasoning. When pursuing such approximate reasoning approaches, however, it is important to be critical not only about appropriate application domains, but also about the quality of the resulting approximate reasoning procedures.

19.1 Linguistic Variables

The concept of a linguistic variable was essential to the development of **fuzzy set theory**. The idea is to use fuzzy sets for the quantification of imprecise terms that appear in our languages [3].

For example, the everyday-life statement "the completion date is late" implies that a linguistic variable COMPLETION DATE takes on the linguistic value *late*. Other values of this linguistic variable might be linguistic values such as *early, normal,* and *late*. Also more complicated linguistic expressions such as *normal or late, very late,* and *not late* are often used.

In order to formalize the concept of a linguistic variable and to make a computerization possible, a formal language (with a syntax) of these values is needed. Furthermore, we have to add meaning to its values (a semantic). For that purpose, we assign to every linguistic expression a suitable fuzzy set.

A linguistic variable is defined as a quintuple:

$$(V, T(V), X, G, M),$$

where V is the name of the variable; $T(V)$ is the set of linguistic expressions of V, also called the term set; X is the universe of discourse; G is a grammar with which the linguistic expressions are formed; and M is a mapping that assigns to each linguistic expression its meaning, which is a fuzzy set on the universe of discourse.

© Springer Nature Switzerland AG 2022
R. Kruse et al., *Computational Intelligence*, Texts in Computer Science,
https://doi.org/10.1007/978-3-030-42227-1_19

The number of different values determines the "granularity" of the representation. It is recommended to assign fuzzy sets to the linguistic expressions in such a way that the fuzzy sets form a fuzzy partition.

Example 19.1 Assume that we want to analyze subjective weather observations. Linguistic variables are always used in a special context. In this example, the climatic attribute is used. Let V be the linguistic variable CLOUDING. The universe of discourse $X = [0, 1]$ describes the possible clouding degrees, i.e., the percentage to which the sky is covered with clouds. The term set $T(V)$ includes elementary linguistic values (*cloudless, clear, fair, cloudy*, ...), and parameterized linguistic expressions (*approximately A, exactly A, between A and B*). It is also allowed to use operators for combining linguistic expressions and to use the values *fair or cloudy, very clear*, and *not cloudless*.

For that purpose, we define the syntax by using a grammar G that specifies the formal language $T(V)$ of all linguistic expressions. Typically a grammar consists of a set of rules with which more complex expressions can be generated, e.g.,

$$A, B \in T(V) \implies (A \text{ or } B) \in T(V),$$
$$A \in T(V) \implies (\text{very } A) \in T(V).$$

The semantic of an elementary linguistic expression A is defined by fuzzy sets μ_A of the universe of discourse X, i.e., $\mu_A : X \to [0, 1]$. For defining the meaning of a composed linguistic expressions we use fuzzy operators. For the linguistic operator *or* we use the fuzzy union operator max: $\mu_{A \text{ or } B}(x) := \max\{\mu_A(x), \mu_B(x)\}$. For modeling the "linguistic hedge" *very A* the membership degrees of μ_A are squared: $\mu_{\text{very } A}(x) := \mu_A(x)^2$. Note that $\mu_{\text{very } A}(x) < \mu_A(x)$ wherever $\mu_A(x) < 1$.

19.2 Computing with Words

Linguistic variables play a central role for computing with words and perceptions [4,5]. In this approach, the objects of computation are words, perceptions, and propositions drawn from a natural language. One way to handle this problem is to transform it from the word/perception level by using linguistic variables to a fuzzy representation. Having solved the problem with fuzzy methods on the fuzzy level, the result is transformed back to the linguistic level. An advantage of this procedure is the interpretability of the results, which is important in several areas.

Example 19.2 A typical application of this method is the analysis of linguistic data. Suppose an observer described the clouding degree for several years by using linguistic expressions, and we want to analyze this sample with descriptive statistics methods. For this purpose, we use the linguistic variable $V = $ CLOUDING introduced in Example 19.1. The given sample is the annual linguistic values $A_1, \ldots, A_n$ and their associated fuzzy sets $\mu_1, \ldots, \mu_n$.

Suppose we are interested in the mean of the linguistic sample. It is impossible to calculate this quantity on the linguistic level. However, using the corresponding fuzzy values this can be easily done via the extension principle, see Chap. 16. The arithmetic mean of real numbers is a function from $\mathbb{R}^n$ to $\mathbb{R}$ defined by

$$\text{mean}(x_1, \ldots, x_n) := \frac{x_1 + \cdots + x_n}{n}.$$

The extension of this function to the fuzzy level gives a function from $\mathcal{F}(\mathbb{R})^n$ to $\mathcal{F}(\mathbb{R})$ that assigns the value $\text{mean}(\mu_1, \ldots, \mu_n)$ to the input $(\mu_1, \ldots, \mu_n)$. This value is a fuzzy set of the real line defined as follows:

$$\text{mean}(\mu_1, \ldots, \mu_n)(y) :=$$

$$\sup \left\{ \min \left\{ \mu_1(x_1), \ldots, \mu_n(x_n) \right\} \;\middle|\; (x_1, \ldots, x_n) \in \mathbb{R}^n : \frac{x_1 + \ldots + x_n}{n} = y \right\}.$$

The mean is described by a fuzzy set. It can be approximated by a suitable linguistic expression of the term set $T(V)$. This "linguistic approximation" can be automated by heuristic search algorithms.

19.3 Generalized Logical Inference

Logical inference can be understood as the process of deriving some facts from other fact by using special rules which imitate rules of human thinking. This formal process is called deduction. The most important deduction method is called *modus ponens*. Formally, it is often written as follows:

$$\begin{array}{ll} implication & A \rightarrow B \\ fact & A \\ \hline conclusion & B \end{array}$$

This reasoning scheme says the following: if we know that the fact A holds true and it is known also the fact B follows from the fact A, then we can conclude that the fact B is also true. Such reasoning schemes (so-called syllogisms) can be traced back to the Greek philosopher Aristotle (384–322 BC). Often the implication is written in the form of the rule "if A then B".

Example 19.3 Let WEATHER be a linguistic variable with the four values *heavy rain, rain, little rain,* and *no rain* and let MOISTURE OF ROAD be another linguistic variable with the two linguistic expressions *wet* and *dry*.

Suppose we know for sure that the rule "If WEATHER = *rain*, then MOISTURE OF ROAD = *wet*" holds, and that the fact WEATHER = *rain* is true. In that situation we would conclude that MOISTURE OF ROAD = *wet*.

The task of an automatic deduction system is to answer queries on the basis of facts and rules. A simple query could be "Is is true that the road is wet?". This query can be easily answered by logical programming systems (such as PROLOG) with classical propositional logic.

In real applications, the experts often have to handle imprecise facts and imprecise if-then rules. Suppose two linguistic variables CAR SPEED and NOISE LEVEL IN THE CAR, the imprecise rule

$$If\ the\ CAR\ \ SPEED\ is\ very\ high,$$
$$then\ the\ NOISE\ \ LEVEL\ \ IN\ \ THE\ \ CAR\ is\ high.$$

and the fact

$$The\ CAR\ \ SPEED\ is\ high.$$

are known. How to answer the query

$$What\ is\ the\ NOISE\ \ LEVEL\ \ IN\ \ THE\ \ CAR?$$

A typical answer of a human is

$$The\ NOISE\ \ LEVEL\ \ IN\ \ THE\ \ CAR\ is\ medium.$$

This result is plausible, but it cannot be derived easily with classical logic. With a fuzzy logic it is very easy to formalize an "approximate" reasoning scheme that solves this task. Users tend to like that, because it is often better to have a good approximate answer than to have no answer.

The idea for solving the problem is to represent imprecise if-then rules by fuzzy relations. With this relational representation we can use the techniques already presented in Chap. 17 to support approximate reasoning tasks.

Let $(U, T(U), X, G_U, M_U)$ and $(V, T(V), Y, G_V, M_V)$ denote two linguistic variables. Consider the approximate rule

$$if\ U = A\ then\ V = B.$$

This rule is interpreted as an approximate logical implication between the fuzzy sets $\mu = M_U(A)$ of X and $\nu = M_V(B)$ of Y that can be represented by a fuzzy relation on $X \times Y$. In most cases, the Gödel implication, the Łukasiewicz implication, or the Zadeh implication is used for defining the fuzzy relation ρ. In Example 17.13, the Gödel implication is chosen.

In order to answer the query, we have to calculate the image of μ under ρ according to Definition 17.2. This is the fuzzy set $\rho[\mu]$ of Y, defined by

$$\rho[\mu](y) := \sup\big\{\min\{\rho(x, y), \mu(x)\} \mid x \in X\big\}.$$

The answer to the query is a fuzzy set, which could additionally be interpreted by a suitable linguistic expression of the term set $V(Y)$ by linguistic approximation.

19.4 Approximation of Functions Using Linguistic If-Then Rules

Linguistic if-then rules are often used to approximate the graph of an unknown function $f : X_1 \times \cdots \times X_n \rightarrow Y$. Typical examples can be found in control theory, such as the following rule with three linguistic variables:

If the TEMPERATURE *is very high and the* PRESSURE *is low*,
then the INFLOW *is high*.

By substituting the linguistic expression in all rules by the corresponding fuzzy sets we obtain a fuzzy if-then rule system. For simplicity, we restrict ourselves in this chapter to the approximation of functions with one input and one output. Then, a fuzzy rule system looks like this:

$$R_1 : \text{If } x \text{ is } \mu_1 \text{ then } y \text{ is } \nu_1.$$

$$\vdots$$

$$R_r : \text{If } x \text{ is } \mu_r \text{ then } y \text{ is } \nu_1.$$

There are several interpretations of such a fuzzy rule base which lead to different fuzzy relations.

19.4.1 Approximation of Functions by Using Rules as Constraints

Consider the imprecise rule "if $x \in [3, 4]$ then $y \in [5, 6]$" as a constraint (Fig. 19.1).
  Then the gray area is considered as a forbidden area for the unknown function. In case of more rules of the form "if $x \in M_i$ then $y \in N_i$" the intersection of these constraints forms a corridor

$$R = \bigcap_{i=1}^{r}(M_i \times N_i \cup (M_i^C \times Y))$$

for the function from X to Y as can be seen in Fig. 19.2.
  This constraint based view of a rule systems can be easily generalized to fuzzy rules of the form "if $x \in \mu$ then $y \in \nu$", since this view is consistent with the logical approach described in the Chaps. 17.4 and 19.3: Each rule is represented by a fuzzy

Fig. 19.1 A rule considered
as a constraint

Fig. 19.2 A set of rules as a constraint

implication relation, and the rule systems are represented by the intersection of k fuzzy relations.

A single rule provides a (fuzzy) constraint on the set of possible values. Since all rules are considered to be correct (true), all fuzzy constraints specified by the rules have to be satisfied, i.e., the fuzzy sets resulting from the single rules have to be intersected with each other. If r rules of the form

$$R_i: \text{ If } x \text{ is } \mu_{R_i}, \text{ then } y \text{ is } \nu_{R_i} \qquad (i = 1, \dots, r)$$

are given and the input is $x = a$, the output fuzzy set of a logic-based controller is

$$\mu_{\mathcal{R},a}^{\text{out, logic}} : Y \to [0, 1], \qquad y \mapsto \min_{i \in \{1, \dots, r\}} \{\llbracket a \in \mu_{R_i} \to y \in \nu_{R_i} \rrbracket\}.$$

Here, we still have to choose a truth function for the implication $\to$. With the Gödel implication we obtain

$$\llbracket a \in \mu_{R_i} \to y \in \nu_{R_i} \rrbracket = \begin{cases} \nu_{R_i}(y) & \text{if } \nu_{R_i}(y) < \mu_{R_i}(a) \\ 1 & \text{otherwise,} \end{cases}$$

and the Łukasiewicz implication leads to

$$\llbracket a \in \mu_{R_i} \to y \in \nu_{R_i} \rrbracket = \min\{1 - \nu_{R_i}(y) + \mu_{R_i}(a), 1\}.$$

In contrast to the Gödel implication, which can lead to discontinuous output fuzzy sets, the output fuzzy sets of the Łukasiewicz implication are always continuous, provided that the involved fuzzy sets are continuous (as real-valued functions).

So far, we have only considered one input variable. When we have to deal with more than one input variable in the rules, we have to replace the value $\mu_{R_i}(a)$ by

$$\llbracket a_1 \in \mu_{R_i}^{(1)} \wedge \dots \wedge a_n \in \mu_{R_i}^{(n)} \rrbracket$$

for the input vector $(a_1, \dots, a_n)$. For the conjunction, we have to choose a suitable t-norm as its truth function, for example, the minimum, the Łukasiewicz t-norm, or the algebraic product.

19.4.2 Approximation of Functions by Solving Fuzzy Relational Equations

We interpret a rule "μ implies ν" as a fuzzy relation ρ that satisfies the fuzzy relational equation

$$\mu \circ \rho = \nu.$$

Which fuzzy relation should we choose? We know from the treatment of fuzzy relational equations: If there is a solution for ρ, then the greatest solution is the Gödel implication. In case of the existence of a solution, the Cartesian product of μ and ν is also a (rather small) solution, and that there are, in general, several minimal solutions.

If a suitable fuzzy relation ρ is chosen, then it can be applied by using the composition to other inputs as well, especially to crisp inputs (real numbers or intervals) whose characteristic functions may be viewed as fuzzy sets. As a result, we obtain a controller with (fuzzy) inputs and fuzzy outputs.

Now consider the following system of m rules with n inputs and one output:

$$\text{If } x_1 \text{ is } \mu_1^1 \text{ and } x_2 \text{ is } \mu_2^1 \text{ and } \ldots \text{ and } x_n \text{ is } \mu_n^1 \text{ then } y \text{ is } \nu^1.$$
$$\vdots$$
$$\text{If } x_1 \text{ is } \mu_1^m \text{ and } x_2 \text{ is } \mu_2^m \text{ and } \ldots \text{ and } x_n \text{ is } \mu_n^m \text{ then } y \text{ is } \nu^m.$$

We interpret this set of rules as a system of fuzzy relational equations:

$$\min \{\mu_1^i, \ldots, \mu_n^i\} \circ \rho = \nu^i, \quad i = 1, \ldots, m.$$

From Sect. 17.6, we know already that if a solution ρ exists, then the minimum of all Gödel implications is the greatest solution. The basic principle to solve a system of relational equation consists in determining the Gödel implication separately for each relational equation, and subsequently finding a combined solution for the entire system by calculating the minimum of the solutions for the single equations.

Suppose that a solution exists. Then *assembling* the overall solution from *small* solutions of the single relational equations is another way to find a convenient solution. A reasonable approach to derive such an assembled solution is to calculate the maximum of the Cartesian products of the single solutions. We obtain as the general solution

$$\max \left\{ \min \left\{ \mu_1^i, \ldots, \mu_n^i \right\} \times \nu^i \right\}.$$

The resulting fuzzy relation corresponds exactly to a Mamdani fuzzy controller (see Chap. 20). In this sense, the Mamdani controller can also be motivated on the basis of fuzzy relational equations.

19.4.3 Approximation of Functions by Interpolation Between Fuzzy Points

Fuzzy rules can be used to give an approximate description of the course of an unknown function f. The idea is to characterize the behavior of f in some regions by rules.

Consider the imprecise rule "if $x \in [3, 4]$ then $y \in [5, 6]$" as an "imprecise" point $[3, 4] \times [5, 6]$ (Fig. 19.3).

In this interpretation, the gray area describes a "patch" where the graph of the unknown function from X to Y "passes". In case of more rules "if $x \in M_i$ then $y \in N_i$" the union

$$S = \bigcup_{i=1}^{r} (M_i \times N_i)$$

of these patches form a patch rug for the function (Fig. 19.4).

This view of a rule systems can be easily generalized to fuzzy rules of the form "if x is μ then y is ν", since this view is consistent with the similarity-based approach described in Chap. 18: each rule is represented by a fuzzy point, and we are trying to find a function that follows the patches. This is in essence an interpolation view that turns out to be useful for control tasks that are studied in the next chapter.

Example 19.4 Suppose we have five rules

if μ_1 then ν_1, if μ_2 then ν_2, if μ_3 then ν_3, if μ_4 then ν_4, if μ_5 then ν_5.

Fig. 19.3 A rule as an imprecise point

Fig. 19.4 A rule set as an imprecise description of the graph of a function

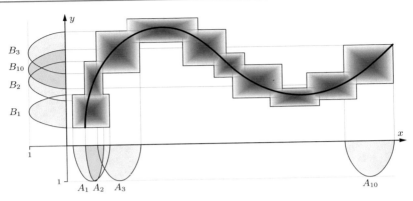

Fig. 19.5 Sketch of a fuzzy function F, represented by a fuzzy function plot approximately describing a crisp function f. Below the x-axis and left of the y-axis the fuzzy sets A_i and B_i, $i = 1, \ldots, k$ are shown

Each of the rules $\mu \to \nu$ is interpreted as a fuzzy relation

$$\rho : X \times Y \to [0, 1], \quad (x, y) \mapsto \min\{\mu(x), \nu(y)\}.$$

The fuzzy sets used in these rules are shown next to the x-axis and y-axis in Fig. 19.5. The union (with t-norm max) of the five fuzzy relations is sketched as a gray-level image. It shows the approximate graph of the unknown function. A plausible function is also drawn. This method is used for several applications, e.g., for fuzzy control.

References

1. Nguyen HT, Walker CL, Walker EA (2019) A first course in fuzzy logic. CRC Press, Boca Raton
2. Novák V, Perfilieva I, Dvorák A (2016) Insight into fuzzy modeling. Wiley, New York
3. Zadeh LA (1975) The concept of a linguistic variable and its application to approximate reasoning. Inf Sci 8(3):199–249
4. Zadeh LA (1999) From computing with numbers to computing with words. From manipulation of measurements to manipulation of perceptions. IEEE Trans Circuits Syst I: Fundam Theory Appl 46(1):105–119
5. Zadeh LA (2001) A new direction in AI: toward a computational theory of perceptions. AI Mag 22(1):73

Fuzzy Control

20

Control theory deals with the control of dynamical systems in engineered processes and machines. The objective is to develop a model or an algorithm governing the application of system inputs to drive the system to a desired state, while minimizing any delay, overshoot, or steady-state error and ensuring a level of control stability, often with the aim to achieve a degree of optimality. To do this, a controller with the requisite corrective behavior is required. This controller monitors the controlled process variable, and compares it with the reference or set point. The difference between actual and desired value of the process variable, called the error signal, is applied as feedback to generate a control action to bring the controlled process variable to the same value as the set point. Other aspects which are also studied are controllability, observability, and performance. Control theory provides reliable solutions if the process can be modeled in an appropriate way, in most cased by using differential equations. Finding realistic models is often complicated, sometimes the description is complex and difficult to understand.

On the other hand, control of processes can often manually realized by experts who know from their experience how the process should be controlled. There are lots of applications in industry, where the results are satisfactory, despite of the fact that the knowledge about the behavior of the process is only approximate. A popular example is the "stabilization of an inverted pendulum": the model of the process can be described by nonlinear differential equations. On the other hand, children can balance a stick with a hand without knowing the exact mathematical model.

Fuzzy control is an effective tool to deal with complex processes. It was originally introduced as a "model-free control" approach based on expert's experience and knowledge. In real applications, we see a growing interest in fuzzy-model-based approaches, since their design methods allow to analyze stability, performance, and robustness of the resulting systems.

© Springer Nature Switzerland AG 2022
R. Kruse et al., *Computational Intelligence*, Texts in Computer Science,
https://doi.org/10.1007/978-3-030-42227-1_20

20.1 Mamdani Fuzzy Controller

Fuzzy control was initiated in [16]. A Mamdani fuzzy controller uses approximate reasoning to control a technical process. Figure 20.1 shows the general architecture.

In a control loop, the controlled system iteratively receives precise inputs and produces precise outputs. The fundamental method in control theory is to use observed errors for control. At each time t, the desired output of the system, the *set-point* $w(t)$, is compared with the actual output $y(t)$ of the process. Then the error $e(t) = w(t) - y(t)$ is calculated, in more complex controllers also the change of the error $\dot{e}(t)$, and also the second derivation of the error $\ddot{e}(t)$ are calculated or at least approximated. Then the output $u(t)$ of the fuzzy controller is defined as a function of the error information. The control action $u(t)$ altered by a disturbance $d(t)$ is the input to the controlled system.

The knowledge base contains the necessary information about the measured and manipulated variables, the used fuzzy sets (parameters, partitions), fuzzy operators (t-norms), the association of fuzzy sets to linguistic terms, and the fuzzy rules (inference type).

The fuzzy controller is a classical function where the input is the error and, depending on the type of the controller, additionally the change of the error and the change of the change of the error. The output is the control action. This function is defined according to the approximate reasoning method described in Sect. 19.4.3. The basic task is to find a function that fits to a linguistic rule system.

In analogy to classical controllers, there are several types of fuzzy controllers. A fuzzy PD controller uses only linguistic rules of the form

> If the error e is μ_e and the change $\dot{e}$ of the error is $\mu_{\dot{e}}$,
> then the output u is ν_u.

Whereas a fuzzy PI controller uses rules of the form

Fig. 20.1 Fuzzy control loop

If the error e is μ_e and the change $\dot{e}$ of the error is $\mu_{\dot{e}}$,
then the change $\dot{u}$ of the output is $\nu_{\dot{u}}$.

Other fuzzy controller types such as the P fuzzy controller and PID fuzzy controller
are defined in the same way.

Formally, a fuzzy controller is a function that approximates linguistic rules. For
that purpose the input is *fuzzified*, approximate inferences based on the fuzzy rules
are performed, and finally the results of the inference process are aggregated via
defuzzification into a crisp output.

The **Mamdani controller** is based on a finite set $\mathcal{R}$ of if-then rules $R \in \mathcal{R}$ of the
form

$$R : \qquad \text{If } x_1 \text{ is } \mu_R^{(1)} \text{ and } \ldots \text{ and } x_n \text{ is } \mu_R^{(n)}$$
$$\text{then } y \text{ is } \mu_R. \tag{20.1}$$

$x_1, \ldots, x_n$ are input variables of the controller and y is the output value. Usually,
the fuzzy sets $\mu_R^{(i)}$ or μ_R stand for linguistic values, that is, for imprecise concepts
like "about zero", "of average height" or "negative small" which, for their part, are
represented by fuzzy sets. In order to simplify the notation, we do not distinguish
between membership functions and linguistic values that they represent.

How to precisely interpret the rules is essential for understanding the Mamdani
controller. Although the rules are formulated in terms of if-then statements, they
should not be understood as logical implications but in the sense of a piecewise
defined function. If the rule base $\mathcal{R}$ consists of the rules $R_1, \ldots, R_r$, we should
understand it as a piecewise definition of a fuzzy function, that is,

$$f(x_1, \ldots, x_n) \approx \begin{cases} \mu_{R_1} \text{ if } x_1 \approx \mu_{R_1}^{(1)} \text{ and } \ldots \text{ and } x_n \approx \mu_{R_1}^{(n)}, \\ \vdots \\ \mu_{R_r} \text{ if } x_1 \approx \mu_{R_r}^{(1)} \text{ and } \ldots \text{ and } x_n \approx \mu_{R_r}^{(n)}. \end{cases} \tag{20.2}$$

This is similar to the pointwise specification of a crisp function presented in
Sect. 19.4.3. This function is defined over a product space of finite sets in the form

$$f(x_1, \ldots, x_n) \approx \begin{cases} y_1 \text{ if } x_1 = x_1^{(1)} \text{ and } \ldots \text{ and } x_n = x_1^{(n)}, \\ \vdots \\ y_r \text{ if } x_1 = x_r^{(1)} \text{ and } \ldots \text{ and } x_n = x_r^{(n)}. \end{cases} \tag{20.3}$$

We obtain the graph of this function by

$$\text{graph}(f) = \bigcup_{i=1}^{r} \left(\hat{\pi}_1(\{x_i^{(1)}\}) \cap \ldots \cap \hat{\pi}_n(\{x_i^{(n)}\}) \cap \hat{\pi}_Y(\{y_i\}) \right). \tag{20.4}$$

"Fuzzifying" this equation using the minimum for the intersection and the maximum
(supremum) for the union, the fuzzy graph of the function described by the rule set $\mathcal{R}$
is the fuzzy set

$$\mu_{\mathcal{R}} : X_1 \times \ldots \times X_n \times Y \to [0, 1],$$

$$(x_1, \ldots, x_n, y) \mapsto \sup_{R \in \mathcal{R}} \{\min\{\mu_R^{(1)}(x_1), \ldots, \mu_R^{(n)}(x_n), \mu_R(y)\}$$

or

$$\mu_{\mathcal{R}} : X_1 \times \ldots \times X_n \times Y \to [0, 1],$$

$$(x_1, \ldots, x_n, y) \mapsto \max_{i \in \{1, \ldots, r\}} \{\min\{\mu_{R_i}^{(1)}(x_1), \ldots, \mu_{R_i}^{(n)}(x_n), \mu_{R_i}(y)\}$$

in the case of a finite rule base $\mathcal{R} = \{R_1, \ldots, R_r\}$.

If a concrete input vector $(a_1, \ldots, a_n)$ for the input variables $x_1, \ldots, x_n$ is given, then we obtain the fuzzy set

$$\mu_{\mathcal{R},a_1,\ldots,a_n}^{\text{output}} : Y \to [0, 1], \qquad y \mapsto \mu_{\mathcal{R}}(a_1, \ldots, a_n, y)$$

as the fuzzy "output value."

The fuzzy set $\mu_{\mathcal{R}}$ can be interpreted as fuzzy relation over the sets $X_1 \times \ldots \times X_n$ and Y. Therefore, the fuzzy set $\mu_{\mathcal{R},a_1,\ldots,a_n}^{\text{output}}$ corresponds to the image of the one-element set $\{(a_1, \ldots, a_n)\}$ or its characteristic function under the fuzzy relation $\mu_{\mathcal{R}}$. So, in principle, instead of a sharp input vector we could also use a fuzzy set as input. For this reason, it is very common to call the procedure of feeding concrete input values to a fuzzy controller as **fuzzification**, i.e., the input vector $(a_1, \ldots, a_n)$ is transformed into a fuzzy set which is nothing else than a representation of a one-element set in terms of its characteristic function.

Fuzzification can also be interpreted in another way. In Chap. 17, we have seen that we can obtain the image of a fuzzy set under a fuzzy relation computing the cylindrical extension of the fuzzy set, intersecting the cylindrical extension with the fuzzy relation and projecting the result to the output space. In this sense, we can understand the cylindrical extension of the measured tuple or the corresponding characteristic function as a the fuzzification procedure which is necessary for intersecting it with the fuzzy relation.

Figure 20.2 shows this procedure. In order to make a graphical representation possible, we consider only one input and output variable. Three rules are shown in the figure:

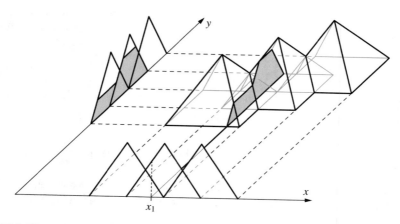

Fig. 20.2 The projection of the input value x_1 onto the output axis y

$$\text{If } x \text{ is } A_i, \text{ then } y \text{ is } B_i, \qquad (i = 1, 2, 3).$$

The fuzzy relation $\mu_{\mathcal{R}}$ is represented by the three pyramids. If the input value x is given, the cylindrical extension of $\{x\}$ defines a plane cutting through the pyramids. The projection of this cutting plane onto the y-axis (pointing from the front to the back) yields us the fuzzy set $\mu_{\mathcal{R},x}^{\text{output}}$ which describes the desired output value (in fuzzy terms).

We can illustrate the computation of the output value by the following scheme. Figure 20.3 shows two rules of a Mamdani controller with two input variables and one variable. At first, we consider only one of the two rules, say the rule R. The degree to which the antecedent is satisfied for the given input values is determined by taking the minimum of the membership degrees of the corresponding fuzzy sets. Then, the fuzzy set in the consequent of the rule is "cutoff" at this level. This means that the membership degree to the actual output fuzzy set is the minimum of the membership degrees to the fuzzy set in the antecedent of the rule and the firing strength of the rule.

If the firing strength of rule is 1, then we exactly obtain the fuzzy set in the consequent as result, i.e., $\mu_R = \mu_{R,a_1,\dots,a_n}^{\text{output}}$. When the rule is not applicable, meaning that the firing strength is 0, we obtain $\mu_{R,a_1,\dots,a_n}^{\text{output}} = 0$, i.e., based on this rule nothing can be said about the output value.

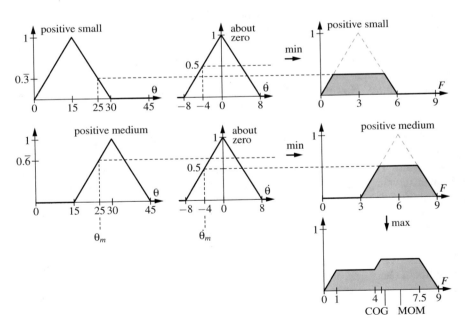

Fig. 20.3 Illustration of the Mamdani controller computation scheme

All other rules are treated in the same way—Fig. 20.3 shows just two rules—so, for every rule R we obtain a fuzzy set $\mu_{R,a_1,\ldots,a_n}^{\text{output}}$. In real applications, usually only a few rules have a non-zero firing strength at the same time and thus contribute to the final output fuzzy set. In the second step, these output fuzzy sets derived from single rules have to be joint to a single fuzzy set which characterizes the described output value.

In order to explain in which way this aggregation is carried out, we reconsider the interpretation of the rule base of a fuzzy controller in the sense of a piecewise definition of a function (cf. Eq. 20.2). For a crisp piecewisely defined function, the r different cases have to be disjoint or, if they overlap they must provide the same output because otherwise the function value is not well defined. We assume that each single case describes a "function value" in the form of a set for every input value: if a considered input value satisfies the condition of the corresponding case, then we obtain a one-element set with the specified function value. Otherwise, we obtain the empty set for this case. With this interpretation the actual function value or the one-element set which contains this function value is given by the union of the sets resulting from the single cases.

For this reason, the (disjoint) union of the fuzzy sets $\mu_{R,a_1,\ldots,a_n}^{\text{output}}$ resulting from the single rules should be computed for the overall fuzzy output. In order to compute the union of fuzzy sets we use a t-conorm, e.g., the maximum. Thus, we obtain

$$\mu_{\mathcal{R},a_1,\ldots,a_n}^{\text{output}} = \max_{R \in \mathcal{R}} \{\mu_{R,a_1,\ldots,a_n}^{\text{output}}\} \tag{20.5}$$

as the final output fuzzy set given both the rule base $\mathcal{R}$ and the input vector $(a_1, \ldots, a_n)$. In this way, the two rules shown in Fig. 20.3 lead to the output fuzzy set which is also shown there.

In order to obtain a single crisp output value, the output fuzzy set has to be *defuzzified*. A large variety of heuristic defuzzification strategies has been proposed in the literature. However, without specifying more precisely how the fuzzy sets and rules should be interpreted, defuzzification remains a matter of pure heuristics. Therefore, we restrict our considerations to a few defuzzification strategies.

In order to understand the fundamental idea of defuzzification applied in the case of Mamdani controllers, we consider the output fuzzy set determined in Fig. 20.3 once again. The fuzzy sets in the consequents of the two rules are interpreted as imprecise values. In the same sense, the resulting output fuzzy set is an imprecise description of the desired output value. Intuitively, the output fuzzy set in Fig. 20.3 can be understood in the sense that we should favor a larger output value, but should also consider a smaller output value with a lower preference. This interpretation is justified by the fact that the first rule, voting for a larger output value, fires with a higher degree than the second rule that points to a lower output value. Therefore, we should choose an output value that is large, but not too large, i.e., a compromise of the proposed outputs of the two rules, however, putting a higher emphasis on the first rule.

A defuzzification strategy which satisfies this criterion is the center of gravity (COG) method. The output value of this method is the center of mass (or, to be precise, its projection onto the output axis) of the area under the output fuzzy set, i.e.,

$$COG(\mu_{\mathcal{R},a_1,...,a_n}^{output}) = \frac{\int_Y \mu_{\mathcal{R},a_1,...,a_n}^{output} \cdot y \, dy}{\int_Y \mu_{\mathcal{R},a_1,...,a_n}^{output} dy}. \tag{20.6}$$

This method requires the implicit condition that the functions $\mu_{\mathcal{R},a_1,...,a_n}^{output} \cdot y$ and $\mu_{\mathcal{R},a_1,...,a_n}^{output}$ are integrable which is always satisfied, provided that the membership functions appearing in the consequents of the rules are chosen in a reasonable way, e.g., when they are continuous.

20.2 Design of a Mamdani Fuzzy Controller

The basic idea of Mamdani fuzzy control is to represent the process states by means of linguistic variables and to exploit these variables as inputs to control rules. Hence, this controller enables to incorporate the expert's skills and experience through a set of fuzzy if-then rules [1,18,24,33]. Thanks to this particular feature, the effectiveness of this type of fuzzy control has been clearly proven if no acceptable mathematical model is available for the controlled system and human operators play a crucial role in the control process and can provide qualitative control rules in terms of fuzzy logic sentences. The idea for stability analysis has been to consider the Mamdani fuzzy controller as a nonlinear controller. In essence, fuzzy control design is recast as a nonlinear control approach using absolute stability theory, sliding mode control, and adaptive fuzzy control [4,23]. This allows Mamdani fuzzy control to easily achieve multi-objective goals by simply setting some fuzzy control rules under one criterion and others under a different performance criterion. The coordination between different control objectives can be performed by fuzzy reasoning. For these reasons, to this date, Mamdani controller has successfully been applied to a large number of industrial processes [21,24].

When choosing the fuzzy sets for the input variables, one should make sure that the domain of each input variable is completely covered, this means, for every possible value there exists at least one fuzzy set to which it has a non-zero membership degree. Otherwise, the fuzzy controller cannot determine an output value for this input value.

The fuzzy sets should represent imprecise values or ranges. Therefore, convex fuzzy sets are preferable. Triangular and trapezoidal membership functions are especially suitable, because they have a simple parametric representation and determining the membership degrees can be achieved with low computational effort. In ranges where the controller has to react very sensitively to small changes of an input value we should choose very narrow fuzzy sets in order to distinguish between the values well enough. We should, however, take into account that the number of possible rules grows very fast with the number of fuzzy sets. If we have k_i fuzzy sets for the ith input variable, then a complete rule base that assigns to all possible combinations of input fuzzy sets an output fuzzy set contains $k_1 \cdot \ldots \cdot k_n$ rules when we have n input variables. If we have four input values with only five fuzzy sets for each of them, then we already get 625 rules.

Concerning the choice of the fuzzy sets for the output variable similar constraints as for the input variables should be considered. They should be convex and in the ranges, where a very exact output value is important, narrow fuzzy sets should be used. Additionally, the choice of fuzzy sets for the output value strongly depends on the defuzzification strategy. It should be noted that the defuzzification of asymmetrical triangular membership functions of the form $\Lambda_{x_0-a,x_0,x_0+b}$ with $a \neq b$ does not always correspond to what we might expect. If only one single rule fires with the degree 1 and all others with degree 0, we obtain the corresponding fuzzy set in the consequent of this rule as the output fuzzy set of the controller before defuzzification. If this is an asymmetrical triangular membership function $\Lambda_{x_0-a,x_0,x_0+b}$, we have $\mathrm{COG}(\Lambda_{x_0-a,x_0,x_0+b}) \neq x_0$, i.e., not the point where the triangular membership function reaches the membership degree 1.

Another problem of the center of gravity method is that it can never return a boundary value of the interval of the output variable. This means, the minimum and maximum value of the output domain are not accessible for the fuzzy controller. A possible solution of this problem is to extend the fuzzy sets beyond the interval limits for the output variable. However, in this case, we have to ensure that the controller will never yield an output value outside the range of permitted output values.

For the design of the rule base, completeness is an important issue. We have to make sure that for any possible input vector there is at least one rule that fires. This does not mean that for every combination of fuzzy sets of input values we have to formulate a rule with these fuzzy sets in the antecedent. On the one hand, a sufficient overlapping of the fuzzy sets guarantees that there will still be rules firing, even if we have not specified a rule for all possible combinations of input fuzzy sets. On the other hand, there might be combinations of input values that correspond to a state which the system can or must never reach. For these cases, it is not necessary to specify rules. We should also avoid contradicting rules. Rules with identical antecedents and different consequents should be avoided.

The Mamdani controller we have introduced here is also called max-min controller because of Eq. 20.5 for computing the output fuzzy set $\mu^{\mathrm{output}}_{\mathcal{R},a_1,\ldots,a_n}$. Maximum and minimum were used to calculate the union and the intersection, respectively, in Eq. 20.4.

Of course, also other t-norms and t-conorms can be considered instead of minimum or maximum. In applications the product t-norm is often preferred and the bounded sum $s(\alpha, \beta) = \min\{\alpha + \beta, 1\}$ is sometimes used as the corresponding t-conorm. The disadvantage of minimum and maximum is the idempotency property. The output fuzzy set $\mu^{\mathrm{output}}_{R,a_1,\ldots,a_n}$ of a rule R depends only on the input variable for which the minimum membership degree to the corresponding fuzzy set in the antecedent is obtained. A change of another input variable will not affect the output of this rule, unless the change is large enough to let the membership degree of this input value to its corresponding fuzzy set drop below the membership degree of the other input values.

If the fuzzy sets $\mu^{\mathrm{output}}_{R,a_1,\ldots,a_n}$ of different rules support a certain output value to some degree, the aggregation based on the maximum will only take the largest of

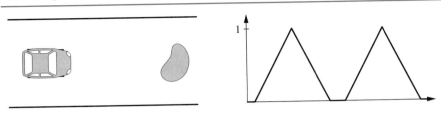

Fig. 20.4 A car driving towards an object (left) and possible output fuzzy set of a steering controller consisting of two disjoint fuzzy (right)

these membership degrees into account. It might be desirable that such an output value obtains a higher support than another one that is supported to the same degree, but only by a single rule. In this case, a t-conorm like the bounded sum should be preferred to the maximum.

In some commercial programs, weighted rules are allowed. The resulting output fuzzy set of a rule is then multiplied by the assigned weight. Such weights increase the number of the free parameters of a fuzzy controller. But the same effect can be achieved directly by an adequate choice of the fuzzy sets in the antecedent or the consequent of the rule without any weights. In most cases, weights make it more difficult to interpret a fuzzy controller.

Another problem of the disjoint fuzzy cases is illustrated in Fig. 20.4 which shows an output fuzzy set where the defuzzification is difficult.

Should we interpolate between the two fuzzy values represented by the triangles like it would be done by the center of gravity method? That would mean that a defuzzification yields a value whose membership degree to the output fuzzy set is 0 which surely does not fit to our intuition. Or do the two triangles represent two alternative output values from which we have to choose one? For example, the illustrated fuzzy set could be the output fuzzy set of a controller which is supposed to drive a car around an obstacle. Then, the fuzzy set says that we have to evade the obstacle either to the left or to the right, but not keep on driving straight ahead. The latter solution would be proposed by the center of gravity method, leading to the effect that the car will bump into the obstacle. The interpretation of two alternative outputs is contradictory to the underlying philosophy of the Mamdani controller as a piecewise definition of a fuzzy function, because, in this case, the function is not well defined, since two fuzzy values are assigned at the same time to only one input.

A general defuzzification strategy has to fulfill two tasks at the same time. It has to turn a fuzzy set into a set and it has to choose one (fuzzy) value from a set of (fuzzy) values. It is not obvious in which order these two steps should be carried out. For example, the fuzzy set from Fig. 20.4 could be defuzzified by first choosing one of the two fuzzy values it represents, i.e., by choosing one of the two triangles. In the second step, the corresponding triangle, representing only a single fuzzy value must be turned into a suitable crisp value. We could also exchange the order of these two steps. Then we would first turn the fuzzy set into a crisp set that in this case would contain two elements. Afterwards, we have to pick one of the values from this crisp set.

At this point, we explain some other defuzzification strategies and their properties in order to better understand the issue of defuzzification [26]. The *mean of maxima* (MOM) method is a very simple defuzzification strategy which chooses as output value the mean values of the values with maximum membership degree to the output fuzzy set. This method is rarely applied in practice because for symmetrical fuzzy sets it leads to a discontinuous controller behavior. Given the input values, the output value depends only on the output fuzzy set which belongs to the rule with the highest firing strength—provided that there are not two or more rule which by chance fire to the same degree and which have different fuzzy sets in their consequents. When we use symmetrical fuzzy sets in the consequents of the rules in the context of the mean of maxima method, the output value will always be on of the centers of the fuzzy sets, except in the rare case when two or more rules fire with the same maximum degree. Therefore, the rule with the highest firing strength will determine the constant output until another rule takes over. Then the controller output will jump directly or with one intermediate step to the output of the rule that has now the maximum firing strength.

Both the center of area method and MOM will result in the possibly non-desired mean value in the defuzzification problem shown in Fig. 20.4. Therefore, they always choose the right-most value of the values with maximum membership degree to the output fuzzy set. (Alternatively, one can always take the left-most one.) But similar to MOM, it can also lead to discontinuous changes of the output value.

The center of gravity method requires relatively high computational costs and does not have the interpolation properties we would expect. To illustrate this, we consider a Mamdani controller with the following rule base:

If x is "about 0", then y is "about 0"
If x is "about 1", then y is "about 1"
If x is "about 2", then y is "about 2"
If x is "about 3", then y is "about 3"
If x is "about 4", then y is "about 4"

The terms "about 0", ..., "about 4" are represented by fuzzy sets in the form of symmetrical triangular membership functions of width two, i.e., $\Lambda_{-1,0,1}$, $\Lambda_{0,1,2}$, $\Lambda_{1,2,3}$, $\Lambda_{2,3,4}$, and $\Lambda_{3,4,5}$, respectively. It seems that these rules describe the function $y = x$, i.e., a straight line. Applying the center of gravity method the resulting function matches the straight line only at the values $0, 0.5, 1, 1.5, \ldots, 3.5$ and 4. At all of the other points, the control function differs a little bit from the simple straight line as shown in Fig. 20.5.

These and other undesirable effects, which, for example, can appear using asymmetrical membership functions in the consequent, can be avoided by using rules with a crisp value in the consequent. For describing the input values we still use fuzzy sets, but the outputs are specified not in terms of fuzzy sets, but as a single value for each rule. In this case, the defuzzification is also very simple. We take the weighted mean of the rule outputs, i.e., each rule assigns its firing strength as weight to its

Fig. 20.5 Interpolation of a straight line using the center of gravity method

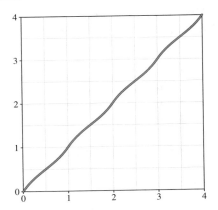

crisp output value in the consequent. In this way, we obtain

$$y = \frac{\sum_R \mu_{R,a_1,\ldots,a_n}^{\text{output}} \cdot y_R}{\sum_R \mu_{R,a_1,\ldots,a_n}^{\text{output}}} \tag{20.7}$$

as the output of the fuzzy controller. The rules have the form

$$R: \quad \text{If } x_1 \text{ is } \mu_R^{(1)} \text{ and } \ldots \text{ and } x_n \text{ is } \mu_R^{(n)}, \text{ then } y \text{ is } y_R$$

with crisp output values y_R. $a_1, \ldots, a_n$ are the measured input values of the input variables $x_1, \ldots, x_n$ and $\mu_{R,a_1,\ldots,a_n}^{\text{output}}$ denotes—as it used to do so—the firing strength of the rule R with these input values.

20.3   Mamdani Controller and Similarity Relations

When we introduced the Mamdani controllers we have already seen that the fuzzy rules there represent fuzzy points on the graph of the control or transfer function which is described by the controller. Based on the concept of similarity relations as discussed in Chap. 18, fuzzy sets, such as the ones appearing in the Mamdani controller, can be interpreted as fuzzy points. Here we discuss this interpretation of the Mamdani controller and its consequences in detail.

20.3.1   Interpretation of a Mamdani Controller Using Similarity Relations

At first, we consider a given Mamdani controller. Let us assume that the fuzzy sets defined for the input and output domains satisfy the conditions of Theorem 18.2 or even better of Theorem 18.3. In this case, we can derive similarity relations such that the fuzzy sets can be interpreted as extensional hulls of single points.

Example 20.1 For a Mamdani controller with two input variables x and y and one output variable z, we use the left fuzzy partition from Fig. 20.9 for the input variables and the right one for the output variable. The rule base consists of four rules.

R_1: If x is *small*   and y is *small*   then z is *positive*
R_2: If x is *medium* and y is *small*   then z is *zero*
R_3: If x is *medium* and y is *big*   then z is *zero*
R_4: If x is *big*   and y is *big*   then z is *negative*

These fuzzy partitions satisfy the conditions of Theorem 18.3 such that suitable scaling functions can be found. For the left fuzzy partition in Fig. 20.9, the scaling function is

$$c_1 : [0, 6] \to [0, \infty), \qquad x \mapsto \begin{cases} 0.25 \text{ if } 0 \leq x < 4 \\ 0.5 \ \text{ if } 4 \leq x \leq 6, \end{cases}$$

for the right fuzzy partition, it is

$$c_2 : [-3, 3] \to [0, \infty), \qquad x \mapsto \frac{1}{3}.$$

The fuzzy sets *small, medium, big, negative, zero,* and *positive* correspond to the extensional hulls of the points $0, 4, 6, -3, 0,$ and 3, respectively, if the similarity relations induced by the above scaling functions are considered.

Then, the four rules say that the graph of the function described by the controller should pass through the corresponding points $(0, 0, 3)$, $(4, 0, 0)$, $(4, 6, 0)$, and $(6, 6, -3)$.   □

The interpretation on the basis of similarity relations in the example above specifies four points on the graph of the control function as well as additional information encoded in the similarity relations. The construction of the whole control function is therefore an interpolation task. We want to find a function which passes through the given points and maps similar values to similar values in the sense of the similarity relations.

If we, for example, want to find a suitable output value for the input $(1, 1)$, we can see that $(0, 0)$ is the most similar input pair for which we know an output value, i.e., 3 according to rule R_1. The similarity degree of 1 to 0 is nothing but the membership degree of the value 1 to the extensional hull of 0, that is, to the fuzzy set *small*, which is 0.75. The input $(1, 1)$ also similar to the input pair $(4, 0)$ in rule R_2, however, much less similar than to $(0, 0)$. The similarity degree of 1 to 4 is 0.25, the similarity of 1 to 0 is again 0.75. Thus, the output value for $(1, 1)$ should be quite similar to the output value 3 for the input $(0, 0)$ (rule R_1) but also a little bit similar to the output value 0 of the input $(4, 0)$ (rule R_2).

So far we have left the question open, how the two similarity degrees, which we obtain for the two components x and y of the input pair, should be aggregated. A t-norm is a suitable choice in this case, e.g., the minimum. As an example, let us determine how well the output value 2 fits to the input $(1, 1)$. In order to answer this question, we have to compute the similarity degrees to the points defined by the four rules.

Each of these similarity degrees is simply the minimum of the membership degrees of the three components 1, 1, and 2 to the three fuzzy sets in the corresponding rule. In this way, for the point induced by rule R_1 we obtain a similarity degree of $2/3 = \min\{3/4, 3/4, 2/3\}$. For R_2 the result is $0.25 = \min\{1/4, 3/4, 2/3\}$. For the two rules R_3 and R_4, the similarity degree is 0 because already the considered input values do not fit to these rules at all. The similarity degree according to the four given points or rules corresponds to the best possible value, that is, $2/3$. Repeating the calculation for any value z in the input/output tuple $(1, 1, z)$, we obtain a function

$$\mu : [-3, 3] \rightarrow [0, 1]$$

for the given input $(1, 1)$. This function can be interpreted as a fuzzy set over the output range. When we compare this computation scheme with the calculations carried for the Mamdani controller, we obtain exactly the output fuzzy set (cf. Eq. 20.5) of the corresponding Mamdani controller.

20.3.2 Construction of a Mamdani Controller Using Similarity Relations

Instead of determining the scaling functions or similarity relations and the corresponding interpolation points indirectly from a Mamdani controller we can also specify them directly and then derive the corresponding Mamdani controller. The advantage of this procedure is on the one hand that we can no longer specify arbitrary fuzzy sets, but only fuzzy sets which are consistent in a certain way. And on the other hand the interpretation of the scaling functions and especially of the interpolation points that have to be specified is very simple. The scaling functions can be interpreted in the sense of Example 18.2 on page 432. In ranges where the control is very sensitive to a change of the value, we should distinguish very exactly between the single values. So we should choose a larger scaling factor. For ranges where the exact values are less important for the control, a small scaling factor is sufficient. This leads to very narrow fuzzy sets in ranges where very precise control actions have to be taken, while larger fuzzy sets are used in regions where a rough control is sufficient. In this way we can explain, why fuzzy sets near the operating point of a controller are usually chosen to be very narrow in contrast to the other ranges. The operating point requires very exact control actions in most cases. In contrast to this, if the process is far from the operating point, first of all rough and strong actions have to be carried out to force the process closer to its operating point.

Using scaling function it is also obvious which additional hidden assumptions are used for the design of a Mamdani controller. The fuzzy partitions are defined for the single domains and are then used in the rules. In the sense of scaling functions this means that the scaling functions for the different domains are assumed to be independent from each other. The similarity of the values in a domain does not depend on the values in other domains. In order to illustrate this issue, we consider a simple PD controller which uses as input variable the error—the difference from the reference variable—and the change of the error. For a very small error value, it is obviously very important for the controller whether the change of the error is slightly positive or slightly negative. Therefore, we would choose a larger scaling factor near zero in the domain of the change of error, resulting in narrow fuzzy sets. On the other hand, if the error value is large, it is not very important whether the change of error tends to be slightly positive or slightly negative. First of all we have to reduce the error. This fact speaks in favor for a small scaling factor near zero in the domain of the change of error, i.e., we should use wider fuzzy sets. In order to solve this problem, there are three possibilities:

1. We specify a similarity relation in the product space of the domains error and change of error which models the dependence described above. But this seems to be quite difficult because the similarity relation in the product space cannot be specified using scaling functions.
2. We choose a high scaling factor near zero in the domain of the change of error. However, in this case, it might be necessary to specify many almost identical rules for the case of a large error value. The rules differ only in the fuzzy sets for the change of the error, like

> If error is *big* and change is *positive small*, then *y* is *negative*.
> If error is *big* and change is *zero*,   then *y* is *negative*.
> If error is *big* and change is *negative small*, then *y* is *negative*.

3. We define rules that do not use all input values, for instance

> If error is *big*, then *y* is *negative*.

The interpretation of the Mamdani controller in the sense of similarity relations also explains why it is very convenient that adjacent sets of a fuzzy partition meet each other at the level of 0.5. A fuzzy set is a imprecise value which will be used for the interpolation points. If a certain value has been specified, this value provides also some information about similar values, where similar should be interpreted in the sense of the corresponding similarity relation. So, as long as the similarity degree does not drop to zero, there is still some information available. Therefore, once the similarity degree is zero, new information is needed, i.e., a new interpolation point must be introduced. Following this principle, the fuzzy sets will exactly overlap at the level of 0.5. Of course, the interpolation points could be chosen closer, provided sufficiently detailed knowledge about the process is available. This would lead to strongly overlapping fuzzy sets. However, this does not make sense, when we desire a representation of the experts knowledge on the process that is as compact as possible. Therefore, new interpolation points are only introduced, when they are needed.

Even if a Mamdani controller does not satisfy the conditions of one of the Theorems 18.2 or 18.3, calculating the similarity relations from Theorem 18.1 will still provide important pieces of information. The corresponding similarity relations always exist and make the fuzzy sets extensional, even though they might not be interpreted as extensional hulls of points. The following results are taken from [11,12]:

1. The output of a Mamdani controller does not change if we use instead of a crisp input value its extensional hull as input.
2. The output fuzzy set of a Mamdani controller is always extensional.

This means that we cannot overcome the indistinguishability that is inherently coded in the fuzzy partitions.

20.4 Takagi–Sugeno Controller

Takagi–Sugeno controllers (TS models) [29,31] use rules of the form

$$R: \quad \text{If } x_1 \text{ is } \mu_R^{(1)} \text{ and } \dots \text{ and } x_n \text{ is } \mu_R^{(n)}, \text{ then } y = f_R(x_1, \dots, x_n). \quad (20.8)$$

In the same manner as in the case of the Mamdani controller (cf. Eq. 20.1), the input values in the rules are described by fuzzy sets. However, using a TS model, the consequent of a single rule consists no longer of a fuzzy set, but determines a function with the input variables as arguments. The basic idea is that the corresponding function is a good local control function for the fuzzy region that is described by the antecedent of the rule. For instance, if we use linear functions, the desired input/output behavior of the controller is described locally (in fuzzy regions) by linear models. At the boundaries between single fuzzy regions, we have to interpolate in a suitable way between the corresponding local models. This is done by

$$y = \frac{\sum_R \mu_{R,a_1,\dots,a_n} \cdot f_R(x_1, \dots, x_n)}{\sum_R \mu_{R,a_1,\dots,a_n}}, \quad (20.9)$$

where $a_1, \dots, a_n$ are the measured input values of the input variables $x_1, \dots, x_n$, and $\mu_{R,a_1,\dots,a_n}$ denotes the firing strength of rule R which results given these input values.

A special case of the TS models is the variant of the Mamdani controller where the fuzzy sets in the consequent of the rules are replaced by constant values and the output value is calculated by Eq. 20.7. In this case, the functions f_R are constant.

These singleton-type fuzzy systems are also called piecewise multi-affine (PMA) systems, since its input–output relation with respect to an affine-in-control system is found to be a multi-affine function. This type of fuzzy systems is also called piecewise bilinear because in the two-dimensional case, the output can be expressed as $y = ax_1 + bx_2 + cx_1x_2 + d$, for some scalars a, b, c, and d.

Mamdani-type fuzzy systems are linguistically understandable, since fuzzy variables are used in both the premises and the consequents. However, TS fuzzy systems do not use linguistic variables as only functional membership functions are used without labels. PMA systems are a simplified special case of both previous types of fuzzy systems. Note that all three types of fuzzy systems are known to have general approximation capability for any nonlinear function.

In order to maintain the interpretability of a TS controller in terms of local models f_R for fuzzy regions, a strong overlap of these regions should be avoided, since otherwise the interpolation formula given in Eq. 20.9 can completely blur the single models and mix them together into one complex model that might have a good control behavior, but looses interpretability completely. As an example, we consider the following rules:

$$\text{If } x \text{ is "very small," then } y = x$$
$$\text{If } x \text{ is "small," then } y = 1$$
$$\text{If } x \text{ is "large," then } y = x - 2$$
$$\text{If } x \text{ is "very large," then } y = 3$$

First, the terms "very small," "small," "large," and "very large" are modeled by the four non-overlapping fuzzy sets in Fig. 20.6. In this case, the four functions or local models $y = x$, $y = 1$, $y = x - 2$, and $Y = 3$ defined in the rules are reproduced exactly as shown in Fig. 20.6. If we choose only slightly overlapping fuzzy sets, the TS model yields the control function in Fig. 20.7. Finally, Fig. 20.8 shows the result of the TS model which uses fuzzy sets that overlap even more.

We can see that the TS model can lead to slight overshoots as shown in Fig. 20.7, even if the fuzzy sets overlap just slightly. If we choose an overlap of the fuzzy sets

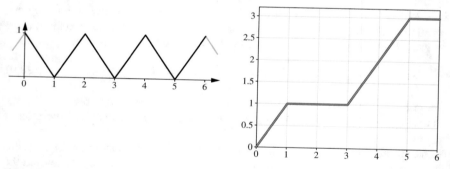

Fig. 20.6 Four non-overlapping fuzzy sets: exact reproduction of the local models

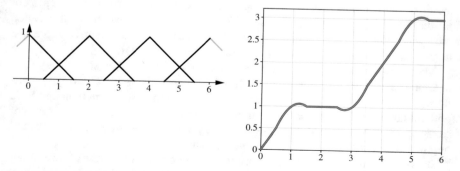

Fig. 20.7 Four slightly overlapping fuzzy sets: slight mixture of the local models

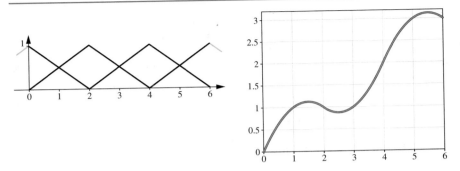

Fig. 20.8 Four strongly overlapping fuzzy sets: almost complete mix of the local models

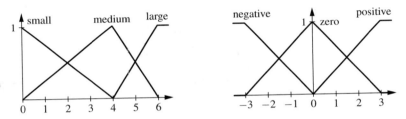

Fig. 20.9 Two fuzzy partitions

typical for Mamdani controllers, then the single local model is not visible anymore as shown in Fig. 20.8.

A suitable way to avoid this undesirable effect is to use trapezoidal membership functions instead of triangular ones, when working with TS models. When we choose trapezoidal membership functions in such a way that an overlap occurs only at the edges of the trapezoidal functions, the corresponding local models are reproduced exactly in the regions where membership degree is 1.

The TS fuzzy model enables a systematic framework to deal with the control design of nonlinear dynamical systems:

Example 20.2 Consider the continuous-time system

$$\dot{x}(t) = f\big(x(t), u(t)\big),$$

where $x(t) \in \mathbb{R}^n$ is the state vector, $u(t) \in \mathbb{R}^m$ is the control input, and the origin $0_n \in \mathbb{R}^n$ is an equilibrium of the unforced system, i.e., $f(0_n, 0_m) = 0_n$ vanishes. Let the vector field f be sufficiently smooth.

This continuous-time nonlinear system can be represented by a TS systems using the following rules:

$$R_i\colon \text{If } z_1(t) \text{ is } \mu_1^i \text{ and } \ldots \text{and } z_p(t) \text{ is } \mu_p^i,$$
$$\text{then } \dot{x}(t) = A_i\, x(t) + B_i\, u(t),$$

where R_i denotes the ith fuzzy inference rule, r is the number of inference rules, μ_j^i with $1 \leq i \leq r$ and $1 \leq j \leq p$ are the fuzzy sets, and (A_i, B_i) the state-space matrices of appropriate dimensions of the ith local model. The vector of premise variables is defined as $z(t) := \big(z_1(t), \ldots, z_p(t)\big)$. Using the center of gravity method for defuzzification, the TS fuzzy model can be represented in the compact form

$$\dot{x}(t) = \sum_{i=1}^{r} h_i\big(z(t)\big) \cdot \big(A_i \, x(t) + B_i \, u(t) \big),$$

where the normalized membership functions $h_i\big(z(t)\big)$ are defined as

$$h_i\big(z(t)\big) := \frac{\omega_i\big(z(t)\big)}{\sum_{i=1}^{r} \omega_i\big(z(t)\big)}, \quad \omega_i\big(z(t)\big) := \prod_{j=1}^{p} \mu_j^i\big(z_j(t)\big).$$

Note that these functions satisfy the following convex sum property:

$$0 \leq h_i(z) \leq 1, \quad \sum_{i=1}^{r} h_i(z) = 1, \quad \sum_{i=1}^{r} \dot{h}_i(z) = 0.$$

Model-based stability analysis can be performed for TS fuzzy systems using conventional Lyapunov-based approaches [32].

There are many methods for choosing the shape of the membership functions used in the TS rule systems, see, for instance, [14]. Nevertheless, more powerful methodologies could be created with the use of interval type-2 TS fuzzy modeling techniques [5, 17]. These fuzzy models allow to deal with imprecise grades of membership. Hence, type-2 TS fuzzy systems are very useful in the cases where exact MFs are difficult to be chosen and/or there is a need to cope with large amounts of imprecision/uncertainty.

References

1. Babuška R (2012) Fuzzy modeling for control. Springer Science
2. Barto AG, Sutton RS, Anderson CW (1983) Neuronlike adaptive elements that can solve difficult learning control problems. IEEE Trans Syst Man Cybern 13(5):834–846. IEEE Press, Piscataway, NJ, USA
3. Berenji HR, Khedkar P (1992) Learning and tuning fuzzy logic controllers through reinforcements. IEEE Trans Neural Netw 3(5):724–740. IEEE Press, Piscataway, NJ, USA
4. Feng G (2006) A survey on analysis and design of model-based fuzzy control systems. IEEE Trans Fuzzy Syst 14(5):676–697
5. Hagras H (2007) Type-2 FLCs: a new generation of fuzzy controllers. IEEE Comput Intell Mag 2(1):30–43
6. Hopf J, Klawonn F (1994) Learning the rule base of a fuzzy controller by a genetic algorithm. In: Kruse R, Gebhardt J, Palm R (eds) Fuzzy systems in computer science. Vieweg, Braunschweig, Germany, pp 63–74

7. Jang J-SR (1993) ANFIS: adaptive-network-based fuzzy inference system. IEEE Trans Syst Man Cybern 23(3):665–685. IEEE Press, Piscataway, NJ, USA

8. Kaelbling LP, Littman MH. Moore AW (1996) Reinforcement learning: a survey. J Artif Intell Res 4:237–285. AI Access Foundation and Morgan Kaufman Publishers, El Segundo/San Francisco, CA, USA

9. Kinzel J, Klawonn F, Kruse R (1994) Modifications of genetic algorithms for designing and optimizing fuzzy controllers. In: Proceedings of IEEE conference on evolutionary computation (ICEC'94, Orlando, FL). IEEE Press, Piscataway, NJ, USA, pp 28–33

10. Klawonn F (1992) On a Lukasiewicz logic based controller. In: Proceedings of international seminar on fuzzy control through neural interpretations of fuzzy sets (MEPP'92). Åbo Akademi, Turku, Finland, pp 53–56

11. Klawonn F, Castro JL (1995) Similarity in fuzzy reasoning. Mathware Soft Comput 2:197–228. University of Granada, Granada, Spain

12. Klawonn F, Kruse R (2004) The inherent indistinguishability in fuzzy systems. In: Lenski W (ed) Logic versus Approximation: Essays Dedicated to Michael M. Richter on the Occasion of his 65th Birthday. Springer, Berlin, Germany, pp 6–17

13. Klawonn F, Novák V (1996) The relation between inference and interpolation in the framework of fuzzy systems. Fuzzy Sets Syst 81:331–354. Elsevier, Amsterdam, Netherlands

14. Lam HK (2018) A review on stability analysis of continuous-time fuzzy-model-based control systems: from membership-function-independent to membership-function-dependent analysis. Eng Appl Artif Intell 67:390–408

15. Lee M, Takagi H (1993) Integrating design stages of fuzzy systems using genetic algorithms. In: Proceedings of IEEE international conference on fuzzy systems (San Francisco, CA). IEEE Press, Piscataway, NJ, USA, pp 612–617

16. Mamdani EH, Assilian S (1975) An experiment in linguistic synthesis with a fuzzy logic controller. Int J Man-Mach Stud 7:1–13. Academic Press, Waltham, MA, USA

17. Mendel JM (2007) Type-2 fuzzy sets and systems: an overview. IEEE Comput Intell Mag 2(1):20–29

18. Michels K, Klawonn F, Kruse R, Nürnberger A (2006) Fuzzy control: fundamentals, stability and design of fuzzy controllers. In: Studies in fuzziness and soft computing, vol. 200. Springer, Berling/Heidelberg, Germany

19. Nauck DD, Klawonn F, Kruse R (1997) Foundations of neuro-fuzzy systems. Wiley, Chichester

20. Nauck D, Kruse R (1993) A fuzzy neural network learning fuzzy control rules and membership functions by fuzzy error backpropagation. In: Proceedings of IEEE international conference on neural networks (ICNN'93, 1993, San Francisco, CA) . IEEE Press, Piscataway, NJ, USA, pp 1022–1027

21. Nguyen AT, Taniguchi T, Eciolaza L, Campos V, Palhares R, Sugeno M (2019) Fuzzy control systems: past, present and future. IEEE Comput Intell Mag 14(1):56–68

22. Nürnberger A, Nauck DD, Kruse R (1999) Neuro-fuzzy control based on the NEFCON-model: recent developments. Soft Comput 2(4):168–182. Springer, Berlin/Heidelberg, Germany

23. Palm R, Driankov D, Hellendoorn H (1997) Model based fuzzy control: fuzzy gain schedulers and sliding mode fuzzy controllers. Springer, Berlin

24. Precup R-E, Hellendoorn H (2011) A survey on industrial applications of fuzzy control. Comput Ind Eng 62(3):213–226

25. Riedmiller M, Spott M, Weisbrod J (1999) FYNESSE: A hybrid architecture for selflearning control. In: Cloete I, Zurada J (eds) Knowledge-Based Neurocomputing. MIT Press, Cambridge, MA, USA, pp 291–323

26. Runkler TA (1997) Selection of appropriate defuzzification methods using application specific properties. IEEE Trans Fuzzy Syst 5(1):72–79

27. Runkler TA (2012) Kernel based defuzzification. In: Moewes C, Nürnberger A (eds) Computational intelligence in intelligent data analysis. Springer, Berlin/Heidelberg, Germany, pp 61–72

28. Runkler TA, Glesner M (1993) A set of axioms for defuzzification strategies—towards a theory of rational defuzzification operators. In: Proceedings of 2nd IEEE international conference on fuzzy systems (FUZZ-IEEE'93, San Francisco, CA). IEEE Press, Piscataway, NJ, USA, pp 1161–1166

29. Sugeno M (1985) An introductory survey of fuzzy control. Inf Sci 36:59–83. Elsevier, New York, NY, USA

30. Sutton RS, Barto AG (1998) Reinforcement learning: an introduction. MIT Press, Cambridge, MA, USA

31. Takagi T, Sugeno M (1985) Fuzzy Identification of systems and its applications to modeling and control. IEEE Trans Syst Man Cybern 15:116–132. IEEE Press, Piscataway, NJ, USA

32. Tanaka K, Sugeno M (1990) Stability analysis of fuzzy systems using Lyapunov's direct method. In: Proceedings of NAFIPS 90, vol 1. Toronto, Canada, pp 133–136

33. Yager RR, Filev DP (1994) Essentials of fuzzy modeling and control. Wiley, New York

34. Zadeh LA (1971) Towards a theory of fuzzy systems. In: Kalman RE, de Claris N (eds) Aspects of networks and system theory. Rinehart and Winston, New York, USA, pp 469–490

35. Zadeh LA (1972) A Rationale for Fuzzy Control. J Dyn Syst Meas Control 94(1):3–4. American Society of Mechanical Engineers (ASME), New York, NY, USA

36. Zadeh LA (1973) Outline of a new approach to the analysis of complex systems and decision processes. IEEE Trans Syst Man Cybern 3:28–44. IEEE Press, Piscataway, NJ, USA

Hybrid Systems for Tuning and Learning Fuzzy Systems

21

In applications, often a combination of different methods is used. In this Chapter, we study such hybrid systems. In the first part of the chapter, we combine the advantages of fuzzy systems (its interpretability) with one of the neural networks (its ability to learn from data). These systems are called neuro-fuzzy systems. There are lots of different cooperative and integrative architectures [15,22], especially for control, classification, and function approximation. There is a very high number of successful applications in different areas [21]. We focus on architectures that are often used for tuning/learning fuzzy controller from data. In the second part of this chapter, we study the application of evolutionary algorithms to the design of fuzzy systems. In such a hybridization, the evolutionary algorithms are used as meta-heuristic based optimization techniques for fuzzy systems. The challenge is to use the learning and adaptation capabilities of evolutionary optimization, but at the same time to keep the inherent interpretability of fuzzy rule system [7]. In this chapter, we focus on evolving/tuning fuzzy controllers by using a genetic algorithm.

21.1 Neuro-Fuzzy Control

Promising approaches for optimization of existing fuzzy controllers and to learn a fuzzy system from scratch are techniques that combine fuzzy systems and learning methods of artificial neural networks. These methods are called neuro-fuzzy systems.

Fuzzy systems provides the ability of interpreting a fuzzy controller and to introduce a priori knowledge whereas and neural networks contribute its capability of learning and therefore the ability of automatic optimization or automatic generation of the whole controller. Due to the possibility to introduce a priori knowledge by fuzzy rules into the system, we expect to strongly reduce both the optimization time and the amount of training data required to train the system in comparison with pure neural network-based controllers. If we have only some training data available, the

© Springer Nature Switzerland AG 2022
R. Kruse et al., *Computational Intelligence*, Texts in Computer Science,
https://doi.org/10.1007/978-3-030-42227-1_21

prior knowledge might be even necessary in order to be able to create a controller. Furthermore, with neuro-fuzzy systems we are—in principle—able to learn a controller and then to interpret its control strategy by analyzing the learned fuzzy rules and fuzzy sets and, if necessary, revise them.

Essentially, neuro-fuzzy controllers can be divided in cooperative and hybrid models. In cooperative models, the neural network and the fuzzy controller operate separately. The neural network generates (offline) or optimizes (online, i.e., during control) some parameters Hybrid models try to unite the structures of neural networks and fuzzy controllers. Thus, a hybrid fuzzy controller can be interpreted as neural network and can even be implemented with the help of a neural network. Hybrid models have the advantage of an integrated structure which does not require any communication between the two different models. Therefore, the system is principally able to learn online as well as offline and thus these approaches have become much more accepted than the cooperative models [16].

The idea of hybrid methods is to map fuzzy sets and fuzzy rules to a neural network structure. This principle is explained in the following. For that, we reconsider the fuzzy rules R_i of a Mamdani controller defined in Eq. 20.1, i.e.,

$$R_i : \quad \text{If } x_1 \text{ is } \mu_i^{(1)} \text{ and } \ldots \text{ and } x_n \text{ is } \mu_i^{(n)}$$
$$\text{then } y \text{ is } \mu_i,$$

or the fuzzy rules R_i' of a TS controller (cf. Eq. 20.8), i.e.,

$$R_i' : \quad \text{If } x_1 \text{ is } \mu_i^{(1)} \text{ and } \ldots \text{ and } x_n \text{ is } \mu_i^{(n)}, \text{ then } y = f_i(x_1, \ldots, x_n).$$

The activation $\tilde{a}_i$ of these rules can be calculated by a t-norm. With given input values $x_1, \ldots, x_n$, we obtain for $\tilde{a}_i$ using the minimum t-norm

$$\tilde{a}_i(x_1, \ldots, x_n) = \min\{\mu_i^{(1)}(x_1), \ldots, \mu_i^{(n)} x_n\}.$$

One way to represent such a rule with a neural network is to replace each real-valued connection weight w_{ji} from an input neuron u_j to an inner neuron u_i by a fuzzy set $\mu_i^{(j)}$. Therefore, an inner neuron represents a rule and the connections from the input units represent the fuzzy sets of the antecedents of the rules. In order to calculate the rule activation of the inner neurons, we just have to modify their network input functions. If we, for example, choose the minimum as t-norm, then we define

$$\text{net}_i = \min\{\mu_i^{(1)}(x_1), \ldots, \mu_i^{(n)} x_n\}$$

as network input function.

If we finally replace the activation function of the neuron by the identity, then the activation of the neuron corresponds to the rule activation $\tilde{a}_i$. Therefore, the neuron can be used directly to compute the rule activity of any fuzzy rule. A graphical representation of such a structure for a rule with two inputs is shown in Fig. 21.1 (on the left).

We obtain another representation if the fuzzy sets of the antecedent are modeled as separate neurons. For that, the network input function is the identity and the activation function of the neuron is the membership function of the fuzzy set. Thus, the neuron calculates for each input the degree of membership to the fuzzy set represented by the

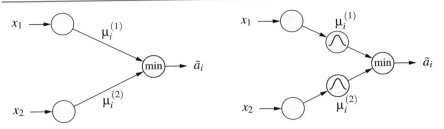

Fig. 21.1 Example of a neural network for calculating the activation of the antecedent of a fuzzy rule: Modeling of the fuzzy sets as weights (left) and as activation functions of a neuron (right)

activation function. For this representation, we need two layers of neurons in order to model the antecedent of a fuzzy rule (see Fig. 21.1 (on the right)). The advantage of this representation is that the fuzzy sets can be directly used in several rules in order to ensure the interpretability of the entire rule base. In this representation, the network weights w_{ij} in the connections from the fuzzy sets to the rule neuron are initialized with 1 and are regarded as constant. The weights of the input values to the fuzzy sets can be used for scaling the input values.

If the evaluation of the entire rule base shall to be modeled as well, then we have to decide whether a Mamdani or a TS controller is needed. For the TS controller, various realizations are possible. But, in principle, for each rule we get one more unit for evaluating the output function f_i—which will then be implemented as network input function—and it will be connected to all of the input units $(x_1, \ldots, x_n)$. In an output neuron, the outputs of these units will be combined with the rule activations $\tilde{a}_i$ which are calculated by the rule neurons. This output neuron will finally calculate the output of the TS controller with the help of the network input function

$$\text{out} = \frac{\sum_{i=1}^{r} \tilde{a}_i \cdot f_i(x_1, \ldots, x_n)}{\sum_{i=1}^{r} \tilde{a}_i}.$$

The connection weights between the neurons are again constantly 1 and the identity is used as activation function.

For the Mamdani controller, the concrete implementation depends on the chosen t-conorm and the defuzzification method. In any case, a common output neuron combines the activations of the rule neurons and calculates a crisp output value with the help of a modified network input function based on the corresponding fuzzy sets in the consequents of the rules.

The transfer of a fuzzy rule base into a network structure can be summarized by the following steps:

1. For every input variable x_i, we create a neuron of the same denotation in the input layer.
2. For every fuzzy set $\mu_i^{(j)}$, we create a neuron of the same denotation and connect it to the corresponding x_i.
3. For every output variable y_i, we create a neuron of the same denotation.

4. For every fuzzy rule R_i, we create an inner (rule) neuron of the same denotation and we specify a t-norm for calculating the rule activation.
5. Every rule neuron R_i is connected according to its fuzzy rule to the corresponding neurons that represent the fuzzy sets of the antecedent.
6. Mamdani controller: Every rule neuron is connected to the output neuron according to the consequent of its fuzzy rule. As connection weight, we have to choose the consequent of the corresponding fuzzy set. Furthermore, a t-conorm and the defuzzification method have to be integrated adequately into the output neurons. TS controller: For every rule unit, one more neuron is created to calculate the output function. These neurons are connected to the corresponding output neuron. Furthermore, all of the input units are connected to the neurons for the calculation of the output function and all of the rule neurons are connected to the output neuron.

Having described the mapping of a rule base as a network, learning algorithms of artificial neural networks can be applied to this structure. However, the learning methods usually have to be modified due to several reasons. First of all, the network input and activation functions changed. Second, not the real-valued network weights but the parameter of the fuzzy sets have to be learned. In the following sections, we discuss two hybrid neuro-fuzzy systems in more detail. Furthermore, we explain the principles and problems of neuro-fuzzy architectures, especially in terms of applications in system control.

21.1.1 Models with Supervised Learning Methods

Neuro-fuzzy models with supervised learning methods try to optimize the fuzzy sets and—for a TS model—the parameters of the output function of a given rule base with the help of known input–output tuples. Therefore, applying supervised models is convenient if we already have a description of the system to be control but need the control behavior—and thus the rule base—to be more exact. If measured data of the system to be modeled exist (tuple of state, output and control variables), they can be used to retrain the system.

Neuro-fuzzy models with supervised learning methods are also convenient if an existing controller is to be replaced by a fuzzy controller, that is, that measured data of the control behavior of the real controller are available. Here, an existing rule base is also presupposed. The learning methods can then be used in order to optimize the approximation of the original controller. If no initial fuzzy rule base is available describing the system that should be approximated and if an approximate rule base also cannot be created manually, we might use fuzzy clustering methods (see Sect. 22.2) or evolutionary algorithms in order to obtain an initial rule base if measured data for training is available.

In the following, we discuss a typical example for a neuro-fuzzy system with supervised learning, i.e., the ANFIS model. Beside this one, there are several other approaches which are based on similar principles.

The ANFIS Model

In [10], the neuro-fuzzy system *ANFIS* (Adaptive-Network-based Fuzzy Inference System) was developed which by now has been integrated in many controllers and simulation tools. The ANFIS model is based on a hybrid structure, i.e., it can be interpreted as neural network and as fuzzy system. The model uses the fuzzy rules of a TS controller. Figure 21.2 shows an example for a model with the three fuzzy rules

$$R_1: \quad \text{If } x_1 \text{ is } A_1 \text{ and } x_2 \text{ is } B_1 \text{ then } y = f_1(x_1, x_2)$$
$$R_2: \quad \text{If } x_1 \text{ is } A_1 \text{ and } x_2 \text{ is } B_2 \text{ then } y = f_2(x_1, x_2)$$
$$R_3: \quad \text{If } x_1 \text{ is } A_2 \text{ and } x_2 \text{ is } B_2 \text{ then } y = f_3(x_1, x_2)$$

where A_1, A_2, B_1, and B_2 are linguistic expressions which are assigned to the corresponding fuzzy sets $\mu_j^{(i)}$ in the antecedents. The functions f_i in the consequent of the ANFIS model are defined by linear combination of the input variables, i.e., in the example above by

$$f_i = p_i x_1 + q_i x_2 + r_i.$$

Here, we use the product t-norm for the evaluation of the antecedent, i.e., the neurons in the layer 2 calculate activation $\tilde{a}_i$ of rule R_i by

$$\tilde{a}_i = \prod_j \mu_i^{(j)}(x_j).$$

In the ANFIS model, the evaluation of the consequent and the calculation of the output value is split up to layers 3–5. Layer 3 calculates the contribution $\bar{a}_i$ each rule made to creating the total output by taking into account the activations $\tilde{a}_i$. Therefore, the neurons of layer 3 calculate

$$\bar{a}_i = a_i = \text{net}_i = \frac{\tilde{a}_i}{\sum_j \tilde{a}_j}.$$

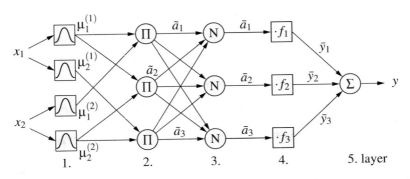

Fig. 21.2 Structure of an ANFIS network with three rules

Then, the neurons of layer 4 calculate the weighed control outputs based on the input variables x_k and the relative control activations $\bar{a}_i$ of the previous layer:

$$\bar{y}_i = a_i = \text{net}_i = \bar{a}_i f_i(x_1, \ldots, x_n).$$

Finally, the output neuron u_{out} of layer 5 calculates the total output of the network and the fuzzy system, respectively:

$$y = a_{\text{out}} = \text{net}_{\text{out}} = \sum_i \bar{y}_i = \frac{\sum_i \bar{a}_i f_i(x_1, \ldots, x_n)}{\sum_i \bar{a}_i}.$$

For learning, the ANFIS model uses supervised learning. Therefore, for training, it is necessary that a sufficient amount of input/output tuples exists. Based on these training data, model parameters are determined, i.e., the parameters of the fuzzy sets and the parameters of the output function f_i.

As learning method, different approaches are suggested in [10]. Besides the pure gradient descent method which is analogous to the backpropagation method for neural networks (see Sect. 5.6), also combinations with methods for solving overdetermined linear equation systems (i.e., the method of least square (estimate)) are suggested. Here, the parameters of the antecedents (i.e., fuzzy sets) are determined with a gradient descent method and the parameters of the consequents (i.e., linear combination of the input variables) with the least squares method. The learning occurs in several separated steps where the parameters of the antecedents and of the consequents are alternatingly optimized by fixing the other ones.

In the first step, all input vectors are propagated through the network to layer 3. For every input vector, the control activations are stored. Based on these values, for the parameters of the functions f_i in the consequent, a overdetermined equation system is created.

Let r_{ij} be the parameters of the output function f_i, $x_i(k)$ the input values, $y(k)$ the output value of the kth training pair, and $\bar{a}_i(k)$ the relative control activation. Then, we obtain

$$y(k) = \sum_i \bar{a}_i(k) y_i(k) = \sum_i \bar{a}_i(k) \left(\sum_{j=1}^n r_{ij} x_j(k) + r_{i0} \right), \quad \forall i, k.$$

Therefore, with $\hat{x}_i(k) := [1, x_1(k), \ldots, x_n(k)]^T$, we obtain the overdetermined linear equation system

$$\mathbf{y} = \bar{a}\mathbf{RX}$$

for a sufficient number m of training data ($m > (n+1) \cdot r$, where r is the number of rules and n the number of input variables).

Therefore, the parameters of the linear equation system, which is built in this way,—the parameters of the output function f_i in the matrix $\mathbf{R}$—can be determined with the method of least squares after the propagation of all of the training data. Finally, the error is determined in the output units based on the new calculated output functions and with the help of a gradient descent method the parameters of the fuzzy sets are optimized. The combination of the two methods leads to an improved convergence because the method of least squares already has an optimal solution (i.e., the one with the least error squares) for the parameters of the output function with regard to the initial fuzzy sets.

Unfortunately, the ANFIS model has no restrictions for the optimization of the fuzzy sets in the antecedents, i.e., it is not ensured that the input range will still be covered completely with fuzzy sets after the optimization. Thus, definition gaps can appear. This has necessarily to be checked after optimizing. Fuzzy sets can also can change, independently form each other and can also exchange their order and thus their importance. We have to pay attention to this, especially if an initial rule base was set manually and the controller has to be interpreted afterwards.

21.1.2   Models with Reinforcement Learning

The fundamental idea of the models of reinforcement learning [2] is to determine a controller with knowing as little as possible about the system. The aim is that the learning process can be managed with a minimum of information about the aim of the control. In the extreme case, the learning method gets merely the information whether the system is still stable or the controller had failed.

The main problem of these approaches is that to rework the control action such that it can be used for learning and optimizing the controller. In general, we cannot assume that the last control action has the greatest influence on the current system state. This problem is also called *credit assignment problem* [2], i.e., the problem of assigning to a control action the (long-term) effect which it has on the system.

By now, many models have been suggested in the field of reinforcement learning. All of these are essentially based on the principle of dividing the learning problem into two systems: A criticizing system (critic) and a system which stores a description of the control strategy and applies this to the system (actor). The critic evaluates the current state considering the previous states and control action, evaluates as well the output of the actor based on these information and, if necessary, adapts its control strategy.

The methods suggested by now for using the principles of reinforcement learning are mostly based on the combination with neural nets [11]. Very promising are the methods which use dynamic programming in order to determine an optimal control strategy. For a more detailed discussion about this topic, see e.g., [20].

In the field of neuro-fuzzy systems, there are many approaches. Examples of such approaches are GARIC [3], FYNESSE [18], and the NEFCON model [17] which is shortly presented below.

The NEFCON Model

The principal aim of the NEFCON model (neuro-fuzzy controller) is to detect online a suitable and interpretable rule base with a the least amount of training cycles. Furthermore, it should be possible to bring previous knowledge into the training process in the easiest way in order to speed up the learning process. This is the difference to the most reinforcement learning approaches which try to generate the most optimal controller and therefore lose much time in the long learning phases. Furthermore, the NEFCON model has also heuristic approaches for learning a rule base. This point is different to the most of the other neuro fuzzy systems which, in general, can only be used for optimizing a rule base.

The NEFCON model is a hybrid model of a neuro-fuzzy controller. We assume the definition of a Mamdani controller and obtain the network structure if we—analogously to the description in Sect. 21.1—interpret the fuzzy sets as weights and the measuring and regulating variables as well as the rules as operating units. Then, the net has the structure of a multilayer perceptron and can be interpreted as three-layer fuzzy perceptron. The fuzzy perceptron arises from a perceptron by modeling the weights, the network inputs and the activation of the output unit as fuzzy sets. An example for a fuzzy controller with five rule units, two measuring variables, and one regulating variable is shown in Fig. 21.3.

The inner units $R_1, \ldots, R_5$ represent the rules, the units x_1, x_2, and y the measuring and regulating variables and $\mu_r^{(i)}$ and ν_r the fuzzy sets for the antecedents and the consequents, respectively. The connection with common weights denote equal fuzzy sets. A changing of these weights makes it necessary that all connections with this weight have to be adapted in order to ensure that the same fuzzy sets keep represented by the same weights. Thus, the rule base defined by the network structure can also be formulated in the form of the fuzzy rules listed in Table 21.1.

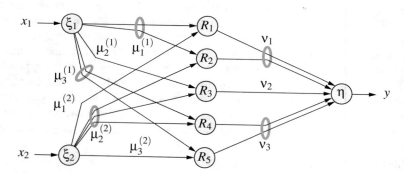

Fig. 21.3 A NEFCON system with two input variables and five rules

Table 21.1 The rule base of the NEFCON system shown in Fig. 21.3

R_1: if x_1 is $\mu_1^{(1)}$ and x_2 is $\mu_1^{(2)}$ then y is v_1

R_2: if x_1 is $\mu_1^{(1)}$ and x_2 is $\mu_2^{(2)}$ then y is v_1

R_3: if x_1 is $\mu_2^{(1)}$ and x_2 is $\mu_2^{(2)}$ then y is v_2

R_4: if x_1 is $\mu_3^{(1)}$ and x_2 is $\mu_2^{(2)}$ then y is v_3

R_5: if x_1 is $\mu_3^{(1)}$ and x_2 is $\mu_3^{(2)}$ then y is v_3

The learning process of the NEFCON model can be divided into two main phases. The first phase is designed to learn an initial rule base, if no prior knowledge about the system is available. Furthermore, it can be used to complete a manually defined rule base. The second phase optimizes the rules by shifting or modifying the fuzzy sets of the rules. Both phases use a fuzzy error e, which describes the quality of the current system state, to learn or to optimize the rule base. The fuzzy error plays the role of the critic element in reinforcement learning models. In addition, the sign of the optimal output value must be known. Thus, the extended fuzzy error E is defined as

$$E(x_1, ..., x_n) = \mathrm{sgn}(y_{opt}) \cdot e(x_1, ..., x_n),$$

with the input $(x_1, ..., x_n)$. At the end of this section, we briefly discuss some methods to describe the system error.

Learning a Rule Base

If for the system which has to be controlled an adequate rule base does not exist or cannot be determined manually, it has to be created by an appropriate rule learning method. Methods to learn an initial rule base can be divided into three classes: Methods starting with an empty rule base, methods starting with a "full" rule base (combination of every fuzzy set in the antecedents with every consequent) and methods starting with a random rule base. In the following, we briefly present algorithms of the first two classes. The algorithms do not require a given fixed learning problem but they try to determine a suitable rule base based on the extended system error E (see also [17]). However, both methods require an appropriate fuzzy partitioning of the input and output variables

If for the system which has to be controlled does not exist an adequate rule base, or it cannot be determined, it has to be created by an appropriate rule learning method. Following we present some algorithms for learning a rule base, they do not need a given fixed learning exercises but based on the system error they try to determine a suitable rule base (see also [17]). Both of the methods need suitable fuzzy partitioning of the measuring and regulating variable ranges (see also the discussion in Sect. 20.3.2).

An Elimination method for Learning a Rule Base

The elimination method starts with a complete overdetermined rule base, that is, the rule base consists of all rules which can be defined by the combination of all fuzzy sets in the initial fuzzy partitions of the input and output variables.

The algorithm can be divided into two phases which are executed during a fixed period of time or a fixed number of iteration steps. During the first phase, rules with an output sign different from that of the optimal output value are removed. During the second phase, a rule base is constructed for each control action by selecting randomly one rule from every group of rules with identical antecedents. The error of each rule (the output error of the whole network weighted by the activation of the individual rule) is accumulated. At the end of the second phase from each group of rule nodes with identical antecedents, the rule with the least error value remains in the rule base. All other rule nodes are deleted. In addition, rules used very rarely are removed from the rule base.

A disadvantage of the elimination method is that it starts with a very large rule base and, thus, for systems with many state variables or fine grained partitions, i.e., many fuzzy sets are used to model the input and output values, it requires much memory and is computationally expensive. Therefore, it is advisable to use the incremental rule learning procedure for larger rule bases.

Incremental Learning the Rule Base

This learning methods starts with an empty rule base. However, an initial fuzzy partitioning of the input and output intervals must be given. The algorithm can be divided into two phases. During the first phase, the rules' antecedents are determined by classifying the input values, i.e., finding that membership function for each variable that yields the highest membership value for the respective input value. Then the algorithm tries to "guess" the output value by deriving it from the current fuzzy error with a heuristics. It is assumed that input patterns with similar error values require similar output (control) values. The so-found rule is inserted into the rule base. In the second phase, we try to optimize the rule consequent by applying the learned rule base to the system and, based on the detected error values, exchanging the fuzzy sets in the consequents, if necessary.

The used heuristics maps the extended error E merely linearly to the interval of the control variable. Thus, we assume a direct dependency between the error and the control variable. This is might be critical, especially for systems which would require an integrator for control, that is, which require a control value unequal 0 in order to achieve the control goal or to keep it. This heuristic also assumes that the error in not determined only from the control deviation but that the error also take the following system states into account.

Through the incremental learning method, it is easily possible to introduce previous knowledge into the rule base. Missing rules are added during learning. However, because of the already discussed problems, both of the presented heuristics cannot provide a rule base which is appropriate for all systems.

The rule bases learned with the methods presented above should be checked manually for its consistence—at least if the optimization method, which we discuss in the following paragraph, is unable to provide a satisfying solution. In any case, the possibility to introduce prior knowledge should be considered, that is, known rules should be inserted into the rule base before learning and the rule learning methods should not change the manually defined rules afterwards.

Optimization of the Rule Base

The NEFCON learning algorithm for optimizing a rule base is based on the idea of the backpropagation algorithm for the multilayer perceptron. The error is, beginning with the output unit, propagated backwards trough the network, and used locally for adapting the fuzzy sets.

The optimization of the rule base is done by changing the fuzzy sets in the antecedents and consequents. Depending on the contribution to the control action and the resulting error the fuzzy sets of a rule are 'rewarded' or 'punished'. Thus, the principle of the reinforcement learning is applied. A 'reward' or 'punishment' can be done by displacing or reducing/enlarging the support of the fuzzy set. These adaptations are made iteratively, that is, during the learning process the controller is used for controlling the system, and after every control action an evaluation of the new state is made as well as a incremental adaption of the controller.

The principal problem of this approach is that the system error has to be determined very carefully in order to avoid the problems of evaluating a control action which we discussed at the beginning of this section. In many cases, this can be very difficult or even impossible. Nevertheless, the discussed methods can be applied to simple systems, and they can very helpful for designing fuzzy controllers for more complex systems. We have to pay attention to the fact that then such a controller should be checked very carefully for its stability.

21.2 Evolutionary Fuzzy Control

The objective of an evolutionary fuzzy system is to support tuning and/or learning of a fuzzy controller. Tuning is targeted at optimizing the performance of an already existing system. The tuning process involves mainly the adaptation of the database, especially parameters of membership functions. The evolutionary learning process is concerned with the automatic derivation of fuzzy rules. Learning faces a much more difficult task as it tries to establish the proper relationship between input and output states from scratch, rather than optimizing the performance of a fuzzy system that already operates at least approximately correct.

Designing a fuzzy rule-based system is equivalent to finding the optimal configuration of fuzzy sets and/or rules, and in that sense can be regarded as an optimization problem. The search space is the set of parameters that encode the membership functions, the operators, and the fuzzy rules. We already pointed out that fuzzy controller exploit the imprecise knowledge by aggregation of similar states into coarse clusters/granules. So the main task for the evolutionary optimization algorithm is to learn the appropriate local relationships between input and output values established by fuzzy if—then rules.

21.2.1 Structure of an Evolutionary Fuzzy Controller

The major components of an evolutionary fuzzy system are shown in Fig. 21.4. The fuzzy controller lies at the core of the hybrid structure, it fuzzifies the input state, performs the inference based on the fuzzy rules, and aggregates the result of the inference process into a crisp output. Depending on the context, the environment can be a plant to be controlled, or a system to be modeled. An external instance (a so-called trainer) evaluates the performance of the fuzzy controller with regard to

Fig. 21.4 Evolutionary
Fuzzy Control

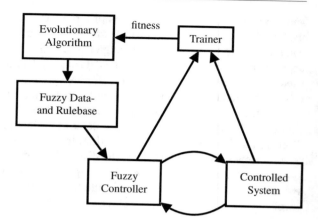

the control task. The performance is aggregated into a fitness value on which basis the evolutionary algorithm works.

If a fuzzy controller should be improved or generated by an evolutionary algorithm, at first, we have to define an objective function which has to be optimized by the evolutionary algorithm. If measure data of a controller are given—e.g., measuring data obtained from observing a human operator—the fuzzy controller should approximate the underlying control function as good as possible. As the fitness function, which has to be minimized, we could use the mean squared error or the absolute error as well as the maximum deviation of the control function from the measured output. When the measured data come from different operators, an approximation can lead to a very bad behavior of the controller. If the single operators use effective, but different control strategies and the controller is forced to approximate the data as good as possible, it will interpolate between the different strategies at each point. In the worst case, this mixed or interpolated control function might not work at all. If, for example, a car shall drive around an obstacle, and in the data for half of the cases an operator (driver) chose to avoid the obstacle to the right side and for the other half to the left side, interpolation will result in a strategy that keeps on driving straight ahead. If possible, the observed data should always be checked for consistency.

When we have a simulation model of the system or process to be controlled, we can define various criteria that a good controller should satisfy, e.g., the time or the energy the controller needs to bring the process from different initial states to the desired state, some kind of evaluation of overshoots, etc. If the evolutionary algorithm uses a simulation model with such an objective function, it is often better to slowly tighten the conditions of the objective function. In a random initial population it is very likely that no individual (controller) at all will be able to bring the process very close to a desired state. Therefore, at first, the objective function might only measure the time how long a controller is able to keep the process in a larger vicinity of the desired state [9]. With an increasing number of generations, the objective function is chosen more strict until it reflects the actual criterion.

From an engineering point of view, the most important optimization criterion is of course the *accuracy* of the problem solution. Additional requirements have been considered for fitness such as *compactness* (small number of rules and fuzzy sets), *completeness* (coverage of relevant areas of input space), *consistency* (no rules with similar antecedents and different consequents).

Nowadays, and especially in some AI applications, there is a need for interpretable solutions. These systems must be optimized not only for accuracy but also for other criteria such as *fairness* or unbiasedness, *privacy, reliability, robustness*, and/or *trust*, among others [7]. This goal is not easy to achieve, as these criteria are usually in conflict. A solution is to carry out the learning procedure by means of multi-objective optimization [4]. The strength of these multi-objective evolutionary algorithms [5] resides in their ability to approximate a wide part of the Pareto front, thus providing multiple solutions.

21.2.2 Optimizing Parameters of an Evolutionary Fuzzy Controller

Evolutionary algorithms can be used for tuning and learning the database and the rule base of the fuzzy controller. A crucial question is whether the rule base and the fuzzy sets should be optimized at the same time or separately. The rule base functions serve as the skeleton of a fuzzy controller. As long as the rule base can change drastically, it does not make sense to fine-tune the fuzzy sets. In order to keep the number of parameters to be optimized by the evolutionary algorithm small, it is often better to learn at first the rule base on the basis of standard fuzzy partitions and then optimize the fuzzy sets with a fixed rule base. The parameters of the fuzzy controller are divided into two groups, the database parameters and the rule base parameters.

21.2.2.1 Database Parameters

Usually, the fuzzy sets are represented in a parametrized form like triangular, trapezoidal or Gaussian membership functions. For example, an asymmetric triangular function can be coded by three parameters: the two extreme points and one central point, cf. (a, b, c) in Fig. 15.5. Instead of the coordinates of the extreme points we might also give distances from the central point, i.e., $(b, b - a, c - b)$. The corresponding real parameters are suitable for an optimization based on evolution strategies.

However, giving complete freedom to the evolution to optimize these parameters, seldom leads to meaningful results. The optimized fuzzy controller might perform perfectly but the fuzzy sets overlap completely so that it is hardly possible to assign meaningful linguistic terms to them and to formulate interpretable rules. In this case, the fuzzy controller corresponds to a black box, without any chance to explain or interpret its control strategy.

It is recommended to choose the parameter set in such a way that the interpretability of the fuzzy controller is always guaranteed. One possible way would be the restriction to triangular functions which are chosen in such a way that the extreme point of one fuzzy set is the central point of the next, i.e., the left and the right neighbor of a fuzzy set get the value 1 at the point where the membership degree of the fuzzy set in the middle drops to 0. In this case, the evolution strategy would have for each input or output variable as many real parameters as fuzzy sets are used for the corresponding domain. The respective real parameters indicate where the corresponding triangular functions assume the value 1.

Even with this parametrization undesired effects can occur, for example, if the fuzzy set "approximately zero", because of mutation, overtakes the fuzzy set "positive small". A simple change in the coding of the parameters can avoid this effect: the value of the parameter k determines no longer the point where the triangular membership function is 1, but how far it is shifted to the right relative to its left neighbor. The disadvantage of this coding is that a change (mutation) of the first value leads to a new position for all fuzzy sets and therefore a rather big change of the total behavior of the controller. If the fuzzy sets are parametrized independently, a mutation only has local effect. Therefore, we should stick to the independent parametrization, but prohibit mutations which lead to overtaking fuzzy sets. In this way, mutations cause small changes and the interpretability of the fuzzy controller is preserved.

21.2.2.2   Rule Base Parameters

Let us first assume that the fuzzy sets are specified in advance or are optimized at simultaneously by another method. For instance, if the controller has two input variables for which n_1, respectively, n_2 fuzzy sets are defined, for each of the possible $n_1 \cdot n_2$ combinations an output can be defined. For a Mamdani controller with n_o fuzzy sets for the output variables, we could use a chromosome with $n_1 \cdot n_2$ parameters (genes), where each of these genes can take one of n_o values. But the coding of the rule table as a linear vector with $n_1 \cdot n_2$ components, which is required for the genetic algorithm, is not optimal for the crossover. Crossover should enable the genetic algorithm to combine to solutions (chromosomes) that have optimized different parameters (genes) to one good solution.

For the optimization of a rule base of a fuzzy controller, the conditions needed to benefit from crossover are satisfied, i.e., that the parameters are independent to a certain degree. Adjacent rules in a rule base operate on overlapping regions so that they contribute to the output of the controller at the same time, meaning that they interact and are dependent. Non-adjacent rules do not interact and never fire simultaneously. In this sense, they are independent. If there are two fuzzy controllers in a population, which found, each one in a different part of the rule table, well-performing entries for the output of the rules, the combination of the two parts of the table will result in a better controller. However, a (fuzzy) region does not correspond to a row or a column in a rule table, but to a rectangular region in the table. A traditional genetic algorithm would only exchange linear parts in the crossover process. In the case of optimizing the rule base of a controller with two input variables, it makes sense to deviate from the linear chromosome structure and use a planar structure for the chromosome. Crossover should then exchange smaller rectangles within the rule table [13]. Here, we discussed the case of only two input variables. The idea can be generalized to more input variables in a straight forward manner. For k input variables, we would use a k-dimensional hyperbox as the chromosome structure.

In order to guarantee small changes in the mutation process, an output fuzzy set should not be replaced by an arbitrary other fuzzy set, but by an adjacent one.

For a TS model for the rule base, we have to determine output functions instead of output fuzzy sets. Usually, these functions are given in parametrized form, e.g.,

$$f(x, y; a_R, b_R, c_R) = a_R + b_R x + c_R y$$

for input variables x and y as well as three parameters a_R, b_R and c_R that have to be determined for each rule R. For a rule table with—like above—$n_1 \cdot n_2$ entries we had to determine, in total, $3 \cdot n_1 \cdot n_2$ real parameters for the rule table. In this case, we should apply an evolution strategy (see Sect. 13.2) since the parameters are not discrete but continuous.

If we do not want to fill the rule table completely and want to limit the number of rules, we could assign to each rule an additional binary gene (parameter) which tells us whether the rule of the controller is used or not. For a TS model, we have a proper evolutionary algorithm, because we have to deal with continuous and discrete parameters at the same time. The number of active rules can be fixed in advance, where we have to ensure that this number is not changed by mutation or crossover. Mutation could always activate one rule and deactivate another one at the same time, so that the number of active rules is not changed by mutation. For crossover, a repairing algorithm is needed. If we have too many active rules after crossover, then, for example, rules could be deactivated randomly, until the desired number of active rule is reached again.

A better strategy is not to fix the number of active rules in advance. Since fuzzy controllers with a smaller number of rules are to be preferred due to better interpretability, we could introduce an additional term in the objective function penalizing higher numbers of rules. This additional term should have a suitable weight. If the weight is too large, the emphasis is put more or less completely on keeping the number of rules small, ignoring the performance of the controller. A weight chosen too small will not contribute enough to keep the rule base small.

21.2.3 Example

In order to illustrate the principle of parameter coding, in the following section, we present a genetic algorithm for learning a TS controller. The algorithm tries to optimize all of the parameters of the controller, that is, the rule base, the form of the fuzzy sets and the parameters of the consequents at the same time.

In order to learn the rules

$$R_r : \text{If } x_1 \text{ is } \mu_R^{(1)} \text{ and } \ldots \text{ and } x_n \text{ is } \mu_R^{(n)} \text{ then } y = f_r(x_1, \ldots, x_n),$$

of a Takagi–Sugeno controller with

$$f_r(x_1, \ldots, x_n) = p_0^r + x_1 \cdot p_1^r + \ldots + x_n \cdot p_n^r,$$

we have to encode the fuzzy sets of the input values and the parameters $p_0, \ldots, p_n$ of each rule.

In this approach, a triangular fuzzy set is described by three binary coded parameters (**membership function chromosome (MFC)**):

leftbase	center	rightbase
10010011	10011000	11101001

The parameters *leftbase*, *rightbase* and *center* are no absolute quantities but denote the distances to the reference point. *leftbase* and *rightbase* refer to the *center* of a fuzzy set and the *center* refers to the distance between the center and the left neighbor of the fuzzy set. If these parameters are positive, passing and abnormal fuzzy sets can be avoided.

The parameters $p_0, \ldots, p_n$ of a rule are encoded directly by binary numbers and result in the **rule-consequent parameters chromosome (RPC)**:

p_0	$\ldots$	p_n
10010011	$\ldots$	11101001

Based on these parameter encodings, the complete rule base of a TS controller is encoded in the form of a bit string:

variable 1	$\ldots$	variable n	parameters of the consequents
$MFC_{1\ldots m_1}$	$\ldots$	$MFC_{1\ldots m_n}$	$RPC_{1\ldots(m_1\cdot\ldots\cdot m_n)}$

Besides the parameter optimization, the algorithm tries to minimize the amount of fuzzy sets assigned to a variable and thus also tries to minimize the amount of rules in a rule base. For that, we assume a maximum amount of fuzzy sets. We eliminate fuzzy sets which are no longer in the permitted domain of a variable. Furthermore, among controllers of the same performance the selection process prefers the ones with less rules.

References

1. Babuška R (2012)Fuzzy modeling for control. Springer Science
2. Barto AG, Sutton RS, Anderson CW (1983) Neuronlike adaptive elements that can solve difficult learning control problems. IEEE Trans Syst Man Cybern 13(5):834–846. IEEE Press, Piscataway, NJ, USA 1983
3. Berenji HR, Khedkar P (1992) Learning and tuning fuzzy logic controllers through reinforcements. IEEE Trans Neural Networks 3(5):724–740. IEEE Press, Piscataway, NJ, USA 1992
4. Coello-Coello A, Lamont G, van Veldhuizen D (2007) Evolutionary algorithms for solving multi-objective problems. Springer, Berlin
5. Fazzolari M, Alcala R, Nojima Y, Ishibuchi H, Herrera F (2013) A review of the application of multi-objective evolutionary systems: current status and further directions. IEEE Trans Fuzzy Syst 21(1):45–65

6. Feng G (2006) A survey on analysis and design of model-based fuzzy control systems. IEEE Trans Fuzzy Syst 14(5):676–697
7. Fernandez A, Cordon O, Herrerra F, Marcelloni F (2019) Evolutionary fuzzy systems for explainability artificial intelligence: why, when, what for, and where to? IEEE Comput Intell Mag
8. Hoffmann F (2001) Evolutionary algorithms for fuzzy control system design. Proc IEEE 89(9):1318–1333
9. Hopf J, Klawonn F (1994) Learning the rule base of a fuzzy controller by a genetic algorithm. In: Kruse R, Gebhardt J, Palm R (eds) Fuzzy Systems in Computer Science, 63–74. Vieweg, Braunschweig, Germany
10. Jang J-SR (1993) ANFIS: adaptive-network-based fuzzy inference system. IEEE Trans Syst Man Cybern 23(3):665–685. IEEE Press, Piscataway, NJ, USA 1993
11. Kaelbling LP, Littman MH, Moore AW (1996) Reinforcement learning: a survey. J Artif Intell Res 4:237–285. AI Access Foundation and Morgan Kaufman Publishers, El Segundo/San Francisco, CA, USA
12. Karr C (1991) Genetic algorithms for fuzzy controllers. Artif Intell Expert 6(2):26–33
13. Kinzel J, Klawonn F, Kruse R (1994) Modifications of genetic algorithms for designing and optimizing fuzzy controllers. In: Proceedings of the IEEE Conference on Evolutionary Computation (ICEC'94, Orlando, FL), 28–33. IEEE Press, Piscataway, NJ, USA
14. Lee M, Takagi H (1993) Integrating design stages of fuzzy systems using genetic algorithms. In: Proceedings IEEE International Conference on Fuzzy Systems (San Francisco, CA)
15. Nauck DD, Klawonn F, Kruse R (1997) Foundations of neuro-fuzzy systems. Wiley, Chichester
16. Nauck D, Kruse R (1993) A fuzzy neural network learning fuzzy control rules and membership functions by fuzzy error backpropagation. In: Proceedings IEEE nternational Conference on Neural Networks (ICNN'93, 1993, San Francisco, CA)
17. Nurnberger A, Nauck DD, Kruse R (1999) Neuro-fuzzy control based on the NEFCON-model: recent developments. Soft Comput 2(4):168–182. Springer-Verlag, Berlin/Heidelberg, Germany
18. Riedmiller M, Spott M, Weisbrod J (1999) FYNESSE: a hybrid architecture for selflearning control. In: Cloete I, Zurada J (eds) Knowledge-Based Neurocomputing, 291–323. MIT Press, Cambridge, MA, USA
19. Shi Y, Eberhart R, Chen Y (1999) Implementation of evolutionary fuzzy systems. IEEE Trans Fuzzy Syst 7(2):109–119
20. Sutton RS, Barto AG (1998) Reinforcement learning: an introduction. MIT Press, Cambridge, MA, USA
21. Tiruneh GG, Fayek AR, Sumati V (2015) Neuro-fuzzy systems in construction engineering and management research. Autom Constr 119:1–23
22. Viharos ZJ, Kis KB (2015) Survey on neuro-fuzzy systems and thier applications in technical diagnostics and measurement. Automation in Construction 119, Measurement, vol 67, pp 126–136

Fuzzy Data Analysis

22

In the first part of this chapter we present fuzzy methods for the analysis of classical data. A typical example is fuzzy cluster analysis. In the second part we study (statistical) methods for the analysis of fuzzy data.

22.1 Fuzzy Methods in Data Analysis

Purely knowledge-driven and purely data-driven approaches for constructing models can be seen as the extreme ends of a range of strategies. It is possible, however, to combine them on any level: to provide a knowledge-based model and adjust only a few parameters—for instance, the exact location of triangular fuzzy sets—or to use domain knowledge only to choose a very general model class from which a specific model with concrete parameters is chosen based on the data. Neural networks are a typical example for the latter strategy where only the type or structure of the neural network is fixed and then the neural network must be trained with data.

With a data-driven approach it is no wonder that fuzzy techniques are also used in the context of data analysis, data mining, and machine learning. Note that here the term **fuzzy data analysis** [19] may be interpreted in two fundamentally different ways:

- **Fuzzy Techniques for the Analysis of (Crisp) Data**
  In this approach, the data are considered to be non-fuzzy, and fuzzy techniques are applied to analyze these data. As two examples for the application of fuzzy methods in data analysis, we focus on (1) fuzzy cluster analysis as the oldest fuzzy approach to data analysis and (2) on possibilistic reasoning. A large variety of other fuzzy techniques have been developed. For example, there is a vast collection of

© Springer Nature Switzerland AG 2022
R. Kruse et al., *Computational Intelligence*, Texts in Computer Science,
https://doi.org/10.1007/978-3-030-42227-1_22

methods to tune parameters or to learn the rules of a fuzzy system based on data, and there are lots of applications in the context of data mining and machine learning, see, e.g., [19].

- **Analysis of Fuzzy Data**
  Measured data are often imprecise or under the influence of noise. Answers in questionnaires usually involve terms like "strongly agree", "agree", "neither agree or disagree", "disagree", and "strongly disagree", which are essentially fuzzy terms. One way to handle this inherent fuzziness in the data is to model the data by fuzzy sets and then to analyze the fuzzy data. There are two main interpretations of fuzzy sets in the context of statistical data analysis [28]: in the **epistemic view** fuzzy sets are used to represent incomplete knowledge about an underlying object or a precise quantity [52,57]. In the **ontic view**, fuzzy sets are considered as real complex humped entities [11,68]. Note that due to the different semantics of the data also the statistical methods have to be different. An interested reader can find an abundance of information on this topic in the textbooks [3,53,73] and the proceedings of the conference series Soft Methods in Probability and Statistics [54].

22.2 Fuzzy Clustering

After a brief overview of clustering, this chapter focuses on fuzzy cluster analysis as the oldest fuzzy approach to data analysis. Fuzzy clustering comprises a family of prototype-based clustering methods that can be formulated as the problem of minimizing an objective function. These methods can be seen as "fuzzifications" of, for example, the classical c-means algorithm, which strives to minimize the sum of the (squared) distances between the data points and the cluster centers to which they are assigned. However, in order to "fuzzify" such a crisp clustering approach, it is not enough to merely allow values from the unit interval for the variables encoding the assignments of the data points to the clusters: the minimum is still obtained for a crisp data point assignment. As a consequence, additional means have to be employed in the objective function in order to obtain actual degrees of membership. This chapter surveys the most common fuzzification means and examines and compares their properties.

22.2.1 Clustering

The general objective of *clustering* or *cluster analysis* [5,31,39,44,47] is to group given objects in such a way that objects from the same cluster are as similar as possible, while objects from different clusters are as dissimilar as possible. In order to formalize the notion of similarity, so that it becomes mathematically treatable, it is usually expressed as a *distance measure* between points (or vectors) representing the objects in a metric space, usually $\mathbb{R}^m$. Two objects are then seen as the more similar, the smaller the distance between the data points that represent them.

A common approach to describe the clusters is to use *prototypes* that capture the location and possibly also the shape and size of the clusters in the data space. With such an approach the general objective of clustering can be reformulated as the task to find a set of cluster prototypes together with an assignment of the data points to them, so that the data points are as close as possible to their assigned prototypes. By formalizing this approach, and using for the prototypes only points in the data space that represent the *cluster centers*, one obtains immediately the objective function of classical c-means clustering [2,38,60]: simply sum the (squared) distances of the data points to the center of the cluster to which they are assigned. The c-means algorithm then strives to minimize this objective function by iteratively updating the assignment of the data points to the clusters and recomputing the cluster centers.

Unfortunately, c-means clustering always partitions the data, that is, each data point is assigned to one cluster and one cluster only. This is often inappropriate, as it can lead to somewhat arbitrary cluster boundaries and certainly does not treat points properly that lie between two (or more) clusters without belonging to any of them unambiguously. Solutions to this problem consist in either using a probabilistic approach, like applying the expectation maximization (EM) algorithm to a mixture of Gaussians (see, for example, [10,22,32]), or to employ one of the different "fuzzifications" of the classical crisp scheme (see, for instance, [6,8,12,30,39,69]).

In this chapter, we focus on the latter approach, that is, on how the objective function of classical c-means clustering can be modified in order to obtain graded cluster memberships (so-called *fuzzy clustering*). We survey different methods that have been suggested in the literature and examine and compare their properties.

22.2.2 Presuppositions and Notation

We are given a data set $\mathbf{X} = \{x_1, \ldots, x_n\}$ with n data points, each of which is an m-dimensional real-valued vector, that is, $\forall j; 1 \leq j \leq n : x_j = (x_{j1}, \ldots, x_{jm}) \in \mathbb{R}^m$. These data points are to be grouped into c clusters, each of which is described by a prototype $c_i, i = 1, \ldots, c$. The set of all prototypes is denoted by $\mathbf{C} = \{c_1, \ldots, c_c\}$. We confine ourselves here to cluster prototypes that consist merely of a cluster center, that is, $\forall i; 1 \leq i \leq c : c_i = (c_{i1}, \ldots, c_{im}) \in \mathbb{R}^m$. The assignment of the data points to the cluster centers is encoded as a $c \times n$ matrix $\mathbf{U} = (u_{ij})_{1 \leq i \leq c; 1 \leq j \leq n}$, which is often called the *partition matrix*. In the crisp case, a matrix element $u_{ij} \in \{0, 1\}$ states whether data point x_j belongs to cluster c_i or not. In the fuzzy case, $u_{ij} \in [0, 1]$ states the degree to which x_j belongs to c_i (degree of membership).

Furthermore, we confine ourselves to the (squared) Euclidean distance as the measure for the distance between a data point x_j and a cluster center c_i, that is,

$$d_{ij}^2 = d^2(c_i, x_j) = (x_j - c_i)^\top (x_j - c_i) = \sum_{k=1}^{m} (x_{jk} - c_{ik})^2.$$

A common alternative is the (squared) Mahalanobis distance with a cluster specific covariance matrix Σ_i [34,37], that is, $d_{ij}^2 = (x_j - c_i)^\top \Sigma_i^{-1} (x_j - c_i)$. However, this choice adds at least a shape parameter and in some approaches also a size parameter to the cluster prototypes (see, for example, [8,12,39]). Nevertheless, extending the approaches to this distance measure is usually fairly straightforward. An extension to the L_1-distance[45], that is, to $d_{ij} = \sum_{k=1}^{m} |x_{jk} - c_{ik}|$, or to other Minkowski metrics, although certainly useful in specific cases, is less simple to achieve and clearly beyond the scope of this chapter.

22.2.3 Classical c-Means Clustering

As already stated, classical c-means clustering strives to find, for a given data set $\mathbf{X}$, a set $\mathbf{C}$ of cluster centers and a partition matrix $\mathbf{U}$, such that the objective function

$$J(\mathbf{X}, \mathbf{C}, \mathbf{U}) = \sum_{i=1}^{c} \sum_{j=1}^{n} u_{ij}\, d_{ij}^2$$

is minimized under the constraints $\forall i; 1 \leq i \leq c : \forall j; 1 \leq j \leq n : u_{ij} \in \{0, 1\}$ and $\forall j; 1 \leq j \leq n : \sum_{i=1}^{c} u_{ij} = 1$. These constraints ensure that each data point is assigned to one cluster and to one cluster only (crisp partition of the data set).

Since the minimum cannot be found directly using analytical means, an *alternating optimization* scheme is employed. At the beginning, the cluster centers are initialized randomly, for example, by selecting c data points arbitrarily or by sampling c points from some distribution on the data space. Then the two steps of *partition matrix update* (data point assignment) and *cluster center update* are iterated until convergence, that is, until the cluster centers do not change anymore.

In the partition matrix update, each data point x_j is assigned to the cluster c_i, the center of which is closest to it. That is, the partition matrix is updated according to

$$u_{ij} = \begin{cases} 1, & \text{if } i = \text{argmin}_{i=1}^{c}\, d_{ij}^2, \\ 0, & \text{otherwise.} \end{cases}$$

In the cluster center update, each cluster center is recomputed as the mean of the data points that were assigned to it (hence the name c-means clustering), that is,

$$c_i = \frac{\sum_{j=1}^{n} u_{ij}\, x_j^2}{\sum_{j=1}^{n} u_{ij}}.$$

This update process is guaranteed to converge and usually does so after fairly few steps. However, it is fairly sensitive to the initial conditions (i.e., the initial cluster centers), due to which it can yield undesired results, which are caused by local minima of the objective function. In order to handle this drawback, it is usually recommended to execute the clustering algorithm multiple times and take the best result, that is, the result that yields the smallest value of the objective function.

As an illustration of the problem of local minima, consider the simple two-dimensional data set shown in Fig. 22.1. Visual inspection clearly tells us that there are three clusters (ignore the additional data point marked with a star for now) and

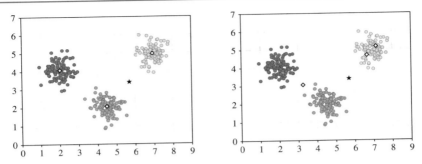

Fig. 22.1 A data set with three clusters and an additional point (marked with a star). Result of a successful c-means clustering (left) and a local optimum (right); diamonds mark cluster centers

hence we expect a clustering result as it is shown on the left: the diamonds mark the cluster centers. Although this result is obtained for a suitable initialization, another initialization yields the result shown on the right. Since two centers are located in the same cluster (the one on the top right), while the other two clusters are captured by a single center that is located between them, this result is clearly undesirable. It corresponds to a local optimum of the objective function, though, and thus the update process may converge to it (depending on the initialization).

However, even if c-means clustering correctly identifies the three clusters (as in Fig. 22.1 on the left), it is not quite clear to which cluster the additional point (marked with a star) should be assigned, because it lies in the middle between two clusters. Although it is clearly counterintuitive to assign this point uniquely to one cluster, c-means clustering has to do so, because it always yields a (crisp) partition of the data set. Clearly, it would be more appropriate to be able to assign this point with degrees of membership to more than one cluster, for example, with a degree of about 0.5 to both of the clusters the centers of which are closest to it.

To obtain such *degrees of membership*, it may seem, at first sight, to be sufficient to simply extend the allowed range of values of the u_{ij} from the set $\{0, 1\}$ to the real interval $[0, 1]$, but to make no changes to the objective function itself. However, this is not the case, the optimum of the objective function is obtained for a crisp assignment, regardless of whether we enforce a crisp assignment or not.

This can easily be demonstrated as follows: let $k_j = \text{argmin}_{i=1}^{c} d_{ij}^2$, that is, let k_j be the index of the cluster center closest to the data point x_j. Then it is

$$
J(X, C, U) \;=\; \sum_{i=1}^{c}\sum_{j=1}^{n} u_{ij}\, d_{ij}^2 \;\geq\; \sum_{i=1}^{c}\sum_{j=1}^{n} u_{ij}\, d_{k_j j}^2 \;=\; \sum_{j=1}^{n} d_{k_j j}^2 \underbrace{\sum_{i=1}^{c} u_{ij}}_{=1 \;\text{(due to the constraints)}}
$$

$$
=\; \sum_{j=1}^{n}\left(1 \cdot d_{k_j j}^2 + \sum_{\substack{i=1 \\ i\neq k_j}}^{c} 0 \cdot d_{ij}^2 \right).
$$

Therefore, it is best to set $\forall j; 1 \leq j \leq n : u_{k_j j} = 1$ and $u_{ij} = 0$ for $1 \leq i \leq c, i \neq k_j$. In other words, the objective function is minimized by assigning each data point crisply to the closest cluster, even though we allowed for degrees of membership.

22.2.4 Fuzzification by Membership Transformation

Since we cannot obtain degrees of membership by merely expanding the range of values of the u_{ij}, we have to modify the objective function if we desire graded assignments. The most common approach is to apply a transformation to the membership degrees, that is, to use an objective function of the form

$$J(\mathbf{X}, \mathbf{C}, \mathbf{U}) = \sum_{i=1}^{c} \sum_{j=1}^{n} h(u_{ij}) \, d_{ij}^2,$$

where h is a convex function on the real interval $[0, 1]$. This general form was first studied by [49], where the convexity of h was derived as follows: for simplicity, we confine ourselves to two clusters c_1 and c_2 and consider the terms of the objective function that refer to a single data point x_j. That is, we consider $J(x_j, c_1, c_2, u_{1j}, u_{2j}) = h(u_{1j}) \, d_{1j}^2 + h(u_{2j}) \, d_{2j}^2$ and study how it behaves for different values u_{1j} and u_{2j}. Note that a crisp assignment should not be ruled out categorically, namely, if the distances d_{1j} and d_{2j} differ significantly. Hence, we assume that d_{1j} and d_{2j} differ only slightly, so that a graded assignment is desired.

$J(x_j, c_1, c_2, u_{1j}, u_{2j})$ is minimized by choosing u_{1j} and u_{2j} appropriately. Exploiting $\sum_{i=1}^{c} u_{ij} = 1$ yields $J(x_j, c_1, c_2, u_{1j}) = h(u_{1j}) \, d_{1j}^2 + h(1 - u_{1j}) \, d_{2j}^2$. A necessary condition for a minimum is $\frac{\partial}{\partial u_{1j}} J(x_j, c_1, c_2, u_{1j}) = h'(u_{1j}) \, d_{1j}^2 - h'(1 - u_{1j}) \, d_{2j}^2 = 0$, where $'$ denotes taking the derivative w.r.t. the argument of the function. This leads to $h'(u_{1j}) \, d_{1j}^2 = h'(1 - u_{1j}) \, d_{2j}^2$, which yields another argument that a graded assignment cannot be optimal without any function h: if h is the identity, we have $h'(u_{1j}) = h'(1 - u_{1j}) = 1$ and thus the equation cannot hold if the distances differ.

For the further analysis let us assume, without loss of generality, that $d_{1j} < d_{2j}$, which implies $h'(u_{1j}) > h'(1 - u_{1j})$. In addition, we know that $u_{1j} > u_{2j} = 1 - u_{1j}$, because the degree of membership should be higher for the cluster that is closer. In other words, the function h must be the steeper, the greater its argument. Therefore, it must be a convex function on the unit interval [49].

Since we confine ourselves to the Euclidean distance (see Sect. 22.2.2), we can already derive the update rule for the cluster centers, namely, by exploiting that a necessary condition for a minimum of the objective function J is that the partial derivatives w.r.t. the cluster centers vanish. Therefore, we have $\forall k; 1 \leq k \leq c$:

$$\nabla_{c_k} J(\mathbf{X}, \mathbf{C}, \mathbf{U}) = \nabla_{c_k} \sum_{i=1}^{c} \sum_{j=1}^{n} h(u_{ij}) \, (x_j - c_i)^\top (x_j - c_i) = -2 \sum_{j=1}^{n} h(u_{ij})(x_j - c_i) \overset{!}{=} 0.$$

Independent of the function h, it follows immediately

$$c_i = \frac{\sum_{j=1}^{n} h(u_{ij}) x_j}{\sum_{j=1}^{n} h(u_{ij})}.$$

This update rule already shows one of the core drawbacks of a fuzzification by membership transformation, namely, that the transformation function enters the update of the cluster centers. It would be more intuitive to use the membership degrees directly as the weights for the mean computation, which would also ensure that all data points enter with the same total unit weight (since $\sum_{i=1}^{c} u_{ij} = 1$ by definition). However, the weights are rather the transformed membership degrees $h(u_{ij})$, which gives unequal weight to the data points as they need not sum to 1.

It may be argued, though, that this effect can actually be desirable: due to the convexity of the function h the total weight $\sum_{i=1}^{c} h(u_{ij})$ of data points x_j with a less ambiguous assignment is higher than that of more ambiguously assigned data points. (The maximum 1 is obtained for a crisp assignment.) Hence, in this scheme, the locations of the cluster centers depend more strongly on the data points that are "typical" for the clusters. Such an effect is very much in the spirit of, for instance, robust regression techniques, in which data points receive a lower weight if they do not fit well to the regression function. This connection to robust statistical methods was explored in more detail, for example, in [21].

In order to derive the update rule for the partition matrix (and thus for the membership degrees u_{ij}), we need to know the exact form of the function h. The most common choice is $h(u_{ij}) = u_{ij}^2$, which leads to the standard objective function of fuzzy clustering [30]. The more general form $h(u_{ij}) = u_{ij}^w$ was introduced by [6]. The exponent w, $w > 1$, is called the *fuzzifier*, since it controls the "fuzziness" of the data point assignments: the higher w, the softer the boundaries between the clusters. This leads to the commonly used objective function [6,8,12,39]

$$J(\mathbf{X}, \mathbf{U}, \mathbf{C}) = \sum_{i=1}^{c} \sum_{j=1}^{n} u_{ij}^w d_{ij}^2.$$

The update rule for the membership degrees is now derived by incorporating the constraints $\forall j; 1 \leq j \leq n : \sum_{i=1}^{c} u_{ij} = 1$ with Lagrange multipliers into the objective function. This yields the Lagrange function

$$L(\mathbf{X}, \mathbf{U}, \mathbf{C}, \Lambda) = \underbrace{\sum_{i=1}^{c} \sum_{j=1}^{n} u_{ij}^w d_{ij}^2}_{=J(\mathbf{X},\mathbf{U},\mathbf{C})} + \sum_{j=1}^{n} \lambda_j \left(1 - \sum_{i=1}^{c} u_{ij}\right),$$

where $\Lambda = (\lambda_1, \ldots, \lambda_n)$ are the Lagrange multipliers, one per constraint.

Since a necessary condition for a minimum of the Lagrange function is that the partial derivatives w.r.t. the membership degrees vanish, we obtain

$$\frac{\partial}{\partial u_{kl}} L(\mathbf{X}, \mathbf{U}, \mathbf{C}, \Lambda) = w\, u_{kl}^{w-1} d_{kl}^2 - \lambda_l \overset{!}{=} 0 \quad \text{and thus} \quad u_{kl} = \left(\frac{\lambda_l}{w\, d_{kl}^2}\right)^{\frac{1}{w-1}}.$$

Summing these equations over the clusters (in order to be able to exploit the corresponding constraints on the membership degrees, which are recovered from the fact that it is a necessary condition for a minimum that the partial derivatives of the Lagrange function w.r.t. the Lagrange multipliers vanish), we get

$$1 = \sum_{i=1}^{c} u_{ij} = \sum_{i=1}^{c} \left(\frac{\lambda_j}{w \, d_{ij}^2} \right)^{\frac{1}{w-1}} \quad \text{and thus} \quad \lambda_j = \left(\sum_{i=1}^{c} \left(w \, d_{ij}^2 \right)^{\frac{1}{1-w}} \right)^{1-w}.$$

Therefore we finally have for the membership degrees $\forall i; \, 1 \le i \le c \colon \forall j; \, 1 \le j \le n$:

$$u_{ij} = \frac{d_{ij}^{\frac{2}{1-w}}}{\sum_{k=1}^{c} d_{kj}^{\frac{2}{1-w}}} \quad \text{and thus for} \, w = 2 : \quad u_{ij} = \frac{d_{ij}^{-2}}{\sum_{k=1}^{c} d_{kj}^{-2}}.$$

This rule is fairly intuitive, as it updates the membership degrees according to the relative inverse squared distances of the data points to the cluster centers.

The effect of the fuzzifier w is illustrated in Fig. 22.2, which shows the membership degrees of points on the x-axis for two clusters, which are located at $x = 0$ and $x = 1$, for the fuzzifiers $w = 2$ (left) and $w = 1.5$ (right). Clearly, the larger fuzzifier $w = 2$ yields a smoother transition between the two clusters as the degree of membership rises (or falls) less steeply than for the fuzzifier $w = 1.5$.

Unfortunately, the above rule has the disadvantage that it necessarily yields a graded assignment. Regardless of how far a data point is from a cluster center, it always receives a non-vanishing degree of membership to the corresponding cluster. Even worse: the farther a data point is from all clusters, the more the membership degrees become equal. This effect can also be seen in Fig. 22.2: the farther we get to the right in the diagrams, the more the degrees of membership to the two clusters approach each other. The undesirable results that can be caused by this property in the presence of clusters with fairly uneven numbers of members have been demonstrated clearly by [49].

In addition, it was revealed in [49] that the reason lies essentially in the fact that $h'(u_{ij}) = \frac{d}{du_{ij}} u_{ij}^{w} = w \, u_{ij}^{w-1}$ vanishes at $u_{ij} = 0$. This suggests the idea to use a transformation function that does not have this property and thus allows, at least for sufficiently large distance relationships, a crisp assignment of data points to cluster centers. In [49], the function $h(u_{ij}) = \alpha u_{ij}^2 + (1 - \alpha) u_{ij}$, $\alpha \in (0, 1]$, or, with a more easily interpretable parametrization, $h(u_{ij}) = \frac{1-\beta}{1+\beta} u_{ij}^2 + \frac{2\beta}{1+\beta} u_{ij}$, $\beta \in [0, 1)$, was suggested as such a transformation. It relies on the standard function $h(u_{ij}) = u_{ij}^2$ and

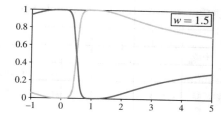

Fig. 22.2 Membership degrees for two clusters with fuzzifiers $w = 2$ (left) and $w = 1.5$ (right)

mixes it with the identity to avoid a vanishing derivative at zero. The parameter β is, for two clusters, the ratio of the smaller to the larger squared distance, at and below which we get a crisp assignment [49]. It therefore takes the place of the fuzzifier w: the smaller β, the softer the boundaries between the clusters.

The update rule for the membership degrees is derived in essentially the same way as for $h(u_{ij}) = u_{ij}^w$, although one has to pay attention to the fact that crisp assignments are now possible and thus some membership degrees may vanish. The detailed derivation, which we omit here, can be found in [49] or in [12]. It yields

$$u_{ij} = \frac{u_{ij}'}{\sum_{k=1}^{c} u_{kj}'} \quad \text{with} \quad u_{ij}' = \max\left\{0, \ d_{ij}^{-2} - \frac{\beta}{1 + \beta(c_j - 1)} \sum_{k=1}^{c_j} d_{\varsigma(k)j}^{-2}\right\},$$

where $\varsigma : \{1, \ldots, c\} \to \{1, \ldots c\}$ is a mapping function for the cluster indices such that $\forall i; 1 \leq i < c : d_{\varsigma(i)j} \leq d_{\varsigma(i+1)j}$ (that is, ς sorts the distances ascendingly) and

$$c_j = \max\left\{k \ \Big| \ d_{\varsigma(k)j}^{-2} > \frac{\beta}{1 + \beta(k - 1)} \sum_{i=1}^{k} d_{\varsigma(i)j}^{-2}\right\}$$

is the number of clusters to which the data point x_j has a non-vanishing membership. This update rule is fairly interpretable, as it still assigns membership degrees essentially according to the relative inverse squared distances to the clusters, but subtracts an offset from them, which makes crisp assignments possible.

22.2.5 Fuzzification by Membership Regularization

We have seen that transforming the membership degrees in the objective function has the disadvantage that the transformation function appears in the update rule for the cluster centers. In order to avoid this drawback, one may try to achieve a fuzzification by leaving the membership degrees in their weighting of the (squared) distances untouched. Graded memberships are rather achieved by adding a regularization term to the objective function, which pushes the minimum away from a crisp assignment. Most commonly, the objective function then takes the form

$$J(\mathbf{X}, \mathbf{C}, \mathbf{U}) = \sum_{i=1}^{c} \sum_{j=1}^{n} u_{ij} d_{ij}^2 + \gamma \sum_{i=1}^{c} \sum_{j=1}^{n} f(u_{ij}),$$

where f is a convex function on the real interval $[0, 1]$. The parameter γ takes the place of the fuzzifier w: the higher γ, the softer the boundaries between the clusters.

To analyze this objective function, we use the same basic means as in the preceding section: we confine ourselves to two clusters c_1 and c_2 and consider the terms of the objective function that refer to a single data point x_j, that is, we consider $J(x_j, c_1, c_2, u_{1j}, u_{2j}) = u_{1j} d_{1j}^2 + u_{2j} d_{2j}^2 + \gamma f(u_{1j}) + \gamma f(u_{2j})$. Since $u_{2j} = 1 - u_{1j}$, it is $J(x_j, c_1, c_2, u_{1j}) = u_{1j} d_{1j}^2 + (1 - u_{1j}) d_{2j}^2 + \gamma f(u_{1j}) + \gamma f(1 - u_{1j})$. A necessary condition for a minimum is $\frac{\partial}{\partial u_{1j}} J(x_j, c_1, c_2, u_{1j}) = d_{1j}^2 - d_{2j}^2 + \gamma f'(u_{1j}) - \gamma f'(1 - u_{1j}) = 0$, where $'$ denotes taking the derivative w.r.t. the argument of the function. This leads to the simple condition $d_{1j}^2 + \gamma f'(u_{1j}) = d_{2j}^2 + \gamma f'(1 - u_{1j})$.

We now assume again, without loss of generality, that $d_{1j} < d_{2j}$, which implies $f'(u_{1j}) > f'(1 - u_{1j})$. In addition we know $u_{1j} > u_{2j} = 1 - u_{1j}$, because the degree of membership should be higher for the cluster that is closer. In other words, the function f must be the steeper, the greater its argument. Hence, it must be a convex function on the unit interval in order to allow for graded memberships.

More concretely, we obtain $(d_{2j}^2 - d_{1j}^2)/\gamma = f'(u_{1j}) - f'(1 - u_{1j})$ as a condition for a minimum. Since f is a convex function on the unit interval, the maximum value of the right-hand side is $f'(1) - f'(0)$. If $f'(1) - f'(0) < \infty$, we have the possibility of crisp assignments, because in this case there exist values for d_{1j}^2, d_{2j}^2 and γ such that the minimum of the function $J(x_j, c_1, c_2, u_{1j})$ w.r.t. u_{ij} either does not exist or lies outside the unit interval. In such a situation, the best choice is the crisp assignment $u_{1j} = 1$ and $u_{2j} = 0$ (still assuming that $d_{1j} < d_{2j}$).

To obtain the update rule for the cluster centers, we can simply transfer the result from the preceding section, since the regularization term does not refer to the cluster centers. Therefore, we have the simple rule (because here $h(u_{ij}) = u_{ij}$)

$$c_i = \frac{\sum_{j=1}^n u_{ij} x_j}{\sum_{j=1}^n u_{ij}}.$$

This demonstrates the advantage of a membership regularization approach, because the membership degrees are directly the weights with which the data points enter the mean computation that yields the new cluster center.

In order to derive the update rule for the membership degrees, we have to respect the constraints $\forall j; 1 \leq j \leq n : \sum_{i=1}^c u_{ij} = 1$. This is achieved in the usual way (cf. the preceding section) by incorporating them with Lagrange multipliers into the objective function. The resulting Lagrange function is

$$L(\mathbf{X}, \mathbf{U}, \mathbf{C}, \Lambda) = \underbrace{\sum_{i=1}^c \sum_{j=1}^n u_{ij} d_{ij}^2 + \gamma \sum_{i=1}^c \sum_{j=1}^n f(u_{ij})}_{=J(\mathbf{X},\mathbf{C},\mathbf{U})} + \sum_{j=1}^n \lambda_j \left(1 - \sum_{i=1}^c u_{ij}\right),$$

where $\Lambda = (\lambda_1, \ldots, \lambda_n)$ are the Lagrange multipliers, one per constraint.

Since a necessary condition for a minimum of the Lagrange function is that the partial derivatives w.r.t. the membership degrees vanish, we obtain

$$\frac{\partial}{\partial u_{kl}} L(\mathbf{X}, \mathbf{U}, \mathbf{C}) = d_{kl}^2 + \gamma f'(u_{kl}) - \lambda_l \overset{!}{=} 0 \quad \text{and thus} \quad u_{kl} = f'^{-1}\left(\frac{\lambda_l - d_{kl}^2}{\gamma}\right),$$

where $'$ denotes taking the derivative w.r.t. the argument of the function and f'^{-1} denotes the inverse of the derivative of the function f. In analogy to Sect. 22.2.4, the constraints on the membership degrees are now exploited to obtain $1 = \sum_{i=1}^c u_{kj} = \sum_{k=1}^c f'^{-1}((\lambda_j - d_{kj}^2)/\gamma)$. This equation has to be solved for λ_j and the result has to be used to substitute λ_l in the expression for the u_{kl} derived above. However, in order to do so, we need to know the exact form of the regularization function f.

The regularization functions f that have been suggested in the literature (concrete examples are studied below) can be seen as derived from a maximum entropy approach. That is, the term of the objective function that forces the u_{ij} to minimize the weighted sum of squared distances is complemented by a term that forces them to maximize the entropies of the distributions over the clusters, the u_{ij} describe for each data point. Thus, the u_{ij} are pushed away from a crisp assignment, which has minimum entropy, and towards a uniform assignment, which has maximum entropy. Generally, such an approach starts from the objective function

$$J(\mathbf{X}, \mathbf{C}, \mathbf{U}) = \sum_{i=1}^{c}\sum_{j=1}^{n} u_{ij}d_{ij}^{2} - \gamma \sum_{j=1}^{n} H(\mathbf{u}_j),$$

where $\mathbf{u}_j = (u_{1j}, \ldots, u_{cj})$ comprises the degrees of membership the data point $\mathbf{x}_j$ has to the different clusters. H computes their entropy, as $\mathbf{u}_j$ is, at least formally, a probability distribution, since it satisfies $\forall i; 1 \leq i \leq c : u_{ij} \in [0, 1]$ and $\sum_{i=1}^{c} u_{ij} = 1$.

In order to develop the maximum entropy approach in more detail, we consider the generalized entropy proposed by [20]. Let $\mathbf{p} = (p_1, \ldots, p_r)$ be a probability distribution over r values. Then *Daróczy entropy* is defined as

$$H_\beta(\mathbf{p}) = \frac{2^{\beta-1}}{2^{\beta-1} - 1} \sum_{i=1}^{r} p_i(1 - p_i^{\beta-1}) = \frac{2^{\beta-1}}{2^{\beta-1} - 1}\left(1 - \sum_{i=1}^{r} p_i^{\beta}\right).$$

From this general formula, the well-known *Shannon entropy* [70] can be derived as the limit for $\beta \rightarrow 1$, that is, as

$$H_1(\mathbf{p}) = \lim_{\beta \to 1} H_\beta(\mathbf{p}) = -\sum_{i=1}^{r} p_i \log_2 p_i.$$

Employing it in the entropy-regularized objective function leads to

$$J(\mathbf{X}, \mathbf{C}, \mathbf{U}) = \sum_{i=1}^{c}\sum_{j=1}^{n} u_{ij}d_{ij}^{2} + \gamma \sum_{i=1}^{c}\sum_{j=1}^{n} u_{ij} \ln u_{ij},$$

where the factor $1/\ln 2$ (which stems from the relation $\log_2 u_{ij} = \ln u_{ij}/\ln 2$) is incorporated into the factor γ, as the natural logarithm allows for easier mathematical treatment. That is, we have $f(u_{ij}) = u_{ij} \ln u_{ij}$ [16,46,59,62] and therefore obtain $f'(u_{ij}) = 1 + \ln u_{ij}$ and $f'^{-1}(y) = e^{y-1}$. Using the latter in the formulas obtained above for deriving the update rule for the membership degrees yields

$$u_{ij} = \frac{e^{-d_{ij}^2/\gamma}}{\sum_{k=1}^{c} e^{-d_{kj}^2/\gamma}}.$$

As was pointed out in [40,65], this update rule relates the approach very closely to the expectation maximization (EM) algorithm for Gaussian mixtures [10,22,32], since by setting $\gamma = 2\sigma^2$, we obtain exactly the formula for the expectation step. As a consequence, this update rule can be interpreted as computing the probability that a data point $\mathbf{x}_j$ was sampled from a Gaussian distribution centered at $\mathbf{c}_i$ and having the variance σ^2. In addition, since the update rule for the cluster centers coincides with the maximization step, this form of fuzzy clustering is actually indistinguishable from the expectation maximization algorithm for a mixture of Gaussians.

It should be noted that $f'(u_{ij}) = 1 + \ln u_{ij}$ implies $f'(1) - f'(0) = \infty$ and thus Shannon entropy regularization always yields graded assignments. However, this drawback is less harmful here, because $e^{-d_{ij}^2/\gamma}$ is much "steeper" than d_{ij}^{-2} and thus is less prone to produce undesired results (cf. the discussion in [24]).

Another commonly used special case of Daróczy entropy is so-called *quadratic entropy*, which results if we set the parameter $\beta = 2$, that is,

$$H_2(\boldsymbol{p}) = 2 \sum_{i=1}^{r} p_i(1 - p_i) = 2 - 2 \sum_{i=1}^{r} p_i^2.$$

Employing it in the entropy-regularized objective function leads to

$$J(\mathbf{X}, \mathbf{C}, \mathbf{U}) = \sum_{i=1}^{c} \sum_{j=1}^{n} u_{ij} d_{ij}^2 + \gamma \sum_{i=1}^{c} \sum_{j=1}^{n} u_{ij}^2,$$

as the constant term 2 has no influence on the location of the minimum and thus can be discarded, and the factor 2 can be incorporated into the factor γ. That is, we have $f(u_{ij}) = u_{ij}^2$ [63] and therefore obtain $f'(u_{ij}) = 2u_{ij}$ and finally $f'^{-1}(y) = \frac{y}{2}$ for the needed inverse function of the derivative.

In order to derive the update rule for the memberships, one has to pay attention to the fact that $f'(1) - f'(0) = 2$. Therefore, crisp assignments are possible and some membership degrees may vanish. However, the detailed derivation can easily be found by following, for example, the same lines as for the analogous approach in the preceding section, which also allowed for vanishing membership degrees.

The resulting membership degree update rule is $\forall i : 1 \leq i \leq c : \forall j : 1 \leq j \leq n$:

$$u_{ij} = \max\left\{0, \frac{1}{c_j}\left(1 + \sum_{k=1}^{c_j} \frac{d_{\varsigma(k)j}^2}{2\gamma}\right) - \frac{d_{ij}}{2\gamma}\right\},$$

where $\varsigma : \{1, \ldots, c\} \rightarrow \{1, \ldots c\}$ is a mapping function for the cluster indices such that $\forall i; 1 \leq i < c : d_{\varsigma(i)j} \leq d_{\varsigma(i+1)j}$ (that is, ς sorts the distances ascendingly) and

$$c_j = \max\left\{k \ \middle| \ \sum_{i=1}^{k} d_{\varsigma(i)j}^2 > k\, d_{kj} - 2\gamma\right\}$$

is the number of clusters to which the data point $\boldsymbol{x}_j$ has a non-vanishing membership. In this update rule, 2γ can be interpreted as a reference distance relative to which all distances are judged. For two clusters, 2γ is the difference between the distances of a data point to the cluster centers, at and above which a crisp assignment is used. Clearly, this is equivalent to saying that the distances, if measured in 2γ units, must differ by less than 1 in order to obtain a graded assignment.

A disadvantage of this update rule is that it refers to the difference of the distances rather than their ratio, which seems more intuitive. As a consequence, a data point that has distance x to one cluster and distance y to the other is assigned in exactly the same way as a data point that has distance $x + z$ to the first cluster and distance $y + z$ to the second, regardless of the value of z (provided $z \geq -\min\{x, y\}$).

Alternatives to the discussed approaches modified the Shannon entropy term, using, for instance, $f(u_{ij}) = u_{ij} \ln u_{ij} + (1 - u_{ij}) \ln(1 - u_{ij})$ [79], or replaced it with Kullback–Leibler information divergence [55] to the (estimated) cluster probability distribution [43], that is, $f(u_{ij}) = u_{ij} \ln \frac{u_{ij}}{p_i}$ with $p_i = \frac{1}{n} \sum_{j=1}^{n} u_{ij}$.

It has also been tried to use $f(u_{ij}) = u_{ij}^w$ [66,78], but combined with $h(u_{ij}) = u_{ij}^w$ (to avoid technical complications), so that the objective function is effectively

$$J(\mathbf{X}, \mathbf{C}, \mathbf{U}) = \sum_{i=1}^{c} \sum_{j=1}^{n} u_{ij}^w (d_{ij}^2 + \gamma).$$

Hence, this is actually a hybrid approach that combines membership transformation and regularization. Another hybrid approach, proposed in [75], combines $h(u_{ij}) = u_{ij}^w$ and Shannon entropy regularization $f(u_{ij}) = u_{ij} \ln u_{ij}$. Finally, a generalized objective function was presented in [9] and analyzed in more detail in [80].

It should be noted, though, that the approach of [33], which is covered by the generalized objective function of [9] and based on

$$J(\mathbf{X}, \mathbf{C}, \mathbf{U}) = \sum_{i=1}^{c} \sum_{j=1}^{n} u_{ij}^w d_{ij}^2 - \gamma \sum_{i=1}^{c} p_i^2 \quad \text{with} \quad p_i = \frac{1}{n} \sum_{j=1}^{n} u_{ij},$$

is *not* a membership regularization scheme, as it yields crisp assignments unless $w > 1$. In this approach, the entropy term (which is added rather than subtracted) serves the purpose to choose the number of clusters automatically.

A closely related approach is *possibilistic clustering* [50,51], which eliminates the membership constraints $\forall j; 1 \leq j \leq n : \sum_{i=1}^{c} u_{ij} = 1$ and is based on the objective function

$$J(\mathbf{X}, \mathbf{C}, \mathbf{U}) = \sum_{i=1}^{c} \sum_{j=1}^{n} u_{ij}^w d_{ij}^2 + \sum_{i=1}^{c} \eta_i \sum_{j=1}^{n} (1 - u_{ij})^w.$$

Here the η_i are suitable positive numbers (one per cluster c_i, $1 \leq i \leq c$) that determine the distance at which the membership degree of a point to a cluster is 0.5. They are usually initialized, based on the result of a preceding run of standard fuzzy clustering, as the average fuzzy intra-cluster distance $\eta_i = \sum_{j=1}^{n} u_{ij}^w d_{ij}^2 / \sum_{j=1}^{n} u_{ij}^w$ and may or may not be updated in each iteration [50].

Although this approach is useful in certain applications, it should be noted that the objective function of possibilistic clustering is truly optimized only if all clusters are identical [72], because the missing constraints decouple the clusters. Thus, it actually *requires* that the optimization process gets stuck in a local optimum in order to yield useful results, which is a somewhat strange property.

22.2.6 Comparison

Since classical c-means clustering does not yield graded data point assignments, even if one allows the membership variables to take values in the unit interval, the objective function has to be modified if graded assignments are desired. There are two fundamental approaches to this: transforming the membership degrees or adding a membership regularization term. In both cases, variants can be derived that allow partially crisp assignments, that is, allow for vanishing membership degrees, as well as variants that enforce graded assignments regardless of the data. All of these variants have advantages and disadvantages: membership transformation suffers generally from the fact that the transformation function enters the cluster center update, but uses a fairly intuitive relative inverse squared distance scheme for the membership updates. Quadratic entropy regularization allows for vanishing membership degrees, but refers to distance differences rather than more intuitive distance ratios. Shannon entropy regularization leads to a procedure that is equivalent to the expectation maximization (EM) algorithm for a mixture of Gaussian and thus is not a specifically "fuzzy" approach anymore. However, judging from the discussion in [24] due to which the forced graded assignment is unproblematic, its practical advantages make it, in our opinion, the most recommendable approach.

22.3 Analysis of Precise Data with Possibility Theory

Probability theory is a well-known calculus for handling uncertainty. Possibility theory, as an alternative with close links to fuzzy sets, has been aired for quite some time, but it is still much less well known. As we already mentioned in Sect. 15.5, a possibility distribution π aims to quantify the state of knowledge of an agent. The aim is to distinguish between what is surprising from what is expected, and what is plausible from what is less plausible. It represents a flexible restriction on what is the actual state with the following convention: $\pi(u) = 0$ means that u is rejected as impossible, $\pi(u) = 1$ means that u is totally possible, and the larger $\pi(u)$ is, the more plausible u is.

Definition 22.1 A possibility distribution π on X is a mapping from the reference set X into the unit interval $[0, 1]$, $\pi : X \mapsto [0, 1]$, for which there exist an $x_0 \in X$ such that $\pi(x_0) = 1$.

A possibility distribution π is said to be at least as specific as another π', if and only if for each x the inequality $\pi(x) \leq \pi'(x)$ holds. Possibility theory is driven by the so-called *principle of minimal specificity*. It states that any hypothesis not known for certain to be impossible cannot be ruled out. Given a piece of information in the form of a fuzzy set μ, this principle leads to represent the knowledge by $\pi \leq \mu$. The minimal specificity principle enforces $\pi = \mu$ if no other piece of information is given (Fig. 22.3).

Fig. 22.3 Fuzzy set for *cloudy*

Example 22.1 Consider the fuzzy set $\mu_{cloudy} : X \mapsto [0, 1]$, where $X = [0, 100]$, as the description of the imprecise concept cloudy. Here $x \in X$ denotes the clouding degree in percent, and $\mu_{cloudy}(x)$ is the membership degree of x to the fuzzy set μ_{cloudy}.

Suppose we are interested in modeling our knowledge about the degree of clouding observable at Brunswick Airport on September 12, 1982, at 2 p.m. There exists exactly one true clouding value at that time, and the only piece of information is a remark of a colleague that the weather was "cloudy" on that day. In order to model our uncertainty about the true but unknown value, a reasonable estimate is to use the possibility distribution π defined by $\pi(x) := \mu(x)$, for all $x \in \mathbb{R}$. Here 40 is rejected as impossible, 70 is totally possible for a possibility distribution, while 60 is possible with a degree of 0.7.

Given a possibility distribution for elementary events, we can also give answers to question such as "Does the true value lie in the set A?". For example, choosing $A = [60, 70]$, we obtain by inspecting the possibility distribution in Fig. 22.4 that A is fully possible, because the true, but unknown value could be 65. But it is neither certain nor necessary that the event A occurs. Using measure theory one can answer such questions in a formalized way.

Definition 22.2 Let $\pi : X \mapsto [0, 1]$ be a possibility distribution. Then the possibility degree of a subset A of X with respect to π is defined as

$$\Pi(A) := \sup \{\pi(x) : x \in A\},$$

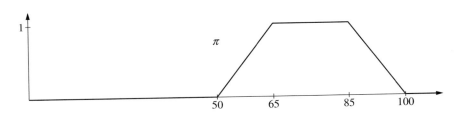

Fig. 22.4 Possibility distribution

and the necessity degree of a subset A of X with respect to π is defined as

$$N(A) := \inf\{1 - \pi(x) : x \in \overline{A}\}.$$

$\Pi(A)$ evaluates to what extent A is consistent with π, while $N(A)$ evaluates to what extent A is certainly implied. The duality is expressed by $N(A) = 1 - \Pi(\overline{A})$ and is valid for all A. Generally

$$\Pi(X) = N(X) = 1$$
$$\Pi(\emptyset) = N(\emptyset) = 0, \text{ and}$$
$$\Pi(A \cup B) = \max\{\Pi(A), \Pi(B)\} \text{ for all } A \text{ and } B$$
$$\Pi(A \cap B) \leq \min\{\Pi(A), \Pi(B)\} \text{ for all } A \text{ and } B$$

holds, but there are cases for which the strict inequality holds.

Example 22.2 Five observers A, B, C, D, E provide information about the same, but unknown airplane. The plane could be of type 1, 2, or 3. The observers are allowed to mark with an × in a table (see Table 22.1), whether it is possible that the observed plane is of a specific type. They are allowed to provide precise information (e.g., A states that the airplane is of type 1) or imprecise information (e.g., B states that it was of type 1 or 2). We can summarize the information given by the observers by using the (normalized) relative frequencies as degrees of possibility, and we obtain as a result $\pi : \{1, 2, 3\} \mapsto [0, 1], \pi(1) = 1, \pi(2) = 0.6, \pi(3) = 0.4$.

There are several interpretations for a possibility degree [29,35]. In the above example, we used the so-called context model. In this interpretation, a set of contexts is presumed from each of which an actual situation can be described. These contexts could be, e.g., several observers that have to state their individual observations. These need not necessarily be disjoints. By these contexts a probability distribution is defined which gives the probability of each context to be used for a description. Each context contains the possible values or (with metric attributes) an interval. A possibility degree is then attributed to each value from the base set, which is equal to

Table 22.1 Information given by the five observers about possible plane types

Observer	Observation		
	1	2	3
A	×		
B	×	×	
C	×		×
D	×	×	×
E	×	×	
Total	5	3	2
Rel. Frequency	1.0	0.6	0.4

the sum of probabilities of the contexts under which the value is possible. Therefore, with this model a possibility degree states the probability that a value is possible (and *not* the probability of the value). Formally the possibility degree can be defined with the help of a random set (i.e., a set-valued random variable) and its envelope. From this point of view, it follows immediately that possibility degrees can be interpreted as upper bounds for probabilities since with the mapping of a context's complete probability to each of its values the upper bound of a conditional probability of 1.0 has been accounted for. Possibility measures belong to the class of fuzzy measures.

Definition 22.3 Let X be a finite set, and let g be a function from 2^X to $[0, 1]$. The function g is called a fuzzy measure if and only if

(i) $g(\emptyset) = 0$, $g(X) = 1$ and
(ii) for all $A, B \subseteq X$, $A \subseteq B$ the inequality $g(A) \leq g(B)$ holds.

Probability measures $p : 2^X \mapsto [0, 1]$ on a finite set X have the additivity property, i.e., if $A \cap B$ is empty, then $P(A \cup B) = P(A) + P(B)$, so they are monotone set functions and therefore fuzzy measures. Another type of fuzzy measures was defined by Sugeno: A Sugeno λ-measure g_λ has the property

$$g_\lambda(A \cup B) = g_\lambda(A) + g_\lambda(B) + \lambda g_\lambda(A) g_\lambda(B).$$

For $\lambda > -1$, a Sugeno measure is a fuzzy measure [56,71]. Note that there are several notions for integrals that can be obtained by starting from the concept of a fuzzy measure [4].

In agreement with other uncertainty theories, possibility and necessity measures represent degrees of plausibility and belief [23,77]. Notions of conditioning and several variants of independence between events exist in possibility theory. For example, possibilistic graphical models as an alternative to probabilistic graphical models [13] have been developed. Possibility theory has not been the main framework for engineering applications of fuzzy sets, but it has significant potential for further applied developments.

22.4 Analysis of Imprecise Data Using Random Sets

Standard statistical data analysis is formally based on random variables and tries to derive properties of these random variables with the help of probability theory. However, most commonly the considered random variables map to the real numbers $\mathbb{R}$ and thus are restricted to precise values. A natural extension that allows for imprecision are so-called *random sets*, which are set-valued random variables and thus can map, for example, to subsets of a finite set or to intervals of the real numbers. While in the case of a random variable $X : \Omega \mapsto A$ the result of an experiment is an element of the set A, in the case of a random set $X : \Omega \mapsto 2^A$ the result of an experiment is a subset of A, i.e., an element of the power set of A.

A random set generalizes the notion of a random variable: In the case of a random variable $X : \Omega \mapsto A$, the result of an experiment is an element of the set A, whereas in case of a random set $X : \Omega \mapsto 2^A$, the result of an experiment is a subset of A, i.e., an element of the power set of A.

The modern theory of random sets was initiated in the 70s by [48] and [61], and it has been applied in different fields such as economy and stochastic geometry. Let us illustrate the main ideas and techniques by studying two examples.

Example 22.3 Let us consider the following set of languages $L = \{$ English, German, French, Spanish $\}$. Suppose that we randomly select a person ω from the set Ω of employees of a working group and we ask about the languages the employee can speak. We can model the outcome of this experiment by means of a set-valued function $\Gamma : \Omega \mapsto 2^{L \setminus \{\emptyset\}}$, where $\Gamma(\omega) \subseteq L$ is the collection of languages the person ω can speak. We suppose that each person is able to speak at least one language in L. If every employee has the same probability to be selected, we use the uniform probability distribution P defined on the finite set Ω. The triplet (Ω, P, Γ) describes the random set adequately. Typical questions that are of interest in such a context are the following:

- What is the proportion P_1 of employees that can speak German and English and cannot speak any other language?
- What is the proportion P_2 of employees that can speak German or English but no other language?
- What is the proportion P_3 of employees that can speak German or English?
- What is the proportion P_4 of employees that can speak at least three languages?

Such questions can be answered by a formal analysis of the corresponding random set, respectively:

$$P_1 = P(\{\omega \in \Omega : \Gamma(\omega) = \{\text{English, German}\}\})$$
$$P_2 = P(\{\omega \in \Omega : \Gamma(\omega) \subseteq \{\text{English, German}\}\})$$
$$P_3 = P(\{\omega \in \Omega : \Gamma(\omega) \cap \{\text{English, German}\} \neq \emptyset\})$$
$$P_4 = P(\{\omega \in \Omega : |\Gamma(\omega)| \geq 3\}).$$

In the above example, we consider the outcome of the random set as a precise piece of information. In the scientific literature, this is called the conjunction interpretation of a set or the *ontic* view of a set [26].

Let $(\Omega, 2^{\Omega}, P)$ be a finite probability space and $\Gamma : \Omega \mapsto 2^U$ be a multi-valued mapping with non-empty images. In most applications, we are interested in the proportion of elements of Ω whose images "touch" a given subset A of U or are completely included in A. These proportions are called upper probability $P^\star(A)$ and lower probability $P_\star(A)$ and can be calculated by

$$P^\star(A) = P(\{\omega \in \Omega : \Gamma(\omega) \cap A \neq \emptyset\}), \quad A \subseteq U, \quad \text{and}$$
$$P_\star(A) = P(\{\omega \in \Omega : \Gamma(\omega) \neq \emptyset \wedge \Gamma(\omega) \subseteq A\}), \quad A \subseteq U.$$

According to the ontic view, the random set is seen as a "random object", i.e., a particular measurable mapping within the framework of canonical probability theory. Hence, it induces a probability measure of the corresponding σ-algebra, and its relevant parameters are calculated as a function of such a probability measure [64]. It is possible to formulate extensions of classical statistics to random sets. One can prove, for example, the strong law of large numbers for random sets, which is the key to inductive statistics [67].

Example 22.4 Let us consider the random experiment that consists in randomly selecting a day in the year 1982 and recording the interval given by the lowest and the highest recorded temperature at Brunswick Airport. The task is to study the distribution of the daily mean temperatures in Brunswick in 1982. We assume that we are just provided with the pair of minimum and maximum temperature and we have no further information about the remaining records during the day.

We can model the experiment again by using a (Laplace) probability P on Ω and the set-valued mapping $\Gamma : \Omega \mapsto 2^{\mathbb{R}}$, $\omega \mapsto [T_{min}(\omega), T_{max}(\omega)]$, where the intervals represent the possible temperatures on a date. In this experiment, the sets have a different semantics, because we know that for each day there is exactly one true mean value. There is an "original" random variable $X_0 : \Omega \to \mathbb{R}$ that assigns to each day its true mean temperature value. We know that for all $\omega \in \Omega$ the inequalities $T_{min}(\omega) \leq X_0(\omega) \leq T_{max}(\omega)$ hold, i.e., $X_0(\omega) \in \Gamma(\omega)$ for all $\omega \in \Omega$. But we do not know which one is the underlying original X_0. All random variables X with the property $X(\omega) \in \Gamma(\omega)$ are candidates for the original. These candidates are called "admissible" random variables or selectors. The random set is regarded as the representation of the perception of an otherwise point-valued random variable. Therefore, the information given by the random set is regarded the same as the information given in the set of admissible selectors. This identification can be used to define statistical parameters of a random set.

Let $(\Omega, 2^{\Omega}, P)$ be a probability space and $\Gamma : \Omega \to 2^{\mathbb{R}}$ a multi-valued (measurable) mapping with non-empty images. Then the expected value is defined as the set

$$E(\Gamma) = \{E(X) \mid X \text{ is random variable such that } E(X) \text{ exists and }$$
$$X(\omega) \in \Gamma(\omega) \text{ for all } \omega \in \Omega\}$$

and the variance as the set

$$\mathrm{Var}(\Gamma) = \{\mathrm{Var}(X) \mid X \text{ is random variable such that } \mathrm{Var}(X) \text{ exists and }$$
$$X(\omega) \in \Gamma(\omega) \text{ for all } \omega \in \Omega\}.$$

$E(\Gamma)$ was studied in [1], $\mathrm{Var}(\Gamma)$ in [52]. Descriptive statistics can be extended using these ideas to random sets [53].

22.5 Analysis of Fuzzy Data with Fuzzy Random Variables

In this section, we study a generalization of the concept of a random set. We assume that the uncertainty handling is modeled by a probability space. The data are modeled by fuzzy sets, where there are two different interpretations in the context of statistical data analysis [18]: In the *epistemic* view, fuzzy sets are used to represent incomplete knowledge about an underlying object or a precise quantity [52,57]. In the *ontic* view, fuzzy sets are considered as real complex humped entities [11,68].

Example 22.5 Let us consider the languages $S = \{$English, German, French, Spanish$\}$. Suppose the set Ω of employees of a working group have different language skills, and we want to model the information about the language skills of the persons.

If we know the persons CEFR levels (European Language Test), then we assign for each person and for each language a membership value on a $[0, 1]$-scale, e.g., by using normalized CEFR levels. The fuzzy sets $\mu : S \mapsto [0, 1]$ describe the language skills of a person. Note that the fuzzy sets are considered as precise pieces of information. We use an ontic view of a fuzzy set [19].

In order to use the information for statistical purposes, it is convenient to model the uncertainty of selecting a person by a probability space on Ω and the information about the persons by fuzzy sets. We obtain a mapping $\Gamma : \Omega \mapsto F(S)$, which is called a fuzzy random variable. The statistical analysis of a fuzzy random variable $\Gamma : \Omega \mapsto F(X)$ with ontic fuzzy sets on a probability space (Ω, B, P) is mathematically demanding. The standard way is to define a suitable σ-algebra over a class of fuzzy sets on X. Then a fuzzy random variable is a measurable mapping, and the probability distribution induced by this generalized random variable can be used to analyze the probabilistic information. A typical question in our example could be "What is the probability of the people that speak both English and German to a degree of at least 0.8?". This question could be answered by calculating the probability $P(\{\omega \in \Omega : \Gamma(\omega) \leq \mu\}$, where $\mu : S \mapsto [0, 1]$ and $\mu(\text{English}) = \mu(\text{German}) = 0.8$, $\mu(\text{French}) = \mu(\text{Spanish}) = 1$.

In applications with objects described as fuzzy sets of the real line, a restriction to fuzzy sets μ with compact α-cuts $[\mu]_\alpha, \alpha > 0$, is mathematically convenient. In that case we can define metrics on this subset of fuzzy sets of the real line by generalizing the Hausdorff pseudometric defined for subsets of the real line:

$$d(\mu, \nu) = \int_0^1 d_H([\mu]_\alpha, [\nu]_\alpha) \, d\alpha,$$

where

$$d_H(A, B) = \max \left\{ \sup_{a \in A} \inf_{b \in B} \{|a - b|\}, \sup_{b \in B} \inf_{a \in A} \{|a - b|\} \right\}$$

is the Hausdorff pseudometric defined for non-empty sets $A, B \subseteq \mathbb{R}$.

Besides the equipment of the space of fuzzy (compact) objects with topological properties we can additionally equip the space with algebraic operations such as an addition and a scalar multiplication by extending the Minkowski operations $A + B = \{a + b : a \in A, b \in B\}$ by using Zadeh's extension principle to fuzzy compact objects. Under these structures one can prove the generalized strong law of large numbers [67], which gives the key to inductive statistics with random fuzzy objects.

Let us now consider a random experiment where the data are described by fuzzy sets with an epistemic interpretation. Then our model consists of a probability space (Ω, B, P) and a mapping $\Gamma : \Omega \mapsto F(X)$ with possibility distributions as values. The mapping is again called a fuzzy random variable, but in this situation the mapping is used for modeling a situation in which a second-order uncertainty occurs.

Example 22.6 Suppose we want to analyze statistical observations on a database, in which the clouding degree at noon for several days are stored. The information about the true clouding degrees is sometimes precise, but in some cases interval valued (using min-max information) and for some days only described by subjective information (using possibility distributions).

In order to explain the procedure, we use the very small data set of Fig. 22.5 consisting only of three data points $\Gamma(\omega_1)$, $\Gamma(\omega_2)$, $\Gamma(\omega_3)$. The first observation is exactly 60, and the second observation is between 50 and 60, but we have no further information about the true value, the underlying "original". Our information about the third datum is encoded by using a possibility distribution: We have some background knowledge about the location of the unknown original. In order to perform descriptive statistics with this data set, we want to determine suitable location and range parameters such as the mean value and the variance. How to define such parameters? This can be done by

(1) defining a possibility distribution on the set of all random variables that describes the possibility of a random set being the original and
(2) applying the extension principle to the mapping that assigns to each random variable its expected value.

Fig. 22.5 Expected value of a random sample of fuzzy sets

Let $U : \Omega \to X$ be a random variable. We know the possibility degree that $U(\omega)$ is the original of $\Gamma(\omega)$ is $\Gamma(\omega)\big(U(\omega)\big)$, and according to the insufficient reason principle we conclude that the possibility that U is the original on Γ is

$$\pi_\Gamma(U) := \inf_{\omega \in \Omega} \left\{ \Gamma(\omega)\big(U(\omega)\big) \right\}$$

Let $E(U)$ denote the expected value of a random variable U with existing expected value, and π_Γ the possibility distribution on the set of all random variables defined above. Consider the mapping E from the set of all random variables to the real numbers that assigns to each random variable U its expected value $E(U)$. Using the extension principle for defining the expected value for fuzzy subsets of the set of random variables, we obtain a reasonable definition for the expected value of a fuzzy random variable Γ.

Definition 22.4 Let $\Gamma : \Omega \to F(X)$ be a fuzzy random variable. Then the expected value $E(\Gamma) : X \to [0, 1]$ is a fuzzy set of X defined by

$$x \mapsto \sup_{\substack{U : \Omega \to X, \\ E(U) = x}} \left\{ \min_{\omega \in \Omega} \left\{ \Gamma(\omega)\big(U(\omega)\big) \right\} \right\}.$$

Parameters such as the variance can be defined in the same way. Note that for finite probability spaces $\Omega = \{\omega_1, \ldots, \omega_n\}$ and possibility distributions on the real line $\mathbb{R}$ the calculation of the parameters can be facilitated by using α-cuts. Then for each $\alpha > 0$ the α-cut $[E(\Gamma)]_\alpha$ is equal to the weighted Minkowski sum of the α-cuts of the data, i.e.,

$$[E(\Gamma)]_\alpha = \sum_{\omega \in \Omega} P\{\omega\} \cdot [\Gamma(\omega)]_\alpha \text{ for } \alpha > 0.$$

If the α-cuts are closed intervals, then the calculation of the expected value can be efficiently done by using simple interval arithmetic.

Using this technique, we can extend classical statistical notions such as the distribution function to the case of fuzzy random variables. Conceptually, the approach can be seen as a built-in "sensitivity analysis".

References

1. Aumann RJ (1965) Integrals of set-valued functions. J Math Anal Appl 23(1):1–12
2. Ball GH, Hall DJ (1967) A clustering technique for summarizing multivariate data. Behav Sci 12(2):153–155. J Wiley & Sons, Chichester, United Kingdom
3. Bandemer H, Näther W (1992) Fuzzy data analysis. Kluwer, Dordrecht, Netherlands
4. Bede B (2013) Mathematics of fuzzy sets and Fuzzy Logic. Springer, Berlin Heidelberg
5. Berthold MR, Borgelt C, Höppner F, Klawonn F, Silipo R (2020) Guide to intelligent data science. Springer

6. Bezdek JC (1981) Pattern recognition with fuzzy objective function algorithms. Plenum Press, New York, NY, USA

7. Bezdek JC, Pal N (1992) Fuzzy models for pattern recognition. IEEE Press, New York, NY, USA

8. Bezdek JC, Keller JM, Krishnapuram R, Pal N (1999) Fuzzy Models and Algorithms for Pattern Recognition and Image Processing. Kluwer, Dordrecht, Netherlands

9. Bezdek JC, Hathaway RJ (2003) Visual Cluster Validity (VCV) displays for prototype generator clustering methods. In: Proceedingsof the 12th IEEE Iternational Conference on Fuzzy Systems (FUZZ-IEEE 2003, Saint Louis, MO), 2:875-880. IEEE Press, Piscataway, NJ, USA 2003

10. Bilmes J (1997) A gentle tutorial on the EM algorithm and its application to parameter estimation for Gaussian mixture and hidden Markov models. Tech Report ICSI-TR-97-021. University of Berkeley, CA, USA

11. Blanco-Fernández A, Casals MR, Colubi A, Coppi R, Corral N, Rosa de Sáa S, D'Urso P, Ferraro MB, García-Bárzana M, Gil MA, Giordani P, González-Rodríguez G, López MT, Lubiano MA, Montenegro M, Nakama T, Ramos-Guajardo AB, Sinova B, Trutschnig W (2012) Arithmetic and distance-based approach to the statistical analysis of imprecisely valued data. In: Borgelt et al. 1–18

12. Borgelt C (2005) Prototype-based classification and clustering. Otto-von-Guericke-University of Magdeburg, Germany, Habilitationsschrift

13. Borgelt C, Steinbrecher M, Kruse R (2009) Graphical models, 2nd edn. Wiley, Chichester

14. Borgelt C, Gil MA, Sousa JMC, Verleysen M (eds) (2012/2013) Towards advanced data analysis by combining soft computing and statistics. Studies in fuzziness and soft computing, vol 285. Springer-Verlag, Berlin/Heidelberg, Germany

15. Borgelt C, Kruse R (2015) Bedeutung von Fuzzy Zugehörigkeitsgraden in der Fuzzy Technologie. Informatik Spektrum, Band 38, Heft 6, 490–499

16. Boujemaa N (2000) Generalized competitive clustering for image segmentation. In: Proceedings of the 19th International Meeting North American Fuzzy Information Processing Society (NAFIPS (2000) Atlanta, GA), 133–137. IEEE Press, Piscataway, NJ, USA

17. Couso I, Dubois D (2009) On the variability of the concept of variance for fuzzy random variables. IEEE Trans Fuzzy Syst 17:1070–1080. IEEE Press, Piscataway, NJ, USA

18. Couso I, Dubois D, Sanchez L (2014) Random sets and random fuzzy variables as ill-perceived random variables, Springer-Verlag, Cham

19. Couso I, Borgelt C, Hullermeier E, Kruse R (2019) Fuzzy sets in data analysis: from statistical foundations to machine learning. IEEE Comput Intell Mag 14(1):31–44

20. Daróczy Z (1970) Generalized information functions. Inf Control 16(1):36–51. Academic Press, San Diego, CA, USA

21. Davé RN, Krishnapuram R (1997) Clustering robust, methods: a unified view. IEEE Trans Fuzzy Syst 5:270–293. IEEE Press, Piscataway, NJ, USA

22. Dempster AP, Laird N, Rubin D (1977) Maximum Likelihood from incomplete data via the EM algorithm. J Roy Stat Soc: Ser B 39:1–38. Blackwell, Oxford, United Kingdom

23. Destercke S, Denoeux T, Cuzzolin F, Martin A (2018) Belief functions: theory and applications. Springer lecture notes in Artificial Intelligence, vol 11069

24. Döring C, Borgelt C, Kruse R (2005) Effects of irrelevant attributes in fuzzy clustering. In: Proceedings of the 14th IEEE International Conference on Fuzzy Systems (FUZZ-IEEE'05, Reno, NV)

25. Dubois D, Prade H (eds) (2000) Fundamentals of fuzzy sets kluwer, Boston

26. Dubois D, Prade H (2012) Gradualness, uncertainty, bipolarity: making sense of fuzzy sets. Fuzzy Sets Syst 192:3–24

27. Dubois D, Nguyen HT, Prade H (2000) Fuzzy sets and probability: misunderstandings, bridges, gaps. In: Dubois and Prade (eds) 343–438

28. Dubois D (2012) Statistical reasoning with set-valued information: Ontic vs. Epistemic views. In: Borgelt et al. (eds) 119–136

29. Dubois D, Prade H (2015) Possibility theory and its applications: where do we stand? In: Kacprzyk J, Pedrycz W (eds) Springer Handbook of computational intelligence. Springer, Dordrecht, pp 31–60

30. Dunn JC (1973) A fuzzy relative of the ISODATA process and its use in detecting compact well-separated clusters. J Cybern 3(3):32–57. American Society for Cybernetics, Washington, DC, USA 1973. Reprinted in [Bezdek and Pal 1992], 82–101

31. Everitt BS (1981) Cluster analysis. Heinemann, London, United Kingdom

32. Everitt BS, Hand DJ (1981) Finite mixture distributions. Chapman & Hall, London, United Kingdom

33. Frigui H, Krishnapuram R (1997) Clustering by competitive agglomeration. Pattern Recognit 30(7):1109–1119. Pergamon Press, Oxford, United Kingdom

34. Gath I, Geva AB (1989) Unsupervised optimal fuzzy clustering. IEEE Trans Pattern Anal Mach Intell (PAMI) 11:773–781. IEEE Press, Piscataway, NJ, USA 1989. Reprinted in [Bezdek and Pal 1992], 211–218

35. Gebhardt J, Kruse R (1993) The context model: an integrating view of vagueness and uncertainty. Int J Approximate Reasoning 9:283–314

36. Grabisch M, Murofushi T, Sugeno M (eds) (2000) Fuzzy measures and integrals: theory and applications ,Springer, Studies in Fuzziness and Soft Computing

37. Gustafson EE, Kessel WC (1979) Fuzzy clustering with a fuzzy covariance matrix. In: Proceedings of the IEEE Conference on Decision and Control (CDC 1979, San Diego, CA)

38. Hartigan JA, Wong MA (1979) A k-Means clustering algorithm. Appl Stat 28:100–108. Blackwell, Oxford, United Kingdom

39. Höppner F, Klawonn F, Kruse R, Runkler T (1999) Fuzzy cluster analysis. Wiley, Chichester, United Kingdom

40. Honda K, Ichihashi H (2005) Regularized linear fuzzy clustering and probabilistic PCA mixture models. IEEE Trans Fuzzy Syst 13(4):508–516. IEEE Press, Piscataway, NJ, USA

41. Hüllermeier E (2005) Fuzzy-methods in machine learning and data mining: status and prospects. Fuzzy Sets Syst 156(3):387–407. Elsevier, Amsterdam, Netherlands

42. Hüllermeier E (2011) Fuzzy sets in machine learning and data mining. Appl Soft Comput 11(2):1493–1505, Elsevier, Amsterdam, Netherlands

43. Ichihashi H, Miyagishi K, Honda K (2001) Fuzzy c-means clustering with regularization by K-L information. In: Proceedings 10th IEEE International Conference on Fuzzy Systems (FUZZ-IEEE 2001, Melbourne, Australia), 924–927. IEEE Press, Piscataway, NJ, USA

44. Jain AK, Dubes RC (1988) Algorithms for clustering data. Prentice Hall, Englewood Cliffs, NJ, USA

45. Jajuga K (2003) L_1-norm based fuzzy clustering. Fuzzy Sets Syst 39(1):43–50. Elsevier, Amsterdam, Netherlands

46. Karayiannis NB (1994) MECA: maximum entropy clustering algorithm. In: Proceedings of the 3rd IEEE International Conference on Fuzzy Systems (FUZZ-IEEE 1994, Orlando, FL)

47. Kaufman L, Rousseeuw P (1990) Finding groups in data: an introduction to cluster analysis. Wiley, New York, NY, USA

48. Kendall DG (1974) Foundations of a theory of random sets. In: Harding EF, Kendall DG (eds) Stochastic geometry, 322–376. Wiley, New York

49. Klawonn F, Höppner F (2003) What is fuzzy about fuzzy clustering? understanding and improving the concept of the fuzzifier. In: Proceedings of the 5th International Symposium on Intelligent Data Analysis (IDA 2003, Berlin, Germany)

50. Krishnapuram R, Keller JM (1993) A possibilistic approach to clustering. IEEE Trans Fuzzy Syst 1(2):98–110. IEEE Press, Piscataway, NJ, USA

51. Krishnapuram R, Keller JM (1996) The possibilistic c-means algorithm: insights and recommendations. IEEE Trans Fuzzy Syst 4(3):385–393. IEEE Press, Piscataway, NJ, USA

52. Kruse R (1987) On the variance of random sets. J Math Anal Appl 122:469–473. Elsevier, Amsterdam, Netherlands

53. Kruse R, Meyer KD (1987) Statistics with vague data. D. Reidel Publishing Company, Dordrecht, Netherlands

54. Kruse R, Berthold MR, Moewes C, Gil MA, Grzegorzewski P, Hryniewicz O(eds) (2012) Synergies of soft computing and statistics for intelligent data analysis. Advances in intelligent systems and computing, vol 190. Springer-Verlag, Heidelberg/Berlin, Germany

55. Kullback S, Leibler RA (1951) On information and sufficiency. Ann Math Stat 22:79–86. Institute of Mathematical Statistics, Hayward, CA, USA

56. Kruse R (1982) A note on λ-additive fuzzy measures. Fuzzy Sets Syst 8(2):219–222. Elsevier, Amsterdam, Netherlands

57. Kwakernaak H (1978) Fuzzy random variables—I. Definitions and theorems. Inf Sci 15:1–29. Elsevier, Amsterdam, Netherlands

58. Kwakernaak H (1979) Fuzzy random variables—II. Algorithms and examples for the discrete case. Inf Sci 17:252–278. Elsevier, Amsterdam, Netherlands

59. Li RP, Mukaidono M (1995) A maximum entropy approach to fuzzy clustering. In: Proceedings of the 4th IEEE International Conference on Fuzzy Systems (FUZZ-IEEE 1994, Yokohama, Japan)

60. Lloyd S (1982) Least squares quantization in PCM. IEEE Trans Inf Theor 28:129–137. IEEE Press, Piscataway, NJ, USA

61. Matheron G (1975) Random sets and integral geometry. Wiley, New York

62. Miyamoto S, Mukaidono M (1997) Fuzzy c-means as a regularization and maximum entropy approach. In: Proceedings of the 7th International Fuzzy Systems Association World Congress (IFSA'97, Prague, Czech Republic)

63. Miyamoto S, Umayahara K (1998) Fuzzy clustering by quadratic regularization. In: Proceedings of the IEEE International Conference on Fuzzy Systems/IEEE World Congress on Computational Intelligence (WCCI 1998, Anchorage, AK)

64. Molchanov I (2006) Theory of random sets. Springer Science & Business Media

65. Mori Y, Honda K, Kanda A, Ichihashi H (2003) A unified view of probabilistic PCA and regularized linear fuzzy clustering. In: Proceedings of the International Joint Conference on Neural Networks (IJCNN, (2003) Portland, OR) I:541–546. IEEE Press, Piscataway, NJ, USA

66. Özdemir D, Akarun L (2002) A fuzzy algorithm for color quantization of images. Pattern Recogn 35:1785–1791. Pergamon Press, Oxford, United Kingdom

67. Puri M, Ralescu D (1983) Strong law of large numbers for Banach space valued randomsets. Ann Probab 11:22–224

68. Puri M, Ralescu D (1986) Fuzzy random variables. J Math Anal Appl 114:409–422. Elsevier, Amsterdam, Netherlands

69. Ruspini EH (1969) A new approach to clustering. Inf Control 15(1):22–32. Academic Press, San Diego, CA, USA. Reprinted in [Bezdek and Pal 1992], 63–70

70. Shannon CE (1948) The mathematical theory of communication. Bell Syst Tech J 27:379–423. Bell Laboratories, Murray Hill, NJ, USA

71. Sugeno M, Murofushi T (1987) Pseudo-additive measures and integrals. J Math Anal Appl 122:197–222

72. Timm H, Borgelt C, Döring C, Kruse R (2004) An extension to possibilistic fuzzy cluster analysis. Fuzzy Sets Syst 147:3–16. Elsevier Science, Amsterdam, Netherlands

73. Viertl R (2011) Statistical methods for fuzzy data. Wiley, Chichester, UK

74. Walley P (1991) Statistical reasoning with imprecise probabilities. Chapman and Hall, London

75. Wei C, Fahn C (2002) The multisynapse neural network and its application to fuzzy clustering. IEEE Trans Neural Networks 13(3):600–618. IEEE Press, Piscataway, NJ, USA

76. Wolkenhauer O (2004) Data engineering: fuzzy mathematics in systems theory and data analysis. Wiley

77. Yager RR, Liu L (eds) (2008) Classic works of the Dempster-Shafer theory of belief functions. vol 219. Springer

78. Yang MS (1993) On a class of fuzzy classification maximum likelihood procedures. Fuzzy Sets Syst 57:365–375. Elsevier, Amsterdam, Netherlands

79. Yasuda M, Furuhashi T, Matsuzaki M, Okuma S (2001) Fuzzy clustering using deterministic annealing method and its statistical mechanical characteristics. In: Proceedings of the 10th IEEE International Conference on Fuzzy Systems (FUZZ-IEEE 2001, Melbourne, Australia)

80. Yu J, Yang MS (2007) A generalized fuzzy clustering regularization model with optimality tests and model complexity analysis. IEEE Trans Fuzzy Syst 15(5):904–915. IEEE Press, Piscatway, NJ, USA

81. Zadeh L (1978) Fuzzy sets as a basis for a theory of possibility. Fuzzy Sets Syst 1:3–28

Part IV
Bayes and Markov Networks

Relational database systems are among the most widespread data management systems in today's businesses. A database typically consists of several tables that contain data about business objects such as customer data, sales orders, or product information. Each table row represents a description of a single object with each table column representing an attribute of that object. Relations between these objects are also modeled via tables. Please note that we use the notions *table* and *relation* interchangeably. A major part of database theory is concerned with the task to represent data with as little redundancy as possible.

A Fictitious Example
Before we introduce the theoretical background, we will illustrate the concepts in the setting of a fictitious car manufacturer. For each part of a car, a table with different suppliers is maintained. For the sake of simplicity, let us assume there are only three different parts: engine, transmission, and brakes.

Tables 23.1, 23.2, and 23.3 show example values for these parts. The first column denotes the primary key, that is, the identifying table entry. Let us assume further that all attributes' value combinations are possible and hence there are 36 different unique car models. The above-mentioned decomposition for redundancy reduction is not shown here. For example, in real life, there would be another table containing the address and contact data for each supplier which in turn are referenced in Tables 23.1, 23.2, and 23.3 by their name.

Apart from just querying the database, we would like to go further than this and draw *inferences* like "Supplier X can only deliver transmission t_4 at the moment. Which consequences have to be inferred for the procurement of brakes and engines?". Answering such questions involves historical information (e.g., installation rates of single car parts from the past) and expert knowledge (e.g., the allowed technical combinations of parts) that are modeled and exploited via probability theory. Such a probability distribution is exhaustively (together with all marginal distributions)

© Springer Nature Switzerland AG 2022
R. Kruse et al., *Computational Intelligence*, Texts in Computer Science,
https://doi.org/10.1007/978-3-030-42227-1_23

Table 23.1 Table engines

EID	Power	Type	...
e_1	100 kW	Diesel	...
e_2	150 kW	Otto	...
e_3	200 kW	Otto	...

Table 23.2 Table brakes

BID	Material	Manufacturer	...
b_1	Steel	Firm 1	...
b_2	Steel	Firm 2	...
b_3	Ceramic	Firm 2	...

Table 23.3 Table transmissions

TID	Gears	Automatic	...
t_1	4	n	...
t_2	5	n	...
t_3	5	y	...
t_4	6	y	...

displayed in Fig. 23.1. Table 23.4 sketches how they are stored inside the database. The above question how the distribution of brakes and engines changes if only transmission t_4 is available is shown in Fig. 23.2. Obviously, all combinations with $T \neq t_4$ become impossible. Since the result needs to be another probability distribution, all remaining entries of the "slice" $T = t_4$ were normalized to sum up to one. This was done by dividing all entries by the marginal installation rate $P(T = t_4) = 0.280$. Now, let dom(E) be the domain of attribute E, that is, the set of all possible values E can assume. To determine, e.g., the new installation rates for brake b_1, the following implicit steps are required:

$$
P(B = b_1 \mid T = t_4) = \frac{\displaystyle\sum_{e \in \mathrm{dom}(E)} P(E = e, B = b_1, T = t_4)}{\displaystyle\sum_{e \in \mathrm{dom}(E)} \sum_{b \in \mathrm{dom}(B)} P(E = e, B = b, T = t_4)}
$$

$$
= \frac{8 + 17 + 9}{80 + 17 + 3 \ + \ 72 + 68 + 6 \ + \ 8 + 17 + 9} \approx 0.122.
$$

These summations clearly show what effort it takes to compute the conditional distributions directly. Dealing with a three-dimensional database, such a procedure is still feasible. In real-world applications, however, one has to deal with hundreds of

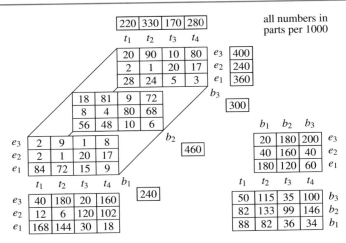

Fig. 23.1 Three-dimensional probability distribution over the attributes Transmission, Engine, and Brakes. Additionally, the marginal distributions (sums over rows, columns, or both) are shown

Table 23.4 Sketch of the ternary relation with the relative frequencies of all 36 possible car combinations. The full relation can be found in Fig. 23.1

TID	EID	BID	$P(\cdot)$
t_1	e_1	b_1	0.084
t_1	e_1	b_2	0.056
⋮	⋮	⋮	⋮
t_4	e_3	b_2	0.072
t_4	e_3	b_3	0.080

attributes having considerably larger domains. Assuming a more realistic number of attributes of 200 and only three values per attribute, the state space has a cardinality of $3^{200} \approx 2.6 \cdot 10^{95}$ and thus contains more unique combinations than there are elementary particles in the universe, that is, an estimated number of 10^{87} elementary particles. Even a summation over subsets of the distribution is impossible. Another challenge is to efficiently store the distribution since an enumeration of all possible combinations is infeasible and most of them are likely never used at all. This can be seen as follows: assume that all vehicles on the planet were built by our manufacturer and no two vehicles are identical (w.r.t. their attribute values). Even then the estimated number of just below a billion of those vehicles is negligible compared to the number 3^{200} of theoretically possible ones.

The main idea to efficiently store and use a high-dimensional probability distribution p (containing the knowledge about a certain area of application) is to decompose it into a set $\{p_1, \ldots, p_s\}$ of lower dimensional and possibly overlapping distributions. If this is achieved one can infer from it the same conclusions as from the original

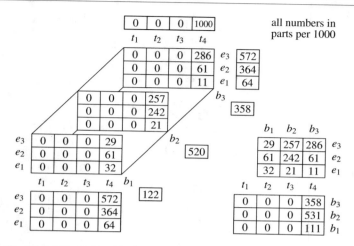

Fig. 23.2 Conditional distribution(s) given the condition $T = t_4$

Fig. 23.3 Graphical model of the relation shown in Fig. 23.1

distribution p. As one can easily verify, the following criterion is valid inside the example distribution $P(T, E, B)$ from Fig. 23.1:

$$P(T = t, E = e, B = b) = \frac{P(T = t, E = e) \cdot P(E = e, B = b)}{P(E = e)}. \qquad (23.1)$$

Obviously, it is sufficient to store the two-dimensional distributions over attributes B and E, and T and E in order to correctly reconstruct the original three-dimensional distribution since the one-dimensional distribution over E can be computed via (affordable) summation over one of the two two-dimensional distributions. The attribute E apparently plays an important role w.r.t. the decomposability since it is contained in both two-dimensional distributions. We will later refer to the criterion in Eq. 23.1 as *conditional independence* (of the attributes T and B given attribute E). Such conditional independences will later be intuitively represented by (directed or undirected) graphs. The objective is to be able to infer all valid stochastic statements from the underlying probability distribution by only using graph-theoretic criteria. The undirected graph for our small example is depicted in Fig. 23.3. The attributes for which we need a joint distribution are connected by an edge. Another advantage of a graphical representation will be the ability to use it to determine the paths and steps needed to update the remaining attributes in the presence of evidence (like known attribute values such as the value t_4 of attribute T in our example) without the need to reconstruct the original distribution.

The graph in Fig. 23.3 describes a path over which information about attribute T can be transferred via attribute E to infer a changed probability distribution of attribute B. Thus, only those two lower dimensional distributions are necessary that are represented by the edges:

$$P(B = b_1 \mid T = t_4) = \frac{1}{P(T = t_4)} \cdot \sum_{e \in \mathrm{dom}(E)} \frac{P(T = t_4, E = e) \cdot P(E = e, B = b_1)}{P(E = e)}$$

$$= \frac{1000}{280} \cdot \left(\frac{18 \cdot 180}{360 \cdot 1000} + \frac{102 \cdot 40}{240 \cdot 1000} + \frac{160 \cdot 20}{400 \cdot 1000} \right) = \frac{34}{280} \approx 0.122.$$

One can easily understand that in a scenario with a reasonable (large) number of attributes, only exploiting the previously sketched concepts allows for a feasible evidence propagation. Concluding the above ideas, the following list of questions will be answered in the next chapters:

1. How can expert knowledge about complex domains be efficiently represented? We will look into knowledge representations based on directed and undirected Bayes and Markov network.
2. How can inferences be drawn inside these representations? The graphical representations will provide the paths over which the evidence needs to be propagated inside the respective Bayes or Markov networks.
3. How can such graphical representations be automatically learned from data? We will discuss learning principles for Bayes and Markov networks.

Elements of Probability and Graph Theory

24

This chapter introduces required theoretical concepts for the definition of Bayes and Markov networks. After important elements of probability theory—especially (conditional) independences—are discussed, we present relevant graph-theoretic notions with emphasis on so-called separation criteria. These criteria will later allow us to capture probabilistic independences with an undirected or directed graph.

24.1 Probability Theory

The classical notion of probability and its interpretation in terms of relative frequencies are deeply embedded in our intuition. Modern mathematics embraced an axiomatic methodology which abstracts from specific meanings of objects. Rather, it assumes all objects to be given with no further property than their own identity (i.e., the objects are mutually distinguishable) and studies the relations amongst these objects resulting from postulated axioms.

Probability theory is formed in that axiomatic way by employing the so-called Kolmogorov axioms [3]. An event in this nomenclature is simply a set of elementary events that are distinguishable, i.e., that have an identity. A probability is then a number assigned to an event satisfying certain criteria that are defined by the above-mentioned axioms. Let us first define the fundamental notions of an *event algebra* and a σ-*algebra*.

Definition 24.1 (*Event algebra*) Let Ω be an event space (i.e., a universal set of elementary events). A system of subsets S over Ω is called *event algebra* if and only if the following conditions hold:

- The certain event Ω and the impossible event $\emptyset$ are in S.
- For every $A \in S$ the complement $\overline{A} = \Omega \backslash A$ is also contained in S.

© Springer Nature Switzerland AG 2022
R. Kruse et al., *Computational Intelligence*, Texts in Computer Science,
https://doi.org/10.1007/978-3-030-42227-1_24

- If A and B are in S, then $A \cup B$ and $A \cap B$ are also in S.

The following condition may also hold:

- If for all $i \in \mathbb{N}$ the event A_i is in S,
  then the events $\bigcup_{i=1}^{\infty} A_i$ and $\bigcap_{i=1}^{\infty} A_i$ are also in S.

In that case S is called a σ-*algebra* .

The semantic of $A \cup B$ is the event that occurs if A or B occurs. The intersection $A \cap B$ occurs if and only if A and B occur. The complement $\overline{A}$ occurs if and only if A does not occur. Two events A and B are *disjoint* if and only if they cannot occur simultaneously, i.e., if their intersection is the impossible event: $A \cap B = \emptyset$.

To assign a probability to an event, the so-called Kolmogorov axioms are used:

Definition 24.2 (*Kolmogorov axioms*) Let S be an event algebra over a finite event space Ω.

- The *probability* $P(A)$ of an event $A \in S$ is a uniquely defined non-negative number of value at most one, i.e., if $0 \le P(A) \le 1$ holds.
- The certain event Ω has probability one: $P(\Omega) = 1$
- *Addition axiom*: If the events A and B are disjoint $(A \cap B = \emptyset)$, then $P(A \cup B) = P(A) + P(B)$ holds.

In event spaces Ω containing infinitely many elementary events, S has to be a σ-algebra and the addition axiom has to be replaced by:

- *Extended addition axiom*: If $A_1, A_2, \ldots$ are countably infinitely many pairwise disjoint events, then

$$P\left(\bigcup_{i=1}^{\infty} A_i\right) = \sum_{i=1}^{\infty} P(A_i).$$

These three axioms already implies the following (incomplete) list of properties:

- $\forall A \in S : P(\overline{A}) = 1 - P(A)$.
- $P(\emptyset) = 0$.
- For pairwise disjoint events $A_1, \ldots, A_n$ holds:

$$P\left(\bigcup_{i=1}^{n} A_i\right) = \sum_{i=1}^{n} P(A_i).$$

- For any (not necessarily disjoint) events A and B holds:

$$P(A \cup B) = P(A) + P(B) - P(A \cap B).$$

The Kolmogorov axioms are consistent since there exist systems that satisfy all axioms. Kolmogorov's axioms allow for an embedding of probability theory into measure theory and to interpret the probability as an non-negative normalized additive set function, i.e. as a measure.

Since the definitions of event algebra and Kolmogorov axioms are not unique but represent a class of set systems and functions, respectively, one has to specify precisely for every application the underlying objects. This is done with the notion of a *probability space*.

Definition 24.3 (*Probability space*) Let Ω be an event space, $\mathcal{S}$ a σ-algebra over Ω and P a probability on $\mathcal{S}$. The the triple $(\Omega, \mathcal{S}, P)$ is called a *probability space*.

Up to now, we only computed the probabilities of events without discussing the change of this probability when new information (in form of, again, events) becomes known. That is, we now ask for the probability of an event given the knowledge that one or more other events have (or have not) occurred.

Definition 24.4 (*Conditional probability*) Let A and B be events with $P(B) > 0$. Then

$$P(A \mid B) = \frac{P(A \cap B)}{P(B)}$$

is called the *conditional probability* of A given (the condition) B.

The following theorem directly follows:

Theorem 24.1 (Product theorem/Multiplication theorem)   *For any two events A and B holds:*

$$P(A \cap B) = P(A \mid B) \cdot P(B) = P(B \mid A) \cdot P(A).$$

For a finite set U of events together with a total ordering $\prec$ on this set, we can generalize the product theorem by induction over the events:

$$P\left(\bigcap_{A \in U} A \right) = \prod_{A \in U} P\left(A \; \middle| \; \bigcap_{B \prec A} B \right)$$

If there is no $B \in U$ with $B \prec A$, then the intersection in the condition of the right-hand side is not empty but it is not computed at all, leading to an implicit Ω:

$$P\left(A \; \middle| \; \bigcap_{B \prec A} B \right) = P\left(A \; \middle| \; \Omega \cap \bigcap_{B \prec A} B \right) = P(A \mid \Omega) = P(A)$$

For $U = \{A, B\}$ with $B \prec A$ the above Theorem 24.1 follows. Further, multiple events can make up the condition.

Theorem 24.2 *Let U, V and W be non-empty sets of events with $U = V \cup W$ and $V \cap W = \emptyset$. Then the following statement holds:*

$$P\left(\bigcap_{A \in U} A\right) = P\left(\bigcap_{A \in V} A \,\Big|\, \bigcap_{A \in W} A\right) \cdot P\left(\bigcap_{A \in W} A\right).$$

A conditional probability satisfies all Kolmogorov axioms. With this we get the following theorem.

Theorem 24.3 *For any fixed event B with $P(B) > 0$ the function P_B defined as*

$$P_B(A) = P(A \mid B)$$

constitutes a probability function that satisfies the condition $P_B(\overline{B}) = 0$.

Definition 24.5 (*Event partition*) Let U be a set of events. The events in U form an *event partition* if all events are pairwise disjoint (that is, if $\forall A, B \in U : A \neq B \Leftrightarrow A \cap B = \emptyset$ holds) and if $\bigcup_{A \in U} = \Omega$ holds (that is, they cover the entire event space).

Theorem 24.4 (Total probability) *Let U be a set of events that form an event partition. Then the probability of any event B can be written as*

$$P(B) = \sum_{A \in U} P(B \mid A) P(A).$$

If one replaces the $P(B)$ in the right-hand side of the equation in Definition 24.4 by an event partition we arrive at the Bayes theorem.

Theorem 24.5 (Bayes theorem) *Let U be a set of events that form an event partition. Further let B be an event with $P(B) > 0$. Then the following equality holds:*

$$\forall A \in U : P(A \mid B) = \frac{P(B \mid A) P(A)}{P(B)} = \frac{P(B \mid A) P(A)}{\sum_{A' \in U} P(B \mid A') P(A')}.$$

This equation is also known as the equation on the probability of hypotheses because it is possible to compute the probability of hypotheses (e.g. diseases), given the knowledge about the probabilities with which the respective hypotheses (here: A) lead to the events $B \in U$ (e.g. symptoms).

24.1.1 Random Variables and Random Vectors

Until now, we combined elements of the event space to make up events without discussing a specific way how to determine the elements of an event A. Further, we still lack the ability to specify properties of the elementary events. Let's assume the event space Ω be the set of all students at the University of Magdeburg. We are interested in, say, the attributes **Gender**, **Year** and **Course**. To assign values to these attributes we use functions defined on Ω having a reasonable domain (e.g. {**male**, **female**} or {**CS**, **Math**, **Econ**, ...}). The preimage of such a function is a set of elementary events (e.g. the set of computer science students or all female students). If these preimages constitute proper events (w.r.t. an underlying event algebra) we will refer to these functions as *random variables*.

Definition 24.6 (*(Discrete) Random variable*) A function X defined on an event space Ω with domain dom(X) is called a *random variable* if the preimage of any subset of its domain has a probability. A subset $W \subseteq$ dom(X) has w.r.t. X the following preimage:

$$X^{-1}(W) = \{\omega \in \Omega \mid X(\omega) \in W\} \overset{\text{abbr.}}{=} \{X \in W\}$$

Note that despite the name random *variable* and the traditional uppercase letter we are dealing with a function here! In the remainder, we refer to the domain of any function X (including random variables, of course) by dom(X). The concept of a random variable can be generalized to sets of random variables.

Definition 24.7 (*Random vector*) Let $X_1, \ldots, X_n$ be random variables over the same event space Ω and the same event algebra $\mathcal{S}$. Then the vector $X = (X_1, \ldots, X_n)$ is called a *random vector*.

We will use the following conjunctive interpretation for the computation of the probability of the value of a random vector. Simultaneously, we will introduce some shorthand notations:

$$\forall x \in \underset{i=1}{\overset{n}{\bigtimes}} \text{dom}(X_i): \quad P(X = x)$$

$$\equiv P(x)$$

$$\equiv P(x_1, \ldots, x_n)$$

$$\equiv P(X_1 = x_1, \ldots, X_n = x_n)$$

$$\equiv P\left(\bigwedge_{i=1}^{n} X_i = x_i\right)$$

$$:= P\left(\bigcap_{i=1}^{n} \{X_i = x_i\}\right)$$

$$= P\left(X_1^{-1}(x_1) \cap \cdots \cap X_n^{-1}(x_n)\right)$$

Given a set of random variables, we can consider the probabilities of all (domain) value combinations a structured representation of the underlying probability space and will refer to them as a probability distribution. We first define distributions for a single random variable.

Definition 24.8 (*(Probability) distribution*) A random variable X with a finite or countably infinite domain $\text{dom}(X)$ is called discrete. The entirety p_X of all pairs

$$\left(x_i, P(X = x_i)\right) \quad \text{with} \quad x_i \in \text{dom}(X)$$

is called the *(probability) distribution* of the discrete random variable X. We use the notation

$$p_X(x_i) = P(X = x_i) \quad \text{for all } x_i \in \text{dom}(X).$$

The generalization of this notion to sets of random variables is straightforward and not given here.

Until now, we used vectors to represent higher dimensions (of random variables). The (domain) value combinations thus were vectors, i.e. elements of the Cartesian product of the domains of the respective random variables. In order to simplify notations later on, we need to do away with the implicit order that underlies a Cartesian product. We choose to use tuples (instead of a vector) as functions on the *set* of random variables. With this rationale, the order of the random variables becomes irrelevant.

Definition 24.9 (*Tuple*) Let $V = \{A_1, \ldots, A_n\}$ be a finite set of random variables with the respective domains $\text{dom}(A_i)$, $i = 1, \ldots, n$. An *instantiation* of the random variables in V or a *tuple* over V is a mapping

$$t_V : V \to \bigcup_{A \in V} \text{dom}(A),$$

that satisfies the following condition:

$$\forall A \in V : t_V(A) \in \text{dom}(A).$$

The vector notation illustrates that the tuple assigns values to multiple random variables. A tuple that assigns a value to only a single random variable will be denoted in its scalar form: t. The index V is dropped if the set V is clear from context. A tuple over the set $\{A, B, C\}$ of random variables which assigns to A the value a_1, to B the value b_2 and to C the value c_2, is denoted as

$$t = (A \mapsto a_1, B \mapsto b_2, C \mapsto c_2),$$

or shorter (if one can infer the attribute from its value):

$$t = (a_1, b_2, c_2)$$

For two tuples to be equal, they must be declared on the same sets of random variables and map to identical values:

$$t_V = t'_U \quad \Leftrightarrow \quad V = U \,\wedge\, \forall A \in V : t(A) = t'(A)$$

The domain of a tuple is restricted using a projection that is defined as follows:

Definition 24.10 (*Projection (of a tuple)*) Let t_X be a tuple over a set X of random variables and $Y \subseteq X$. Then $\text{proj}_Y^X(t_X)$ be the *projection* of tuple t_X to Y. That is, the mapping $\text{proj}_Y^X(t_X)$ assigns values only to elements of Y.

We, again, drop the index X if it is clear from context.

Up to now, we always considered all random variables for computation of probabilities. If only fewer random variables are required, we marginalize (sum) over all value combinations of the variables to be eliminated.

Definition 24.11 (*Marginalization, Marginal distribution*) Let $V = \{X_1, \ldots, X_n\}$ be a set of random variables over the same probability space and p_V a probability distribution over V. For any subset $M \subset V$ the *marginalization over M* is defined as the distribution $p_{V \setminus M}$ that results when summing over all values of all random variables in M, i.e., if the following holds:

$$\forall x_1 \in \text{dom}(X_1) : \cdots \forall x_n \in \text{dom}(X_n) :$$

$$p_{V \setminus M}\left(\bigwedge_{X_i \in V \setminus M} X_i = x_i \right) = \sum_{\substack{\forall X_j \in M: \\ \forall x_j \in \text{dom}(X_j)}} p_V\left(\bigwedge_{X_j \in M} X_j = x_j, \bigwedge_{X_i \in V \setminus M} X_i = x_i \right)$$

For $V \setminus M = \{X\}$ p_X is called the *marginal distribution* of X.

Using this definition, we can easily verify the values of the marginal distributions in Table 24.1:

$$P(\text{G} = \text{m}) = 0.5 \qquad\qquad P(\text{Sm} = \text{sm}) = 0.3$$
$$P(\text{G} = \text{f}) = 0.5 \qquad\qquad P(\text{Sm} = \overline{\text{sm}}) = 0.7$$

$$P(\text{Sm} = \text{sm}, \text{Pr} = \text{pr}) = 0.01 \qquad P(\text{Sm} = \text{r}, \text{Pr} = \overline{\text{pr}}) = 0.29$$
$$P(\text{Sm} = \overline{\text{sm}}, \text{Pr} = \text{pr}) = 0.04 \qquad P(\text{Sm} = \overline{\text{sm}}, \text{Pr} = \overline{\text{pr}}) = 0.66$$

In the last paragraphs we used the notion random variable and attribute interchangeably. Additional synonyms that we will use in the remainder, are random variable, property and dimension. Further, we will from now on only use probability statements w.r.t. random variables; no directly specified events (as subsets of Ω) will be used anymore. Therefore, we need to emphasize the difference between the probability statements: $P(A)$ together with a specific event $A \subseteq \Omega$ stands for a specific probability, i.e., $P(A) \in [0, 1]$. If the object A is a random variable (as it will be the case from now on) the proposition $P(A)$ represents an all-quantified statement over all values of the domain of A. For two random variables A and B the equation

$$P(A \mid B) = \frac{P(A, B)}{P(B)}$$

is a shorthand notation for the following verbose statement:

$$\forall a \in \text{dom}(A) : \forall b \in \text{dom}(B) : P(A = a \mid B = b) = \frac{P(A = a, B = b)}{P(B = b)}.$$

24.1.2 Independences

In Chap. 23 we motivated the decomposition of a high-dimensional distribution into several lower-dimensional distributions. The exploited property for that process is the (conditional) independence between attributes. Let us start with the so-called marginal (i.e. unconditional) independence. As the name suggests, we need a criterion which tells that for two attributes A and B it is irrelevant for the probabilities of A to know anything about B. Formally, the marginal probability distribution of A shall not be any different from the conditional distribution of A given B:

Definition 24.12 (*Independence of random variables*) The random variable A is (stochastically) independent of random variable B with $0 < P(B) < 1$ if and only if

$$P(A \mid B) = P(A)$$

or, equivalently, if

$$P(A, B) = P(A) \cdot P(B).$$

The last proposition is obtained by applying Definition 24.4 which was declared on events to $P(A \mid B)$ and resolve for $P(A, B)$ where we use the implicitly all-quantified version over all values of the attributes. Note that the (stochastic) independence is symmetric, i.e., if A is (stochastically) independent of B, then B is (stochastically) independent of A. The notion of (stochastic) independence is easily generalized to more than two events:

Definition 24.13 (*Full (stochastic) independence*) Let U be a set of random variables. The random variables in U are fully (stochastically) independent, if the following holds:

$$\forall V \subseteq U : \quad P\left(\bigcap_{A \in V} A \right) = \prod_{A \in V} P(A).$$

As none of the probability statements had a condition, we refer to the above concept as *unconditional* or *marginal* independence.

Table 24.1 shows a three-dimensional example distribution over the Boolean attributes **Gender**, **Pregnant** and **Smoker**. The distribution after marginalizing over

Table 24.1 Example distribution with conditional independence

P_{orig}	G = m		G = f	
	Sm = sm	Sm = $\overline{sm}$	Sm = sm	Sm = $\overline{sm}$
Pr = pr	0	0	0.01	0.04
Pr = $\overline{pr}$	0.2	0.3	0.09	0.36

Table 24.2 Distributions to illustrate conditional independence of "Pregnant is conditionally independent of Smoker given Gender."

p_1	Sm = sm	Sm = $\overline{\text{sm}}$	
Pr = pr	0.01	0.04	0.05
Pr = $\overline{\text{pr}}$	0.29	0.66	0.95
	0.30	0.70	

(a) Distribution $p_1 = P(\text{Sm}, \text{Pr})$

p_2	Sm = sm	Sm = $\overline{\text{sm}}$	
Pr = pr	0.015	0.035	0.05
Pr = $\overline{\text{pr}}$	0.285	0.665	0.95
	0.300	0.700	

(b) Distribution $p_2 = P(\text{Sm}) \cdot P(\text{Pr})$

p_3	Sm = sm	Sm = $\overline{\text{sm}}$	
Pr = pr	0	0	0
Pr = $\overline{\text{pr}}$	0.4	0.6	1.0
	0.4	0.6	

(c) Distribution $p_3 = P(\text{Sm}, \text{Pr} \mid G = m)$

p_4	Sm = sm	Sm = $\overline{\text{sm}}$	
Pr = pr	0.02	0.08	0.1
Pr = $\overline{\text{pr}}$	0.18	0.72	0.9
	0.20	0.80	

(d) Distribution $p_4 = P(\text{Sm}, \text{Pr} \mid G = f)$

the attribute Gender is depicted in Table 24.2a. Marginalizing further to the one-dimensional distributions for Smoker and Pregnant, we can constructing the two-dimensional joint distribution via multiplication as depicted in Table 24.2b. Even though the values are close, both two-dimensional distributions over Smoker and Pregnant (the original and the reconstructed one) are different. Hence, attributes Smoker and Pregnant are *not* independent.

However, if we consider the columns for the values of attribute Gender in Table 24.1 separately and renormalize the probabilities to one, we arrive at the conditional distributions as shown in Table 24.2c and Table 24.2d, respectively.

Testing these distributions for independence reveals that in both cases the attributes Pregnant and Smoker are independent. We observe an independence under the condition that the value of the third attribute Gender is known. Hence, we refer to this concept as *conditional independence of* Pregnant *and* Smoker *given (the condition)* Gender.

The mathematical formulation results by inserting into the probability functions of Definition 24.12 one or more additional conditions:

$$P(A \mid B, C) = P(A \mid C) \quad \Leftrightarrow \quad P(A, B \mid C) = P(A \mid C)\, P(B \mid C).$$

Note that the independence has to hold for all conditions, i.e. for all attribute values of C, in order to conclude a conditional independence given C:

$$\forall a \in \text{dom}(A) : \forall b \in \text{dom}(B) : \forall c \in \text{dom}(C) :$$
$$P(A = a, B = b \mid C = c) = P(A = a \mid C = c)\, P(B = b \mid C = c)$$

In the remainder we will use the generalized notion of conditional independence which is extended to sets of attributes.

Definition 24.14 (*Conditional independence of random variables*) Let $X = \{A_1, \ldots, A_k\}$, $Y = \{B_1, \ldots, B_l\}$ and $Z = \{C_1, \ldots, C_m\}$ be three pairwise disjoint sets of random variables. X and Y are conditionally independent given Z (w.r.t. a given distribution p) — written as $X \perp\!\!\!\perp_p Y \mid Z$ — if and only if the following holds:

$\forall a_1 \in \text{dom}(A_1) : \cdots \forall a_k \in \text{dom}(A_k) :$

  $\forall b_1 \in \text{dom}(B_1) : \cdots \forall b_l \in \text{dom}(B_l) :$

   $\forall c_1 \in \text{dom}(C_1) : \cdots \forall c_m \in \text{dom}(C_m) :$

$$P(A_1 = a_1, \ldots, A_k = a_k \mid B_1 = b_1, \ldots, B_l = b_l, C_1 = c_1, \ldots, C_m = c_m)$$
$$= P(A_1 = a_1, \ldots, A_k = a_k \mid C_1 = c_1, \ldots, C_m = c_m).$$

Or in shorthand notation:

$$P(A_1, \ldots, A_k \mid B_1, \ldots, B_l, C_1, \ldots, C_m) = P(A_1, \ldots, A_k \mid C_1, \ldots, C_m).$$

24.2 Graph Theory

In order to represent Bayes and Markov networks, we will need *directed, acyclic graphs* (short: *DAGs*) and undirected graphs. This section introduces the necessary graph-theoretic notions.

24.2.1 Background

Definition 24.15 (*(Simple) graph*) A *simple graph*—in the remainder just referred to as *graph*—is a tuple $G = (V, E)$ where $V = \{A_1, \ldots, A_n\}$ is a finite set of n *vertices* or *nodes* and $E \subseteq (V \times V) \setminus \{(A, A) \mid A \in V\}$ is a set of *edges*.

Such a graph is called *simple* because no multiple edges or loops (edges from a node to itself) are allowed.

Definition 24.16 (*Directed edge*) Let $G = (V, E)$ be a simple graph. An edge $e = (A, B) \in E$ is called a *directed edge* if

$$(A, B) \in E \Rightarrow (B, A) \notin E.$$

Such an edge points from A to B which is denoted as $A \to B$. The node A is called *parent node* of B while B is called the *child node* of A.

Definition 24.17 (*Undirected edge*) Let $G = (V, E)$ be a (simple) graph. Two pairs (A, B) and (B, A) from E comprise a single *undirected edge* between nodes A and B if

$$(A, B) \in E \Rightarrow (B, A) \in E.$$

We denote such an edge as $A - B$ or $B - A$.

Definition 24.18 (*Adjacency set*) Let $G = (V, E)$ be a graph. The set of nodes that are reachable from a given node A is called *adjacency set* of A:

$$\text{adj}(A) = \{B \in V \mid (A, B) \in E\}$$

Definition 24.19 (*Path*) Let $G = (V, E)$ be a graph. A series ρ of r pairwise different nodes

$$\rho = \langle A_{i_1}, \ldots, A_{i_r} \rangle$$

is called a *path* from A_i to A_j if

- $A_{i_1} = A_i$
- $A_{i_r} = A_j$
- $(A_{i_k}, A_{i_{k+1}}) \in E$ or $(A_{i_{k+1}}, A_{i_k}) \in E$, $1 \le k < r$

The symmetric formulation of the last item allows paths that run against the edge direction (which will be important later on). A path consisting only of undirected edges is called *undirected path* and is denoted as

$$\rho = A_{i_1} - \cdots - A_{i_r}$$

while paths consisting only of directed edges that run exclusively in edge direction are called *directed paths* and denoted as

$$\rho = A_{i_1} \to \cdots \to A_{i_r}.$$

Paths with only directed edges that also might run in opposite edge directed are called *mixed paths* and are denoted according to the contained edges. The left graph of Fig. 24.1 contains e.g. the mixed path $F \leftarrow D \to G \leftarrow E$. If two nodes A and B are connected via a directed path ρ in a graph G we denote it as $A \xrightarrow{\rho}_{G} B$. In case of an undirected path, we write $A \xleftrightarrow{\rho}_{G} B$.

A graph with only undirected edges is called *undirected graph*. A graph with only directed edges is called *directed graph*. Note that the path Definition 24.19 allows edges from the last node to the first one. In that case we call them cycles or circles for directed or undirected graphs, respectively.

Definition 24.20 (*Cycle*) Let $G = (V, E)$ be a directed graph. A path

$$\rho = X_1 \to \cdots \to X_k$$

with $X_k \to X_1 \in E$ is called a *cycle*.

Fig. 24.1 The adjacency sets of node D w.r.t. the directed and undirected case are shaded in gray. The closed path $A - D - F - C$ represents a cycle and circle, respectively

 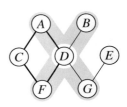

Definition 24.21 (*Circle*) Let $G = (V, E)$ be an undirected graph. A path

$$\rho = X_1 - \cdots - X_k$$

with $X_k - X_1 \in E$ is called *circle*.

In Fig. 24.1, the path $A - D - F - C$ is a cycle in the directed graph and a circle in the undirected graph.

Definition 24.22 (*Tree*) An undirected graph in which any pair of nodes is connected by exactly one path is called a *tree*.

Definition 24.23 (*Minimum spanning tree*) Let $G = (V, E)$ be an undirected graph and w a function assigning to each edge in E a weight:

$$w : E \to \mathbb{R}$$

A graph $G' = (V, E')$ is called *minimum spanning tree* if

- G' is a tree.
- $E' \subseteq E$
- $\sum_{e \in E'} w(e) = \min$

That is, there is no other tree over all nodes V with a smaller edge weight sum. There may be, however, multiple trees with equal minimal edge weight sums.

Along these lines *maximal spanning trees* can be defined: The sum of the edge weights has then to be maximal. Two widely known algorithms for constructing minimal or maximal spanning trees are:

- KRUSKAL algorithm [4]
- PRIM algorithm [6]

We now introduce notions for directed graphs.

Definition 24.24 (*Parent nodes, child nodes, family*) Let $G = (V, E)$ be a directed graph and $A \in V$ a node.

$$\text{pa}(A) = \{B \in V \mid B \to A \in E\}.$$

Analogous, the set of *child nodes* of A is defined as:

$$\text{ch}(A) = \{B \in V \mid A \to B \in E\}.$$

The *family* of a node A consists of the node A itself together with its parent nodes:

$$\text{fa}(A) = \{A\} \cup \text{pa}(A).$$

Definition 24.25 (*Directed acyclic graph (DAG)*) A directed graph $G = (V, E)$ is called *acyclic* if for each path $X_1 \to \cdots \to X_k$ in G holds:

$$X_k \to X_1 \notin E$$

Definition 24.26 (*Ancestors, Descendants, Non-descendants*) Let $G = (V, E)$ be a directed acyclic graph and $A \in V$ a node. In contrast to Definition 24.24 we require acyclicity. Otherwise, nodes could be ancestors or descendants of themselves. The set of *ancestors* of A is defined as:

$$\text{ancs}(A) = \{B \in V \mid \exists \rho : B \overset{\rho}{\underset{G}{\leadsto}} A\}$$

The set of *descendants* of A is defined as:

$$\text{descs}(A) = \{B \in V \mid \exists \rho : A \overset{\rho}{\underset{G}{\leadsto}} B\}$$

The set of *non-descendants* of node A is defined as:

$$\text{non-descs}(A) = V \setminus \{A\} \setminus \text{descs}(A)$$

Example 24.1 The left graph in Fig. 24.2 shows a directed acyclic graph (DAG) together with the parents, children and the family of node F. The right graph shows the DAG together with the ancestors, descendants and non-descendants of node F. Table 24.3 contains the ancestors, descendants and non-descendants of all nodes of the DAG in Fig. 24.2. □

Definition 24.27 (*(Induced) subgraph*) Let $G = (V, E)$ be an undirected graph and $W \subseteq V$ a set of nodes. Then $G_W = (W, E_W)$ with

$$E_W = \{(u, v) \in E \mid u, v \in W\}$$

is called *subgraph of G induced by W*.

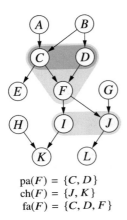

pa(F) = {C, D}
ch(F) = {J, K}
fa(F) = {C, D, F}

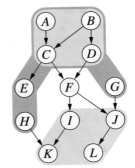

ancs(F) = {A, B, C, D}
descs(F) = {J, K, L, M}
non-descs(F) = {A, B, C, D, E, G, H}

Fig. 24.2 Node relations in directed graphs

Table 24.3 Properties of the example graph in Fig. 24.2

	Ancestors	Descendants	Non-descendants
A	$\emptyset$	$\{C, E, F, J, K, L, M\}$	$\{B, D, G, H\}$
B	$\emptyset$	$\{C, D, E, F, J, K, L, M\}$	$\{A, G, H\}$
C	$\{A, B\}$	$\{E, F, J, K, L, M\}$	$\{A, B, D, G, H\}$
D	$\{B\}$	$\{F, J, K, L, M\}$	$\{A, B, C, E, G, H\}$
E	$\{A, B, C\}$	$\emptyset$	$\{A, B, C, D, F, G, H, J, K, L, M\}$
F	$\{A, B, C, D\}$	$\{J, K, L, M\}$	$\{A, B, C, D, E, G, H\}$
G	$\emptyset$	$\{K, M\}$	$\{A, B, C, D, E, F, H, J, L\}$
H	$\emptyset$	$\{L\}$	$\{A, B, C, D, E, F, G, J, K, M\}$
J	$\{A, B, C, D, F\}$	$\{L\}$	$\{A, B, C, D, E, F, G, H, K, M\}$
K	$\{A, B, C, D, F, G\}$	$\{M\}$	$\{A, B, C, D, E, F, G, H, J, L\}$
L	$\{A, B, C, D, F, H, J\}$	$\emptyset$	$\{A, B, C, D, E, F, G, H, J, K, M\}$
M	$\{A, B, C, D, F, G, K\}$	$\emptyset$	$\{A, B, C, D, E, F, G, H, J, K, L\}$

Definition 24.28 (*Minimal ancestral graph*) Let $G = (V, E)$ be a directed acyclic graph and $M \subseteq V$ a set of nodes. The smallest subgraph of G that contains all ancestors of all nodes of M is called *minimal ancestral graph*. That is, it is the subgraph of G induced by the following set:

$$M \cup \bigcup_{A \in M} \text{ancs}(A)$$

Definition 24.29 (*Moral graph, Moralization*) Let $G = (V, E)$ be a directed acyclic graph. Its moral graph G' is an undirected graph with the same node set that is obtained by first adding (arbitrarily directed) edges between unconnected parent nodes of all families and then replacing all edges by undirected ones. This transformation is known as *moralization*.

This notion goes back to [5]. The intuition being that "unmarried" parents of a common child node are "married". This rather conservative naming is unfortunate as child nodes may have more than two parents that by definition have to be married to each other. The notion, however, caught on.

Example 24.2 (*Moralized minimal ancestral graph*) We first consider the minimal ancestral graph induced by the node set $\{E, F, G\}$ in Fig. 24.2. It is drawn black in Fig. 24.3. The dropped descendants of the nodes of the inducing set are drawn in gray. This graph shall now be moralized: All unconnected parents have to be connected after which all edge directions are dropped. This applies to the parents $\{C, D\}$ of node F and parents $\{A, B\}$ of node C. Note that in case of more than two parents, all possible parent pairs have to be connected! Fig. 24.4 illustrates this. The edge $A - D$ must not be overlooked here.   □

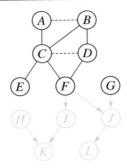

Fig. 24.3 Moral graph of the minimal ancestral graph induced from Fig. 24.2 by the node set $\{E, F, G\}$. Edges inserted during moralization are drawn dashed

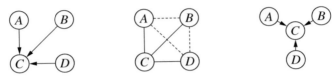

Fig. 24.4 All pairs of parents have to be connected by an edge during moralization. It might be helpful to change the graph layout to observe all missing edges more easily. The edge $A - D$ may be overlooked in the left graph. An equivalent graph is shown on the right which clearly shows all needed edges for moralization

Definition 24.30 (*Complete graph*) An undirected graph $G = (V, E)$ is called *complete* if and only if each pair of (different) nodes from V is connected by an edge.

Definition 24.31 (*Complete set, Clique*) Let $G = (V, E)$ be an undirected graph. A set $W \subseteq V$ is called *complete* if and only if it induces a complete subgraph. Additionally, W is called a *clique* if and only if it is maximal, i.e., if it is impossible to add a node to W without violating the completeness.

We will later apply the notion of a clique to subgraphs. We then refer to the node set of the respective subgraph.

Example 24.3 (*Cliques*) The three graphs in Fig. 24.5 contain the following cliques:

left: $\{A, B, C, D\}$ and $\{B, D, E\}$
middle: $\{A, B, C\}$ and $\{B, E\}$
right: $\{A, B, C, D\}$

In a tree (e.g. the graph in Fig. 24.2 without edge (B, C) and without edge directions) each edge represents a clique.   □

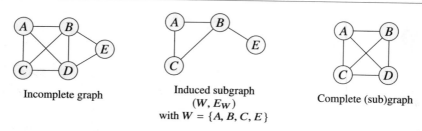

Incomplete graph

Induced subgraph
(W, E_W)
with $W = \{A, B, C, E\}$

Complete (sub)graph

Fig. 24.5 Induction of two subgraphs from the left graph

Definition 24.32 (*Ordering*) Let $G = (V, E)$ be an (arbitrary) graph and α a bijective function with

$$\alpha : V \to \{1, \ldots, |V|\}.$$

Then α is called an *ordering*.

Definition 24.33 (*Topological ordering*) Let α be an ordering on a directed acyclic graph $G = (V, E)$. α is called a *topological ordering* if

$$\forall A \in V : \forall B \in \text{descs}(A) : \alpha(A) < \alpha(B).$$

A directed acyclic graph may have multiple topological orderings.

Definition 24.34 (*Perfect ordering*) Let $G = (V, E)$ be an undirected graph with n nodes and a total ordering $\alpha = \langle v_1, \ldots, v_n \rangle$ on V. α is called *perfect* if the sets

$$\text{adj}(v_i) \cap \{v_1, \ldots, v_{i-1}\}, \quad i = 1, \ldots, n$$

are complete.

An undirected graph can have multiple perfect orderings or none at all.

Figure 24.6 depicts an undirected graph and a node ordering α that is perfect w.r.t. the graph. The table shows the perfectness criterion for each node. The intersections on the right side show that for each step the criterion from Definition 24.34 is satisfied: the two-element sets correspond to edges that are contained in the graph. All single-element sets are nodes and thus trivially complete.

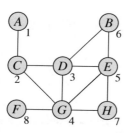

i	$\text{adj}(v_i) \cap \{v_1, \ldots, v_{i-1}\}$	
1	$\{C\} \cap \emptyset$	$= \emptyset$
2	$\{A, D, G\} \cap \{A\}$	$= \{A\}$
3	$\{C, B, E, G\} \cap \{A, C\}$	$= \{C\}$
4	$\{F, C, D, E, H\} \cap \{A, C, D\}$	$= \{C, D\}$
5	$\{B, D, G, H\} \cap \{A, C, D, G\}$	$= \{D, G\}$
6	$\{D, E\} \cap \{A, C, D, G, E\}$	$= \{D, E\}$
7	$\{G, E\} \cap \{A, C, D, G, E, B\}$	$= \{G, E\}$
8	$\{G\} \cap \{A, C, D, G, E, B, H\}$	$= \{G\}$

Fig. 24.6 α is a perfect ordering

Definition 24.35 (*Chord of a circle*) A chord of a circle is an edge between two nodes of the circle which is not contained in the circle itself.

The circle $B - D - G - H - E$ of the graph in Fig. 24.6 has two chords: $D - E$ and $G - E$. Obviously, only circles of length greater than tree can have chords.

Chords subdivide larger circles into smaller ones. It will be beneficial in the later sections not to have chord-less circles with more than three nodes.

Definition 24.36 (*Triangulated graph*) An undirected graph is called *triangulated* if and only if every simple circle (i.e. a path with its nodes occurring at most once (except for the start/end node, of course)) with more than three nodes has a chord.

Note that a triangulated graph not necessarily needs to consists solely of triangles. For example, the graph in the center of Fig. 24.5 is triangulated: the edge $C - E$ need not be inserted. Contrary, not every graph that consists of triangles is also triangulated: The top part of Fig. 24.7 depicts an obvious triangulation. The same insertion, however, must also be applied in the bottom part. The circle $A - B - E - C$ has no chord in both scenarios.

Definition 24.37 (*Maximum cardinality search (MCS)*) Let $G = (V, E)$ be an undirected graph. A node ordering α by maximum cardinality search is constructed as follows:

1. Choose an arbitrary start node from V and assign it number 1.
2. Assign the next higher order number to the node that is adjacent to the largest number of already numbered nodes.

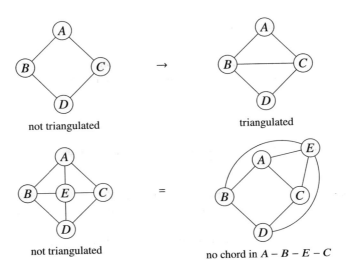

Fig. 24.7 Top: triangulation is achieved by insertion of edge $B - C$ (or $A - D$). Bottom: graph is not triangulated as can be easier seen in the alternative drawing

Fig. 24.8 Example of a
maximum cardinality search
starting with node A.
Number 3 can be assigned to
nodes D or G, for number 6
we have the choice between
nodes H and B

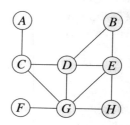

Obviously, MCS is not unique. Figure 24.8 shows an example. Node A gets assigned number 1, node C then has to follow with number 2 as it is the only neighbor. For number 3, we have the choice of node D and G because both are adjacent to one already numbered node (C). Let's choose D as the third node which leads immediately to node G with number 4 as it is the only one that is now connected to two already numbered nodes (C and D). The same arguments leads to number 5 of node E. For number 6, we again have two choices: node B or H. Setting $\alpha(B) = 6$ leads to $\alpha(H) = 7$ and finally to $\alpha(F) = 8$.

24.2.2 Join Graphs

The (undirected) graph concept discussed so far provided an algebraic structure to describe the relations amongst single objects (nodes). It is a relation in the algebraic sense, because the edge set is a subset of $V \times V$. The relations coincide with the edges. We now want to extend this concept to relations between *sets of nodes* and introduce so-called join graphs (or cluster graphs). Since we only define notions that we need later on, we refrain from an exhaustive introduction of the matter. For more details the reader is referred to [1]. Instead allowing arbitrary subsets of V to be the new nodes, we restrict them to be cliques of a graph.

Definition 24.38 (*Join graph*) Let $G = (V, E)$ be an undirected graph and $C = \{C_1, \ldots, C_p\}$ its cliques. $G' = (C, E')$ is called a *join graph* if E' only contains edges between non-disjoint nodes, i.e., if holds:

$$(C_i, C_j) \in E' \implies C_i \cap C_j \neq \emptyset$$

Given an undirected graph, we can easily derive its associated join graph: After identifying all cliques we connect them by an edge if their intersection is non-empty. Figure 24.9 shows an example. The undirected graph on the left contains six cliques: $\{A, C\}$, $\{C, D, G\}$, $\{B, D\}$, $\{B, E\}$, $\{G, F\}$ and $\{E, G, H\}$. These form the nodes of the join graph depicted on the right. The seven edges follow from the mutual intersection conditions.

We will later use join graphs to propagate evidence about certain attributes (which coincide with the nodes in V, obviously) to all other attributes. The underlying algorithm will, however, require a special form of a join graph. First, it must be

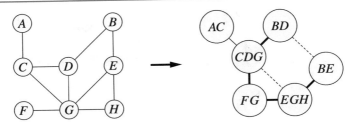

Fig. 24.9 Corresponding join graph of an undirected graph for which there is no join tree. If we delete e.g. the two dashed edges to form a tree, the RIP will not be satisfied: The marked path from BD to BE does not contain the attribute B. It is not possible here to achieve the RIP by deleting any other edges

guaranteed that information is transferred from clique to clique across a unique path. And second, any change of an attribute contained in a clique must be transferable to any other clique containing that attribute. The first requirement can be achieved by using so-called join trees, i.e. join graphs with tree structure. Since in a tree there is exactly one path between any two nodes, this will be the unique path for the evidence propagation. The second requirement can be formulated as follows: If two cliques share some attributes (i.e., have a non-empty intersection), then these attributes must be contained in each clique of the path connecting the two cliques. In that way, no "gaps" on the path will block the evidence propagation. This latter property is known as the so-called running intersection property.

Definition 24.39 (*Running intersection property (RIP)*) Let $G = (V, E)$ be an undirected graph with r cliques. An ordering of these cliques has *running intersection property* if for each $j > 1$ there exists an $i < j$ for which the condition

$$C_j \cap (C_1 \cup \cdots \cup C_{j-1}) \subseteq C_i$$

holds.

Before illustrating the running intersection property, we need to introduce the notion of a join tree.

Definition 24.40 (*Join tree*) A join graph with tree structure whose cliques satisfy the running intersection property, is called a *join tree*.

Let us discuss the RIP with an example. Consider the graph in Fig. 24.10. Sorting the cliques in ascending order w.r.t. their index will yield an ordering satisfying RIP which can be validated in Table 24.4. This allows us to use the RIP to construct the tree: For each clique we immediately know that there is at least one preceding clique with at least one common attribute. Therefore, each clique (except the first one, of course) has a neighbor candidate. We do not need to worry about circles as we only consider preceding cliques (w.r.t. the ordering). Hence we can use the following scheme to form a join tree: Starting with the last clique of the ordering

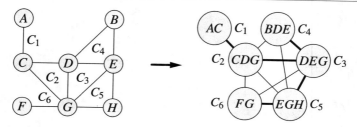

Fig. 24.10 Corresponding join graph of an undirected, triangulated graph

Table 24.4 The clique ordering $C_1, \ldots, C_6$ of the graph in Fig. 24.10 has RIP

j				i
2	$C_2 \cap C_1$	$= \{C\}$	$\subseteq C_1$	1
3	$C_3 \cap (C_1 \cup C_2)$	$= \{D, G\}$	$\subseteq C_2$	2
4	$C_4 \cap (C_1 \cup C_2 \cup C_3)$	$= \{D, E\}$	$\subseteq C_3$	3
5	$C_5 \cap (C_1 \cup C_2 \cup C_3 \cup C_4)$	$= \{E, G\}$	$\subseteq C_3$	3
6	$C_6 \cap (C_1 \cup C_2 \cup C_3 \cup C_4 \cup C_5)$	$= \{G\}$	$\subseteq C_5$	5

(that satisfies the RIP), we connect it to that preceding clique that shares the largest number of attributes. Ties are resolved arbitrarily. Figure 24.10 shows a possible join tree which was created as follows:

1. Starting with the last clique C_6 of the ordering, we look for preceding cliques satisfying the RIP (see also Table 24.4). There are three candidates: C_5, C_3 and C_2. All three have an equally large (single-element) intersection $\{G\}$ with C_6. We (arbitrarily) choose a clique to be the neighbor in the join tree: C_5.
2. C_5 forms the largest intersection $\{E, G\}$ with C_3 and hence is connected to it.
3. C_4 also forms the largest intersection $\{D, E\}$ with C_3 and is connected to it.
4. Clique C_3 forms a non-empty intersection with only C_2 and is therefore connected to that clique.
5. C_2 is connected trivially to C_1 as it is the last remaining clique with non-empty intersection.

The created join tree is shown on the right of Fig. 24.10 by the bold edges.

To create a tree structure from a given join graph is rather simple as only some edges need to be deleted. The question arises whether the RIP can always be achieved. This is not the case as it can be seen from Fig. 24.9. Two edges need to be deleted to arrive at a tree structure. However, none of the two resulting trees satisfies the RIP.

Table 24.5 Generating a clique ordering with RIP from a perfect node ordering α of Fig. 24.6

Clique	Rank		
$\{A, C\}$	$\max\{\alpha(A), \alpha(C)\}$	$= 2$	$\to C_1$
$\{C, D, G\}$	$\max\{\alpha(C), \alpha(D), \alpha(G)\}$	$= 4$	$\to C_2$
$\{D, E, G\}$	$\max\{\alpha(D), \alpha(E), \alpha(G)\}$	$= 5$	$\to C_3$
$\{B, D, E\}$	$\max\{\alpha(B), \alpha(D), \alpha(E)\}$	$= 6$	$\to C_4$
$\{G, E, H\}$	$\max\{\alpha(G), \alpha(E), \alpha(H)\}$	$= 7$	$\to C_5$
$\{F, G\}$	$\max\{\alpha(F), \alpha(G)\}$	$= 8$	$\to C_6$

If we delete e.g. the edges $CDG - EGH$ and $BD - BE$, the cliques BD and BE both contain the attribute B but this does not hold for the cliques on their connecting path.

The question whether there is a structural property of a join graph (or its underlying undirected graph) that guarantees the existence of a corresponding join tree was positively answered in [2]: An undirected graph G has a join tree if and only if G is triangulated. Unfortunately, the definition of the RIP is not constructive, i.e., it gives us a criterion to test a given clique ordering for RIP but it does not produce an algorithm to find such a clique ordering. We exploit the following relation to solve the problem: If an undirected graph has a perfect ordering and if we sort its cliques in ascending order w.r.t. the largest perfect number of its contained attributes, then the resulting clique ordering will satisfy the RIP. The clique ordering from Table 24.4 was generated from the perfect ordering in Fig. 24.6. Table 24.5 illustrates the assignment.

We just reduced the problem of finding an appropriate clique ordering to the problem of finding a perfect node ordering, which might seem little beneficial at first sight since there exist more nodes than cliques. The objective thus is to find a perfect ordering on the nodes of an undirected graph. For that task, the following relationship is helpful: The nodes of an undirected graph have at least one perfect ordering if and only if the graph is triangulated.

The last building block in this chain is the construction of such a perfect ordering. We exploit the following statement: A node ordering induced by a maximum cardinality search on a triangulated graph is perfect. Let us summarize this chain of reasoning:

1. Undirected, triangulated graph $G = (V, E)$
2. Node ordering induced by MCS on V is perfect
3. Clique ordering w.r.t. largest perfect node number has RIP
4. Construct join tree using RIP

24.2.3 Separations

One objective of graphical models is to capture as many (conditional) independences of a high-dimensional probability distribution as possible in an undirected or directed graph in order to infer probabilistic statements solely by exploiting graph-theoretic criteria. The fact that there exist certain (conditional) independences amongst a set of attributes shall be represented by means of separation criteria of the corresponding nodes inside a graph. We will discuss such separation criteria for both directed and undirected graphs.

Let us begin with an undirected graph $G = (V, E)$ and three mutually disjoint node subsets X, Y and Z. The set Z shall u-separate X and Y (the u stands for *u*ndirected) if and only if each path from a node in X to a node in Y is blocked. A path is blocked if and only if it contains a blocking node. And, finally, a node is blocking if and only if it is contained in Z. This somewhat artificial definition of u-separation will help us later to define the corresponding concept of d-separation for directed graphs.

Definition 24.41 (*u-separation*) Let $G = (V, E)$ be an undirected graph and X, Y and Z three disjoint subsets of V. The sets X and Y are *u-separated* by Z in G, written $X \perp\!\!\!\perp_G Y \mid Z$, if and only if each path from a node in X to a node in Y contains at least one node of Z. A path that contains at least one node of Z is called *blocked*, otherwise it is *active*.

Let us consider an example in Fig. 24.11. The set $Z = \{E, F\}$ separates the node sets $\{A, B, C, D\}$ and $\{G, H, I\}$ since all paths from the one set to the other are running through Z as is shown exemplary with path $A - B - E - G - H$.

An alternative but equivalent way to test for u-separation is as follows: the nodes (and all adjacent edges) of set Z are deleted from the graph. If no path is left that connects the sets X and Y, we can conclude the u-separation of them by the set Z. Figure 24.12 illustrates this with the example taken from Fig. 24.11. Because of semantic reasons discussed later, for directed graphs we have to take the edge di-

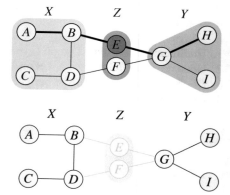

Fig. 24.11 The node sets $\{A, B, C, D\}$ and $\{G, H, I\}$ are u-separated by $\{E, F\}$. The emphasized path $A - B - E - G - H$ is blocked by node E

Fig. 24.12 After deleting the nodes of set Z from the graph there is no path left from X to Y: both sets are u-separated by Z

rections into account when deciding whether a path is blocked or not. From now on, we will consider mixed paths, i.e. paths that might run against edge directions. Again, let X, Y and Z be three disjoint subset of nodes of a directed graph. We use the same reasoning as above for u-separation, only the blocking criteria are modified for the directed case. Again, the sets X and Y are d-separated by Z (the d stands for directed) if and only if each path from a node in X to a node in Y is blocked (by Z). A path is blocked if and only if it contains at least one blocking node. A node is blocking if its edge directions *along the path* are

- serial or diverging and the node itself lies in Z, or
- converging and neither the node itself nor any of its descendants lies in Z.

The four possible edge directions at a node are grouped as follows:

The type of a node (i.e. serial, converging or diverging) depends on the path in which it is contained. In Fig. 24.13 the node E is serial w.r.t. the path $C \to E \to G$ while it is converging w.r.t. the path $C \to E \leftarrow D$. The same node is finally diverging in the path $F \leftarrow E \to G$.

Definition 24.42 (*d-separation*) Let $G = (V, E)$ be a directed graph and X, Y and Z three disjoint subsets of V. X and Y are *d-separated* by Z in G, written $X \perp\!\!\!\perp_G Y \mid Z$, if and only if there is no path from a node in X to a node in Y along which the following criteria are satisfied:

1. Every node with converging edges (along the path) is in Z or has a descendant in Z.
2. Every other node is not in Z.

A path satisfying both above criteria is called *active*, else *blocked* (by Z).

Note the two equivalent but mutually negated definitions of d-separation. The first description defines the blockade of a path and requires for a d-separation every path to be blocked. The second description defines d-separation as the absence of any active path. We mentioned both version here as both are found in literature.

Let us consider some examples. In the following figures, the sets X and Y will be shaded light and medium gray, while Z will be drawn in dark gray.

Example 24.4 In Fig. 24.13 both X and Y contain just one node and Z is empty. Since we deal with a tree, only the path $A \to C \to E \leftarrow D$ needs to be checked.

- C is a serial node in the path and not in Z. It is therefore non-blocking.

Fig. 24.13 Node E blocks the only path from A to D: $A \perp\!\!\!\perp_G D \mid \emptyset$.

Fig. 24.14 The path from A to E is active because of $E \in Z$: $A \not\perp\!\!\!\perp_G D \mid E$

Fig. 24.15 I activates the path from A to D: $A \not\perp\!\!\!\perp_G D \mid I$

- E is a converging node in the path and not in Z. Also, none of its descendants F, H, G and I is in Z. Thus, E is blocking.

The existence of one blocking node is sufficient for the blockade of the entire path. Hence we can conclude $A \perp\!\!\!\perp_G D \mid \emptyset$.   □

Example 24.5 In contrast to the Example 24.4, the set Z now contains the node E. Figure 24.14 depicts this situation. Again, only the path $A \rightarrow C \rightarrow E \leftarrow D$ needs to be checked.

- C is a serial node in the path and not in Z. Therefore, it is non-blocking.
- E is a converging node in the path and contained in Z. Thus, it is non-blocking as well.

All nodes of the only path are non-blocking which renders the path active. Hence, A and D are not d-separated and therefore $A \not\perp\!\!\!\perp_G D \mid E$.   □

Example 24.6 Instead of node E we now add one of its descendants to the set Z, see Fig. 24.15. Again, only the path $A \rightarrow C \rightarrow E \leftarrow D$ needs to be checked.

- C is a serial node in the path and not in Z. Therefore, it is non-blocking.
- E is a converging node in the path but not in Z. However, one of its descendants, namely I, is in Z. Hence, E is non-blocking.

Fig. 24.16 Graph for
example 24.7

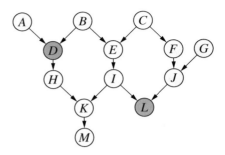

All nodes of the only path are non-blocking which renders the path active. A and D are not d-separated by I. Therefore $A \not\perp_G D \mid I$. □

Example 24.7 Let us consider Fig. 24.16 as an example with more than one path between the (still single element) sets $X = \{D\}$ and $Y = \{L\}$. We will successively add nodes to the set Z to check different scenarios. Following paths exist between the nodes D and L:

1. $D \to H \to K \leftarrow I \to L$
2. $D \leftarrow B \to E \to I \to L$
3. $D \leftarrow B \to E \leftarrow C \to F \to J \to L$

- $Z = \emptyset$

  Path 1: blocked $K, M \notin Z$
  Path 2: active $B, E, I \notin Z$ $\Rightarrow D \not\perp_G L \mid \emptyset$
  Path 3: blocked $E, I, K, M \notin Z$

- $Z = \{E\}$

  Path 1: blocked $K, M \notin Z$
  Path 2: blocked $E \in Z$ $\Rightarrow D \not\perp_G L \mid E$
  Path 3: active $E \in Z$

- $Z = \{E, J\}$

  Path 1: blocked $K, M \notin Z$
  Path 2: blocked $E \in Z$ $\Rightarrow D \perp_G L \mid E, J$
  Path 3: blocked $J \in Z$

□

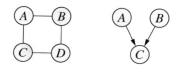

Fig. 24.17 Undirected graph without equivalent directed graph and directed graph without equivalent undirected graph

Note that u-separation (in contrast to d-separation) is monotonic. If a set Z u-separates two node sets X and Y, then X and Y are also u-separated by all supersets of Z. Formally we have:

$$X \perp\!\!\!\perp Y \mid Z \quad \Rightarrow \quad X \perp\!\!\!\perp Y \mid Z \cup W, \qquad (24.1)$$

where W is a fourth disjoint node set. This is immediately clear as paths can only be blocked by nodes in Z. Adding more nodes to Z may block further paths but it can never activate a blocked path. This is completely different to d-separation where adding nodes to Z may indeed activate a path and thus render previously d-separated sets X and Y connected. In Example 24.7 the set $Z = \{E, J\}$ d-separates the nodes D and L. If we added the node K to Z, it would activate the path $D \to H \to K \leftarrow I \to L$ and thus prevents the d-separation of D and L:

$$D \perp\!\!\!\perp L \mid E, J \quad \text{but} \quad D \not\!\perp\!\!\!\perp L \mid E, J, K.$$

Also, u-separation and d-separation have different expressiveness. Consider the undirected graph in Fig. 24.17. Via u-separation we can easily conclude the following separations:

$$A \perp\!\!\!\perp C \mid B, D \quad \text{and} \quad B \perp\!\!\!\perp D \mid A, C.$$

However, there exists no directed graph encoding exactly the same separations. We need to require an exact match here to prevent any additional separations. Otherwise, an edge-less graph would encode all possible separations and thus also both above-mentioned ones. If we directed the edges starting at A via B and D to C, we would actually get the separation $A \perp\!\!\!\perp C \mid B, D$, but this would imply $B \not\!\perp\!\!\!\perp D \mid A, C$ because C would be a converging node which activates the path $D \to C \leftarrow B$.

Likewise, there is not necessarily for every directed graph an equivalent undirected one (w.r.t. the encoded separations) as illustrated on the right of Fig. 24.17. It encodes the d-separation

$$A \perp\!\!\!\perp B \mid \emptyset.$$

Again, we cannot represent this separation in an undirected graph without introducing new separations (which are not valid in the directed graph).

Finally, we recall the running intersection property (RIP) which defines two other important node sets:

Definition 24.43 *(Residual set, Separator set)* Let $C_1, \ldots, C_n$ be a clique ordering satisfying the RIP. The sets

$$S_i = C_i \cap (C_1 \cup \cdots \cup C_{i-1}), \quad i = 1, \ldots, n, \quad S_1 = \emptyset$$

C_i	R_i	S_i
C_1	$\{A, C\}$	$\emptyset$
C_2	$\{D, F\}$	$\{C\}$
C_3	$\{E\}$	$\{D, F\}$
C_4	$\{B\}$	$\{D, E\}$
C_5	$\{H\}$	$\{E, F\}$
C_6	$\{G\}$	$\{F\}$

$\longrightarrow$

$$D, F \perp\!\!\!\perp_G A \mid C$$
$$E \quad \perp\!\!\!\perp_G A, C \mid D, F$$
$$B \quad \perp\!\!\!\perp_G A, C, F \mid D, E$$
$$H \quad \perp\!\!\!\perp_G A, B, C, D \mid E, F$$
$$G \quad \perp\!\!\!\perp_G A, B, C, D, E, H \mid F$$

Fig. 24.18 Residual and separator sets of the graph in Fig. 24.10 together with the encoded u-separations

are called *separator sets* and the sets

$$R_i = C_i \setminus S_i, \quad i = 1, \ldots, n$$

are called *residual sets*.

With these definitions together with the RIP we can easily conclude the following separation criteria inside join graphs:

$$R_i \perp\!\!\!\perp (R_1 \cup \cdots \cup R_{i-1}) \setminus S_i \mid S_i, \quad i = 2, \ldots, n \tag{24.2}$$

Example 24.8 Let us clarify these concepts with the graphs in Fig. 24.10. Figure 24.18 lists all residual and separator sets of all cliques. The u-separations that are derivable via Eq. 24.2 are given as well. □

References

1. Castillo E, Gutierrez JM, Hadi AS (1997) Expert Systems and Probabilistic Network Models. Springer-Verlag, New York, NY, USA
2. Jensen FV (1988) Junction Trees and Decomposable Hypergraphs. Research Report, JUDEX Data Systems, Aalborg, Denmark
3. Kolmogorov AN, der Wahrscheinlichkeitsrechnung Grundbegriffe, Springer-Verlag Heidelberg (1933) English edition: Foundations of the Theory of Probability. Chelsea, New York, NY, USA, p 1956
4. J.B. Kruskal. On the Shortest Spanning Subtree of a Graph and the Traveling Salesman Problem. *Proc. American Mathematical Society* 7(1):48–50. American Mathematical Society, Providence, RI, USA 1956
5. S.L. Lauritzen and D.J. Spiegelhalter. Local Computations with Probabilities on Graphical Structures and Their Application to Expert Systems. *Journal of the Royal Statistical Society, Series B*, 2(50):157–224. Blackwell, Oxford, United Kingdom 1988
6. R.C. Prim. Shortest Connection Networks and Some Generalizations. *The Bell System Technical Journal* 36:1389-1401. Bell Laboratories, Murray Hill, NJ, USA 1957

Decompositions

25

The objective of this chapter is to connect the concepts of conditional independence with the separation in graphs. Both can be represented by a ternary relation $(\cdot \perp\!\!\!\perp \cdot \mid \cdot)$ on either the set of attributes or nodes and it seems to be promising to investigate how to represent the probabilistic properties of a distribution by the means of a graph. The idea then is to use only graph-theoretic criteria (separations) to draw inferences about (conditional) independences because it is them what enables us to decompose a high-dimensional distribution and propagate evidence.

We first introduce an axiomatic approach to the concept of conditional independence (or separation) that goes back to [2] and [5]. With these axioms, we can syntactically derive new separations or independences from a given set without checking for graph-theoretic nor probabilistic conditions.

Definition 25.1 (*Semi-graphoid and Graphoid Axioms*)
Let V be a set of (mathematical) objects and $(\cdot \perp\!\!\!\perp \cdot \mid \cdot)$ a ternary relation of subsets of V. Furthermore, let W, X, Y, and Z be four disjoint subsets of V. The four statements

(a) symmetry: $\quad (X \perp\!\!\!\perp Y \mid Z) \implies (Y \perp\!\!\!\perp X \mid Z)$
(b) decomposition: $\quad (W \cup X \perp\!\!\!\perp Y \mid Z) \implies (W \perp\!\!\!\perp Y \mid Z) \wedge (X \perp\!\!\!\perp Y \mid Z)$
(c) weak union: $\quad (W \cup X \perp\!\!\!\perp Y \mid Z) \implies (X \perp\!\!\!\perp Y \mid Z \cup W)$
(d) contraction: $\quad (X \perp\!\!\!\perp Y \mid Z \cup W) \wedge (W \perp\!\!\!\perp Y \mid Z) \implies (W \cup X \perp\!\!\!\perp Y \mid Z)$

are called the *semi-graphoid axioms*. A ternary relation $(\cdot \perp\!\!\!\perp \cdot \mid \cdot)$ that satisfies the semi-graphoid axioms for all W, X, Y, and Z is called a *semi-graphoid*. The above four statements together with

(e) intersection: $\quad (W \perp\!\!\!\perp Y \mid Z \cup X) \wedge (X \perp\!\!\!\perp Y \mid Z \cup W) \implies (W \cup X \perp\!\!\!\perp Y \mid Z)$

© Springer Nature Switzerland AG 2022
R. Kruse et al., *Computational Intelligence*, Texts in Computer Science,
https://doi.org/10.1007/978-3-030-42227-1_25

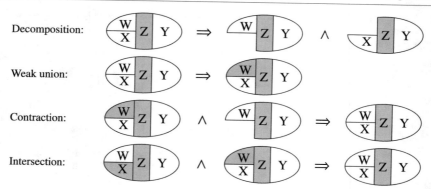

Fig. 25.1 Illustration of the graphoid axioms and separation in graphs

are called the *graphoid axioms*. A ternary relation $(\cdot \perp\!\!\!\perp \cdot \mid \cdot)$ that satisfies the graphoid axioms for all W, X, Y, and Z is called a *graphoid*.

The axioms (b) to (e) are illustrated in Fig. 25.1.

If we talk about sets $\mathcal{I}$ of independence or separation statements, we refer to the same structures in an algebraic way—only the origin is different.

Definition 25.2 Let $V = \{A_1, \ldots, A_n\}$ be a set of attributes. Furthermore, let p be a distribution over the attributes V. Then $\mathcal{I}_p \subseteq 2^V \times 2^V \times 2^V$ be the set of all (conditional) independence statements that are valid in p.

Definition 25.3 Let $V = \{A_1, \ldots, A_n\}$ be as set of nodes. Furthermore, let G be an undirected (directed) graph with node set V. Then $\mathcal{I}_G$ be the set of all separation statements that can be read from G via u-separation (d-separation).

Given a set of axioms $\mathcal{A}$, we can syntactically derive from a set $\mathcal{I}$ of conditional independence statements additional statements. If a statement I can be derived from $\mathcal{I}$ by application of $\mathcal{A}$, we write

$$\mathcal{I} \vdash_{\mathcal{A}} I.$$

If a conditional independence statement follows semantically (w.r.t. an underlying theory $\mathcal{T}$) from a set $\mathcal{I}$ of statements, we denote this with

$$\mathcal{I} \models_{\mathcal{T}} I.$$

In our case the theories $\mathcal{T}$ refer to the probability theory $\mathcal{P}$, d- and u-separation. The semantic deducibility of a conditional independence I from a set of independence statements $\mathcal{I}$ then means that I is valid in all distributions in which all statements of $\mathcal{I}$ are valid:

$$\mathcal{I} \models_{\mathcal{P}} I \quad \Leftrightarrow \quad \forall p : \mathcal{I} \subseteq \mathcal{I}_p \Rightarrow I \in \mathcal{I}_p$$

The semantic deducibility for graphs is defined in an analogous way.

For every system of axioms, it needs to be answered whether the deductions are *sound* and *complete*. That is, whether each deducible statement is correct w.r.t. the theory $\mathcal{T}$ and whether all statements that can be inferred from I via theory $\mathcal{T}$ can also be deduced by applying the axioms.

$$I \vdash_{\mathcal{A}} I \;\Rightarrow\; I \models_{\mathcal{T}} I \quad \text{(Soundness)}$$
$$I \models_{\mathcal{T}} I \;\Rightarrow\; I \vdash_{\mathcal{A}} I \quad \text{(Completeness)}$$

After these theoretical considerations, we turn to specific answers w.r.t. conditional independence and both separation criteria.

Theorem 25.1 *Conditional stochastic independence satisfies the semi-graphoid axioms. For strictly positive probability distributions, it satisfies the graphoid axioms.*

Theorem 25.2 *Both u-separation and d-separation satisfy the graphoid axioms.*

Both theorems obviously contain soundness statements: the semi-graphoid axioms are sound for arbitrary probability distributions, whereas the graphoid axioms are sound for strictly positive probability distributions, d- and u-separation. The axioms are, however, not complete in general [6,7]. Table 25.1 summarizes these facts.

We observed that both u- and d-separation satisfy the same axioms as the notion of conditional independence (of a strictly positive probability distribution). The idea suggests itself to use a (directed or undirected) graph to encode the conditional independence statements that hold in a given distribution such that deducible separations directly correspond to valid conditional independence statements.

Unfortunately, there is in general no isomorphism of both notion (that is, between the conditional independence and one of the separations). One reason for that is that u-separation satisfies stronger axioms than the graphoid axioms. As an example, consider the discrepancy between the (semi-)graphoid axiom of weak union and the monotonicity of the u-separation: On the one hand, the Eq. 24.1 on page 551 resembles the weak union axiom with the difference that the node set W does not have to be u-separated beforehand. The Eq. 24.1 is also called strong union axiom [5].

Table 25.1 Summary of soundness and completeness of the (semi-)graphoid axioms w.r.t. four different independence and separation criteria

		Soundness	Completeness
Conditional	General distributions	Semi-graphoid axioms	
Independence	Strictly positive dists.	Graphoid axioms	No*
u-separation		Graphoid axioms	No†
d-separation		Graphoid axioms	No‡

* Conjectured in [4], refuted in [6,7]
† If weak union is replaced by two other axioms [5]
the resulting axiom set is sound and complete
‡ Counterexample in [1, p. 102]

On the other hand, the conditional independence in general only satisfies the semi-graphoid axioms (see Theorem 25.1). That is, applying the intersection axiom can already lead to incorrect results as the following example illustrates:

Example 25.1 (Cond. indep. does not satisfy intersection axiom in general)
We consider the following (not strictly positive) three-dimensional probability distribution over the binary attributes A, B, and C with $P(A = a_1, B = b_1, C = c_1) = P(A = a_2, B = b_2, C = c_2) = 0.5$ (all other value combinations have probability zero). In this distribution, the following conditional independences hold:

$$A \perp\!\!\!\perp B \mid C, \qquad A \perp\!\!\!\perp C \mid B \quad \text{and} \quad B \perp\!\!\!\perp C \mid A.$$

This can be easily seen as follows: The left table shows the distribution p_{ABC}, while the other two show the conditional distributions given the attribute C. Both distributions also show the marginal distributions $p_{A|C}$ and $p_{B|C}$.

	c_1		c_2	
p_{ABC}	a_1	a_2	a_1	a_2
b_1	$\frac{1}{2}$	0	0	0
b_2	0	0	0	$\frac{1}{2}$

$p_{AB|c_1}$	a_1	a_2	
b_1	1	0	1
b_2	0	0	0
	1	0	

$p_{AB|c_2}$	a_1	a_2	
b_1	0	0	0
b_2	0	1	1
	1	0	

Obviously, the following relationship holds which corresponds to the above conditional independences:

$$\forall a \in \mathrm{dom}(A) : \forall b \in \mathrm{dom}(B) : \forall c \in \mathrm{dom}(C) :$$
$$P(A = a, B = b \mid C = c) = P(A = a \mid C = c) \cdot P(B = b \mid C = c)$$

Applying the intersection axiom (with $Z = \emptyset$) to the above independences results in

W	X	Y	Inference
$\{A\}$	$\{C\}$	$\{B\}$	$AC \perp\!\!\!\perp B \mid \emptyset$
$\{A\}$	$\{B\}$	$\{C\}$	$AB \perp\!\!\!\perp C \mid \emptyset$
$\{B\}$	$\{C\}$	$\{A\}$	$BC \perp\!\!\!\perp A \mid \emptyset$

p_{AB}	a_1	a_2	
b_1	$\frac{1}{2}$	0	$\frac{1}{2}$
b_2	0	$\frac{1}{2}$	$\frac{1}{2}$
	$\frac{1}{2}$	$\frac{1}{2}$	

p'_{AB}	a_1	a_2	
b_1	$\frac{1}{4}$	$\frac{1}{4}$	$\frac{1}{2}$
b_2	$\frac{1}{4}$	$\frac{1}{4}$	$\frac{1}{2}$
	$\frac{1}{2}$	$\frac{1}{2}$	

Applying the decomposition axiom on the previous independences, we finally arrive at the following statements:

$$A \perp\!\!\!\perp B \mid \emptyset, \qquad B \perp\!\!\!\perp C \mid \emptyset \quad \text{and} \quad C \perp\!\!\!\perp A \mid \emptyset.$$

None of these three (here: marginal) independences holds true in p_{ABC} as can be easily verified: The distribution p_{AB} cannot be reconstructed from the two marginal distributions p_A and p_B as the comparison between p_{AB} and p'_{AB} shows. The same holds also for the other attribute combinations.

Since we cannot rely on isomorphism between the separation and independence concepts, we will have to confine ourselves to weaker statements. For that, we first consider the following definitions.

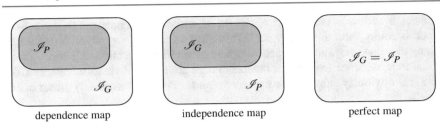

dependence map   independence map   perfect map

Fig. 25.2 Relationships between deducibility of conditional independences in different map types

Definition 25.4 (*Dependence, Independence, and Perfect Map*)
Let $(\cdot \perp\!\!\!\perp_p \cdot \mid \cdot)$ be a ternary relation representing the conditional independences of a given distribution p over the attribute set V. An undirected (directed) graph $G = (V, E)$ is called *conditional dependence graph* or *dependence map* w.r.t. p if and only if for all disjoint subsets $X, Y, Z \subset V$ the implication

$$X \perp\!\!\!\perp_p Y \mid Z \;\Rightarrow\; X \perp\!\!\!\perp_G Y \mid Z$$

holds. That is, if G describes via u-separation (d-separation) all conditional independences of p and thus describes only sound dependences.
An undirected (directed) graph $G = (V, E)$ is called *conditional independence graph* or *independence map* w.r.t. p if and only if for all disjoint subsets $X, Y, Z \subset V$ the implication

$$X \perp\!\!\!\perp_G Y \mid Z \;\Rightarrow\; X \perp\!\!\!\perp_p Y \mid Z$$

holds. That is, if G describes via u-separation (d-separation) only conditional independences that are valid in p. G is called a *perfect map* of the conditional (in)dependences in p if and only if it is both a dependence map and an independence map.

A dependence graph represent all conditional independences that are valid in the distribution p but may contain additional ones that are not valid in p. An independence graph, however, encodes only those conditional independences that are valid in p but maybe not all of them. Figure 25.2 illustrates these relationships.

We saw in Chap. 23 that we can decompose a distribution by exploiting conditional independences. Consequently, we must make sure that no independence can be read from the graph that is not valid inside the distribution. Since we cannot hope for perfect maps in general, we will confine ourselves to independence maps. We may not be able to derive all valid independences from them (and, hence, the decomposition might not be as efficient as it could be) but we can be sure that we do not derive invalid independences (w.r.t. the distribution) and thus that the decomposition will be correct.

Note that any isolated graph $G = (V, \emptyset)$ which consists only of unconnected nodes constitutes a trivial dependence map. Such a graph encodes all possible conditional and marginal independences among the attributes in V and thus, obviously, also those contained in any distribution (over V). Put differently: Definition 25.4 stated that a dependence map encodes only correct dependences. Dependences, however, require connected nodes. In consequence, the set of encoded dependences in an isolated graph is empty. And this empty set is (trivially) a subset of any set of dependences of any distribution. Contrary, each complete graph (that is, a clique V) constitutes a trivial independence map. Since independences are encoded via missing edges, the set $\mathcal{I}_G$ is obviously empty in a complete graph G. Thus, it is a (trivial) subset of any set $\mathcal{I}_p$.

A trivial independence map is—albeit correct—of little use. Therefore, we are interested in graphs that encode as many independences of a given distribution as possible, that is, where $|\mathcal{I}_G|$ is large, but without violating the inclusion $\mathcal{I}_G \subseteq \mathcal{I}_p$. Although we maximize the number of conditional independences that are contained in $\mathcal{I}_G$, they are referred to as *minimal* independence maps in the literature because the number of edges is minimized. We will follow this convention here and define.

Definition 25.5 (*Minimal Independence Map*) Let $G = (V, E)$ be a (directed or undirected) independence map of a probability distribution p_V. G is called a *minimal independence map* or *minimal independence graph* if it is not possible to delete an edge from E without introducing a conditional independence that is invalid in p_V.

An analogous definition for dependence maps is possible as well. We omit it here, as we do not use this notion in the remainder. As we now have defined the type of graphs that are of interest to us, we can now define the notion of decomposition or decomposability (which was used rather colloquially up to now). Again, we distinguish between the two graph types.

Definition 25.6 (*Decomposable w.r.t. an Undirected Graph*)
A probability distribution p_V over an attribute set $V = \{A_1, \ldots, A_n\}$ is called decomposable or factorizable w.r.t. to an undirected graph $G = (V, E)$ if and only if it can be written as a product of non-negative functions that are defined on the cliques of G. Precisely: Let C be a family (that is, a set) of subsets of V such that the subgraphs induced by the sets $C \in C$ coincide with the cliques of G. Let further be $\mathcal{E}_C$ the set of events that can be described by assigning values to the attributes in C. Then p_V is called decomposable or factorizable w.r.t. G if there exist functions $\phi_C : \mathcal{E}_C \to \mathbb{R}_0^+$, $C \in C$, such that the following holds:

$$\forall a_1 \in dom(A_1) : \cdots \forall a_n \in dom(A_n) :$$
$$p_V\left(\bigwedge_{A_i \in V} A_i = a_i \right) = \prod_{C \in C} \phi_C\left(\bigwedge_{A_i \in C} A_i = a_i \right)$$

Definition 25.7 (*Decomposable w.r.t. a Directed Acyclic Graph*)
A probability distribution p_V over an attribute set $V = \{A_1, \ldots, A_n\}$ is called decomposable or factorizable w.r.t. a directed acyclic graph $G = (V, E)$ if and only if it can be written as the product of the conditional probabilities of the attributes

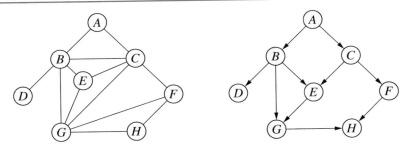

Fig. 25.3 Two graphs to illustrate the decomposition of a distribution

given their parent attributes in G. That is, if the following holds:

$$\forall a_1 \in \text{dom}(A_1) : \cdots \forall a_n \in \text{dom}(A_n) :$$

$$p_V\left(\bigwedge_{A_i \in V} A_i = a_i \right) = \prod_{A_i \in V} P\left(A_i = a_i \ \Big| \ \bigwedge_{A_j \in \text{pa}_G(A_i)} A_j = a_j \right)$$

Let us consider two examples to illustrate the decomposition of a distribution w.r.t. an undirected and a directed acyclic graph.

Example 25.2 (Decomposition w.r.t. an undirected graph)
The undirected graph in Fig. 25.3 contains the following cliques:

$$C_1 = \{B, C, E, G\}, C_2 = \{A, B, C\}, C_3 = \{C, F, G\}, C_4 = \{B, D\}, C_5 = \{F, G, H\}.$$

The decomposition induced by these cliques reads:

$$\forall a \in \text{dom}(A) : \cdots \forall h \in \text{dom}(H) :$$
$$p_V(A = a, \ldots, H = h) = \phi_{C_1}(B = b, C = c, E = e, G = g)$$
$$\cdot \ \phi_{C_2}(A = a, B = b, C = c)$$
$$\cdot \ \phi_{C_3}(C = c, F = f, G = g)$$
$$\cdot \ \phi_{C_4}(B = b, D = d)$$
$$\cdot \ \phi_{C_5}(F = f, G = g, H = h)$$

Example 25.3 (Decomposition w.r.t. a directed acyclic graph)
The directed acyclic graph in Fig. 25.3 induces the following decomposition:

$$\forall a \in \text{dom}(A) : \cdots \forall h \in \text{dom}(H) :$$
$$p_V(A = a, \ldots, H = h) = P(H = h \mid F = f, G = g) \cdot P(G = g \mid B = b, E = e)$$
$$\cdot \ P(F = f \mid C = c) \cdot P(E = e \mid B = b, C = c)$$
$$\cdot \ P(D = d \mid B = b) \cdot P(C = c \mid A = a)$$
$$\cdot \ P(B = b \mid A = a) \cdot P(A = a)$$

As both the notions decomposability w.r.t. a graph and independence map are now defined, we still need the connection between them. Our objective is to infer correct (sound) conditional independences of an underlying distribution from the separation criteria read from an independence map. This connection is established by the following two theorems.

Theorem 25.3 *Let p_V be a strictly positive probability distribution over a set V of (discrete) attributes. p_V is factorizable w.r.t. an undirected graph $G = (V, E)$, if and only if G is a conditional independence map of p_V.*

Theorem 25.4 *Let p_V be probability distribution over a set V of (discrete) attribute. p_V is factorizable w.r.t. a directed acyclic graph $G = (V, E)$ if and only if G is a conditional independence map of p_V.*

Both theorems can be summarized as follows (taking care of the restriction w.r.t. strict positivity of p in Theorem 25.3):

$$G \text{ factorizes } p \quad \Leftrightarrow \quad G \text{ is an independence map of } p.$$

Let us clarify which conclusion we can draw from this equivalence. Assume, we observe that a distribution p_V can be decomposed w.r.t. a given graph G. It is immediately clear that G must be an independence map of p_V (implication from left to right). This in turn allows us to read every separation in G as a valid conditional independence in p_V (implication from right to left).

Until now, most definitions and theorems assumed a graph and a distribution to be given. Usually, this is not the case. Therefore, we will now answer the two questions how to construct a minimal independence map (directed and undirected) given a distribution and how to construct a distribution for which a given (directed and undirected) graph is an independence map. Hence, we will now introduce four algorithms that establish the constructive connection between distributions and the decomposition w.r.t. a graph.

Algorithm 25.1 (Decomposition w.r.t. an Undirected Triangulated Graph)
Input: undirected triangulated graph $G = (V, E)$
Output: decomposition of a distribution p_V with independence map G

1. Determine all cliques of G.
2. Determine a clique ordering $C_1, \ldots, C_r$ with the running intersection property (see Definition 24.39).
3. Determine the sets $S_j = C_j \cap (C_1 \cup \cdots \cup C_{j-1})$ and $R_j = C_j \setminus S_j$.
4. Return

$$\forall a_1 \in \text{dom}(A_1) : \cdots \forall a_n \in \text{dom}(A_n) :$$

$$p_V\Big(\bigwedge_{A_i \in V} A_i = a_i \Big) = \prod_{C_j \in C} P\Big(\bigwedge_{A_i \in R_j} A_i = a_i \ \Big| \ \bigwedge_{A_i \in S_j} A_i = a_i \Big)$$

Example 25.4 (Illustration of Algorithm 25.1) Consider again the undirected triangulated graph in Fig. 25.3. The following clique ordering has running intersection property (see Definition 24.43 on page 551) which allows us to state the following residual and separator sets:

i	C_i	R_i	S_i
1	$\{B, C, E, G\}$	$\{B, C, E, G\}$	$\emptyset$
2	$\{A, B, C\}$	$\{A\}$	$\{B, C\}$
3	$\{C, F, G\}$	$\{F\}$	$\{C, G\}$
4	$\{B, D\}$	$\{D\}$	$\{B\}$
5	$\{F, G, H\}$	$\{H\}$	$\{F, G\}$

This leads to the following decomposition as described by Algorithm 25.1 (for the sake of clarity, we omit the all-quantifiers over the respective attributes domains):

$$
\begin{aligned}
p_V&(A, B, C, D, E, F, G, H)\\
&= P(R_1 \mid S_1) \quad\cdot P(R_2 \mid S_2) \quad\cdot P(R_3 \mid S_3) \quad\cdot P(R_4 \mid S_4) \cdot P(R_5 \mid S_5)\\
&= P(B, C, E, G) \cdot P(A \mid B, C) \cdot P(F \mid C, G) \cdot P(D \mid B) \quad\cdot P(H \mid F, G)\\
&= \frac{P(B, C, E, G)}{1} \cdot \frac{P(A, B, C)}{P(B, C)} \cdot \frac{P(F, C, G)}{P(C, G)} \cdot \frac{P(D, B)}{P(B)} \cdot \frac{P(H, F, G)}{P(F, G)}\\
&= \frac{P(C_1)}{1} \cdot \frac{P(C_2)}{P(S_2)} \cdot \frac{P(C_3)}{P(S_3)} \cdot \frac{P(C_4)}{P(S_4)} \cdot \frac{P(C_5)}{P(S_5)}
\end{aligned}
$$

The last line in this equation shows an alternative representation without explicitly stating conditional probabilities (by just replacing them by their definition). For a general clique set C satisfying RIP, we get

$$\forall a_1 \in \mathrm{dom}(A_1) : \cdots \forall a_n \in \mathrm{dom}(A_n) :$$

$$
p_V\left(\bigwedge_{A_i \in V} A_i = a_i\right) = \frac{\displaystyle\prod_{j=1}^{r} P\left(\bigwedge_{A_i \in C_j} A_i = a_i\right)}{\displaystyle\prod_{j=2}^{r} P\left(\bigwedge_{A_i \in S_j} A_i = a_i\right)}
$$

The start index 2 in the denominator excludes the probability of the formally empty separator set S_1. We will use the above decomposition formula in join trees. The separator sets are the intersections of neighboring cliques. Since there only $n-1$ edges in a tree with n nodes, this explains the "absence" of one separator set.

Algorithm 25.2 (Decomposition w.r.t. a Directed Acyclic Graph)
Input: directed acyclic graph $G = (V, E)$
Output: decomposition of a distribution p_V with independence map G

1. Determine the parent sets $\mathrm{pa}_G(A_i)$.
2. Return

$$\forall a_1 \in \mathrm{dom}(A_1) : \cdots \forall a_n \in \mathrm{dom}(A_n) :$$

$$p_V \left(\bigwedge_{A_i \in V} A_i = a_i \right) = \prod_{A_i \in V} P \left(A_i = a_i \ \Big| \bigwedge_{A_j \in \mathrm{pa}_G(A_i)} A_j = a_j \right)$$

This algorithm basically coincides with Definition 25.7. In contrast to a decomposition by an undirected graph where appropriate clique potentials have to be found, they are immediately clear here. Example 25.3 serves as an illustration.

Algorithm 25.3 (Minimal Undirected Independence Map)
Input: strictly positive distribution p_V over a set $V = \{A_1, \ldots, A_n\}$ of attributes
Output: minimal undirected independence map $G = (V, E)$ of p_V.

1. Start with $G = (V, E)$ a fully connected graph: $E = V \times V \setminus \{(A, A) \mid A \in V\}$.
2. For each edge $A - B \in E$ compute:

$$p_{V \setminus \{A\}} \left(\bigwedge_{A_i \in V \setminus \{A\}} A_i = a_i \right) = \sum_{a \in \mathrm{dom}(A)} p_V \left(\bigwedge_{A_i \in V} A_i = a_i \right)$$

$$p_{V \setminus \{B\}} \left(\bigwedge_{A_i \in V \setminus \{B\}} A_i = a_i \right) = \sum_{b \in \mathrm{dom}(B)} p_V \left(\bigwedge_{A_i \in V} A_i = a_i \right)$$

$$p_{V \setminus \{A,B\}} \left(\bigwedge_{A_i \in V \setminus \{A,B\}} A_i = a_i \right) = \sum_{a \in \mathrm{dom}(A)} \sum_{b \in \mathrm{dom}(B)} p_V \left(\bigwedge_{A_i \in V} A_i = a_i \right)$$

If $p_V \left(\bigwedge_{A_i \in V} A_i = a_i \right) \cdot p_{V \setminus \{A,B\}} \left(\bigwedge_{A_i \in V \setminus \{A,B\}} A_i = a_i \right)$

$= p_{V \setminus \{A\}} \left(\bigwedge_{A_i \in V \setminus \{A\}} A_i = a_i \right) \cdot p_{V \setminus \{B\}} \left(\bigwedge_{A_i \in V \setminus \{B\}} A_i = a_i \right),$

then delete edge $A - B$ from E.
3. Return G.

The independence test in step 2 exploits the following fact of undirected graphs: the absence of an edge $A - B$ in an undirected graph $G = (V, E)$ implies that those two nodes A and B are u-separated by all other nodes:

$$\forall A, B \in V, A \neq B : \quad (A, B) \notin E \ \Rightarrow \ A \perp\!\!\!\perp_G B \mid V \setminus \{A, B\}.$$

Since the graph G shall be an independence map w.r.t. a given distribution p_V, we must ensure that deducible u-separations also hold true in form of conditional independence statements in p_V. The separation $A \perp\!\!\!\perp_G B \mid V \setminus \{A, B\}$ in G has to have the following corresponding probabilistic counterpart in p_V:

$$P(A, B \mid V \setminus \{A, B\}) = P(A \mid V \setminus \{A, B\}) \cdot P(B \mid V \setminus \{A, B\}).$$

Applying equivalence transformations (multiply twice with $P(V \setminus \{A, B\})$), we arrive at the test criterion of step 2:

$$P(A, B \mid V \setminus \{A, B\}) = P(A \mid V \setminus \{A, B\}) \cdot P(B \mid V \setminus \{A, B\})$$
$$P(V) = P(A \mid V \setminus \{A, B\}) \cdot P(V \setminus \{A\})$$
$$P(V) \cdot P(V \setminus \{A, B\}) = P(V \setminus \{B\}) \cdot P(V \setminus \{A\})$$

The resulting independence map is obviously minimal: no further edge can be deleted (because we tested them all) without encoding an invalid independence.

Algorithm 25.4 (Minimal Directed Independence Map)
Input: distribution p_V over a set $V = \{A_1, \ldots, A_n\}$ of attributes
Output: minimal directed independence map $G = (V, E)$ of p_V.

1. Determine an arbitrary attribute ordering $A_1, \ldots, A_n$.
2. Find for each A_i a minimal predecessor set Π_i, that renders A_i conditionally independent of $\{A_1, \ldots, A_{i-1}\} \setminus \Pi_i$.
3. Start with $G = (V, \emptyset)$ and insert for each A_i an edge from each node in Π_i to A_i.
4. Return G.

Example 25.5 Consider the three-dimension distribution of Table 24.1 on page 534 which is repeated as Table 25.2. Let us assume the following node order:

$$G \prec Pr \prec Sm.$$

We know from Sect. 24.1.2 (and the distributions in Table 24.2) that only the following conditional independence holds in the distribution p_{GPrSm} (and, of course, its symmetric counterpart):

$$Pr \perp\!\!\!\perp_{p_{GPrSm}} Sm \mid G.$$

Let us now start to determine the sets Π_i. We always start with all predecessors as the candidate set: $\Pi_i = \{A_1, \ldots, A_{i-1}\}$. We then delete as many attributes from Π_i

Table 25.2 Three-dimensional example distribution

p_{orig}	G = m		G = w	
	Sm = sm	Sm = $\overline{sm}$	Sm = sm	Sm = $\overline{sm}$
Pr = pr	0	0	0.01	0.04
Pr = $\overline{pr}$	0.2	0.3	0.09	0.36

Fig. 25.4 Resulting graph, for Example 25.5

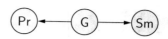

until the conditional independence would be violated. The remaining attributes will become the parents of A_i. It may happen, that we cannot exclude any attribute from the initial set Π_i. We then add all predecessors $A_1, \ldots, A_{i-1}$ of A_i as its parents.

1. G has no predecessor. Hence it is (and remains) $\Pi_G = \emptyset$.
2. For Pr we start with $\Pi_{Pr} = \{G\}$. The first (and only) reduction of Π_{Pr} is the one leading to the empty set. Consequently, we test whether $Pr \perp\!\!\!\perp_{PGPrSm} G \mid \emptyset$. This is not the case as can be easily verified. There are no other options to set Π_S, hence the initial set remains: $\Pi_{Pr} = \{G\}$.
3. We start with $\Pi_{Sm} = \{G, Pr\}$ and test the following three options:

$$R \perp\!\!\!\perp_{PGPrSm} Pr \mid G, \qquad R \perp\!\!\!\perp_{PGPrSm} G \mid Pr \qquad \text{and} \qquad R \perp\!\!\!\perp_{PGPrSm} G, Pr \mid \emptyset$$

The only valid independence is $Sm \perp\!\!\!\perp_{PGPrSm} Pr \mid G$ which leads to $\Pi_{Sm} = \{G\}$.

We arrive at the graph shown in Fig. 25.4, which is not only a minimal independence map but also a perfect map in this case.

Finally, we can define the core structures covered by the term *graphical models*.

Definition 25.8 (*Markov Network*) A *Markov network* is an undirected conditional independence graph $G = (V, E)$ of a probability distribution p_V together with a family of non-negative functions ϕ_M of the factorization induced by the graph.

The undirected graph in Fig. 25.3 together with the decomposition from Example 25.2 (and appropriately selected clique potentials) is a Markov network.

Definition 25.9 (*Bayesian Network*) A *Bayes network* is a directed conditional independence graph of a probability distribution p_V together with a family of conditional probabilities of the factorization induced by the graph.

The directed acyclic graph in Fig. 25.3 together with the decomposition from Example 25.3 (and appropriately selected conditional distributions) is a Bayes network.

References

1. Borgelt C, Steinbrecher M, Kruse R (2009) Graphical models—representations for learning, reasoning and data mining, 2nd edn. Wiley, Chichester, United Kingdom
2. Dawid AP (1979) Conditional independence in statistical theory. J Roy Stat Soc Ser B (Methodological) 41(1):1–31. Blackwell, Oxford, United Kingdom

3. Gebhardt J, Kruse R (2005) Operations knowledge-based, for graphical models in planning. In: Proceedings of the Europian Conference on symbolic and quantitative approaches to reasoning with uncertainty (ECSQARU, (2005) Barcelona, Spain), LNAI 3571:3–14. Springer-Verlag, Berlin, Germany

4. Pearl J (1988) Probabilistic reasoning in intelligent systems: networks of plausible inference. Morgan Kaufmann, San Mateo, CA, USA

5. Pearl J, Paz A (1987) Graphoids: a graph based logic for reasoning about relevance relations. In: Boulay BD, Hogg D, Steels L (eds) Advances in Artificial Intelligence 2, 357–363. North Holland, Amsterdam, Netherlands

6. Studený M (1989) Multiinformation and the problem of characterization of conditional independence relations. Probl Contr Inf Theory 1:3–16

7. Studený M (1990) Conditional independence relations have no finite complete characterization. Kybernetika 25:72–79. Institute of Information Theory and Automation, Prague, Czech Republic

Evidence Propagation

After having discussed efficient representations for expert and domain knowledge, we intent to exploit them to draw inferences when new information (evidence) becomes known. The objective is to propagate the evidence through the underlying network to reach all relevant attributes. Obviously, the graph structure will play an important role.

Given a graphical model (a Markov or Bayes network) $G = (V, E)$ over an attribute set V with underlying distribution p_V and an observed value a_o of the attribute $A_o \in V$, propagating this evidence corresponds formally to the calculation of the following probabilities:

$$\forall A \in V \setminus \{A_o\} : \forall a \in \mathrm{dom}(A) : \quad P(A = a \mid A_o = a_o).$$

Evidence propagation is thoroughly treated in the literature and a multitude of different algorithms exist. They can be distinguished by model type (Markov or Bayes network), by graph topology (tree, polytree, general graph, or completely independent of any graph structure), or by type of computation (exact or approximate), just to name the most important criteria for comparison. A polytree is a directed tree where nodes can have more than one parent. The graph G_1 in Fig. 26.2 on page 567 is a polytree, while G_2 is a simple tree.

It is not our intention to provide a wide coverage over those algorithms. We rather intent to discuss one particular evidence propagation algorithm in greater detail in order to stress the underlying ideas.

We already saw in the introduction Chap. 23 that we can exploit the graph structure to guide the evidence propagation: First, the (undirected) tree structure guaranteed unique paths between the attributes and, second, the tree decomposes into multiple isolated subgraphs when any node is deleted. The latter coincides with the statement that (in an undirected tree) an instantiated attribute (where instantiated here means that, for this attributes, we have observed evidence) separates the attributes in the subtrees as depicted in Fig. 26.1. An alike separation can also be observed in directed trees: Given an attribute A, the attributes of the child trees (i.e., the subtrees having

© Springer Nature Switzerland AG 2022
R. Kruse et al., *Computational Intelligence*, Texts in Computer Science,
https://doi.org/10.1007/978-3-030-42227-1_26

Fig. 26.1 The gray node separates the notes in all four subtrees. We will use that idea to simplify the evidence propagation

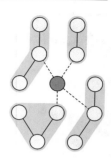

a child of A as the root node) become conditionally independent from each other. However, we have to take into account that, in a polytree, we might have to deal with multiple parent nodes. These form a converging connection with A which is activated when A is instantiated. The matter becomes even more complex when dealing with general graphs, i.e., arbitrary undirected or directed acyclic graphs: Since there might be multiple paths between two nodes, we do not have a unique route for propagating evidence. There exist algorithms that deal with all of the mentioned problems. We will shortly discuss a solution that can deal with all of the above-mentioned network and graph types.

The evidence propagation itself will be carried out on a join tree. The tree structure guarantees the uniqueness of the evidence flow, while the choice of a join tree is justified by the fact that every general (undirected or directed) graph can be transformed into a semantically equivalent join tree. Semantic equivalence means here that the resulting join tree is still a conditional independence map of the underlying distribution.

How any undirected graph can be transformed into a join tree was shown in Sect. 24.2.2. Let us therefore focus on the transformation of a directed acyclic graph into a join tree. The idea to just drop the edge directions in order to get an equivalent undirected graph does unfortunately not work as we can easily see. Consider the directed acyclic graph G_1 in Fig. 26.2. It encodes only the following conditional independence statements (although the symmetric counterparts do, of course, hold but we omit them here for the sake of brevity):

$$\mathcal{I}_{G_1} = \{\quad A \perp\!\!\!\perp_{G_1} D \mid C, \quad B \perp\!\!\!\perp_{G_1} D \mid C, \quad A \perp\!\!\!\perp_{G_1} B \mid \emptyset \quad\}.$$

By dropping the edge directions, we arrive at the undirected graph G_u as depicted in Fig. 26.2. Its set of encoded conditional independence statements is, alas, no subset of those of G_1:

$$\mathcal{I}_{G_u} = \{\quad A \perp\!\!\!\perp_{G_u} D \mid C, \quad B \perp\!\!\!\perp_{G_u} D \mid C, \quad A \perp\!\!\!\perp_{G_u} B \mid C \quad\} \not\subseteq \mathcal{I}_{G_1}.$$

There is one new conditional independence statement (here: $A \perp\!\!\!\perp B \mid C$) that is not valid in the original graph G_1. The fact that we lost a previously valid independence, namely $A \perp\!\!\!\perp_{G_1} B \mid \emptyset$, is unfortunate but does not invalidate the character of the independence map. Hence, G_u is not an equivalent undirected graph for G_1! The goal must be to find a transformation that by no means introduces any new conditional independences—and be it at the loss of existing ones.

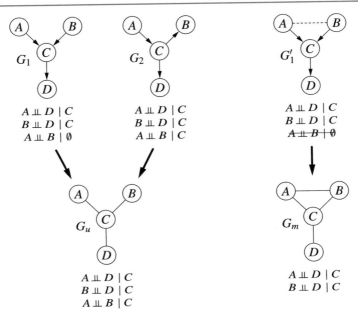

Fig. 26.2 In general, ignoring the edge directions in general does not result in an equivalent (w.r.t. the encoded conditional independence statements) undirected graph

Note, that not every directed acyclic graph suffers from the illustrated phenomenon. Consider graph G_2 in Fig. 26.2. For its undirected pendant G_u, we indeed have

$$\mathcal{I}_{G_u} \subseteq \mathcal{I}_{G_2}.$$

(We even achieve equality here which is not necessarily always the case.)

Obviously, the illustrated problems arise due to the asymmetric definition of the d-separation which cause problems when dealing with converging nodes. It must be prevented that both parent nodes of a common child node become conditionally independent given the child node. This is achieved by connecting parent nodes that are not yet connected (the edge direction can be chosen arbitrarily as we drop it in the next step). This is depicted on the right of Fig. 26.2 and graph G_1'. By connecting the nodes A and B, we loose a (marginal) independence statement, but we prevent the introduction of a new one which leads to $\mathcal{I}_{G_u} \subseteq \mathcal{I}_{G_1'}$.

The transformation of connecting unconnected parent nodes we already know as moralization (see Definition 24.29 on page 540). With these prerequisites, we can transform any given graphical model into a common input format (namely, a join tree): If the given graph is a Markov network, we might need to triangulate the graph before it is transformed into a join tree. If we are given a Bayes network, we moralize the graph, triangulate it (if necessary), and transform it into a join tree afterwards. We will illustrate this with an example that we will reuse later when running the evidence propagation algorithm.

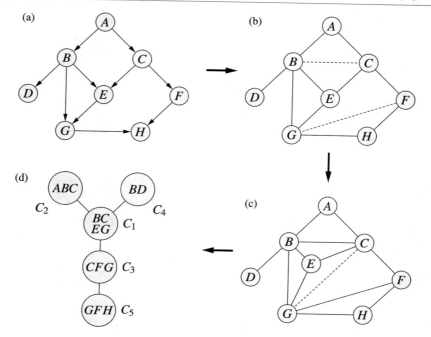

Fig. 26.3 Transformation of a Bayes network into a join tree. **a** Bayes network, **b** Moral graph, **c** Triangulated moral graph, **d** Join tree

Example 26.1 (*Join tree from a Bayes network*) We will shortly consider an propagation of evidence inside the Bayes network depicted in Fig. 26.3a. The result of the moralization step can be seen in Fig. 26.3b. The circle $E - C - F - G$ has no chord which leads to the insertion of one. We decide for chord $C - G$ ($E - F$ would have been possible as well) and arrive at the triangulated graph in Fig. 26.3c. We know this graph already from Fig. 25.3 and Example 25.2 which allows us to directly reuse the cliques identified back then, resulting in the join tree in Fig. 26.3 d. Note, that during triangulation as well as moralization only edges might be added (never deleted) and hence the character of a conditional independence map never changes.   □

The evidence propagation will consist of the following principal steps:

Initialization: Evidences, i.e., the known attribute values will be incorporated into all relevant clique potentials.

Message passing: Neighboring cliques send a message to each other to announce the changes. Hence, there are exactly two messages sent across each edge. Having r cliques in the tree, $2(r - 1)$ messages will be sent.

Update: After all messages have been sent, each clique can update the joint probability of its attributes.

Marginalization: Since we are interested in the marginal distribution of single attributes, we finally marginalize out each attribute from the smallest clique that contains it.

We now investigate these four steps more closely.

26.1 Initialization

We already know Algorithm 25.1 from page 560 to determine the clique potentials. If the join tree resulted from an initial Bayes network (as in our example case), we can easily determine the clique potentials of all cliques $C \in C$ as the product of the conditional probabilities of all node families which are properly contained in C:

$$\phi_C(c) = 1 \cdot \prod_{fa(A) \subseteq C} P\left(A = \text{proj}_{\{A\}}^C(c) \,\Big|\, \text{pa}(A) = \text{proj}_{\text{pa}(A)}^C(c)\right). \qquad (26.1)$$

If the values of one or more attributes become known, it leads to the modification of those cliques that contain these attributes: All attribute value combinations that are incompatible with the evidence get assigned a value of zero by the potential functions. In all other cases, the potential functions remain unchanged:

$$\phi_C'(c) = \begin{cases} 0, & \text{if } \text{proj}_{E \cap C}^C(c) \neq \text{proj}_{E \cap C}^E(e) \\ \phi_C(c), & \text{otherwise.} \end{cases}$$

26.2 Message Passing

After initialization, neighboring cliques exchange messages to inform each other about (a possible) change in their potential functions. Each clique sends exactly one message to each neighbor. The message $M_{B \to C}$ from clique B to clique C is declared on the attributes of the separator set S_{BC}. The values of the message from B to C depend on the potential of the sending clique ϕ_B and the messages of all *other* neighbors of B. The values are multiplied and marginalized on the attributes in S_{BC}:

$$M_{B \to C}(s_{BC}) = \sum_{b \setminus s_{BC}} \left[\phi_B(b) \cdot \prod_{D \in \text{adj}(B) \setminus \{C\}} M_{D \to B}(s_{DB}) \right]. \qquad (26.2)$$

Since we need for the computation of the message from B to C all received messages from all other neighbors of B, it is clear that only cliques with no other neighbors can send its messages first: namely, the outer nodes of the join tree. We will later see how this leads to a message sending cascade across the join tree.

26.3 Update

After all messages have been sent, each clique C can compute is joint probability distribution $P(c)$ as the product of its potential function and all received messages from its neighbors:

$$P(c) \; \propto \; \phi_C(c) \cdot \prod_{B \in \mathrm{adj}(C)} M_{B \to C}\left(\mathrm{proj}_{S_{BC}}^C(c)\right). \qquad (26.3)$$

The $\propto$-sign denotes that the distribution $P(c)$ needs to be normalized in case it does not add up to one (over all c).

26.4 Marginalization

After each clique has updated its joint distribution, we look for each attribute A for the smallest clique C that contains it in order to minimize the marginalization effort (a marginalization from other cliques that contain A would be also possible, of course):

$$P(a) = \sum_{c \backslash a} P(c)$$

Let us now study a full run of the algorithm. The parameters of the Bayes network in Fig. 26.3a are shown in Fig. 26.4.

Example 26.2 (*Initialization, propagating "zero" evidence*) For the join tree in Fig. 26.3d, we use the initial Bayes network from Fig. 26.3a and can compute the following potentials:

$P(A)$	
a_1	0.6
a_2	0.4

$P(B \mid A)$	a_1	a_2
b_1	0.2	0.1
b_2	0.8	0.9

$P(C \mid A)$	a_1	a_2
c_1	0.3	0.7
c_2	0.7	0.3

$P(D \mid B)$	b_1	b_2
d_1	0.4	0.7
d_2	0.6	0.3

$P(F \mid C)$	c_1	c_2
f_1	0.1	0.4
f_2	0.9	0.6

$P(E \mid B, C)$	b_1		b_2	
	c_1	c_2	c_1	c_2
e_1	0.2	0.4	0.3	0.1
e_2	0.8	0.6	0.7	0.9

$P(G \mid B, E)$	b_1		b_2	
	e_1	e_2	e_1	e_2
g_1	0.95	0.4	0.7	0.5
g_2	0.05	0.6	0.3	0.5

$P(H \mid G, F)$	g_1		g_2	
	f_1	f_2	f_1	f_2
h_1	0.2	0.4	0.5	0.7
h_2	0.8	0.6	0.5	0.3

Fig. 26.4 Parameters of the Bayes network in Fig. 26.3a

$$\phi_{C_1}(b, c, e, g) = P(e \mid b, c) \cdot P(g \mid e, b)$$
$$\phi_{C_2}(a, b, c) = P(b \mid a) \cdot P(c \mid a) \cdot P(a)$$
$$\phi_{C_3}(c, f, g) = P(f \mid c)$$
$$\phi_{C_4}(b, d) = P(d \mid b)$$
$$\phi_{C_5}(g, f, h) = P(h \mid g, f)$$

Since each conditional probability occurs in exactly on potential, it is immediately clear that the product of all potentials equals the network's decomposition formula. The clique potentials derived via Algorithm 25.1 based on separator and residual sets would lead to the following potentials:

$$\phi_{C_1}(b, c, e, g) = P(b, e \mid c, g)$$
$$\phi_{C_2}(a, b, c) = P(a \mid b, c)$$
$$\phi_{C_3}(c, f, g) = P(c \mid f, g)$$
$$\phi_{C_4}(b, d) = P(d \mid b)$$
$$\phi_{C_5}(g, f, h) = P(h, g, f)$$

In the remainder of this example, we will use potentials retrieved from the Bayes network. Since we want to compute the a priori distribution of all attributes first, no clique potentials are modified in this run. $\square$

Example 26.3 (*Message passing*)
Figure 26.5 shows the join tree from Fig. 26.3d together with all eight messages which are computed as follows:

$$M_{21}(b, c) = \sum_a \phi_2(a, b, c),$$
$$M_{41}(b) = \sum_d \phi_4(b, d),$$
$$M_{53}(f, g) = \sum_h \phi_5(f, g, h),$$
$$M_{13}(c, g) = \sum_{b,e} \phi_1(b, c, e, g)\, M_{21}(b, c)\, M_{41}(b),$$
$$M_{31}(c, g) = \sum_f \phi_3(c, f, g)\quad M_{53}(f, g),$$
$$M_{12}(b, c) = \sum_{e,g} \phi_2(b, c, e, g)\, M_{31}(c, g)\, M_{41}(b),$$
$$M_{35}(f, g) = \sum_c \phi_3(c, f, g)\quad M_{13}(c, g),$$
$$M_{14}(b) = \sum_{c,e,g} \phi_1(b, c, e, g)\, M_{21}(b, c)\, M_{31}(c, g).$$

As can easily be seen, the messages M_{41}, M_{53}, and M_{21} can be computed (in arbitrary order) first because they do not require other messages for their computation. This is immediately clear as they originate from leaf nodes which do not have any other neighbors except the recipient of the message. After these three messages are sent, we can compute (and send) M_{31} and M_{13}. Only then we are able to compute M_{12}, M_{14}, and M_{35}. Figure 26.5 depicts the just mentioned dependencies as a directed graph with edges directed to the dependent messages. The computed messages are as follows:

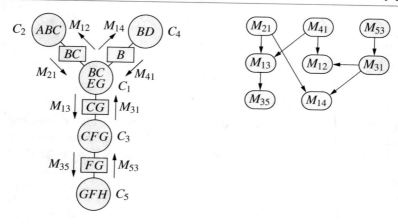

Fig. 26.5 Messages that need to be sent across the join tree together with a dependency graph of the messages

$$M_{21} = \overset{b_1,c_1 \ \ b_1,c_2 \ \ b_2,c_1 \ \ b_2,c_2}{(\ 0.06, 0.10, 0.40, 0.44 \ )} \qquad M_{41} = \overset{b_1 \ b_2}{(\ 1, \ 1 \ )}$$

$$M_{13} = \overset{c_1,g_1 \ \ \ c_1,g_2 \ \ \ c_2,g_1 \ \ \ c_2,g_2}{(\ 0.254, 0.206, 0.290, 0.250 \ )} \qquad M_{35} = \overset{f_1,g_1 \ \ f_1,g_2 \ \ f_2,g_1 \ \ f_2,g_2}{(\ 0.14, 0.12, 0.40, 0.33 \ )}$$

$$M_{53} = \overset{f_1,g_1 \ \ f_1,g_2 \ \ f_2,g_1 \ \ f_2,g_2}{(\ \ 1 \ , \ \ 1 \ , \ \ 1 \ , \ \ 1 \ )} \qquad M_{31} = \overset{c_1,g_1 \ \ c_1,g_2 \ \ c_2,g_1 \ \ c_2,g_2}{(\ \ 1 \ , \ \ 1 \ , \ \ 1 \ , \ \ 1 \ )}$$

$$M_{12} = \overset{b_1,c_1 \ \ b_1,c_2 \ \ b_2,c_1 \ \ b_2,c_2}{(\ \ 1 \ , \ \ 1 \ , \ \ 1 \ , \ \ 1 \ )} \qquad M_{14} = \overset{b_1 \ \ \ b_2}{(\ 0.16, 0.84 \ )} \qquad \qquad \square$$

Example 26.4 (*Update*) After all messages have been sent, the cliques can update their joint probability distribution. For the five cliques of our example, we get

$$
\begin{aligned}
P(c_1) &= P(b, c, e, g) = \phi_1(b, c, e, g) \cdot M_{21}(b, c) \cdot M_{31}(c, g) \cdot M_{41}(b) \\
P(c_2) &= P(a, b, c) \quad \propto \phi_2(a, b, c) \quad \cdot M_{12}(b, c) \\
P(c_3) &= P(c, f, g) \quad \propto \phi_3(c, f, g) \quad \cdot M_{13}(c, g) \cdot M_{53}(f, g) \\
P(c_4) &= P(b, d) \quad \ \ \propto \phi_4(b, d) \quad \ \ \cdot M_{14}(b) \\
P(c_5) &= P(f, g, h) \quad \propto \phi_5(f, g, h) \quad \cdot M_{35}(f, g)
\end{aligned}
$$

The numbers can be found in the P-columns of the potential tables in Table 26.1. In this case, no normalization is necessary. $\qquad \square$

Example 26.5 (*Marginalization*) Finally, we determine the marginal probabilities for each attribute from the clique potentials. To keep the effort low, we choose the smallest clique that contains the attribute. We arrive at the following marginalizations:

$$
\begin{array}{ll}
P(a) = \sum_{b,c} P(a, b, c), & P(b) = \sum_d P(b, d), \\
P(c) = \sum_{a,b} P(a, b, c), & P(d) = \sum_b P(b, d), \\
P(e) = \sum_{b,c,g} P(b, c, e, g), & P(f) = \sum_{c,g} P(c, f, g), \\
P(g) = \sum_{c,f} P(c, f, g), & P(h) = \sum_{g,f} P(g, f, h).
\end{array}
$$

Table 26.1 The potential tables of the join tree from Fig. 26.3d. The first column contains the potentials according to Eq. 26.1, the second column contains the updated joint probabilities of the cliques according to Eq. 26.3 after zero evidence has been propagated. The third column contains the probabilities after the evidence $H = h_1$ has been propagated

ϕ_1:

				ϕ_1	P	P'
b_1	c_1	e_1	g_1	0.19	0.0122	0.0095
			g_2	0.01	0.0006	0.0009
		e_2	g_1	0.32	0.0205	0.0161
			g_2	0.48	0.0307	0.0431
	c_2	e_1	g_1	0.38	0.0365	0.0241
			g_2	0.02	0.0019	0.0025
		e_2	g_1	0.24	0.0230	0.0152
			g_2	0.36	0.0346	0.0443
b_2	c_1	e_1	g_1	0.21	0.0832	0.0653
			g_2	0.09	0.0356	0.0501
		e_2	g_1	0.35	0.1386	0.1088
			g_2	0.35	0.1386	0.1947
	c_2	e_1	g_1	0.07	0.0311	0.0205
			g_2	0.03	0.0133	0.0171
		e_2	g_1	0.45	0.1998	0.1321
			g_2	0.45	0.1998	0.2559

ϕ_2:

			ϕ_2	P	P'
a_1	b_1	c_1	0.036	0.0360	0.0392
		c_2	0.084	0.0840	0.0753
	b_2	c_1	0.144	0.1440	0.1523
		c_2	0.336	0.3360	0.3220
a_2	b_1	c_1	0.028	0.0280	0.0305
		c_2	0.012	0.0120	0.0108
	b_2	c_1	0.252	0.2520	0.2665
		c_2	0.108	0.1080	0.1035

ϕ_3:

			ϕ_3	P	P'
c_1	f_1	g_1	0.1	0.0254	0.0105
		g_2	0.1	0.0206	0.0212
	f_2	g_1	0.9	0.2290	0.1892
		g_2	0.9	0.1850	0.2675
c_2	f_1	g_1	0.4	0.1162	0.0480
		g_2	0.4	0.0998	0.1031
	f_2	g_1	0.6	0.1742	0.1440
		g_2	0.6	0.1498	0.2165

ϕ_4:

		ϕ_4	P	P'
b_1	d_1	0.4	0.0640	0.0623
	d_2	0.6	0.0960	0.0934
b_2	d_1	0.7	0.5880	0.5910
	d_2	0.3	0.2520	0.2533

ϕ_5:

			ϕ_5	P	P'
f_1	g_1	h_1	0.2	0.0283	0.0585
		h_2	0.8	0.1133	0
	g_2	h_1	0.5	0.0602	0.1243
		h_2	0.5	0.0602	0
f_2	g_1	h_1	0.4	0.1613	0.3331
		h_2	0.6	0.2419	0
	g_2	h_1	0.7	0.2344	0.4841
		h_2	0.3	0.1004	0

Table 26.2 Marginal probabilities of the Bayes network (w.r.t. the corresponding join tree) in Fig. 26.3 without evidence present. They represent the a priori distribution of the attributes and were computed by propagating "zero evidence"

$P(\cdot)$	A	B	C	D	E	F	G	H
$\cdot_1$	0.6000	0.1600	0.4600	0.6520	0.2144	0.2620	0.5448	0.4842
$\cdot_2$	0.4000	0.8400	0.4500	0.3480	0.7856	0.7380	0.4552	0.5158

The actual numbers can be found in Table 26.2.   □

Example 26.6 (*Second propagation run, now with evidence*)
We will now do a second propagation run and propagate the evidence $H = h_1$ in order to arrive at the conditional probability distributions $P(A \mid H = h_1)$ to $P(G \mid H = h_1)$. As the steps are identical to the previous run, we will only tell the changed parameters and intermediate values.

For initialization, we set all those clique potential entries to zero that contradict the evidence, i.e., for which $H \neq h_1$. Since H is only contained in clique C_5, we only need to change one potential table.

ϕ_5'			P
f_1	g_1	h_1	0.2
		h_2	**0**
	g_2	h_1	0.5
		h_2	**0**
f_2	g_1	h_1	0.4
		h_2	**0**
	g_2	h_1	0.7
		h_2	**0**

The message computation and passing is equivalent to the previous run. The messages are now as follows:

$$M_{21} = \left(\overset{b_1,c_1 \quad b_1,c_2 \quad b_2,c_1 \quad b_2,c_2}{0.06, 0.10, 0.40, 0.44} \right) \qquad M_{41} = \left(\overset{b_1 \ b_2}{1, \ 1} \right)$$

$$M_{13} = \left(\overset{c_1,g_1 \quad c_1,g_2 \quad c_2,g_1 \quad c_2,g_2}{0.254, 0.206, 0.290, 0.250} \right) \qquad M_{35} = \left(\overset{f_1,g_1 \quad f_1,g_2 \quad f_2,g_1 \quad f_2,g_2}{0.14, 0.12, 0.40, 0.33} \right)$$

$$M_{53} = \left(\overset{f_1,g_1 \quad f_1,g_2 \quad f_2,g_1 \quad f_2,g_2}{0.2, \ 0.5, \ 0.4, \ 0.7} \right) \qquad M_{31} = \left(\overset{c_1,g_1 \quad c_1,g_2 \quad c_2,g_1 \quad c_2,g_2}{0.38, 0.68, 0.32, 0.62} \right)$$

$$M_{12} = \left(\overset{b_1,c_1 \quad b_1,c_2 \quad b_2,c_1 \quad b_2,c_2}{0.527, 0.434, 0.512, 0.464} \right) \qquad M_{14} = \left(\overset{b_1 \quad b_2}{0.075, 0.409} \right)$$

After all messages have been sent, we can again update the joint probability distributions according to Eq. 26.3. These result in the values of the P'-column of Table 26.1. In this case, a normalization is necessary. Finally, we marginalize over the clique joint distributions to get the updated marginal attribute distributions (given $H = h_1$). These are shown in Table 26.3. □

Figure 26.6 contains notations that will help to illustrate the propagation algorithm [2].

Table 26.3 Marginal probabilities of the Bayes network (w.r.t. the corresponding join tree) in Fig. 26.3 given evidence $H = h_1$. They represent the a-posteriori distributions of the attributes

$P(\cdot \mid h_1)$	A	B	C	D	E	F	G	H
$\cdot 1$	0.5888	0.1557	0.4884	0.6533	0.1899	0.1828	0.3916	1.0000
$\cdot 2$	0.4112	0.8443	0.5116	0.3467	0.8101	0.8172	0.6084	0.0000

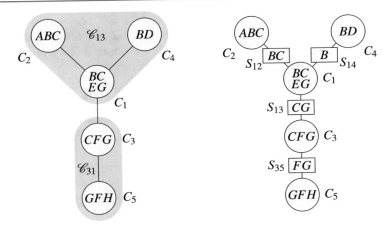

Fig. 26.6 Example for the notation used to derive the evidence propagation steps

1. C_{CB} be the set of cliques of the join subtree containing C that results when edge $C - B$ is removed. If the cliques differ only in the indices of its names, we will use these indices in the notation.
   Example: $C_{C_1,C_3} = C_{13} = \{C_1, C_2, C_4\}$ and $C_{C_3,C_1} = C_{31} = \{C_3, C_5\}$
2. The union of all attributes of all cliques of C_{CB} is denoted as X_{CB}.
   Example: $X_{C_1,C_3} = X_{13} = \{A, B, C, D, E, G\}$ and $X_{C_3,C_1} = X_{31} = \{C, F, G, H\}$
3. We obviously get $V = X_{CB} \cup X_{BC}$.
4. Separator and residual sets are already known. We will now index them with the cliques between which they are defined:

$$S_{CB} = S_{BC} = C \cap B$$

and

$$R_{CB} = X_{CB} \setminus S_{CB}.$$

Example: $R_{C_1,C_3} = R_{13} = \{A, B, D, E\}$, $R_{C_3,C_1} = R_{31} = \{F, H\}$ and $S_{C_1,C_3} = S_{13} = \{C, G\}$

5. From 3. and 4. follows that the entire node set V partitions into the following three disjoint node sets given any edge $(C, B) \in E$:

$$V = R_{CB} \cup S_{CB} \cup R_{BC}.$$

Hence, the attributes R_{CB} are u-separated from the attributes R_{BC} by the attributes S_{BC}.

$$R_{BC} \perp\!\!\!\perp R_{CB} \mid S_{CB}$$

Derivation

We will now derive the origin of the essential Eqs. 26.2 and 26.3 needed for evidence propagation. Let us begin with the update rule for the clique distribution. This can be written as follows (we discuss the single steps afterwards):

$$P(c) \overset{(1)}{=} \underbrace{\sum_{v \backslash c}}_{\text{Marginalization}} \underbrace{\prod_{D \in C} \phi_D(\text{proj}_D^V(v))}_{\text{Decomposition of } P(v)}$$

$$\overset{(2)}{=} \phi_C(c) \sum_{v \backslash c} \prod_{D \neq C} \phi_D(\text{proj}_D^V(v))$$

$$\overset{(3)}{=} \phi_C(c) \sum_{\bigcup_{B \in \text{adj}(C)} r_{BC}} \prod_{D \neq C} \phi_D(\text{proj}_D^V(v))$$

$$\overset{(4)}{=} \phi_C(c) \prod_{B \in \text{adj}(C)} \underbrace{\left(\sum_{r_{BC}} \prod_{D \in C_{BC}} \phi_D(\text{proj}_D^V(v)) \right)}_{M_{B \to C}(s_{BC})}$$

$$= \phi_C(c) \prod_{B \in \text{adj}(C)} M_{B \to C}(s_{BC})$$

(1) The vector v is defined on all attributes. Since we are only interested in attributes in C, we marginalize all remaining attributes' value combinations $v \backslash c$. The product is the application of Definition 25.6 on page 558 about the decomposition of a distribution w.r.t. an undirected graph.

(2) The product in (1) ran over all cliques, including C. This factor is now pulled out of the product and since it is independent of the sum, also pulled out of the sum.

(3) We simplify the marginalization. Instead to run over all value combinations $v \backslash c$, we exploit the following relationship that we can derive from the separations induced by the running intersection property in Eq. 24.2 on page 552. Applied to the join tree, we get:

$$V \backslash C = \left(\bigcup_{B \in \text{adj}(C)} X_{BC} \right) \backslash C = \bigcup_{B \in \text{adj}(C)} (X_{BC} \backslash C) = \bigcup_{B \in \text{adj}(C)} R_{BC}$$

The sum index $\bigcup_{B \in \text{adj}(C)} r_{BC}$ means the iteration over all value combinations of all attributes in all R_{BC}.

(4) The sum is split up by assigning the residuals to the respective neighboring cliques. The resulting factors are the starting point for the messages.

The messages $M_{B \to C}(s_{BC})$ declared on the separators can be further simplified to finally arrive at the Eq. 26.2:

$$M_{B \to C}(s_{BC}) = \underbrace{\sum_{x_{BC} \setminus s_{BC}} \prod_{D \in C_{BC}} \phi_D(d)}_{r_{BC}}$$

$$= \sum_{b \setminus s_{BC}} \phi_B(b) \prod_{D \in \mathrm{adj}(B) \setminus \{C\}} M_{D \to B}(s_{DB})$$

Other Propagation Algorithms

We will sketch two related evidence propagation algorithms at the end of this chapter. Information on other approaches can be found in e.g. [1,2].

The algorithm described above [4–6] allows any directed (but acyclic) graph structure as input. If simpler structures are employed, one can use simpler algorithms, of course. One example are polytree propagation algorithms [7]. Since polytrees are singly connected (i.e., they fall apart into two subgraphs if an edge is removed), the propagation path is already clear without the need to construct a secondary structure. Because of the edge directions, two message types are distinguished: λ-messages from child-to-parent nodes and π-messages from parent-to-child nodes. Of course, one can apply the join tree propagation algorithm on polytrees. This will lead to cliques consisting of nodes together with its parent nodes.

Another technique that not necessarily needs to operate on graph structures is the so-called bucket elimination [3,8]. Given a factorization of a probability distribution, an attribute can be eliminated by summing over all factors containing the attribute. Successive summing allows to isolate target attributes. The efficiency of this approach depends heavily on the order of the summations. A conditional independence graph can be used to suggest an order by means of its edges.

References

1. Borgelt C, Steinbrecher M, Kruse R (2009) Graphical models—representations for learning, reasoning and data mining, 2nd edn. Wiley , Chichester, United Kingdom
2. Castillo E, Gutiérrez JM, Hadi AS (1997) Expert systems and probabilistic network models. Springer-Verlag, New York, NY, USA
3. Dechter R (1996) Bucket elimination: a unifying framework for probabilistic inference. In: Proceedings of the 12th Conference on uncertainty in artificial intelligence (UAI'96, Portland, OR, USA)
4. Jensen FV (1996) An introduction to Bayesian networks. UCL Press, London, United Kingdom
5. Jensen FV (2001) Bayesian networks and decision graphs. Springer-Verlag, Berlin, Germany

6. Jensen FV, Nielsen TD (2007) Bayesian networks and decision graphs, 2nd edn. Springer-Verlag, London, United Kingdom
7. Pearl J (1988) Probabilistic reasoning in intelligent systems: networks of plausible inference. Morgan Kaufmann, San Mateo, CA, USA
8. Zhang NL, Poole D (1996) Exploiting causal independence in Bayesian network inference. J Artif Intell Res 5:301–328. Morgan Kaufmann, San Mateo, CA, USA

Learning Graphical Models

We will now address the third question from Chap. 23, namely how graphical models can be learned from given data. Until now, we were given the graphical structure, now we will introduce heuristics that allow us to induce it.

For the remainder of this chapter, we assume that we are given a database D of samples over an attribute set V (we confine ourselves here to categorical variables) and some prior information ξ, which can be used as a bias towards certain graphical structures. For example, we might prefer graphs that are not too dense or we might have expert knowledge allowing us to impose constraints on the possible graphs.

Approaches for learning graphical models typically fall into one of two categories: score-based approaches and constraint-based approaches. Score-based approaches consist of two elements: a score function to evaluate how well graph candidates fit the database, and some search heuristic (possibly guided by the scores) to traverse the set of graphs. The goal of constraint-based approaches is to use conditional (in)dependence tests to construct a graphical model which is a perfect map (or independence map) of the data-generating distribution.

Note that we typically do not have access to the underlying joint probability distribution $P(V)$. We assume that the cases of database D were independently sampled from $P(V)$, however, we should not fall prey to the assumption to be able to estimate it from D. The reason for this is that even for moderate numbers of attributes and domain sizes, the number of parameters of the joint distribution exceeds practical limits. Assume that V contains only 20 attributes with 5 possible values each. This already results in a state space of size $5^{20} \approx 9.5 \cdot 10^{13}$. Even for this small example, observing each attribute value combination at least once requires more data than will be available in any practical setting (even worse, to get reliable estimates we would need to see each value combination many times). This will turn out to be a problem for constraint-based approaches if they are designed around conditional independence tests which become unreliable if we have only sparse data available.

Before we examine both approaches, we need to mention that we will have to accept suboptimal results when learning graphical models (except for very small

© Springer Nature Switzerland AG 2022
R. Kruse et al., *Computational Intelligence*, Texts in Computer Science,
https://doi.org/10.1007/978-3-030-42227-1_27

numbers of attributes). This is because certain problems related to learning graphical models are known to be NP-hard. For example, it is known from database theory that deciding whether a given relation is decomposable w.r.t. a given family of attribute sets is NP-hard [6]. Likewise, it is known to be NP-hard to find the minimal decomposition of a given relation and it has been conjectured that it is NP-hard to determine whether a given relation can be decomposed even if the size of the attribute sets is restricted to some maximum number [6]. Finally, the specific task of learning a Bayesian network has been shown to be NP-hard if a consistent scoring function[1] is used to evaluate the networks [3].

27.1 Score-Based Approaches

In principle, we seek for a graph G^* that maximizes a quality or score function S. Formally, we iterate through all possible graphs over the attribute set V (denoted as $\mathcal{G}_V$) and pick the one[2] that leads to the best quality score:

$$G^* := \underset{G \in \mathcal{G}_V}{\text{argmax}}\, S(G; D, \xi)$$

We assume that a database D over the attribute set V is given and fixed. The quantity ξ represents any prior information and is also considered a constant. A score function S typically will incorporate (some of) of the following parts [5]:

- Prior information: This part can be thought of a probability distribution over $\mathcal{G}_V$ in order to favor some graphs over others. For example, domain experts may have stipulated constraints on the (in)dependences of attributes which can be encoded in said distribution. If no prior information is available, this part can be omitted (which corresponds to a uniform distribution over $\mathcal{G}_V$).
- Goodness-of-fit: This part describes how well a candidate graph G fits the given database D and prior information ξ. A typical choice is the log-likelihood of the database (see Sect. 27.1.1 below) given the current candidate graph and prior: $\log P(D \mid G, \xi)$
- Model complexity: This part is used to penalize graph structures that are overly complex w.r.t. the number of edges or the size of the factor potentials they would induce.

In addition, there are desirable properties for score functions; some of which can be formalized, while others are phrased in an intuitive manner [8]:

---

[1] Consistency of a score function is described in Sect. 27.1 below.
[2] There may be multiple different graphs that have the same maximal score. We omit this detail here to simplify the formal treatment. One can always introduce additional criteria that would allow to distinguish them.

- Consistency: If two candidate graphs G_1 and G_2 are independence maps of the data-generating distribution p and G_1 encodes more (conditional) independences than G_2, that is, $\mathcal{I}_{G_2} \subset \mathcal{I}_{G_1}$, then the score of G_1 should be asymptotically better than the score of G_2:

$$\mathcal{I}_{G_2} \subset \mathcal{I}_{G_1} \subseteq \mathcal{I}_p \quad \Rightarrow \quad \lim_{|D| \to \infty} S(G_1; D, \xi) > \lim_{|D| \to \infty} S(G_2; D, \xi)$$

If we assume G_1 to be a minimal independence map, then the above inequality claims that minimal independence maps should get a higher score than non-minimal independence maps.

- Similar to consistency, graph candidates which are perfect maps (w.r.t. to the data-generating distribution p) should get higher scores than non-perfect maps.
- Equivalence: Graph candidates which are equivalent w.r.t. to their independence structure should be undistinguishable by the score:

$$\mathcal{I}_{G_1} = \mathcal{I}_{G_2} \quad \Rightarrow \quad S(G_1; D, \xi) = S(G_2; D, \xi)$$

- Decomposability: The score function of a graph $G = (V, E)$ should be the sum of the scores of the subgraphs induced by G's families:

$$S(G; D, \xi) = \sum_{A \in V} S(G_{\text{fa}(A)}; D, \xi)$$

An exhaustive search through all possible graphs excludes itself as the size of the search space grows huge even for small numbers of nodes. For example, the set of directed acyclic graphs grows super-exponentially in the number of nodes [11]: For ten nodes, the graph set has already a cardinality of $4.18 \cdot 10^{18}$. This is why all learning algorithms use some heuristic to limit the number of candidate graphs that need to be evaluated. The general steps of a score-based learning algorithm for graphical models are as follows:

- Fix prior information ξ and score function S.
- Use a heuristic that efficiently traverses the search space $\mathcal{G}_V$ and generates promising graph candidates.
- Use the score function to evaluate each candidate graph. This value is then used to guide the heuristic through the search space.

27.1.1 Likelihood of a Database

As described above, one part of a score function computes the goodness-of-fit of the database D w.r.t. a given graph candidate G. One way to do this is to compute the probability with which the graph might have generated the database. (We will see shortly, that this choice requires a complexity penalization in order to properly work.) However, adaptations of it are used in many learning algorithms, like the K2 algorithm that we will discuss later on. Generating a database based on a graph means that the probability distribution described by the graphical model was used to sample tuples that make up the database.

Fig. 27.1 Induction of the
potential tables structure B_S
of the depicted Bayes
network

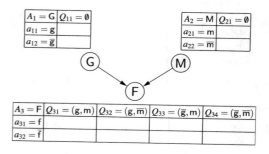

$A_1 = G$	$Q_{11} = \emptyset$
$a_{11} = g$	
$a_{12} = \bar{g}$	

$A_2 = M$	$Q_{21} = \emptyset$
$a_{21} = m$	
$a_{22} = \bar{m}$	

$A_3 = F$	$Q_{31} = (g,m)$	$Q_{32} = (g,\bar{m})$	$Q_{33} = (\bar{g},m)$	$Q_{34} = (\bar{g},\bar{m})$
$a_{31} = f$				
$a_{32} = \bar{f}$				

In the following, we will refer to the graph structure (no matter whether directed or undirected) of a graphical model as B_S. The set of all probability parameters (entries of the potential tables in the undirected case or the conditional probabilities in the directed case) be B_P. Let us consider the component B_P in case of a Bayes network. It contains the specific entries of the conditional probability distributions $P(A \mid \mathrm{pa}(A))$. Each node maintains a potential table which contains for each combination of the parent attributes values of a node A_i the probability of the attribute values of A_i. In general, the potential table of attribute A_i contains the columns $Q_{i1}, \ldots, Q_{iq_i}$. Each of these q_i columns corresponds to a conditional probability distribution of the attribute values $a_{i1}, \ldots, a_{ir_i}$ given the attribute value combination associated with the column.

As an example, consider the simple Bayes network in Fig. 27.1. All attributes be binary and represent **Grippe**, **Malaria** and **Fever**. Note that we use the dated notion *grippe* here as the modern word *flu* would obviously interfere with the variable naming. The node **F** has two-parent attributes (**G** and **M**) which are both binary. Hence, there exist four-parent attributes value combinations: (g, m), $(g, \bar{m})$, $(\bar{g}, m)$, and $(\bar{g}, \bar{m})$ which correspond to the four columns in the potential table for **F** in Fig. 27.1. If the node A_i is a root node, the potential table contains the marginal distribution $P(A_i)$. This artificial column is denoted by $\emptyset$.

The entries of these tables are denoted as θ_{ijk} and represent the probability of attribute A_i assuming value a_{ik}, while its parent attributes $\mathrm{pa}(A_i)$ assume the jth value combination Q_{ij}. Table 27.2 shows a general potential table whose indices we use in the remainder: the index i runs over the attributes ($i = 1, \ldots, n$), j runs over the different parent attributes' value combinations of attribute A_i ($j = 1, \ldots, q_i$) and k runs over all r_i values of A_i ($k = 1, \ldots, r_i$). The values $(\theta_{ij1}, \ldots, \theta_{ijr_i})$ represent a probability distribution: They correspond to the parameters of multinomial distribution of order r_i.

Let us now discuss two examples for computing the parameters B_P from given data D and given structure B_S. With D and B_S provided, we can easily determine the parameters B_P by counting. This is justified because the conditional distributions are multinomial distributions and relative frequencies constitute optimal estimators (Fig. 27.2). Table 27.1 shows an example database with 100 cases.

Fig. 27.2 General potential table of attribute A_i. Each column represents a probability distribution

A_i	Q_{i1}	$\cdots$	Q_{ij}	$\cdots$	Q_{iq_i}
a_{i1}	θ_{i11}	$\cdots$	θ_{ij1}	$\cdots$	θ_{iq_i1}
$\vdots$	$\vdots$	$\ddots$	$\vdots$	$\ddots$	$\vdots$
a_{ik}	θ_{i1k}	$\cdots$	θ_{ijk}	$\cdots$	θ_{iq_ik}
$\vdots$	$\vdots$	$\ddots$	$\vdots$	$\ddots$	$\vdots$
a_{ir_i}	θ_{i1r_i}	$\cdots$	θ_{ijr_i}	$\cdots$	$\theta_{iq_ir_i}$

Table 27.1 An example database with 100 cases

Grippe	$\overline{g}$	$\overline{g}$	$\overline{g}$	$\overline{g}$	g	g	g	g
Malaria	$\overline{m}$	$\overline{m}$	m	m	$\overline{m}$	$\overline{m}$	m	m
Fever	$\overline{f}$	f	$\overline{f}$	f	$\overline{f}$	f	$\overline{f}$	f
#	34	6	2	8	16	24	0	10

In the first part, we assume the network structure B_S to be an edgeless graph, i.e., we deal with three marginal distributions. The "decomposition" of the joint distribution then reads:

$$P(\mathsf{G} = g, \mathsf{M} = m, \mathsf{F} = f) = P(\mathsf{G} = g)\, P(\mathsf{M} = m)\, P(\mathsf{F} = f)$$
$$\text{with } g \in \{\mathsf{g}, \overline{\mathsf{g}}\},\, m \in \{\mathsf{m}, \overline{\mathsf{m}}\},\, f \in \{\mathsf{f}, \overline{\mathsf{f}}\}$$

Consequently, the distributions $P(\mathsf{G})$, $P(\mathsf{M})$ and $P(\mathsf{F})$ have to be estimated from data.

$$P(\mathsf{G} = g) \approx \widehat{P}(\mathsf{G} = g) = \frac{\#(\mathsf{G} = g)}{|D|}$$

where the statements $\#(X = x)$ and $\#(x)$ denote the number of tuples in D having the value x for attribute X, that is, $\#(X = x) = \#(x) \overset{\text{Def}}{=} |\{t \in D \mid t(X) = x\}| = |\{X = x\}|$. Estimated from the example distribution in Table 27.1 we get:

$$\widehat{P}(\mathsf{G} = \mathsf{g}) = {}^{50}/_{100} = 0.50, \qquad \widehat{P}(\mathsf{G} = \overline{\mathsf{g}}) = 1 - \widehat{P}(\mathsf{G} = \mathsf{g}) = 0.50,$$
$$\widehat{P}(\mathsf{M} = \mathsf{m}) = {}^{20}/_{100} = 0.20, \qquad \widehat{P}(\mathsf{M} = \overline{\mathsf{m}}) = 1 - \widehat{P}(\mathsf{M} = \mathsf{m}) = 0.80,$$
$$\widehat{P}(\mathsf{F} = \mathsf{f}) = {}^{48}/_{100} = 0.48, \qquad \widehat{P}(\mathsf{F} = \overline{\mathsf{f}}) = 1 - \widehat{P}(\mathsf{F} = \mathsf{f}) = 0.52.$$

In the second part, we assume the network structure B_S to be as depicted in Fig. 27.1. The decomposition reads

$$P(\mathsf{G} = g, \mathsf{M} = m, \mathsf{F} = f) = P(\mathsf{G} = g)\, P(\mathsf{M} = m)\, P(\mathsf{F} = f \mid \mathsf{G} = g, \mathsf{M} = m).$$

The estimations for $P(\mathsf{G})$ and $P(\mathsf{M})$ are computed as in the previous example, while we use the following formula to estimate the conditional probability distribution: $P(\mathsf{F} \mid \mathsf{G}, \mathsf{M})$:

$$\widehat{P}(f \mid g, m) = \frac{\widehat{P}(f, g, m)}{\widehat{P}(g, m)} = \frac{\dfrac{\#(g, m, f)}{|D|}}{\dfrac{\#(g, m)}{|D|}} = \frac{\#(g, m, f)}{\#(g, m)}.$$

We get the following numbers for the distribution $\hat{P}(\mathsf{F} \mid \mathsf{G}, \mathsf{M})$ from the example data:

$$\hat{P}(\mathsf{F} = \mathsf{f} \mid \mathsf{G} = \mathsf{g}, \mathsf{M} = \mathsf{m}) = \frac{10/100}{10/100} = 1.00,$$

$$\hat{P}(\mathsf{F} = \mathsf{f} \mid \mathsf{G} = \mathsf{g}, \mathsf{M} = \overline{\mathsf{m}}) = \frac{24/100}{40/100} = 0.60,$$

$$\hat{P}(\mathsf{F} = \mathsf{f} \mid \mathsf{G} = \overline{\mathsf{g}}, \mathsf{M} = \mathsf{m}) = \frac{8/100}{10/100} = 0.80,$$

$$\hat{P}(\mathsf{F} = \mathsf{f} \mid \mathsf{G} = \overline{\mathsf{g}}, \mathsf{M} = \overline{\mathsf{m}}) = \frac{6/100}{40/100} = 0.15.$$

Given a database D, a potential network structure B_S and the estimated parameters B_P, we can compute $P(D \mid B_S, B_P)$ when accepting the following assumptions.

1. The data-generating process can be modeled by the Bayes network represented by (B_S, B_P).
2. The tuples in the database occur independently from each other.
3. All tuples are complete, i.e., there are no missing values.

Assumption 1 legitimates the investigation for a Bayes network as the underlying model since in case of a violation of this assumption, the entire search would be pointless. Assumption 2 states that the occurrence of a tuple does not affect the probability of the other tuples. It is not to be confused with the statement of all tuples being equally probable! Assumption 3 finally allows us to use the above-mentioned counting as we do not need to take care of missing values.

The probability of database D (given the graph candidate) can now be computed as follows:

$$P(D \mid B_S, B_P) = \prod_{h=1}^{100} P(c_h \mid B_S, B_P)$$

$$= \underbrace{\overbrace{P(\mathsf{g}, \mathsf{m}, \mathsf{f})}^{\text{case 91}} \cdots \overbrace{P(\mathsf{g}, \mathsf{m}, \mathsf{f})}^{\text{case 100}}}_{10 \text{ times}} \cdots \underbrace{\overbrace{P(\overline{\mathsf{g}}, \mathsf{m}, \mathsf{f})}^{\text{case 43}} \cdots \overbrace{P(\overline{\mathsf{g}}, \mathsf{m}, \mathsf{f})}^{\text{case 50}}}_{8 \text{ times}} \cdots \underbrace{\overbrace{P(\overline{\mathsf{g}}, \overline{\mathsf{m}}, \overline{\mathsf{f}})}^{\text{case 1}} \cdots \overbrace{P(\overline{\mathsf{g}}, \overline{\mathsf{m}}, \overline{\mathsf{f}})}^{\text{case 34}}}_{34 \text{ times}}$$

$$= \underbrace{P(\mathsf{g}, \mathsf{m}, \mathsf{f})^{10}}_{} \cdots \underbrace{P(\overline{\mathsf{g}}, \mathsf{m}, \mathsf{f})^{8}}_{} \cdots \underbrace{P(\overline{\mathsf{g}}, \overline{\mathsf{m}}, \overline{\mathsf{f}})^{34}}_{}$$

$$= \overbrace{P(\mathsf{f}|\mathsf{g}, \mathsf{m})^{10} P(\mathsf{g})^{10} P(\mathsf{m})^{10}} \cdots \overbrace{P(\mathsf{f}|\overline{\mathsf{g}}, \mathsf{m})^{8} P(\overline{\mathsf{g}})^{8} P(\mathsf{m})^{8}} \cdots \overbrace{P(\overline{\mathsf{f}}|\overline{\mathsf{g}}, \overline{\mathsf{m}})^{34} P(\overline{\mathsf{g}})^{34} P(\overline{\mathsf{m}})^{34}}$$

$$= P(\mathsf{f} \mid \mathsf{g}, \mathsf{m})^{10} P(\overline{\mathsf{f}} \mid \mathsf{g}, \mathsf{m})^{0} P(\mathsf{f} \mid \mathsf{g}, \overline{\mathsf{m}})^{24} P(\overline{\mathsf{f}} \mid \mathsf{g}, \overline{\mathsf{m}})^{16}$$
$$\cdot \ P(\mathsf{f} \mid \overline{\mathsf{g}}, \mathsf{m})^{8} P(\overline{\mathsf{f}} \mid \overline{\mathsf{g}}, \mathsf{m})^{2} P(\mathsf{f} \mid \overline{\mathsf{g}}, \overline{\mathsf{m}})^{6} P(\overline{\mathsf{f}} \mid \overline{\mathsf{g}}, \overline{\mathsf{m}})^{34}$$
$$\cdot \ P(\mathsf{g})^{50} P(\overline{\mathsf{g}})^{50} P(\mathsf{m})^{20} P(\overline{\mathsf{m}})^{80}$$

The last equation shows the principle by which the factors were reordered: First, sort by attribute (in the example F, G, then M). Within an attribute we group by parent value combinations (in the example, for F we do (g, m), $(\mathsf{g}, \overline{\mathsf{m}})$, $(\overline{\mathsf{g}}, \mathsf{m})$, and $(\overline{\mathsf{g}}, \overline{\mathsf{m}})$). Finally, we sort by equal attribute values (in the example, for attribute F:

first $\mathbf{f}$, then $\bar{\mathbf{f}}$). The general computation of the probability of a database D is as follows [2]:

$$P(D \mid B_S, B_P) = \prod_{i=1}^{n} \prod_{j=1}^{q_i} \prod_{k=1}^{r_i} \theta_{ijk}^{\alpha_{ijk}}. \tag{27.1}$$

We saw that with known structure B_S we could estimate the parameters B_P from the database. This could lead us to the idea to use the maximum likelihood of the database as the score function S (omitting prior ξ as we assume a uniform distribution over $\mathcal{G}_V$):

$$S\big((B_S, B_P); D\big) = P(D \mid B_S, B_P)$$

That is, the best graph candidate would be defined as

$$\widehat{B}_S = \underset{B_S \in \mathcal{G}_V}{\mathrm{argmax}} \, P(D \mid B_S, B_P) \tag{27.2}$$

This approach, alas, has the drawback that the probability increases with the number of parameters which would lead to a model with maximal parameter number, that is, a fully connected graph B_S. This can be fixed either by using an appropriate prior or complexity penalization which favors graphs with a medium amount of edges or penalizes dense graphs, respectively.

27.1.2   K2 Algorithm

We will now discuss the derivation of a specific score function, the K2 metric, and introduce the corresponding search heuristic which results in the K2 algorithm [5].

Introducing a prior on the graph structures into Eq. 27.2, we arrive at a maximum a posteriori estimator. The approach hence would be as follows:

$$\widehat{B}_S = \underset{B_S}{\mathrm{argmax}} \, P(B_S \mid D) = \underset{B_S}{\mathrm{argmax}} \, \frac{P(B_S, D)}{P(D)} = \underset{B_S}{\mathrm{argmax}} \, P(B_S, D)$$

Consequently, we seek a computational expression for the term $P(B_S, D)$. The result of the following derivation will be the K2 metric. An elaborate treatment can be found in [5].

First, we consider $P(B_S, D)$ to be the marginalization of $P(B_S, B_P, D)$ over all possible parameters B_P which is known as model averaging. The number of parameters is fixed (finite number of entries in the potential tables of the attributes), however, the values of these entries—the θ_{ijk}—are continuous values which requires us to integrate over all models. Indeed, we deal with a multiple integral over all θ_{ijk} which is later made explicit.

$$P(B_S, D) = \int_{B_P} P(B_S, B_P, D)\, dB_P \tag{27.3}$$

$$= \int_{B_P} P(D \mid B_S, B_P)\, P(B_S, B_P)\, dB_P \tag{27.4}$$

$$= \int_{B_P} P(D \mid B_S, B_P)\, f(B_P \mid B_S) P(B_S)\, dB_P \tag{27.5}$$

$$= \underbrace{P(B_S)}_{\text{A-priori prob.}} \int_{B_P} \underbrace{P(D \mid B_S, B_P)}_{\text{Prob. of data}} \underbrace{f(B_P \mid B_S)}_{\text{Parameter densities}}\, dB_P \tag{27.6}$$

The a priori distribution can be used to put a bias on certain classes of graph structures (for example, by assigning a low probability to overly complex structures). Assuming that there is an underlying Bayes network, that the database cases are independent of each other and that the tuples are complete, we can apply Eq. 27.1:

$$P(B_S, D) = P(B_S) \int_{B_P} \left[\prod_{i=1}^{n} \prod_{j=1}^{q_i} \prod_{k=1}^{r_i} \theta_{ijk}^{\alpha_{ijk}} \right] f(B_P \mid B_S)\, dB_P$$

The parameter densities $f(B_P \mid B_S)$ make a statement how probable the respective parameters B_P of a given network structure are. Hence, they are second-order densities as they represent densities about probability distributions. A vector $(\theta_{ij1}, \ldots, \theta_{ijr_i})$ is a probability distribution for fixed i and j (the jth column of the ith potential table, see Table 27.2). Assuming the densities of all columns of all attributes are mutually independent, we can simplify the expression for $f(B_P \mid B_S)$ to:

$$f(B_P \mid B_S) = \prod_{i=1}^{n} \prod_{j=1}^{q_i} f(\theta_{ij1}, \ldots, \theta_{ijr_i})$$

Plugging this term into $P(B_S, D)$, we arrive at the following expression:

$$P(B_S, D)$$

$$= P(B_S) \int \cdots \int_{\theta_{ijk}} \left[\prod_{i=1}^{n} \prod_{j=1}^{q_i} \prod_{k=1}^{r_i} \theta_{ijk}^{\alpha_{ijk}} \right] \cdot \left[\prod_{i=1}^{n} \prod_{j=1}^{q_i} f(\theta_{ij1}, \ldots, \theta_{ijr_i}) \right] d\theta_{111}, \ldots, d\theta_{nq_nr_n}$$

$$= P(B_S) \prod_{i=1}^{n} \prod_{j=1}^{q_i} \int \cdots \int_{\theta_{ijk}} \left[\prod_{k=1}^{r_i} \theta_{ijk}^{\alpha_{ijk}} \right] \cdot f(\theta_{ij1}, \ldots, \theta_{ijr_i})\, d\theta_{ij1}, \ldots, d\theta_{ijr_i}$$

The last simplifying assumption, again, regards the parameter densities. For fixed i and j, the density $f(\theta_{ij1}, \ldots, \theta_{ijr_i})$ be uniform. Hence, we get

$$f(\theta_{ij1}, \ldots, \theta_{ijr_i}) = (r_i - 1)!$$

$$P(B_S, D) = P(B_S) \prod_{i=1}^{n} \prod_{j=1}^{q_i} \int_{\theta_{ijk}} \cdots \int \left[\prod_{k=1}^{r_i} \theta_{ijk}^{\alpha_{ijk}} \right] \cdot (r_i - 1)! \, d\theta_{ij1}, \ldots, d\theta_{ijr_i}$$

$$= P(B_S) \prod_{i=1}^{n} \prod_{j=1}^{q_i} (r_i - 1)! \underbrace{\int_{\theta_{ijk}} \cdots \int \prod_{k=1}^{r_i} \theta_{ijk}^{\alpha_{ijk}} \, d\theta_{ij1}, \ldots, d\theta_{ijr_i}}$$

$$\text{Dirichlet integral} = \frac{\prod_{k=1}^{r_i} \alpha_{ijk}!}{(\sum_{k=1}^{r_i} \alpha_{ijk} + r_i - 1)!}$$

Finally, we get the equation for $P(B_S, D)$ which is referred to as the K2 metric of the network structure B_S given the data D

$$P(B_S, D) = \text{K2}(B_S \mid D) = P(B_S) \prod_{i=1}^{n} \prod_{j=1}^{q_i} \left[\frac{(r_i - 1)!}{(N_{ij} + r_i - 1)!} \prod_{k=1}^{r_i} \alpha_{ijk}! \right]$$

$$\text{with} \quad N_{ij} = \sum_{k=1}^{r_i} \alpha_{ijk} \tag{27.7}$$

The parameter independences are two important properties of the K2 metric. They can be distinguished into *global* and *local* properties (see [9, pp. 13f]):

- *Global*—This property manifests itself in the outer product of the K2 formula: The product runs over all K2 values of the families of the attributes. It originates from the likelihood Eq. 27.1.
- *Local*—The Eq. 27.1 assumes the independence of the child attribute values given the parent attributes values. This is manifested in the product over all q_i different parent attributes' value combinations of the attribute A_i. The equation for the K2 metric also contains this product.

We exploit the global parameter independence to represent the K2 metric as follows:

$$\text{K2}(B_S \mid D) = P(B_S) \prod_{i=1}^{n} \text{K2}_{\text{local}}(A_i \mid D) \quad \text{with}$$

$$\text{K2}_{\text{local}}(A_i \mid D) = \prod_{j=1}^{q_i} \left[\frac{(r_i - 1)!}{(N_{ij} + r_i - 1)!} \prod_{k=1}^{r_i} \alpha_{ijk}! \right]$$

The presented K2 metric computes for a given database D the quality of a network candidate B_S. We saw that the set of candidate graphs $\mathcal{G}_V$ for the given attribute set V becomes way too large to allow to consider each individual graph. Hence, we need a heuristic how to check only a computationally feasible subset of $\mathcal{G}_V$.

The following algorithm uses the K2 metric as its evaluation metric and a greedy search as search heuristic. It further requires a topological ordering (see Definition 24.33 on page 538) on the attributes (nodes).

The search starts with an isolated graph, that is, with n isolated nodes and hence empty parent sets pa(A_i). The indices $1 \le i \le n$ shall represent the topological ordering. The function k_i is defined as the local K2 value of node A_i given the parent set M:

$$k_i(M) = \text{K2}_{\text{local}}(A_i \mid D) \quad \text{with} \quad \text{pa}(A_i) = M$$

The parent sets $\text{pa}(A_i)$ for all nodes A_i are determined incrementally using the following heuristic:

1. Determine for a parentless node A_i the quality measure $k_i(\emptyset)$.
2. Then, all predecessors $\{A_1, \ldots, A_{i-1}\}$ are tested individually as potential parent nodes and the quality measure is recomputed. Let Y be the node that leads to the best quality:

$$Y = \underset{1 \leq l \leq i-1}{\text{argmax}} \, k_i(\{A_l\})$$

This best quality be $g = k_i(\{Y\})$.
3. If g is better than $k_i(\emptyset)$, the node Y is permanently added as a parent of A_i: $\text{pa}(A_i) = \{Y\}$.
4. Steps 2 and 3 are repeated to augment the parent node set until no potential attributes are left, the quality cannot be increased or a certain maximal number of parent nodes is reached.

Algorithm 27.1 shows the K2 algorithm in pseudocode. Figures 27.3 and 27.4 show a specific example.

Algorithm 27.1 (K2 Algorithm)

procedure K2;
begin
    for $i \leftarrow 1 \ldots n$ **do**   (∗ initialization ∗)
        $\text{pa}(A_i) \leftarrow \emptyset$;
    for $i \leftarrow n \ldots 1$ **do begin**   (∗ iteration ∗)
      repeat
        Choose $Y \in \{A_1, \ldots, A_{i-1}\} \setminus \text{pa}(A_i)$ which maximizes $g = k_i(\text{pa}(A_i) \cup \{Y\})$;
        $\delta \leftarrow g - k_i(\text{pa}(A_i))$;
        if $\delta > 0$ **then** $\text{pa}(A_i) \leftarrow \text{pa}(A_i) \cup \{Y\}$; **end**
      until $\delta \leq 0$ **or** $\text{pa}(A_i) = \{A_1, \ldots, A_{i-1}\}$ **or** $|\text{pa}(A_i)| = n_{\max}$;
    end
end

27.2 Constraint-Based Approaches

The underlying idea of constraint-based learning approaches is to use conditional (in)dependence tests to construct a graph that encodes the independence constraints of the data-generating distribution. Typically algorithms in this category first build

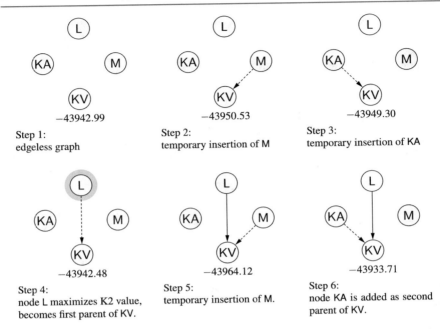

Step 1:
edgeless graph

Step 2:
temporary insertion of M

Step 3:
temporary insertion of KA

Step 4:
node L maximizes K2 value,
becomes first parent of KV.

Step 5:
temporary insertion of M.

Step 6:
node KA is added as second
parent of KV.

Fig. 27.3 An example run of the K2 algorithm (Step 1-6). The topological ordering is L ≺ KA ≺ M ≺ KV

an undirected graph (often referred to as the skeleton) after which the edge directions are determined to arrive at a Bayes network.

One of the most well-known algorithms in this category is SGS (sometimes named like this after its authors in the literature) which uses the following rationale [13]. Under the assumption, there is a perfect map for the domain under consideration, that is, a graph that encodes exactly those conditional independence statements that hold true in the joint distribution of the domain, one can argue as follows: If two attributes A and B are conditionally independent given a (possibly empty) set Z, that is, $A \perp\!\!\!\perp B \mid Z$, then they cannot be connected by an edge. If they were connected by an edge, the graph would no longer encode the above conditional independence statement and thus no longer be a perfect map (because it now encodes at least one conditional independence statement less than the probability distribution). Algorithm 27.2 implements this idea and returns an undirected perfect map (or a skeleton if we plan to induce edge directions).

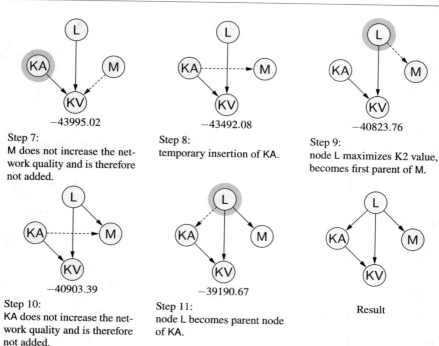

Step 7:
M does not increase the network quality and is therefore
not added.

Step 8:
temporary insertion of KA.

Step 9:
node L maximizes K2 value,
becomes first parent of M.

Step 10:
KA does not increase the network quality and is therefore
not added.

Step 11:
node L becomes parent node
of KA.

Result

Fig. 27.4 An example run of the K2 algorithm (Step 7-12). The topological ordering is L ≺ KA ≺ M ≺ KV

Algorithm 27.2 (SGS part 1: Perfect undirected map by cond. indep. tests)

Input: A sampled distribution $\hat{p}$ over attribute set V
Output: Perfect undirected map $G_u = (V, E)$ for distribution P

begin
$\quad G \leftarrow (V, \emptyset)$
$\quad$**for each** $A, B \in V$ **do**
$\quad\quad$**if** $\nexists S_{AB} \subseteq V \setminus \{A, B\} : A \perp\!\!\!\perp_{\hat{p}} B \mid S_{AB}$ **then**
$\quad\quad\quad E \leftarrow E \cup \{A - B\}$
$\quad\quad$**end**
$\quad$**end**
end

To derive a directed perfect map we need to orient the edges of the undirected output of Algorithm 27.2. This is implemented by Algorithm 27.3. We check all pairs of non-adjacent attributes sharing a common neighbor node C that is not in

S_{AB}.[3] The conditional independence statement of A and B (given S_{AB}) means that every path between A and B is u-separated by S_{AB}. In order for A and B to remain d-separated in a directed graph, the common neighbor C must be a converging node in the path.[4] Finally, the remaining undirected edges are oriented such that the graph remains acyclic and no converging nodes are created. Note that there can be at most one undirected perfect map for a given distribution. In contrast to this, there may be several directed acyclic perfect maps of a distribution.

Algorithm 27.3 (SGS part 2: Edge orientation)

Input: A sampled distribution $\hat{p}$ over attribute set V, undirected perfect map G_u
OutputPerfect directed map $G = (V, E)$ for distribution P

begin
  for each pair $A, B \in V$ of non-adjacent attributes **do**
    if exists a common neighbor $C \notin S_{AB}$ **then**
      Direct the edges towards C, that is, $E \leftarrow E \cup \{A \rightarrow C, B \rightarrow C\}$
    end
  end
  Orient all remaining undirected edges without introducing cycles or converging edges.
end

Despite being simple and intuitive, the presented approach suffers from some drawbacks inherent to constraint-based learning.

Consider the set S_{AB} of Algorithm 27.2. To ascertain that there is no such set, we have to potentially check all subsets of $V \setminus \{A, B\}$ which becomes intractable even for moderate numbers of attributes in V.

Also, the sets S_{AB} themselves may contain a large number of attributes which makes the conditional independence tests hard or even impossible. Note that in the algorithms above, we referred to an underlying probability distribution $P(V)$. As mentioned in the beginning of this chapter, we typically do not know this distribution, but only the database D (which we assume to have been sampled from it). To check for a conditional independence $A \perp\!\!\!\perp_D B \mid S_{AB}$, each instantiation of the attributes A, B and S_{AB} have to be evaluated.[5] For such a conditional independence test to be reliable, the underlying database should contain each tuple (that is, each instantiation of the attributes) multiple times. However, depending on the number of attributes it might be impossible to even see each instantiation at least once.

---

[3] The fact that A and B are not adjacent means they were not connected by Algorithm 27.2 and this means they were determined to be conditionally independent given some S_{AB} (which might be empty).
[4] See Definition 24.42 on page 545.
[5] See Definition 24.14 on page 532.

Another problem arises by assuming that there is a perfect map underlying the distribution. Not every distribution has a (directed or undirected) perfect map. The question arises what the output of the algorithms will look like in such cases (and whether we can at least derive an independence map). Alas, if the perfect map criterion is violated, the resulting graph is not necessarily a dependence or independence map.

Other Constraint-Based Approaches

An early approach to learning graphical models that fits into the constraint-based category are Chow–Liu trees [4]. The algorithm produces a graph that encodes at most second-order potentials which implies that the result is a tree (or a forest, that is, a set of disconnected trees, if the learning process is stopped before a single tree is obtained). The algorithm constructs a maximum-weight spanning tree using information divergence between nodes as edge weights. This results in an optimal approximation of the underlying probability distribution w.r.t. the Kullback–Leibler information divergence [10].

The authors of SGS later proposed the PC algorithm [12] which improves on both stages (skeleton construction and edge orientation) which leads to more efficient execution, especially in situations where the underlying perfect map is sparse. The number of conditional independence tests for determining the skeleton is reduced by starting with empty conditioning sets Z and then grow them by adding neighboring nodes of the tested attributes which limits the largest conditioning set to the (shrinking) neighboring sets. The edge orientation phase is implemented avoiding additional conditional independence tests.

The Max–Min Parents and Children (MMPC) algorithm [14] returns for each attribute the set of parent and children attributes (but without telling whether an attribute would be a parent node or child node) of the perfect map (assuming it exists). This is done by first growing (maximizing) for each attribute T an eponymous candidate parent-and-children set by collecting all attributes that cannot be shown to be independent of T (the conditional independence test is chosen to be a G^2 test). These candidate sets are then shrunk (minimized) by removing attributes if they can be shown to d-separate T given any subset of the candidate set. The authors showed the applicability of the algorithm even for scenarios with large numbers of attributes. The MMPC algorithm serves as the skeleton-constructing phase of the MMHC algorithm (Max–Min Hill Climbing [15]). That is, MMPC is run for each attribute resulting in candidate attributes for parents and children and thus an undirected skeleton. MMHC then performs a greedy hill-climbing search in the space of Bayesian networks, beginning with an empty graph and selecting the graph that leads to the largest increase in BDeu score [1] when applying editing operations such as edge addition, edge deletion, and edge reversal. The search is constrained to only consider adding an edge if it was discovered by MMPC during skeleton construction. Since the second phase uses a scoring function to guide the edge direction (and thus the trajectory through the search space), MMHC is a hybrid algorithm exhibiting constraint-based properties as well as score-based ones. MMHC performs well on very large numbers of attributes and is one of the current state-of-the-art algorithms for structure learning. See [8] and [7] for summaries of other structure learning algorithms.

References

1. Buntine W (1991) Theory refinement on Bayesian networks. In: Proceedings of the 7th Conference on Uncertainty in Artificial Intelligence (UAI'91, Los Angeles, CA, USA), pp 52–60. Morgan Kaufmann, San Mateo, CA, USA
2. Castillo E, Gutiérrez JM, Hadi AS (1997) Expert systems and probabilistic network models. Springer-Verlag, New York, NY, USA
3. Chickering DM, Heckerman D, Meek C (2004) Large-sample learning of Bayesian networks is NP-hard. J Mach Learn Res 5:1287–1330
4. Chow CK, Liu CN (1968) Approximating discrete probability distributions with dependence trees. IEEE Trans Inf Theory IT 14(3):462–467
5. Cooper GF, Herskovits E (1992) A Bayesian method for the induction of probabilistic networks from data. Mach Learn 9:309–347. Kluwer, Dordrecht, Netherlands
6. Dechter R, Pearl J (1992) Structure identification in relational data. Artif Intell 58:237–270. North-Holland, Amsterdam, Netherlands
7. Drton M, Maathuis MH (2017) Structure learning in graphical modeling. Annu Rev Stat Appl 4(1):365–393
8. Gasse M (2017) Probabilistic graphical model structure learning: application to multi-label classification. Université de Lyon, Lyon, France
9. Heckerman D, Geiger D, Chickering DM (1994) Learning Bayesian networks: the combination of knowledge and statistical data, MSR-TR-94-09. Microsoft Research, Advanced Technology Division, Redmond, WA, USA
10. Kullback S, Leibler RA (1951) On information and sufficiency. Ann Math Stat 22:79–86. Institute of Mathematical Statistics, Hayward, CA, USA
11. Robinson RW (1977) Counting unlabeled acyclic digraphs. In: Little CHC (ed) Combinatorial Mathematics V LNMA 622:28–43. Springer-Verlag, Heidelberg, Germany
12. Spirtes P, Glymour C (1991) An algorithm for fast recovery of sparse causal graphs. Soc Sci Comput Rev 9(1):62–72
13. Spirtes P, Glymour C, Scheines R (1993) Causation, prediction, and search (Lecture Notes in Statistics 81). Springer-Verlag, New York, NY, USA
14. Tsamardinos I, Aliferis CF, Statnikov A (2003) Time and sample efficient discovery of Markov blankets and direct causal relations. In: Proceedings of 9th ACM SIGKDD International Conference on knowledge discovery and data mining (Washington, DC, USA)
15. Tsamardinos I, Brown LE, Aliferis CF (2006) The max-min hill-climbing Bayesian network structure learning algorithm. Mach Learn 65:31–78

22=

Belief Revision

In Chap. 25, we discussed graphical models as tools for structuring uncertain knowledge about high-dimensional domains. More precisely, we introduced Bayes networks, which are based on directed graphs and conditional distributions, and Markov networks, which refer to undirected graphs and marginal distributions or factor potentials. Further, we presented an evidence propagation method in Chap. 26 with which it is possible to condition the probability distribution represented by a graphical model on given evidence, i.e., on observed values for some of the variables, a reasoning process that is also called *focusing*. Efficient and user-friendly commercial tools for this task, like [7] and [10], are widely available. In practice, however, the need also arises to revise the probability distribution that is represented by a graphical model so that it satisfies given frame conditions, for example, given marginal distributions. For this task, pure evidence propagation methods like join tree propagation and bucket elimination are unsuited. The available commercial tools, however, only provide these standard methods, and thus do not support the revision operation.

28.1 Introduction

Suppose that we are given a Markov network which represents a joint probability distribution p_U on a set U of variables. In addition, let J be a set of indices of variables together with marginal probability distributions $p^*_{A_j}$ on the domains of the variables A_j for all $j \in J$. The task is to *revise* the Markov network in such a way that the marginal probability distributions p'_{A_j} of the distribution p'_U represented by the new Markov network coincide with the distributions $p^*_{A_j}$ for all $j \in J$.

W.r.t. the *principle of minimal change* [1], p'_U should also preserve as much as possible the properties of the old probability distribution p_U. This means that the structure of the graph underlying the Markov network must remain unchanged, i.e.,

© Springer Nature Switzerland AG 2022
R. Kruse et al., *Computational Intelligence*, Texts in Computer Science,
https://doi.org/10.1007/978-3-030-42227-1_28

the new probability distribution should exhibit (at least) the conditional independences that hold in the old one. Furthermore, we require the (conditional) dependences between variables, as they can be assessed by some dependence measure, to stay constant or to change as little as possible. It is quite easy to see that p'_U in case of existence is uniquely determined: although we may find many distributions that fit the conditions on the new marginals, there can be only one distribution that maximally preserves the available knowledge on qualitative and quantitative dependences adapted from the prior probability distribution.

In mathematical terms, this sort of additional information leads to competing partial or total changes of selected (conditional) low-dimensional probability distributions in the Markov network. They can be interpreted as the basis for a revision operation, where a prior state of knowledge (represented by the initial Markov network) in the light of new information (which is the new set of probability distributions) is revised to a posterior state of knowledge that incorporates the new information in the sense of the principle of minimal change.

Translated into the probabilistic framework, the task is to calculate a posterior Markov network that meets the new distribution conditions, only accepting a minimal change of the qualitative and quantitative interaction structures of the underlying prior distribution.

Preserving the qualitative structure means to leave the network structure (independence relations) unchanged, which, of course, is only possible if the revision conditions themselves do not contradict this assumption. That is, all new distributions fit the clique structure of the Markov network.

Preserving the quantitative structure means to hold the cross-product ratios of the distributions, as far as they are not affected by the revision information.

Using some concepts of multivariate statistics, especially the method of *proportional iterative fitting* [14], and formalizing the concept of revision and the principle of minimal change to Markov networks, it is possible to prove that there is a unique solution of the above-mentioned revision problem. The only natural side condition to get this result is that the new marginal probability distributions do not contradict themselves and do not contradict to the rules system that is integrated in the prior network. An in-depth examination of such a clean knowledge-based solution of the revision problem can be found in [2]. The obtained results have the advantage that they are also useful for the development of decision support algorithms for the recognition, description, avoidance, and goal-oriented correction of inconsistent revision input data.

In the following sections, we present a knowledge-based probabilistic formalization and solution of the basic revision problem for Markov networks, restricted to a set of non-conditional marginals for single variables. This clean probabilistic approach avoids any concepts offered by calculi with deviating semantic foundations, for example, to minimize probabilistic difference measures that could be adapted from information theory.

28.2 Revision Procedure

In order to solve the revision problem specified above, we need an approach that supports the construction of the distribution p'_U that we are searching for. From multivariate statistics, iterative proportional fitting is a well-known algorithm for adapting the marginal distributions of a given joint distribution to desired values. It consists in computing the following sequence of probability distributions:

$$p_U^{(0)}(u) \equiv p_U(u)$$

$$\forall i = 1, 2, \ldots : \qquad p_U^{(i)}(u) \equiv p_U^{(i-1)}(u) \frac{p_{A_j}^*(a)}{p_{A_j}^{(i-1)}(a)},$$

where j is the $((i-1) \mod |J| + 1)$-th element of J, the index set that indicates the variables for which marginal distributions are given. $p_{A_j}^*$ is the desired marginal probability distribution on the domain of the variable A_j, $p_{A_j}^{(i-1)}$ the corresponding distribution as it can be computed from $p_U^{(i-1)}$ by summing over the values of all variables in U except A_j.

In each step, the probability distribution is modified in such a way that the resulting distribution satisfies one given marginal distribution (namely, the distribution $p_{A_j}^*$). However, this will, in general, change the marginal distribution for a variable A_k, which has been processed in a previous step. Therefore, the adaptation has to be iterated, traversing the set of variables several times.

If the revision problem has a solution, it converges to a (uniquely determined) probability distribution that has the desired marginals, has the same structure of (in)dependences as the original distribution as well as the same conditional probability distribution of the non-revised variables given the revised ones. For proofs of these properties, see e.g., [3]. In case a unique solution does not exist due to inconsistencies, experts will rely on techniques that allow to refine and resolve the knowledge base [11, 13].

Up to now we considered revision with iterative proportional fitting on the whole probability distribution p_U represented by the given Markov network. In practice, however, this cannot be done, because one of the main reasons for using a Markov network in the first place is that the distribution p_U cannot be represented as a whole due to its size. Therefore, we have to consider how the revision has to be performed on the parameters of the given Markov network, that is, the factor potentials of the factorization. However, in implementations, usually not the factor potentials are stored with the cliques, but the corresponding marginal distributions, because this has advantages w.r.t. evidence propagation.

The idea of the algorithm is to first assign each attribute (the marginal distribution of which is to be revised) to a maximal clique of the Markov network. As a next phase, steps of iterative proportional fitting are used to adapt the marginal distributions on the maximal cliques. Finally, information added by such a revision is distributed to the other maximal cliques by standard evidence propagation (preferably carried out by join tree propagation). More formally, we split the set J of indices of all variables, the marginal distributions of which are to be revised, into sets J_C of indices for all $C \in C$, where C is the set of sets of variables underlying the maximal cliques of the Markov network. These sets J_C of indices must satisfy the following three conditions:

$$(i) \qquad \bigcup_{C \in C} J_C = J$$

$$(ii) \qquad \forall C_1, C_2 \in C : (C_1 \neq C_2) \Rightarrow (J_{C_1} \cap J_{C_2} = \emptyset)$$

$$(iii) \qquad \forall j \in J_C : A_j \in C$$

Thus each variable A_j is assigned to the clique C, for which $j \in J_C$. Note that there are, in general, several ways to split the set J into sets $J_C, C \in C$, because a variable may be contained in several cliques and thus there may be a choice into which set J_C to put its index.

In pseudocode, the resulting revision algorithm reads

forall $C \in C$ **do**   // traverse the maximal cliques and
$\quad p_C^{(0)}(c) :\equiv p_C(c)$ // initialize marginal distributions
$i := 0;$ // initialize the step counter
repeat   // iterative proportional fitting
$\quad i := i + 1;$ // increment the step counter
$\quad$ **forall** $C \in C$ **do begin**   // traverse the maximal cliques
$\quad\quad$ **forall** $j \in J_C$ **do**   // traverse the assigned attributes

$$p_C^{(i)}(c) :\equiv p_C^{(i-1)}(c) \cdot \frac{p_{A_j}^*(a)}{p_{A_j}^{(i-1)}(a)};$$ // revise the marginal distribution

$\qquad\qquad\qquad\qquad\qquad\qquad\qquad\qquad$ // on the current maximal clique
$\quad\quad$ do evidence propagation;   // adapt the marginal distributions
$\quad$ **end**   // on all other maximal cliques
until convergence;   // until limiting distribution reached

Convergence may be checked, for instance, by determining the maximal change of a marginal probability on a maximal clique: If this maximal change falls below a user-defined threshold, the loop is terminated. It is useful to add a check on the number of iterations done, because there may be no distribution that satisfies the revision settings and then convergence cannot be achieved. That is, if a user-defined threshold on the number of iterations is reached without meeting the convergence criterion, the loop is terminated and a failure reported. In this case, the revision settings (i.e., the desired marginal distributions) have to be reconsidered and adapted.

As an alternative to adapting the revision settings, it is also possible to take one of the competing revision distributions that level off during the iteration process as an approximation that meets at least a part of the desired constraints.

That what is done inside the loop over the maximal cliques in the above algorithm, namely applying some iterative proportional fitting steps only to the marginal distribution on a maximal clique and distributing the revision information afterwards by standard evidence propagation, is indeed equivalent to performing the analogous iterative proportional fitting steps on the whole probability distribution, can easily be seen by considering how join tree propagation [8,9] works. Intuitively, the conditional probability distribution of all variables not contained in the current maximal clique given the variables in this maximal clique is factored out, the marginal distribution of the maximal clique is adapted, and the two factors are put together again.

Note that revision is only one of the many operations on Markov networks needed in order to develop and implement an innovative software system that is focused on, but not restricted to item planning and the prediction of parts demand in the automotive industry. Other necessary functionalities consist in the fusion of historical random samples with rules systems for a future planning interval into a Markov network, different kinds of updating operations, high-speed propagation and information retrieval techniques, explanation functionality for inconsistencies arising in revision [12], and so forth.

28.3   A Real-World Application

This section illustrates knowledge representation with Markov networks using a real-world industrial example. Volkswagen Group favors a marketing policy that provides a maximum degree of freedom in choosing individual specifications of vehicles. That is, considering personal preferences, a customer may select from a large variety of options, each of which is taken from a so-called item family that characterizes a certain line of equipment. Body variants, engines, gearshifts, door layouts, seat coverings, radios, and navigation systems reflect only a small subset of the whole range of item families. In case of the VW Golf—being Volkswagen's most popular car class—there are about 200 families with typically 4–8 values each, and a total range of cardinalities from 2 up to 150.

Of course, not all of the possible instantiations of item variables lead to valid vehicle configurations, since technical rules, restrictions in manufacturing, and sales requirements induce a common rule system that limits the acceptable ways of item combination. Nevertheless, dealing with more than 10,000 technical rules in the Golf class and even more rules delivered by the sales programs for the special needs of the different countries, there remains a giant number of correct vehicle specifications.

The major aim of the project EPL (EigenschaftsPLanung, the German word for item planning) at Volkswagen Group is the development and implementation of a software system that supports item planning, parts demand calculation, and capacity management with the aim of short-term as well as medium-term forecasts up to 24 months of future vehicle production.

In order to reach high quality of planning results, all relevant information sources have to be considered, namely rules for the correct combination of items into complete vehicle specifications, samples of produced vehicles as a reflection of customers' preferences, market forecasts that lead to stipulations of modified item rates for planning intervals, capacity restrictions, and production programs that fix the number of planned vehicles.

W.r.t. the logistics view, the most essential result of the item planning process is to assess the rates of all those item combinations that are known to be relevant for the demand calculation of parts, always related to a certain vehicle class in a certain planning interval. The importance of these item combinations arises from the fact that a vehicle can be interpreted as a large set of installation points, each of which is characterized by a set of alternative built-in parts for the corresponding location. Which one of the alternative parts has to be chosen at an installation point, depends on its installation condition that can be specified by an item combination. Of course, at a selected installation point, all occurring installation conditions have to be disjoint, and their disjunction has to form a tautology. That is, given any correct vehicle specification, for each installation point, we obtain a unique decision which of its alternative parts has to be used.

In the context of the Golf class, a total of about 70,000 different item combinations are required as installation conditions for the whole set of installation points. The data structure that lists all installation points, their installation conditions, and the built-in quantities of the referenced parts, is called a variants-related bill of material. The task of predicting the total demand of a certain part with respect to a future planning interval is to sum up the demands over all of its installation points. The demand at any installation point results from multiplying the rate of the item combination that represents its installation condition with the built-in quantity and the total number of vehicles intended to be produced in the respective planning interval.

Knowledge Representation

The domain knowledge is formally represented by a distribution p_U over the set of relevant item families or attributes U. Conditional independences are used to decompose this distribution into lower dimensional distributions using a Markov network. Figure 28.1 shows a real-world Markov Network from the planning application [5]. The different attributes of a car are described by 204 attributes which are anonymized by numbers in the figure. The Markov network in the figure consists of 123 at most 18-dimensional cliques (51 attributes of the original 204 are independent, that is, they are not connected to the join tree and omitted here for brevity, otherwise the figure would show 174 cliques). Let us make clear the efficiency gain of such a representation by comparing the theoretical necessary storage for the original 204-dimensional distribution p_{orig} with the distribution p_{net} encoded by the Markov network. To simplify the estimation, we assume that the domains of all attributes contain only five values which comes close to the average number of domain sizes in typical industrial projects. Our argumentation w.r.t. the efficiency gain does not change if we assume

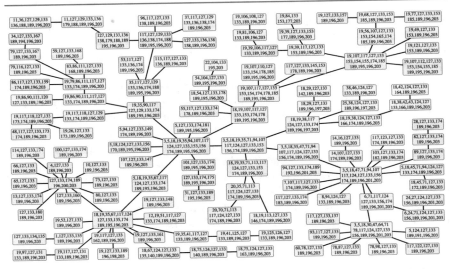

Fig. 28.1 Markov network of a real-world application with 204 attributes (51 of which are independent and therefore not shown). The distribution can be approximated by 123 cliques (or 174 cliques if the independent attributes are taken into account) with at most 18 dimensions

a different domain size. With this assumption we arrive at the following number of parameters for each representation (i.e., single probabilities or clique potential entries):

Distribution	Number of parameters
p_{orig}	$1 \cdot 5^{204} \approx 4 \cdot 10^{142}$
p_{net}	$\sum_{i=1}^{174} 5^{|C_i|} \approx 1 \cdot 10^{13}$

The number of parameters for p_{orig} exceeds any imaginable measure. The size of p_{net}, however, can be handled if certain additional observations are taken into account. The high-dimensional cliques, for example, are usually quite sparse which allows for better representations than a full joint distribution.

For reasons of fairness, we have to admit that this sparseness criterion also applies to the distribution p_{orig}, which would be extremely sparse in reality. Even if every human being on earth would own a unique Volkswagen, then p_{orig} would only contain around 7 billion non-zero probabilities which would be almost negligible in contrast to the 10^{142} possible combinations. Hence, we would most likely use a data structure that accounts for the sparsity which would lead to some kind of list which in turn would grow with the number of cars. The Markov network representation, however, can cater for any car combination without growing as long as the representation is compatible with the conditional (in)dependences encoded by the network structure. It is immediately clear that any new data (in form of adjusted car configuration frequencies) have to satisfy the (conditional) independences because it is these independences that allow for the efficient representation. In practice, the exact decomposability never fully holds. One confines oneself with an approximate notion of decomposability. For details regarding the treatment of approximate independence structures), refer to [4,6].

Once a basic prior Markov network for a certain planning interval has been generated, it becomes subject to a variety of planning operations which involve marketing and sales stipulations (e.g., installation rate of comfort navigation system increases from 20% to 35%) and capacity restrictions from logistics (e.g., maximum availability of seat coverings in leather is 5,000). These quantitative input data are strongly related to the planning interval itself and therefore not learnable, neither from historical data nor from the nonprobabilistic rules system. They typically consist of predicted installation rates or absolute demands for single items, sets of items, or (sets of) item combinations, and are frequently related to refined planning contexts (e.g., VW Golf with all-wheel drive for the US market).

References

1. Gärdenfors P (1988) Knowledge in flux: modeling the dynamics of epistemic states. MIT Press
2. Gebhardt J (2001) The revision operator and the treatment of inconsistent stipulations of item rates. Project EPL: internal report 9. ISC Gebhardt and Volkswagen Group, K-DOB-11, Wolfsburg, Germany
3. Gebhardt J, Borgelt C, Kruse R, Detmer H (2004) Knowledge revision in Markov networks. Mathware Soft Comput 11(2–3):93–107. University of Granada, Granada, Spain
4. Gebhardt J, Detmer H, Madsen AL (2003) Predicting parts demand in the automotive industry—an application of probabilistic graphical models. In: Proceedings of the Bayesian Modelling Applications Workshop at International Joint Conference on Uncertainty in Artificial Intelligence (UAI 2003, Acapulco, Mexico)
5. Gebhardt J, Kruse R (2005) Operations knowledge-based, for graphical models in planning. In: Proceedings of the Europian Conference on Symbolic and Quantitative Approaches to Reasoning with Uncertainty (ECSQARU, (2005) Barcelona, Spain), LNAI 3571:3–14. Springer-verlag, Berlin, Germany
6. Gebhardt J, Klose A, Detmer H, Rügheimer F, Kruse R (2006) Graphical models for industrial planning on complex domains. In: Della Riccia D, Dubois D, Kruse R, Lenz HJ (eds) Decision theory and multi-agent planning, CISM courses and lectures 482:131–143. Springer-Verlag, Berlin, Germany
7. Hugin Expert (2016) HUGIN—Bayesian network software. Aalborg, Denmark 1989–2016. http://www.hugin.com
8. Jensen FV (1996) An introduction to Bayesian Networks. UCL Press, London, United Kingdom
9. Lauritzen SL, Spiegelhalter DJ (1988) Local computations with probabilities on graphical structures and their application to expert systems. J Roy Stat Soc Ser B 2(50):157–224. Blackwell, Oxford, United Kingdom
10. Norsys Software Corp (2016) NETICA—Bayesian network software. Vancouver, BC, Canada 1995–2016. http://www.norsys.com/
11. Schmidt F, Wendler J, Gebhardt J, Kruse R (2013) Handling inconsistencies in the revision of probability distributions. HAIS13, Lecture notes in computer science, vol 8073, Springer, Berlin Heidelberg, pp 598–607

12. Schmidt F, Gebhardt J, Kruse R (2015) Handling revision inconsistencies: towards better explanations. Symbolic and quantitative approaches to reasoning with uncertainty, Springer International Publishing, pp 257–266
13. Schmidt F, Gebhardt J, Kruse R (2015) Handling revision inconsistencies: creating useful explanations. In: Destercke S, Denoeux T (eds) HICSS 2015 Proceedings. Koloa, Kauai, HI, USA
14. Whittaker J (1990) Graphical models in applied multivariate statistics. Whiley, Chichester, United Kingdom

Decision Graphs

29

In this chapter, we will introduce the concepts of decision graphs (also referred to as influence diagrams) that are extensions to Bayes networks targeted for sequential decision-making with dedicated focus on evaluating strategies and expected utilities. Section 29.1 motivates requirements based on a pure Bayes network while Sect. 29.2 formally defines decision graphs. In Sect. 29.3, we will introduce the concepts of decision policies and strategies, Sect. 29.4 discusses how to arrive at optimal strategies and closes with a thorough example.

29.1 Motivation

Let us discuss a pure Bayes network as depicted in Fig. 29.1 to identify additional requirements desired for decision-making. It shows a simplified decision scenario from the medical domain. A certain disease causes different symptoms (here: cough, nausea, or chill) which might incline a physician towards a blood test. The blood test result will then determine whether or not to administer a certain drug. Even though the network structure encodes plausible causal dependencies, there are some drawbacks when it comes to decision-making.

All nodes in a Bayes network are random variables by definition. However, the nodes "Blood Test" and "Drug" should represent a physician's decisions based on evidence and hopefully be not random at all. If we treat all nodes as random variables, we need to specify factor potentials for them. In case of a Bayes network, the factorization is constructed as the product of all conditional probability distributions of all nodes given their parent nodes. While the distributions for "Disease" and the three symptoms can be specified using disease incidence rates and expert knowledge, we will have a harder time parameterizing the "Blood Test" node. One could argue that the probability parameters needed for this node can be extracted from clinical treatment guidelines. However, those guidelines will contain specific

© Springer Nature Switzerland AG 2022
R. Kruse et al., *Computational Intelligence*, Texts in Computer Science,
https://doi.org/10.1007/978-3-030-42227-1_29

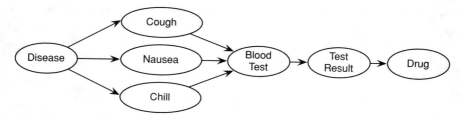

Fig. 29.1 Pure Bayes network illustrating a medical decision scenario

treatment plans or strategies involving a physician's judgment which we cannot guarantee to be representable by pure probabilities. In addition to that, assume a physician observes some symptoms and decides to administer a blood test. This corresponds to initializing the respective symptom nodes and the "Blood Test" node with evidence. Propagating this evidence through the network will likely update the probabilities of all other nodes including the "Disease" node. From the physician's point of view, it is undesirable to have the disease distribution change based on the conduction of the blood test because she wants to determine the state of the "Disease" node rather than altering it with a pure observation decision. One could argue that for an outside observer, however, this effect (of the "Disease" distribution changing based on evidence for "Blood Test" node) is plausible. If a physician conducts a blood test based on observed symptoms, then a disease might be more likely as a blood test costs money and nobody would order one without reasonable cause. A decision-making framework should be able to model and treat those two different points of view separately.

Let us discuss the above-mentioned cost aspects further. A blood test costs money and depending on the severity of the (conjectured) disease a physician might opt against it. The same is true for drug administration. Also, drugs can have side effects that may outweigh the curative properties (if the disease is not too severe). A decision-making framework should have devices to explicitly model costs (or losses) in order to guide the decisions that will be made. The following list summarizes requirements regarding an extension framework of Bayes networks in order to facilitate decision-making.

- *Decisions.* The framework shall offer explicit tools to model decision points.
- *Temporal order.* Decisions shall be explicitly temporally ordered to maintain consistency of sequential decisions.[1]
- *Utilities.* The framework shall offer explicit tools to model utilities (costs) based on decisions and random nodes.

---

[1] We have not covered this requirement explicitly in the medical example as the decision for "Drug" obviously follows the decision for "Blood Test". In more complex networks it might become less obvious in what order to make decisions.

- *Strategies.* The framework shall offer the explicit notion of a strategy in order to evaluate them.
- *Bayes semantics.* While extending the notion of Bayes networks, the underlying probabilistic properties and tools (such as evidence propagation) shall be kept intact in order to drive adoption.

29.2 Definition

Decision graphs introduce the following extensions which are illustrated in Fig. 29.2 and will be summarized in Definition 29.1. Let us assume, we are given a Bayes network with the underlying directed, acyclic graph $G = (V_C, E_C)$.[2] As usual, we depict the random (or chance) nodes as ellipses. We introduce a second set of nodes V_D of so-called *decision nodes* which will be drawn as rectangles. Each decision node $D \in V_D$ has a finite domain dom(D) consisting of all possible decisions that can be made. For example, in our medical scenario above, we would have dom(BloodTest) = {yes, no}. Further, we introduce a third set V_U of so-called *utility nodes*. Each utility node $U \in V_U$ represents a utility function that assigns to each value combination of its parent nodes (both chance or decision nodes are allowed as parent nodes) a real-valued (cost or loss) utility value[3] :

$$U : \mathrm{dom}(\mathrm{pa}(U)) \to \mathbb{R}$$

We impose the restrictions depicted in Fig. 29.3 as to which node types can be connected by a directed edge in the final decision graph:

(a) Probabilistic edge: Normal edge between chance nodes in a Bayes network encoding qualitative knowledge.
(b) Informational edge: The known state of node C might have an influence on the decision D and thus needs to be determined before the decision D is made.
(c) Utility edge: The state of chance node C must be known in order to compute the utility value of U.
(d) Probabilistic edge: The decision of D influences the probability distribution of C. For chance node C, the decision node D is treated as a normal parent node. That is, the conditional probability distribution has the shape $P(C|D)$.
(e) Temporal edge: An edge between decision nodes imposes an order in which to make the decisions. We impose that such edges are inserted to ensure a total ordering of the nodes in V_D. With this property satisfied, we call the network a *perfect recall* decision graph.
(f) Utility edge: The decision of D influences the utility value of U. That is, the decision must be made before the utility value can be computed.

---

[2] The index C is often used in decision graph literature to highlight *chance* nodes, i.e. random variables, of a Bayes network.
[3] The shorthand notation dom(W) with W being a set of nodes shall denote the set of tuples representing all combinations of values in all domains of nodes in W.

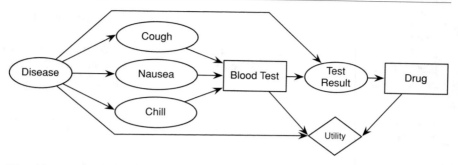

Fig. 29.2 Decision graph that better encodes the requirements motivated in Sect. 29.1 and Fig. 29.1

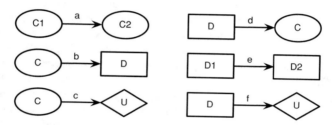

Fig. 29.3 Allowed edge connections in a decision graph

Obviously, utility nodes are sinks, that is, they cannot have outgoing edges.

Figure 29.2 shows a decision graph for the medical scenario illustrated above. The nodes **BloodTest** and **Drug** have been turned into decision nodes and a **Utility** node was added. The value of the **Utility** node represents some combined quantitative measure of the medical and monetary quality of the sequence of the two decisions. That is, a severe disease that goes untested and untreated would lead to a very low utility, while a detected and treated disease would be assigned high utility. Unnecessarily conducted blood tests would be penalized with a low utility. Note, that we added an edge from **Disease** to **TestResult**. Otherwise, the test result would solely depend on whether a blood test is conducted or not. With this edge, we Can, e.g., model the specificity of the test.[4]

Assuming the following domains, we will briefly outline the structure of the conditional probability distributions and utility function that are needed to fully specify the decision graph.

---

[4] We could have added this edge with the same rationale already in Fig. 29.1 but omitted it for brevity.

$$\text{dom(Disease)} = \text{dom(TestResult)} = \{\text{none, malaria, flu}\}$$
$$\text{dom(Cough)} = \text{dom(Nausea)} = \text{dom(Chill)} = \{\text{yes, no}\}$$
$$\text{dom(BloodTest)} = \{\text{yes, no}\}$$
$$\text{dom(Drug)} = \{\text{none, drug1, drug2, drug3}\}$$
$$\text{Utility}: \quad \text{dom(Disease)} \times \text{dom(BloodTest)} \times \text{dom(Drug)} \to \mathbb{R}$$

The structure of the conditional probability distributions is as follows:

$$P(\text{Disease})$$
$$P(\text{Cough}|\text{Disease})$$
$$P(\text{Nausea}|\text{Disease})$$
$$P(\text{Chill}|\text{Disease})$$
$$P(\text{TestResult}|\text{Disease, BloodTest})$$

The following definition formally summarizes the notion of a perfect recall decision graph [1,3]:

Definition 29.1 (*(Perfect recall) Decision Graph*) A (perfect recall) decision graph is a tuple $\mathcal{D} = (V_C \cup V_D, E, P, V_U)$ satisfying the following properties:

- V_C is a set of chance nodes, V_D is a set of decision nodes, and V_U is a set of utility nodes.
- The structure $(V_C \cup V_D \cup V_U, E)$ is a directed, acyclic graph.
- The subgraph induced by $V_C \cup V_D$ together with P is a Bayes network.
- The nodes sets V_C, V_D and V_U are mutually disjoint.
- The graph structure imposes a total ordering on the nodes in V_D.
- Each utility node $U \in V_U$ is assigned a function $U : \text{dom(pa}(U)) \to \mathbb{R}$.

The definition explicitly splits the node set into $V_C \cup V_D$ and V_U to be consistent with the literature where the utility set is explicitly exposed in the algebraic structure.

Given a fully specified decision graph, we can compute the *joint expected utility* for the chance and decision nodes $\text{EU}(V_C \cup V_D)$[5]:

$$\text{EU}(V_C \cup V_D) := \prod_{A \in V_C \cup V_D} P(A | \text{pa}(A)) \sum_{U \in V_U} U(\text{pa}(U)). \tag{29.1}$$

Given all conditional probability distributions and utility function definitions, we can derive from this joint expected utility function all information needed to evaluate decision policies and strategies.

---

[5] Note, that we use the symbol U to denote the node in a graph as well as the utility function associated with it for brevity. This is perfectly fine as nodes in a graph are symbolic objects and a function qualifies as such an object. Further, we have a one-to-one mapping between a utility node and its assigned utility function. In some literature, nodes and functions are distinguished.

29.3 Policies and Strategies

To solve a certain problem for which we are given a decision graph, we need to visit all decision nodes $D_i \in V_D$ in the imposed order and make a decision $d_i^* \in \mathrm{dom}(D_i)$. In our medical example, we first need to decide whether or not to conduct a blood test and then which drug to administer (or not to administer any drug at all). We will use the notion of a *policy* to describe the rules or function for a single decision node and the notion of a *strategy* to describe sequences of policies.

Formally, a policy δ_D for a decision node $D \in V_D$ must assign for each value combination of its parent nodes a decision from the domain $\mathrm{dom}(D)$:

$$\delta_D : \mathrm{dom}(\mathrm{pa}(D)) \rightarrow \mathrm{dom}(D)$$

Technically, all information that is available at the time of decision can influence that decision. In order to formally amount for that prior information, we impose that an edge shall be inserted that connects that prior information and the decision node. For example, the Drug decision node in Fig. 29.2 is only connected to the TestResult chance node and thus a policy δ_{Drug} will only base its decision on the test result outcome. However, the fact whether the patient shows any symptoms (that is, the state of the chance nodes Cough, Nausea, and Chill) are known by the time of the drug decision, too. In order to allow the policy δ_{Drug} to take those observations into account requires to insert a directed edge from, e.g., Cough to Drug. The assumption that all such prior information is always available is known as the *no-forgetting assumption*.

An example for a policy of the decision node BloodTest could be defined or derived as follows (we will discuss how to find optimal sequences of policies below):

$$\delta_{\mathsf{BloodTest}}(\mathsf{Cough}, \mathsf{Nausea}, \mathsf{Chill}) = \begin{cases} \mathsf{yes} & \text{if two or all symptom nodes equal } \mathsf{yes} \\ \mathsf{no} & \text{if no or one symptom node equals } \mathsf{yes} \end{cases}$$

A strategy Δ is a sequence of policies, e.g., $\Delta = \langle \delta_{\mathsf{BloodTest}}, \delta_{\mathsf{Drug}} \rangle$. Let $\mathrm{EU}(\Delta)$ denote the expected utility of the strategy Δ. We are henceforth interested in finding an *optimal strategy* $\hat{\Delta}$ for which holds:

$$\forall \Delta : \mathrm{EU}(\hat{\Delta}) \geq \mathrm{EU}(\Delta)$$

29.4 Finding Optimal Strategies

When dealing with finding optimal strategies, we need to distinguish two modes in which to reason about a strategy:

1. *Execution time.* This is the time when the strategy is actually carried out by a decision-maker. Executing a strategy $\Delta = \{\delta_1, \ldots, \delta_n\}$ entails sequentially executing all policies δ_i. If a policy δ_i contains a chance node in the parent set,

the decision-maker must wait until this chance node has evaluated to a specific value which can be observed and which then will drive the policy decision.

2. *Optimization time.* This is the time when the fully specified decision graph (and hence its joint expected utility function) are assessed in order to determine an optimal strategy. At that time, no chance node has assumed a specific value. Yet we must come up with policies that tell how to act when certain chance nodes will have assumed final values (during execution time).

The underlying idea for finding optimal strategies is to average over chance nodes[6] and maximize over decision nodes [2]. Averaging accounts for the uncertainty about the specific observations of the chance nodes at optimization time. Taking the maximum ensures to choose the best decision to build up an optimal strategy.

Since we have a total order on the decision nodes, we can pick the last one (D_n) and compute the expected utilities for all decision options $\text{dom}(D_n)$ given the past. From this, we pick the decision option with maximal expected utility, essentially creating a policy $\hat{\delta}_n$. We then advance backwards to D_{i-1} and repeat the process until a full sequence of policies $\hat{\Delta} = \left(\hat{\delta}_1, \ldots, \hat{\delta}_n\right)$ is obtained. The averaging and maximization to determine the expected utilities is carried out on the joint expected utility function induced by any fully specified decision graph:

$$\text{EU}(\hat{\Delta}) = \sum_{I_0} \max_{D_1} \sum_{I_1} \max_{D_2} \cdots \sum_{I_{n-1}} \max_{D_n} \sum_{I_n} \text{EU}(V_C \cup V_D).$$

The sets $I_i \subseteq V_C$ comprise chance nodes that need to be observed before the decision D_{i+1} can be made. As mentioned above, the order in which to maximize over the decision nodes is determined by the total order on them. However, the averaging over the chance nodes needs to be carried out in an order that represents the temporal order in which they will (need to) be observed at execution time. Hence, we must derive from the decision graph a partition of the chance nodes that tells which chance nodes need to be observed before the first and in between any subsequent decision.

For the medical scenario above, we would derive the following order of chance node sets and decision nodes:

$$\underbrace{\{\text{Cough, Nausea, Chill}\}}_{I_0} \prec \underbrace{\text{BloodTest}}_{D_1} \prec \underbrace{\{\text{TestResult}\}}_{I_1} \prec \underbrace{\text{Drug}}_{D_2} \prec \underbrace{\{\text{Disease}\}}_{I_2}$$

Let us now discuss a full scenario from the definition of the decision graph, over the joint expected utility function towards the final optimal strategy. We borrow from a network contained in [3] that has fewer chance nodes but two utility nodes and thus still induces a smaller amount of parameters to discuss and present.

---

[6] We stick to the notion of a node since it also represents a chance or decision variable.

29.5 Example Scenario

An oil wildcatter needs to decide whether or not to drill in a specific location. There is uncertainty about the amount of oil beneath that location (Table 29.1). He could decide to test the location first by taking seismic soundings which give an indication

Table 29.1 Original joint expected utility function for the oil wildcatter decision graph $EU(O, S, T, D)$, joint expected utility function $EU(S, T, D)$ after averaging over Oil and expected utility $EU(Test)$ for Test

Oil	Seismic	Test	Drill	EU(O, S, T, D)
dry	diffuse	no	no	0
			yes	-11.67
		yes	no	-3
			yes	-24
	open	no	no	0
			yes	-11.67
		yes	no	-1.5
			yes	-12
	closed	no	no	0
			yes	-11.67
		yes	no	-0.5
			yes	-4
wet	diffuse	no	no	0
			yes	5
		yes	no	-0.9
			yes	3.6
	open	no	no	0
			yes	5
		yes	no	-1.2
			yes	4.8
	closed	no	no	0
			yes	5
		yes	no	-0.9
			yes	3.6
soaking	diffuse	no	no	0
			yes	13.34
		yes	no	-0.2
			yes	3.8
	open	no	no	0
			yes	13.34
		yes	no	-0.8
			yes	15.2
	closed	no	no	0
			yes	13.34
		yes	no	-1
			yes	19

Seismic	Test	Drill	EU(S, T, D)
diffuse	no	no	0
		yes	6.67
	yes	no	-4.1
		yes	-16.6
open	no	no	0
		yes	6.67
	yes	no	-3.5
		yes	8
closed	no	no	0
		yes	6.67
	yes	no	-2.4
		yes	18.6

Test	EU(Test)
no	20
yes	22.5

of the geographical structure of the location (and thus give hints about potential oil quantities). Figure 29.4 depicts the qualitative structure of the scenario. The domains of the chance and decision nodes are as follows:

$$\text{dom(Test)} = \{\text{no, yes}\}$$
$$\text{dom(Drill)} = \{\text{no, yes}\}$$
$$\text{dom(Oil)} = \{\underbrace{\text{dry}}_{\text{no oil}}, \underbrace{\text{wet}}_{\text{some oil}}, \underbrace{\text{soaking}}_{\text{much oil}}\}$$
$$\text{dom(Seismic)} = \{\underbrace{\text{closed}}_{\text{hints much oil}}, \underbrace{\text{open}}_{\text{hints some oil}}, \underbrace{\text{diffuse}}_{\text{hints no oil}}\}$$

Tables 29.2 and 29.3 contain the conditional probabilities for the chance nodes Seismic and Oil. A test costs 10,000, while drilling costs 70,000. The utilities for reaching a dry, wet or soaking oil source are 0, 120,000, and 270,000, respectively. Hence, both utility functions U_1 and U_2 are defined as follows (we divide the costs by 1,000):

$$U_1(\text{Test}) = \begin{cases} 0 & \text{if Test} = \text{no} \\ -10 & \text{if Test} = \text{yes} \end{cases}$$

$$U_2(\text{Drill, Oil}) = \begin{cases} -70 & \text{if Drill} = \text{yes, Oil} = \text{dry} \\ 50 & \text{if Drill} = \text{yes, Oil} = \text{wet} \\ 200 & \text{if Drill} = \text{yes, Oil} = \text{soaking} \\ 0 & \text{if Drill} = \text{no} \end{cases}$$

With these information, the decision graph is fully specified. The objective is now to derive an optimal strategy for conducting the decisions Test and Drill. For this, we will apply the averaging and maximization procedure on the joint expected utilities function. This function is fully specified in Table 29.1 and was computed via Eq. 29.1. As an example, the value for the last entry in the joint expected utility function in Table 29.1 is computed as follows:

$$\begin{aligned} \text{EU}(\text{O} &= \text{soaking, S} = \text{closed, T} = \text{yes, D} = \text{yes}) \\ &= P(\text{O} = \text{soaking}) \cdot P(\text{S} = \text{closed}|\text{O} = \text{soaking, T} = \text{yes}) \cdot \\ & \quad (U_1(\text{yes}) + U_2(\text{yes, soaking})) \\ &= 0.2 \cdot 0.5 * (-10 + 200) = 19 \end{aligned}$$

Fig. 29.4 The Oil Wildcatter example network [3]

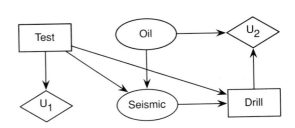

Table 29.2 Conditional probabilities of Seismic node. The uniform distributions for the cases where Test = no model ignorance when no test was conducted

P(Seismic|Test, Oil)	Test = no			Test = yes		
	Oil = dry	Oil = wet	Oil = soaking	Oil = dry	Oil = wet	Oil = soaking
Closed	$1/3$	$1/3$	$1/3$	0.1	0.3	0.5
Open	$1/3$	$1/3$	$1/3$	0.3	0.4	0.4
Diffuse	$1/3$	$1/3$	$1/3$	0.6	0.3	0.1

Table 29.3 Marginal distribution for Oil

P(Oil)	
Dry	0.5
Wet	0.3
Soaking	0.2

The network structure and the total ordering on both decision nodes leads to the following node set order:

$$\emptyset \prec \text{Test} \prec \{\text{Seismic}\} \prec \text{Drill} \prec \{\text{Oil}\}$$

This allows us to write down the averaging and maximization equation[7]:

$$\text{EU}(\hat{\Delta}) = \max_T \sum_S \max_D \sum_O P(O) \cdot P(S|O, T) \cdot \big(U_1(T) + U_2(D, S)\big)$$

That is, in order to determine the policy for Drill, we need to average over Oil. Since the utilities are part of the joint expected utility function, we can just marginalize over Oil.[8] The result is depicted in the upper right of Table 29.1. Maximizing it over Drill allows us to derive the following drilling policy $\hat{\delta}_{\text{Drill}}$:

$$\hat{\delta}_{\text{Drill}}(\text{Seismic, Test}) = \begin{cases} \text{yes} & \text{if Seismic} = \text{closed, Test} = \text{no} \\ \text{yes} & \text{if Seismic} = \text{open, Test} = \text{no} \\ \text{yes} & \text{if Seismic} = \text{diffuse, Test} = \text{no} \\ \text{yes} & \text{if Seismic} = \text{closed, Test} = \text{yes} \\ \text{yes} & \text{if Seismic} = \text{open, Test} = \text{yes} \\ \text{no} & \text{if Seismic} = \text{diffuse, Test} = \text{yes} \end{cases}$$

This policy can be summarized as "Always drill unless the test shows no indication of oil." Continuing this procedure leads us finally to the expected utility for Test as

---

[7] We abbreviate the node names using the first letter to reduce visual clutter.
[8] Since the utility values are already included in the expected utility values, a marginalization effectively represents the mentioned averaging.

depicted in the lower right of Table 29.1. From it, we derive the policy $\hat{\delta}_{\text{Test}} = \text{yes}$ which translates to "Always test." This concludes the task of finding an optimal strategy.

It is obvious that operating according to the averaging and maximization rule in Eq. 29.1 becomes quickly infeasible as the number of nodes and their domains grow. Instead, one can exploit the network structure to simplify computation (similar to exploiting network structure for evidence propagation).

References

1. Howard RA, Matheson JE (1981) Influence diagrams. Readings in decision analysis, strategic decisions group. Chapter 38, pp 763–771. Menlo Park, CA, USA
2. Jensen FV (2001) Bayesian networks and decision graphs. Springer-Verlag, Berlin, Germany
3. Kjærulff UB, Madsen AL (2013) Bayesian networks and influence diagrams: a guide to construction and analysis, 2nd edn. Spinger-Verlag, London, United Kingdom

Causal Networks

We studied how directed and undirected graphs can be used to encode conditional independence statements in Chap. 25. Directed graphs suggest themselves for an additional interpretation. On the one hand, edge directions can be considered a purely structural means which encodes conditional independence statements or dictates the set of conditioning (that is, parent) nodes for each graph node. On the other hand, directed edges between nodes are intuitive and are often used to indicate causal relationships between a cause and an effect. It is a common practice to have domain experts draft causal influences as a directed graph and then transforming it to a directed independence map. This raises the question whether it is possible to use graphical models that were induced by some learning algorithm (see Chap. 27) to reflect (at least some of) the causal relations in the data.

Correlation is perhaps the most frequently used concept in applied statistics. Its standard measure is the correlation coefficient, which assesses what can be called the intensity of linear relationship between two measures [2]. Correlation is closely related to probabilistic dependence, although the two concepts are not identical, because zero correlation does not imply independence. However, since this difference is of no importance for this discussion, we use the term "correlation" in the vernacular sense, that is, as a synonym for (probabilistic) dependence.

Note that neither in the narrower statistical nor in the wider vernacular sense correlation is connected directly to causal relation. We usually do not know why a correlation exists or does not exist, only that it is present or not. It is easily demonstrated that indeed the vast majority of all correlations are, without doubt, noncausal. Consider, for example, the distance between the continents America and Europe over the past 20 years. Due to continental drift, this distance increases a few centimeters every year. Consider also the average price of Swiss cheese in the United States over the same period.[1] The correlation coefficient of these two measures is

---

[1] Or any other quantity that has risen.

© Springer Nature Switzerland AG 2022
R. Kruse et al., *Computational Intelligence*, Texts in Computer Science,
https://doi.org/10.1007/978-3-030-42227-1_30

close to 1, that is, even in the narrow statistical sense they are strongly correlated. But obviously, there is no causal relation whatsoever between them. Of course, we could have used also a lot of other measures that increased over the past years, for example, the distance of Halley's comet (since its last visit in 1986) or the reader's age. The same can be achieved with measures that decreased over the past years. Therefore, causality may neither be inferred from correlation with certainty (since there are counterexamples), nor even inferred with a high probability (since causal correlations themselves are fairly rare).

According to these arguments, it seems to be a futile effort to try to infer causation from observed statistical dependences. Indeed, there is no way to infer causation from a single correlation (that is, a dependence between two variables). But this does not exclude immediately the possibility to infer from a set of (conditional) dependences and independences between several variables something about the underlying causal influences. There could be connections between the causal and the probabilistic structure, which enable us to discover the former at least partly.

30.1   Causal and Probabilistic Structure

Our intuition of causation is perhaps best captured by a binary predicate "X (directly) causes Y" or "X has a (direct) causal influence on Y," where X is the cause and Y the effect. This predicate is usually seen as antisymmetric, that is, if "X (directly) causes Y" holds true, then "Y (directly) causes X" does not. Thus, there is an inherent direction in causal influence, which seems to be a characteristic property of causation. For the most part, it is due to our intuition that a cause precedes its effect in time.

Because of the inherent direction, we can use a directed graph to represent causal influences, which we will call the causal structure. In principle directed cycles, that is, circular causal influences, are possible. (Such cycles are often exploited for control mechanisms, for example Watt's conical pendulum governor of the steam engine.) Nevertheless, we do not consider circular causal structures here, but assume that the causal influences form a directed acyclic graph in order to make them comparable to a probabilistic structure.

The theory of Bayes networks, which are based on directed acyclic graphs, suggests itself as an appropriate framework for a discussion of the connection between the causal and a probabilistic structure of a domain. It is perfectly acceptable to call a Bayes network a causal network as long as the term causal is only meant to indicate that a human domain expert, who "manually" constructed the network, did so by starting from a causal model of the domain under consideration. In this case, the knowledge of the human expert about the causal relations in the modeled domain ensures that the Bayes network represents not only statistical (in)dependences but also causal influences. If, however, the conditional independence graph of a Bayes network is learned automatically from data, calling it a causal network seems to be questionable. What could it be that ensures in an automatically generated network that directed edges show the directions of causal influences (that is, from cause to effect)? Obviously, we need to establish a relation between causal dependence and statistical dependence, since it is only statistical dependence we can test for automatically. The most direct approach is, of course, to identify the two structures, that is, to use the causal structure as a conditional independence graph of a given domain. If the causal and a probabilistic structure of a domain are identified in this way, we should be able to read from the causal structure, using the d-separation criterion (see Definition 24.42 on page 545), certain conditional independences that hold in the domain. This suggests the idea to invert the procedure, that is, to identify the causal structure or at least a part of it by conditional independence tests.

A fundamental problem of an approach that tries to discover the causal structure with conditional independence tests is that the d-separation criterion does not say anything about the dependence or independence of two sets X and Y of attributes given a third set Z, if X and Y are not d-separated by Z. This is sufficient, if a Bayesian network is constructed for a given domain, since for applications it is not essential to find and represent all independences. However, we need more to identify a causal structure, because we must be able to infer something about the causal structure from a conditional independence statement, which we cannot do if we do not know whether this statement is represented by d-separation in the causal structure or not.

In order to cope with this problem, it is assumed that, in a (sampled) probability distribution $\hat{p}$ of the domain under consideration, there exist exactly those (conditional) independences that can be read from the causal structure using d-separation, that is, we assume the causal structure is a perfect map of the conditional independence statements that hold true in the domain. This property is referred to as *stability* in [4].

An important property of d-separation together with the stability assumption is that they distinguish a common effect of two causes on the one hand from the mediating variable in a causal chain and the common cause of two effects on the other. In the structures shown on the left and in the middle of Fig. 30.1, A and C are independent given B, but become dependent if B is not given. In contrast to this, in the structure on the right of Fig. 30.1, A and C are independent unconditionally, but become dependent if B is given. It is this (alleged) fundamental asymmetry of the basic causal structures, which was studied first in [7], that makes statistical inferences about causal relations possible.

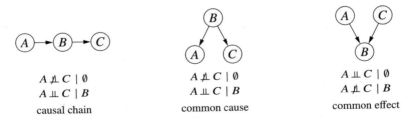

Fig. 30.1 Causal structures and corresponding conditional independence statements

However, even with the d-separation criterion and the stability assumption there are usually several causal structures that are compatible with the observed (conditional) dependences and independences. The main reason is that d-separation and stability cannot distinguish between causal chains and common causes. But in certain situations, all compatible causal structures have a common substructure. The aim of inductive causation is to find these invariant substructures.

With the ingredients discussed above, [4] developed the *inductive causation algorithm* shown in Listing 30.1. It is closely related to the SGS algorithm discussed in Chap. 27 (the first two steps are, in fact, identical to Algorithm 27.2).

Algorithm 30.1 (Inductive causation)

Input: A sampled distribution $\hat{p}$ over attribute set V
Output: A marked hybrid acyclic graph core$(\hat{p})$

1. For each pair of attributes A and B, search for a set $S_{AB} \subseteq V \setminus \{A, B\}$, so that $A \perp\!\!\!\perp_{\hat{p}} B \mid S_{AB}$ holds, that is, so that A and B are conditionally independent given S_{AB}. If there is no such set S_{AB}, connect the attributes by an undirected edge.
2. For each pair of nonadjacent attributes A and B with a common neighbor C (that is, C is adjacent to A as well as to B), check whether $C \in S_{AB}$. If it is not, direct the edges towards C, that is $A \to C \leftarrow B$.
3. Form core$(\hat{p})$ by recursively directing edges according to the following two rules:

   - If for two adjacent attributes A and B there is a strictly directed path from A to B not including the edge connecting A and B, then direct the edge towards B.
   - If there are three attributes A, B, and C with A and B not adjacent, either $A \to C$ or $A \leftarrow C$, and $C - B$, then direct the edge $C \to B$.

4. For each triplet of attributes A, B, and C: If A and B are not adjacent, $C \to B$, and either $A \to C$ or $A \leftarrow C$, then mark the edge $C \to B$.

The algorithms uses conditional independence tests to produce an at least partially directed causal structure. One should not overestimate the power of the algorithm, though. It makes strong assumptions and there are critical arguments about the stochastic interpretation of the induced causal structure. See [1] for more information.

Step 1 of Algorithm 30.1 determines the attribute pairs between which there must exist a direct causal influence or a hidden common cause, because an indirect influence should enable us to find a set S_{AB} that renders the two attributes independent. In step 2, the asymmetry inherent in the d-separation criterion is exploited to direct edges towards a common effect. Part 1 of step 3 ensures that the resulting structure is acyclic. Part 2 uses the fact that $B \to C$ is impossible, since otherwise step 2 would have already directed the edge in this way. Finally, step 4 marks those unidirected links that cannot be replaced by a hidden common cause. The reason is that if the attributes B and C named in this step were connected by a hidden common cause, A and B would not be independent given C, which they must, because otherwise step 2 would have directed both edges towards C.

The output graph $core(\hat{p})$ has four kinds of edges:

1. marked unidirected edges representing genuine causal influences (which must be direct causal influences in a projection),
2. unmarked unidirected edges representing potential causal influences (which may be direct causal influences or brought about by a hidden common cause),
3. bidirected edges representing spurious associations (which are due to a hidden common cause in a projection), and
4. undirected edges representing unclassifiable relations.

Let us illustrate the algorithm with a small example. Assume the domain under consideration can be described by four attributes A, B, C, and D (see left part of Fig. 30.2). We assume further that stochastic tests revealed the following conditional independence statements: $A \perp\!\!\!\perp B \mid \emptyset$, $A \perp\!\!\!\perp D \mid C$, $B \perp\!\!\!\perp D \mid C$, and all statements that can be derived by applying the graphoid axioms (see Definition 25.1 on page 551). From the given independence statements, we can determine that there must be edges between A and C, B and C, as well as C and D. This follows from the fact that we do not have (conditional) independence statements (given or deduced via graphoid axioms) which render those pairs of nodes independent. This results in the graph shown in the middle of Fig. 30.2, which we may refer to as the skeleton of the directed graph, because the edge directions are still missing. Finally, we try to determine the edge directions. We know that $A \perp\!\!\!\perp B \mid \emptyset$ (this marginal independence was assumed to be given or detected), but not $A \perp\!\!\!\perp B \mid C$ (independence statement not given or derivable do not hold). This can only be achieved by making C the common child of A and B, that is, by establishing a converging structure. The edge direction between C and D is determined using a similar rationale. We know that $A \perp\!\!\!\perp D \mid C$, but not $A \perp\!\!\!\perp D \mid \emptyset$ (again, by assuming only those independence statement that were given or derived). The only remaining structure that satisfies these constraints is the causal chain and thus the edge is directed from C to D. Note, we could have derived the same result using the nodes B, C, and D.

It is not always possible to direct all edges of a skeleton graph, which can be illustrated with a simple example as shown in Fig. 30.3. We cannot assign a direction

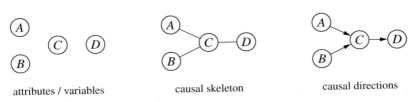

attributes / variables   causal skeleton   causal directions

Fig. 30.2 Deriving a causal structure from conditional independence statements

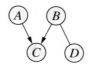

Fig. 30.3 Determining the edge direction is not always possible

for the edge between B and D because both options would satisfy the same set of independence statements. There may even be a hidden, unobserved attribute which is the common cause of both B and D. Such a scenario is consistent with the same set of independence statements. Alone a common effect (that is, a converging structure from B and D toward the hidden variable) can be ruled out.

30.2 The *Do* Operator

We introduced evidence propagation in Chap. 26 which allows us to use the observed values of one or more nodes of a Bayes (or Markov) network to update the probability distributions of all the other nodes. We may refer to this approach as *observational inference*. This comes with some caveats that may be unintuitive when interpreting the results in terms of causality. Think of a simple Bayes network consisting just of two attributes A (for air pressure) and B (for barometer reading) and a single edge $A \rightarrow B$. We can easily think of a conditional and marginal probability distribution that intuitively describe the fact that high air pressure readings cause corresponding barometer readings. Evidence propagation allows us to update the probability distribution of A after observing the value of B. However, the observation of B does not cause the value of A. If we intervened in a way to make the barometer B read a different value, this would not change the air pressure.

A classical view on causality is to interpret the causal relation "A causes B" as the counterfactual conditional proposition "one could cause B, if one could do A", which is known as *interventional inference* [9]. This allows us to cater for the asymmetry between cause and effect we encountered above, that is, we cannot achieve A by doing B, but only B, if A happens (or is done). It also allows to determine scenarios with confounding effects, that is, where two (or more) effects have the same cause. For example, take the above example and add the edge $A \rightarrow S$ with S representing the occurrence of a storm. By not observing air pressure A, the barometer reading B and the storm occurrence S become dependent ($C \not\perp S \mid \emptyset$) and hence a change in the observation of one changes the probability of the other.[2]

---

[2] We assume that the underlying (conditional and marginal) probability distributions exhibit such a behavior, which they do in general unless complete mutual marginal independence between A, B, and S are assumed.

		all seamen				old seamen (A = old)				young seamen (A = young)			
healing H:		no	yes	Σ	%	no	yes	Σ	%	no	yes	Σ	%
treatment C:	no	20	20	40	50	8	2	10	20	12	18	30	60
	yes	24	16	40	40	21	9	30	30	3	7	10	70
	Σ	44	36			29	11			15	25		

Fig. 30.4 Fictitious example for scurvy treatment using lemon juice

An approach to implement interventional inference is the *do* operator by [3]. The term do(C) denotes an external intervention which enforces the truth of (the cause) C. The question is whether the probability of an effect E increases under the intervention do(C), that is whether we have

$$P(E \mid do(C)) > P(E \mid do(\neg C)).$$

We will see below that assuming an intervention corresponds to changing the structure of the network (for the computation) and can be seen as conducting a thought experiment in a hypothetical state space by checking how different interventions would affect the outcomes.

We discuss this by using an example taken from [3] that also illustrates the Simpson paradox [8] which is crucial for this particular scenario. James Lind (1716–1794) discovered that scurvy (a vitamin C deficiency condition) can be treated with lemon juice. Let us assume a fictitious experiment in which half of 80 seamen suffering from scurvy are treated with lemon juice, results in the numbers shown in the left of Fig. 30.4.[3] Counterintuitively, we observe that the vitamin C treatment leads to fewer healings: 50% without treatment, but only 40% with treatment.

This surprising observation is a result of a dependence between age and healing success. If we condition on age, we find a positive effect of the vitamin C treatment in both groups (middle and right of Fig. 30.4):

$$30\% = P(H{=}yes \mid A = \text{old}, C = yes) > P(H{=}yes \mid A = \text{old}, C = no) = 20\%$$
$$70\% = P(H{=}yes \mid A = young, C = yes) > P(H{=}yes \mid A = young, C = no) = 60\%$$

Note that we only used observational inference so far. The underlying Bayes network of the example is depicted on the left of Fig. 30.5 with the middle and the right of Fig. 30.4 representing the parametrization. The above probabilities for treatment success can also be derived by using the decomposition of the Bayes network. For example, for the probability of healing given vitamin C, we derive

---

[3] The attributes for age (A), vitamin C treatment (C) and healing (H) are all binary.

Fig. 30.5 Network
strcutures without (left) and
with *do* operator (right)

$$P(H = \text{yes} \mid C = \text{yes})$$

$$= \frac{P(H = \text{yes}, C = \text{yes})}{P(C = \text{yes})}$$

$$= \frac{1}{P(C = \text{yes})} \sum_{a \in \{\text{old, young}\}} P(A = a, H = \text{yes}, C = \text{yes})$$

$$= \frac{1}{P(C = \text{yes})} \sum_{a \in \{\text{old, young}\}} P(H = \text{yes} \mid C = \text{yes}, A = a) P(C = \text{yes} \mid A = a) P(A = a)$$

$$= \frac{1}{0.5} \left(0.3 \cdot \frac{3}{4} \cdot 0.5 + 0.7 \cdot \frac{1}{4} \cdot 0.5 \right) = 0.4 = 40\%$$

For interventional inference, we apply the *do* operator as follows: By setting variable C (vitamin C treatment), we need to remove the dependence of that variable on age A, which results in the Bayes network on the right of Fig. 30.5 and the following numeric results:

$$P(H = \text{yes} \mid \text{do}(C = \text{yes}))$$

$$= \sum_{a \in \{\text{old, young}\}} P(H = \text{yes} \mid C = \text{yes}, A = a) P(A = a)$$

$$= 0.3 \cdot \frac{1}{2} + 0.7 \cdot \frac{1}{2} = 0.5 = 50\%$$

$$P(H = \text{yes} \mid \text{do}(C = \text{no}))$$

$$= \sum_{a \in \{\text{old, young}\}} P(H = \text{yes} \mid C = \text{no}, A = a) P(A = a)$$

$$= 0.2 \cdot \frac{1}{2} + 0.6 \cdot \frac{1}{2} = 0.4 = 40\%$$

Note that the derivation follows the factorization of the Bayes network on the right of Fig. 30.5, where all terms of the form $P(C = c)$ (with $c \in \{\text{no, yes}\}$ depending on the case of interest) are set to 1.

Given a Bayes network G, the general rule for applying the *do* operator is as follows [3]: If a set of nodes X is the target of the intervention, all attributes in X are set to the desired "true" values. The graph G is modified by removing all incoming edges to all nodes in X. It is then checked for which set of nodes Y and which corresponding attribute values of nodes in Y the probabilities increase. Those attributes Y are then assumed to be caused by X.

We can now close the loop and revisit the above-mentioned interpretation of interventional inference as a thought experiment. The Bayes network constitutes the hypothetical state space. We then fix certain nodes to specific values and check how they change the remaining attributes' probabilities using evidence propagation.

References

1. Borgelt C, Steinbrecher M, Kruse R (2009) Graphical models—representations for learning, reasoning and data mining, 2nd edn. Wiley, Chichester, United Kingdom
2. Everitt BS (1998) The Cambridge dictionary of statistics. Cambridge University Press, Cambridge, United Kingdom
3. Pearl J (2000) Causality: models, reasoning, and inference. Cambridge University Press, Cambridge, United Kingdom
4. Pearl J, Verma TS (1991) A theory of inferred causation. In: Proceedings of 2nd international conference on principles of knowledge representation and reasoning. Morgan Kaufmann, San Mateo, CA, USA
5. Pearl J, Glymour M, Jewell N.P (2016) Causal Inference in statistics—A Primer, John Wiley, Chichester, United Kingdom
6. Peters J, Janzing D, Schölkopf B (2017) Elements of Causal Inference, MIT Press, Cambridge, USA
7. Reichenbach H (1956) The direction of time. University of California Press, Berkeley, CA, USA
8. Simpson EH (1951) The interpretation of interaction in contingency tables. J R Stat Soc Ser B 13:238–241. Blackwell, Oxford, United Kingdom
9. von Wright GH (1974) Explanation and understanding. Cornell University Press, Ithaca, NY, USA

Index

© Springer Nature Switzerland AG 2022
R. Kruse et al., *Computational Intelligence*, Texts in Computer Science,
https://doi.org/10.1007/978-3-030-42227-1